St. Olaf College

MAR 1 1 1993

Science Library

# ICIAM 91

*Proceedings
of the
Second
International
Conference
on
Industrial
and
Applied
Mathematics*

# ICIAM 91

**Proceedings of the Second International Conference on Industrial and Applied Mathematics**

*Edited by*
**Robert E. O'Malley, Jr.**
University of Washington

siam. *Philadelphia*
Society for Industrial and Applied Mathematics

# ICIAM 91

*Proceedings of the Second International Conference on Industrial and Applied Mathematics*

This conference, which took place July 8 – 12, 1991, was partially supported by the National Science Foundation under grant #DMS-9023170, the Department of Energy under grant #DE-FG02-91ER25091, the Office of Naval Research under grant #N00014-91-J-1468, the Air Force Office of Scientific Research, and the Sloan Foundation under grant #B1991-18.

SIAM also thanks the following organizations for their help and support in making ICIAM 91 possible: Bellcore, The Boeing Company, E.I. DuPont de Nemours & Company, Eastman Kodak Company, GTE Laboratories Incorporated, IBM Corporation, NEC Research Institute, and United Technologies Corporation.

**Library of Congress Cataloging-in-Publication Data**

International Conference on Industrial and Applied Mathematics (2nd :
  1991 : Washington, D.C.)
    ICIAM 91 : proceedings of the Second International Conference on
Industrial and Applied Mathematics / edited by Robert E. O'Malley,
Jr.
    p. cm.
  Includes bibliographical references and index.
  ISBN 0-89871-302-1
  1. Mathematics—Congresses.   I. O'Malley, Robert E.   II. Title.
QA1.I73  1991
510—dc20
                                                                                                                                    92-16500

---

**siam** is a registered trademark.

All rights reserved. Printed in the United States of America. No part of this book may be reproduced, stored, or transmitted in any manner without the written permission of the Publisher. For information, write the Society for Industrial and Applied Mathematics, 3600 University City Science Center, Philadelphia, PA 19104-2688.

Copyright © 1992 by the Society for Industrial and Applied Mathematics.

*This ICIAM 91 volume is dedicated to
Gene H. Golub,
Fletcher Jones Professor of Computer Science
at Stanford University,
in recognition of his continuing inspirational
leadership in the international applied
mathematics and scientific computing
community and his special and effective
dedication to the encouragement of young
students and researchers in the field.*

 **PREFACE**

The ICIAM 87 and ICIAM 91 conferences mark the coming of age of the new computationally oriented discipline of applied mathematics as a healthy and technologically important contributor to society in many countries. It's so impressive to me that more than 2200 participants from 49 diverse nations could come together and communicate excitedly about the latest ideas in mathematics and its ever-growing applications. The program had approximately 1500 speakers and offered everyone plenty worth hearing all week long. Certainly, everything couldn't be included in this volume. What was selected, however, gives a rare panorama of our field today. There is no better place for government, industry, and university leadership to learn the immense value and potential of our discipline. Imagine what's in store for us at ICIAM 95!

The introductory material lists some of those individuals and organizations that contributed in many crucial ways to the success of ICIAM 91. It fails to explicitly thank many SIAM staff members who worked harder than ever for months before and after the conference. I'm proud that many observed that SIAM was such a gracious host society. Attendees got used to seeing a busy photographer at all the sessions and social events. He's SIAM's managing director, Dr. I. E. Block, who not only took all the candid photos in this volume, but superbly orchestrated millions of details to make ICIAM 91 the success we all enjoyed.

From Ball to Yserentant, the volume highlights 17 invited talks by world-class applied mathematicians. It also lists the broad array of minisymposia, organized according to subject area. The bibliographies listed after each subject grouping will give readers pointers on where to catch up on what's new in the business. The listing of conference attendees and authors finishes the volume. We hope they are accurate and apologize for any misspelled or missing names. We thank all of the authors for their papers.

Robert E. O'Malley, Jr.<br>
University of Washington

# COMMITTEES

### Committee for International Conferences on Industrial and Applied Mathematics

Roger Temam *(SMAI)*, Université Paris-Sud
*Chair*
Vinicio Boffi *(SIMAI)*, Consiglio Nazionale delle Richerche
C. William Gear *(SIAM)*, NEC Research Institute
*President, ICIAM 91*
Douglas S. Jones *(IMA)*, University of Dundee
Patrick Lascaux *(SMAI)*, Centre d'Etudes de Limeil
Robert M.M. Mattheij *(ECMI)*, Technische Universiteit Eindhoven
James McKenna *(SIAM)*, Bellcore
Jean-Claude Nedelec *(SMAI)*, École Polytechnique
Helmut Neunzert *(GAMM, ECMI)*, Universität Kaiserslautern
Ronald A. Scriven *(IMA)*, Central Electricity Research Laboratory
Albertó Tesei *(SIMAI)*, Istituto per le Applicazioni del Calcolo "Mauro Picone"
Wolfgang Walter *(GAMM)*, Universität Karlsruhe

### Program Committee

Robert E. O'Malley, Jr. *(SIAM)*, University of Washington
*Chair*
Robert Azencott *(SMAI)*, Université Paris-Sud
Franco Brezzi *(SIMAI)*, Università di Pavia
Carlo Cercignani *(SIMAI)*, Polytecnico Institut di Milano
John E. Dennis, Jr. *(SIAM)*, Rice University
Peter J. Deuflhard *(GAMM)*, Konrad-Zuse-Center
Iain S. Duff *(IMA)*, Harwell Laboratory
J.C.R. Hunt *(IMA)*, University of Cambridge
P.A. Raviart *(SMAI)*, Université de Paris VI
Ivar Stakgold *(SIAM)*, University of Delaware
Adriaan H.P. Van der Burgh, Delft University of Technology
Wolfgang L. Wendland *(GAMM)*, Universität Stuttgart

# INTRODUCTORY REMARKS

**Roger Temam**
Chair of the Committee for International Conferences on Industrial and Applied Mathematics

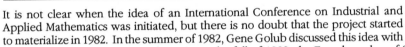

It is not clear when the idea of an International Conference on Industrial and Applied Mathematics was initiated, but there is no doubt that the project started to materialize in 1982. In the summer of 1982, Gene Golub discussed this idea with a group of mathematicians in the U.S. In the fall of 1982, the French analog of SIAM, La Société de Mathématiques Appliquées et Industrielles (SMAI) was constituted. The founders immediately wanted to establish it in the international community by organizing such a meeting. The two ideas merged and created the momentum which led to the first "ICIAM," organized in Paris-La Villette in 1987. ICIAM 87 was sponsored by GAMM, IMA, SIAM, and SMAI and was organized in Paris by SMAI and INRIA.

Organizing such a large meeting is never easy but organizing ICIAM 87 was particularly difficult because of the lack of previous experience. However, the response was tremendous and the number of participants much higher than the best expectations of the organizers.

After the success of the First International Conference on Industrial and Applied Mathematics in Paris-La Villette in 1987, the sponsoring societies decided such an event should take place every four years.

The Steering Committee, formed for ICIAM 87, became a permanent organization during the period 1987-91, with the name CICIAM—Committee for International Conferences on Industrial and Applied Mathematics. As the organization became more structured, eight societies representing more countries (and regions) joined the original four. At the time of ICIAM 91, the membership of CICIAM was:

- Brazilian Society for Applied and Computational Mathematics (SBMAC, f. 1978)
- Canadian Applied Mathematics Society (CAMS, f. 1980)
- China SIAM (CSIAM, f. 1990)
- European Consortium for Mathematics in Industry (ECMI, f. 1987)
- Gesellschaft für Angewandte Mathematik und Mechanik (GAMM, f. 1922)
- Institute of Mathematics and Its Applications (IMA, f. 1965)
- Israel Society for Applied Mathematics (ISAM, f 1978)
- Japan SIAM (JSIAM, f. 1990)
- Nordiska föreningen för Tillämpad och Industriell Matematik (NORTIM, f. 1991)
- Societâ Italiana di Matematica Applicata e Industriale (SIMAI, f. 1988)
- Société de Mathématiques Appliquées et Industrielles (SMAI, f. 1982)
- Society for Industrial and Applied Mathematics (SIAM, f. 1951)

The rapid evolution of computers and scientific computing has continued during the period 1987-91 and with it, applied mathematics has been playing an increasingly important role in many fields of engineering and science. By its size and scope, the ICIAM 91 program reconfirmed the breadth and diversity of applied mathematics and demonstrated its extension into an ever growing number of disciplines.

ICIAM 91 was another milestone in the worldwide recognition of applied mathematics as a new discipline. And, with the remarkable organizational work of SIAM, ICIAM 91 was both a fruitful and enjoyable meeting. There is no doubt that ICIAM has now established itself as an important event in applied mathematics and that applied mathematicians will be looking forward every four years for this picture of their discipline.

I wish much success for the next CICIAM chaired by C. William Gear and for the next ICIAM, which will be organized by GAMM in 1995.

### C. William Gear
President of ICIAM 91

ICIAM 91 is a further step in the evolution of the international applied mathematics community. Great advances have been made in the methodologies of applied mathematics, many of them driven by the increasing capability and performance of computers. Coupled with this have been equally great advances in the application of mathematics to problems of design in engineering and manufacturing, medical research, systems design, and the development of models that increase understanding of the natural and man-made world. In looking to the future, we can expect such progress to accelerate.

The increasing internationalization of research, development, and commerce challenge us to internationalize our science. We now have a quadrennial meeting and future ICIAMs will play an important role in increasing the interaction among the worldwide community of applied and computational mathematicians, as well as mathematically-oriented engineers and scientists.

Future ICIAMs will be important in promoting the use of mathematics in the solution of the grand challenges that face the world in the years ahead — problems, for example, of health, education, environmental protection, and global stability and growth.

At its annual meetings over the next four years the expanded CICIAM will be working to build a base for future international activities of which ICIAM 95 in Germany is but one. A site for ICIAM 99 will be chosen in 1994. Applied mathematics activities will be encouraged in those regions not currently having applied mathematics organizations and the structure of CICIAM will be further strengthened. I urge every one of you to strengthen your international ties and to apply your professional skills to the improvement of life around the world.

### Robert E. O'Malley, Jr.
Chair, ICIAM 91 Program Committee

The success of ICIAM 91 exceeded all our expectations. It's a special pleasure to realize that the high level of invited presentations, despite competing parallel sessions, made ICIAM an effective medium for learning what experts in many fields found exciting and important in applied mathematics today.

We're grateful to many applied and computational mathematicians for their effort in organizing minisymposia. Special thanks go to the SIAM staff for many person-years' work efficiently accomplished under pressure.

Scientists throughout the world can look to future ICIAM conferences as showplaces for the promise applied mathematics brings to science, technology, industry, and society.

# INVITED AND SPECIAL PRESENTATIONS

## Invited Presentations

**Dynamic Energy Minimization and Phase Transformations in Solids**
J. M. Ball

**Intermediate Asymptotics in Micromechanics**
G. I. Barenblatt

**Models of Vision**
Michael Brady

**Large-Scale Nonlinear Constrained Optimization**
A. R. Conn

**Modulation Equations Arising in the Mechanics of Continuous Media**
Wiktor Eckhaus

**Modeling the Solidification of Polymers:
An Example of an ECMI Cooperation**
A. Fasano

**Numerical Solution of Linear Equations Arising
from the Convection-Diffusion Equation**
Gene H. Golub

**Discrete Mathematics in Manufacturing**
Martin Grötschel

**Computational Methods for Chemically Reacting Flow**
Thomas J. R. Hughes

**Interior-Point Methods in Optimization**
Narendra Karmarkar

**Forced Nonlinear Surface Waves**
Klaus W. Kirchgassner

**Optimal Control and Viscosity Solutions**
P. L. Lions

**Wavelets or Wavelet-Packets Applied to Signals and Images:
Theory and Experiences**
Y. Meyer (presented by R. Coifman)

**Dynamics of Patterns, Waves, and Interfaces
from the Reaction-Diffusion Aspect**
Masayasu Mimura

**Complex Pattern Formation in Embryology:
Models, Mathematics, and Biological Implications**
J. D. Murray

**Radar Architectures:
From Microwave Processing to Computational Power**
Gabriel Ruget

**Approximation Algorithms for Packing and Covering Problems**
Éva Tardos

**Applications of Massively Parallel Computing**
D. J. Wallace

**Modeling Heterogeneities, Physical and Chemical,
for Contaminant Transport Problems in Porous Media**
Mary F. Wheeler

**Hierarchical Bases and Related Preconditioners
for Elliptic Partial Differential Equations**
Harry Yserentant

**Special Presentation**

**Science, Technology, and Mathematics in the 1990s**
D. Allan Bromley

 # AWARDS

## Special Award
The Committee for International Conferences on Industrial and Applied Mathematics recognizes outstanding contributions to the international industrial and applied mathematics community and presents its first award to Gene H. Golub for his leadership, which was pivotal in the conception of the first International Conference, and his wisdom, enthusiasm, ideas, and advice, which were major factors in its success.

Roger Temam and Gene H. Golub

## Wilkinson Prize for Numerical Software
The Wilkinson Prize for Numerical Software is awarded in honor of the outstanding contributions of James Hardy Wilkinson to the field of numerical software. The award is presented to the entry that best addresses all phases of the preparation of high quality numerical software. The software must be written within a widely available programming language and must execute on a significant class of computers.

The prize is sponsored by Argonne National Laboratory, the National Physical Laboratory, and the Numerical Algorithms Group.

Heather Wilkinson (right) congratulates Linda Petzold, winner of the first Wilkinson Prize for Numerical Software.

The first Wilkinson Prize was awarded to Linda R. Petzold for her entry "DASSL: A Differential/Algebraic System Solver."

## SIAM Student Paper Competition
Students in good standing who have not yet received PhDs are eligible to submit papers in applied and computational mathematics for this annual competition. To be eligible, a paper must be the work of a single author. Three winners are chosen on the basis of the originality, applicability, and clarity of exposition of their papers.

*The Society for Industrial and Applied Mathematics recognizes the following students as winners of the 1991 SIAM Student Paper Competition.*

Mary Ann Horn, University of Virginia
Uniform Stabilization of the Euler-Bernoulli Plate with Feedback Operator Acting via Bending Moments

Brian C. Morris, University of Massachusetts
Models and Computations for Interstellar Magnetic Gas Clouds: A Variational Method

Nicolas Seube, Université de Paris-Dauphine
Non-Linear Feedback Control Law Learning by Neural Networks: A Viability Approach

*From left to right:* Brian C. Morris, Mary Ann Horn, Margaret Wright, and Nicolas Seube.

## SIAM Award in the Mathematical Contest in Modeling

This prize is awarded to two of the undergraduate student teams judged "Outstanding" in the annual Mathematical Contest in Modeling. Each year two problems are presented and one winning team is selected for each of the two problems. MCM is sponsored by the Consortium for Mathematics and its Applications.

*The Society for Industrial and Applied Mathematics recognizes the following two teams for exceptional performance in the 1991 Mathematical Contest in Modeling:*

*From left to right:* Zvi Margaliot, Professor Henning Rasmussen, Alex Pruss, and Patrick Surry.

*From left to right:* Eiluned Roberts, Anupama Rao, and Anna Baumgartner.

### University of Western Ontario
Zvi Margaliot, Alex Pruss, and Patrick Surry
Faculty Advisor: Professor Henning Rasmussen

### University of Alaska-Fairbanks
Anna Baumgartner, Anupama Rao, and Eiluned Roberts
Faculty Advisor: Professor Robert A. Hollister

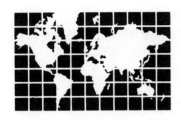

# CONTENTS

| Part I | | INVITED PRESENTATIONS |
|---|---|---|
| 3 | Chapter 1 | **DYNAMIC ENERGY MINIMIZATION AND PHASE TRANSFORMATIONS IN SOLIDS**<br>J. M. Ball |
| 15 | Chapter 2 | **INTERMEDIATE ASYMPTOTICS IN MICROMECHANICS**<br>G. I. Barenblatt |
| 30 | Chapter 3 | **COMPUTER VISION: MATHEMATICS AND COMPUTING**<br>Michael Brady |
| 41 | Chapter 4 | **ADAPTED WAVE FORM ANALYSIS, WAVELET-PACKETS AND APPLICATIONS**<br>R. Coifman, Y. Meyer, and V. Wickerhauser |
| 51 | Chapter 5 | **LARGE-SCALE NONLINEAR CONSTRAINED OPTIMIZATION**<br>A. R. Conn, Nick Gould, and Ph. L. Toint |
| 71 | Chapter 6 | **TIME-SPLITTING METHODS FOR ADVECTION-DIFFUSION-REACTION EQUATIONS ARISING IN CONTAMINANT TRANSPORT**<br>Clint N. Dawson and Mary F. Wheeler |
| 83 | Chapter 7 | **ON MODULATION EQUATIONS OF THE GINZBURG-LANDAU TYPE**<br>Wiktor Eckhaus |
| 99 | Chapter 8 | **MODELLING THE SOLIDIFICATION OF POLYMERS: AN EXAMPLE OF AN ECMI COOPERATION**<br>A. Fasano |
| 119 | Chapter 9 | **DISCRETE MATHEMATICS IN MANUFACTURING**<br>Martin Grötschel |

| | | |
|---|---|---|
| 146 | Chapter 10 | **ANALYSIS OF HYPERSONIC FLOWS IN THERMOCHEMICAL EQUILIBRIUM BY APPLICATION OF THE GALERKIN/LEAST-SQUARES FORMULATION**<br>Frédéric L. Chalot and Thomas J. R. Hughes |
| 160 | Chapter 11 | **INTERIOR-POINT METHODS IN OPTIMIZATION**<br>Narendra Karmarkar |
| 182 | Chapter 12 | **VISCOSITY SOLUTIONS AND OPTIMAL CONTROL**<br>P. L. Lions |
| 196 | Chapter 13 | **DYNAMICS OF PATTERNS, WAVES, AND INTERFACES FROM THE REACTION-DIFFUSION ASPECT**<br>Masayasu Mimura |
| 212 | Chapter 14 | **COMPLEX PATTERN FORMATION IN EMBRYOLOGY: MODELS, MATHEMATICS, AND BIOLOGICAL IMPLICATIONS**<br>J. D. Murray |
| 227 | Chapter 15 | **TRENDS IN RADAR ARCHITECTURES**<br>Gabriel Ruget |
| 241 | Chapter 16 | **MASSIVELY PARALLEL COMPUTING: STATUS AND PROSPECTS**<br>D. J. Wallace |
| 256 | Chapter 17 | **HIERARCHICAL BASES**<br>Harry Yserentant |
| **Part II** | | **MINISYMPOSIA** |
| 279 | Chapter 18 | **APPLIED PROBABILITY AND STATISTICS** |
| 282 | Chapter 19 | **CHEMICAL KINETICS AND COMBUSTION** |
| 283 | Chapter 20 | **COMPUTATIONAL FLUID DYNAMICS** |
| 285 | Chapter 21 | **COMPUTER SCIENCE** |
| 288 | Chapter 22 | **CONTROL AND SYSTEMS THEORY** |
| 292 | Chapter 23 | **DISCRETE MATHEMATICS** |
| 294 | Chapter 24 | **DYNAMICAL SYSTEMS** |
| 296 | Chapter 25 | **ELECTROMAGNETICS AND SEMICONDUCTORS** |
| 299 | Chapter 26 | **ENVIRONMENTAL SCIENCE** |
| 299 | Chapter 27 | **FLUID DYNAMICS** |

| | | |
|---|---|---|
| 304 | Chapter 28 | **GEOMETRIC MODELING, DESIGN, AND COMPUTATION** |
| 307 | Chapter 29 | **GEOPHYSICAL SCIENCES** |
| 307 | Chapter 30 | **INVERSE PROBLEMS** |
| 312 | Chapter 31 | **MANUFACTURING SYSTEMS** |
| 312 | Chapter 32 | **MATERIAL SCIENCE** |
| 316 | Chapter 33 | **MATHEMATICS EDUCATION** |
| 316 | Chapter 34 | **MECHANICS, WAVES, AND SOLIDS** |
| 319 | Chapter 35 | **MEDICINE AND BIOLOGY** |
| 321 | Chapter 36 | **NUMERICAL LINEAR ALGEBRA** |
| 324 | Chapter 37 | **NUMERICAL METHODS IN ORDINARY DIFFERENTIAL EQUATIONS** |
| 326 | Chapter 38 | **NUMERICAL METHODS IN PARTIAL DIFFERENTIAL EQUATIONS** |
| 331 | Chapter 39 | **OCEAN AND ATMOSPHERIC SCIENCE** |
| 332 | Chapter 40 | **OPTIMIZATION** |
| 335 | Chapter 41 | **ORDINARY DIFFERENTIAL EQUATIONS** |
| 337 | Chapter 42 | **PARTIAL DIFFERENTIAL EQUATIONS** |
| 340 | Chapter 43 | **QUANTUM AND STATISTICAL MECHANICS** |
| 341 | Chapter 44 | **REAL AND COMPLEX ANALYSIS** |
| 344 | Chapter 45 | **SCIENTIFIC COMPUTATION** |
| 348 | Chapter 46 | **SIMULATION AND MODELING** |
| 351 | | **LIST OF ATTENDEES** |
| 357 | | **AUTHOR INDEX** |

Part I

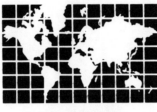

# INVITED PRESENTATIONS

# Chapter 1
# DYNAMIC ENERGY MINIMIZATION AND PHASE TRANSFORMATIONS IN SOLIDS

**J. M. BALL**
Heriot-Watt University
United Kingdom

**1. Introduction.** In this article I discuss the question of how to give a dynamical justification of the variational principles of mechanics and physics. For example, in classical and continuum mechanics it is common to analyse equilibrium problems by seeking a configuration that minimizes an appropriate energy functional; how can this procedure be rationalized on the basis of the behaviour of solutions to the corresponding dynamical equations?

Central to any discussion of these matters is the Second Law of Thermodynamics. A formulation of the Second Law which has earned a fair measure of acceptance by practitioners of continuum thermodynamics, on account of its apparently wide applicability and clear statement, is the *Clausius-Duhem inequality* (for an interesting discussion see Ericksen [25]). Assuming the external volumetric heat supply to be zero, this inequality has the form

$$(1.1) \qquad \frac{d}{dt}\int_\Omega \rho_R \eta \, dx + \int_{\partial\Omega} \frac{q_R \cdot N}{\theta} \, dA \geq 0,$$

where $\Omega \subset \mathbf{R}^3$ denotes the region occupied by a material body in a reference configuration, $\eta$ denotes the entropy density, $q_R$ the material heat-flux vector, $\theta$ the absolute temperature, $\rho_R = \rho_R(x)$ the mass density in the reference configuration, and $N = N(x)$ the unit outward normal to the boundary $\partial\Omega$. In two important cases this inequality endows the governing equations of a material body with a *Lyapunov function*, i.e. a real-valued function of the state of the system which is nonincreasing in time for motions of the body. The first case is that of a thermally isolated body, when $q_R = 0$ on $\partial\Omega$, so that (1.1) implies that

$$(1.2) \qquad \frac{d}{dt}\int_\Omega -\rho_R \eta \, dx \leq 0.$$

The second case is that of a body for which part of the boundary $\partial\Omega_1$ is thermally insulated and the remainder is in contact with a heat-bath at temperature $\theta_0$, where $\theta_0$ does not depend on $x$ but could depend on $t$. Assume for simplicity that the mechanical boundary conditions do no work. Then combining (1.1) with the energy equation

$$\text{(1.3)} \qquad \frac{d}{dt}\int_\Omega \rho_R \left(\frac{1}{2}\mid v\mid^2 +\varepsilon + \psi\right) dx + \int_{\partial\Omega} q_R \cdot N\, dA = 0,$$

where $v$ denotes the velocity, $\varepsilon$ the internal energy density, and $\psi$ the potential energy density of the body-forces, we deduce from (1.1) that

$$\text{(1.4)} \qquad \frac{d}{dt}\int_\Omega \rho_R \left(\frac{1}{2}\mid v\mid^2 +\psi + \varepsilon - \theta_0\eta\right) dx \leq 0,$$

(*cf.* Duhem [21], Ericksen [22], Coleman & Dill [19], for example). Note that the expression $\varepsilon - \theta_0\eta$ in (1.4) is not the same as the Helmholtz free energy $\varepsilon - \theta\eta$.

If $\theta_0$ is allowed to depend on $x$ then in general the existence of a Lyapunov function is in doubt. However, in some special cases when $\theta_0 = \theta_0(x)$ and $q_R = \hat{q}_R(x, \theta, \operatorname{Grad}\theta)$ it has been shown by Ball & Knowles [12] that

$$\text{(1.5)} \qquad \frac{d}{dt}\int_\Omega \rho_R \left(\frac{1}{2}\mid v\mid^2 +\psi + \varepsilon - \varphi(x)\eta\right) dx \leq 0,$$

where $\varphi$ is the solution to the steady-state heat problem

$$\text{(1.6)} \qquad \operatorname{div}\hat{q}_R(x, \varphi, \operatorname{Grad}\varphi) = 0, \quad x \in \Omega,$$

$$\text{(1.7)} \qquad \varphi\mid_{\partial\Omega\setminus\partial\Omega_1} = \theta_0, \quad \hat{q}_R(x, \varphi, \operatorname{Grad}\varphi) \cdot N\mid_{\partial\Omega_1} = 0.$$

There seems to be a wide gulf between modern continuum thermodynamics as expressed, for example, by the balance of energy and the Clausius-Duhem inequality, and nonequilibrium statistical physics and kinetic theory. The Clausius-Duhem inequality is systematically applied with apparent success to a wide range of different materials out of equilibrium, leading to the existence of Lyapunov functions as described above. On the other hand statistical physics provides Lyapunov functions only for very special materials, the main example being the H-theorem for the Boltzmann equation, which models a moderately rarified monatomic gas. One approach to trying to bridge this gulf is to attempt to make precise the idea that in a system of interacting particles local equilibrium is rapidly achieved (*cf.* Guo, Papanicolaou & Varadhan [28], Varadhan [43]).

With the above as motivation, suppose we are given a dynamical system $T(t)_{t\geq 0}$ on some (say, topological) space $X$, i.e. a family of mappings $T(t): X \to X$ satisfying (i) $T(0) = identity$, (ii) $T(s+t) = T(s)T(t)$ for all $s, t \geq 0$, (iii) the mapping $(t, \varphi) \mapsto T(t)\varphi$ is continuous. Thus $T(t)\varphi$ represents the state reached by the system after time $t$ starting with initial data $\varphi$. Let $V: X \to \mathbf{R}$ be a continuous Lyapunov function, so that $V(T(t)\varphi)$ is nonincreasing on $[0, \infty)$ for each $\varphi \in X$. In order to justify dynamically the variational principle

$$\text{(1.8)} \qquad \text{Minimize } V$$

we would like to show that *if $t_j \to \infty$ then $T(t_j)\varphi$ will be a local minimizing sequence* (appropriately defined) *for $V$*. As described in Ball [5] (of which this article is an updated version), there are various obstacles to proving such a result. Some of these are:

(*a*) Exceptional initial data (such as an unstable rest-point) must be excluded.

(*b*) There may be constants of motion $c_i: X \to \mathbf{R}$, so that $c_i(T(t))\varphi = c_i(\varphi)$ for all $i$ and $t \geq 0$; the traditional remedy is then to change the variational principle to

$$\text{(1.9)} \quad \begin{aligned} &\text{Minimize } V(\psi) \\ &c_i(\psi) = \alpha_i \end{aligned}$$

(c) Nonhyperbolic rest-points. (For example, $V(x) = -x$ is a Lyapunov function for the ordinary differential equation $\dot{x} = x^2(1-x^2)$ on $[-1,1]$. This has the three rest-points -1,0,1. The rest-point 0 is not hyperbolic and is not a local minimizer of $V$, but $x(t) \to 0$ whenever $-1 < x(0) \leq 0$.)

(d) $V$ may not attain a minimum.

As far as I am aware, no sufficiently general theorem for dynamical systems is known that overcomes these obstacles. Some ideas which are of relevance are to be found in the study of prolongational limit sets for dynamical systems (*cf.* Ura [41,42], Auslander & Seibert [2], Bhatia & Szego [14]), and the behaviour of solutions to differential equations with small random perturbations (see for the finite-dimensional case Freidlin & Wentzell [26], and for results for parabolic partial differential equations forthcoming work of Freidlin). I intend to discuss this in a future paper. In particular, the infinite-dimensional situation is far from being understood, and the main purpose of this article is to discuss some infinite-dimensional examples arising in models for phase tranformations in solids, in which (d) is an issue. In these examples the minimum of $V$ is not attained, minimizing sequences for $V$ tending weakly to a state that is not a minimizer.

In Section 2 I consider systems of coagulation-fragmentation equations, and how they model the phenomenon of condensation. Here it is shown that the dynamics realize an absolute minimizing sequence for the free energy. Section 3 concerns a variational approach to the formation of microstructure arising from displacive phase transformations in crystals, while in Section 4 I outline some attempts to understand how this microstructure forms dynamically. Some model problems are discussed which e xhibit surprising differences between the asymptotic behaviour of solutions for rather similar systems.

**2. Coagulation-fragmentation dynamics.** Coagulation-fragmentation equations occur frequently in applications to fields such as astrophysics, atmospheric physics, biology, colloidal chemistry, polymer science and the kinetics of phase transformations in alloys. They are appropriate for many systems in which the objects of interest are a large number of clusters of particles, and model the time-evolution of the distribution of cluster sizes as the clusters coalesce to form larger clusters and fragment to form smaller ones. In the case of phase transformations in a binary alloy, the clusters consist of atoms of the minority component of the alloy (or, in more complex cases, atoms of both components in a definite proportion corresponding to a particular phase of the alloy), and the i nteractions between clusters occur by diffusion of atoms on the underlying crystal lattice.

We assume that only binary interactions between clusters occur. For example, a 3-cluster (i.e. a cluster consisting of 3 particles) may coalesce with a 7-cluster to form a 10-cluster, or the reverse may occur, a 10-cluster fragmenting into a 3-cluster and a 7-cluster, but we ignore interactions in which, say, a 4-cluster fragments into a 2-cluster and two 1-clusters. Let $c_j = c_j(t) \geq 0$, $j = 1, 2...$, denote the expected number per unit volume of $j$-clusters in the system at time $t$. Assuming that the rate a t which $j$-clusters coalesce with $k$-clusters to form $(j+k)$-clusters is proportional both to $c_j$ and $c_k$, while the rate at which $(j+k)$-clusters fragment into $j$-clusters and $k$-clusters is proportional to $c_{j+k}$, we are led to the *discrete coagulation-fragmentation equations* for $c = (c_j)$:

$$(2.10) \quad \dot{c}_j = \frac{1}{2}\sum_{k=1}^{j-1}(a_{j-k,k}c_{j-k}c_k - b_{j-k,k}c_j) - \sum_{k=1}^{\infty}(a_{j,k}c_jc_k - b_{j,k}c_{j+k}),$$

$$j = 1, 2, \dots.$$

The rate coefficients $a_{j,k}$, $b_{j,k}$ are assumed to be constant and satisfy the conditions

$$(2.11) \quad a_{j,k} = a_{k,j} \geq 0,$$
$$(2.12) \quad b_{j,k} = b_{k,j} \geq 0.$$

Because each interaction preserves the number of particles, we expect that density will be conserved, i.e.

$$(2.13) \quad \sum_{j=1}^{\infty} jc_j(t) = \rho = constant.$$

While (2.13) holds formally for solutions of (2.10), it is well known that for some rate coefficients $a_{j,k}$, $b_{j,k}$ density conservation breaks down after a finite time due to the formation of an infinite cluster or gel. Mathematically, the situation is analogous to that for nonlinear hyperbolic equations, for which various entropies which are conserved for smooth solutions fail to be conserved for solutions containing shock waves. Here we confine attention to solutions conserving density. It is proved in Ball & Carr [7] that at least one density-conserving solution exists provided $a_{j,k} \leq K(j+k)$ for all $j, k \geq 1$, where $K$ is a constant. Conditions on $a_{j,k}$, $b_{j,k}$ are also given in [7] under which all solutions conserve density, and under which solutions are unique.

We now assume that the rate coefficients satisfy the *detailed balance* condition

$$(2.14) \quad a_{j,k}Q_jQ_k = Q_{j+k}b_{j,k},$$

for positive constants $Q_j$ with $Q_1 = 1$, and we confine attention to the physically interesting case

$$(2.15) \quad 0 < z_s < \infty,$$

where

$$(2.16) \quad z_s^{-1} = \limsup_{j\to\infty} Q_j^{1/j}.$$

It is easily verified that for each $z \geq 0$

$$(2.17) \quad c_j = Q_j z^j, \quad j = 1, 2\dots,$$

is a rest-point of (2.10). In fact, in this case the net rate $W_{j,k} = a_{j,k}c_jc_k - b_{j,k}c_{j+k}$ of conversion of $j$-clusters and $k$-clusters to $(j+k)$-clusters is zero (this being the physical meaning of detailed balancing). To see whether these rest-points have finite density given by (2.13), note that by (2.16) the radius of convergence of the series $F(z) = \sum_{j=1}^{\infty} jQ_j z^j$ is $z_s$. Let

$$(2.18) \quad \rho_s = \sum_{j=1}^{\infty} jQ_j z_s^j.$$

Then for $0 \leq \rho \leq \rho_s$, $\rho < \infty$, there is a unique rest-point $c^\rho$ of the form (2.17) of density $\rho$, namely

(2.19) $$c_j = Q_j z(\rho)^j, \quad j = 1, 2, \ldots,$$

where $F(z(\rho)) = \rho$.

Let

(2.20) $$V(c) = \sum_{j=1}^{\infty} c_j \left( \ln \left( \frac{c_j}{Q_j} \right) - 1 \right),$$

where $c = (c_j)$. A formal calculation shows that

(2.21) $$\frac{dV}{dt} = -\frac{1}{2} \sum_{j,k=1}^{\infty} \left( \ln (a_{j,k} c_j c_k) - \ln (b_{j,k} c_{j+k}) \right) (a_{j,k} c_j c_k - b_{j,k} c_{j+k}),$$

so that $V$ is a Liapunov function (the *free energy*). Note that (2.21) shows formally that all rest-points of (2.10) are of the form (2.17).

Since the density (2.13) is a constant of motion, the appropriate variational principle for $c = (c_j) \geq 0$ is

(2.22) $$\begin{array}{c} \text{Minimize} \quad V(c) \\ \sum_{j=1}^{\infty} j c_j = \rho \end{array},$$

where $\rho \geq 0$. If $0 \leq \rho \leq \rho_s$, $\rho < \infty$, then it is easily verified that $c^\rho$ is the unique minimizer of (2.22), and that any minimizing sequence $c^{(k)}$ converges strongly in

(2.23) $$X \stackrel{\text{def}}{=} \{ c \geq 0 : \| c \| = \sum_{j=1}^{\infty} j \mid c_j \mid < \infty \},$$

to $c^\rho$. However, if $\rho_s < \rho < \infty$ then the minimum of $V$ subject to $\sum_{j=1}^{\infty} j c_j = \rho$ *is not attained*. In fact, in this case minimizing sequences $c^{(k)}$ for (2.22) converge weak* to $c^{\rho_s}$ in $X$ i.e. $c_j^{(k)} \to c_j^{\rho_s}$ as $k \to \infty$ for all $j$. There is therefore a difference $\rho - \rho_s$ between the density of elements $c^{(k)}$ of the minimizing sequence and the density of the limit $c^{\rho_s}$. This difference corresponds to the formation as $k$ increases of clusters of arbitrarily large size. In this way the variational principle (2.22) predicts the macroscopic phenomenon of *condensation*. It is of interest to note (see Ball [4]) that this same variational structure occurs in a number of other problems, for example the Thomas-Fermi model of atoms and molecules, the equilibrium of an incompressible fluid above a surface, and an adiabatic model predicting a finite height for the atmosphere. In each case there is a critical parameter value $\rho_s$ such that for $0 < \rho \leq \rho_s$ the minimum of the energy is attained, while for $\rho > \rho_s$ the minimum is not attained, minimizing sequences converging weakly to the minimizer of maximal parameter value $\rho_s$.

In order to provide a dynamical description of condensation one must show that given initial data $c_0 = (c_{0j})$ of density $\rho$, the solution $c$ of (2.10) with $c(0) = c_0$ generates a minimizing sequence for (2.22), that is that if $0 \leq \rho \leq \rho_s$, $\rho < \infty$ then $c(t) \to c^\rho$ strongly in $X$ as $t \to \infty$, while if $\rho_s < \rho < \infty$ then $c(t) \stackrel{*}{\rightharpoonup} c^{\rho_s}$ in $X$ as $t \to \infty$. This has so far only been established in some special cases. For the *Becker-Döring equations*, for which

(2.24) $$a_{j,k} = b_{j,k} = 0 \quad \text{if both } j > 1, k > 1,$$

the result was proved by Ball, Carr & Penrose [8] (for technical refinements see Ball & Carr [6], Slemrod [38]). When all the $a_{j,k}, b_{j,k}$ are nonzero the only result known is that of Carr [17] under hypotheses which imply that $\rho_s = \infty$, so that condensation does not occur. For the general case, parts of the argument for proving the result are available in [7] (for the continuous a nalogue of (2.10) see Stewart [39]); what is missing is an analogue or substitute for the maximum principle technique used in [8].

## 3. A variational approach to the formation of fine microstructure in crystals.

When cooled below a critical temperature at which a phase transformation involving a change of symmetry occurs, crystals typically develop characteristic patterns of *microstructure*. In the simplest case this consists of many fine parallel bands, in each of which the deformation of the crystal is affine with respect to its original configuration. In some materials (e.g. InTl [13,16]) this microstructure can be observed in optical microscopes, but in others (e.g. NiMn [3]) high resolution electron microscopy is required, and reveals that the bands are sometimes only a few atomic spacings thick.

Why does such microstructure form, and how can its geometric features be predicted? I describe briefly here a variational approach to this question due to Ball & James [10,11] (for related work on the same model see, for example, [18,24,20]). Following Ericksen [23] the crystal is modelled using nonlinear (thermo)elasticity, which allows both for large deformations and nonlinear stress-strain behaviour. While rarely used by metallurgists, this theory is ideally suited to the study of the mechanical properties of crystals since it incorporates in a natural way, and without the inappropriate approximations of linear elasticity, the symmetries arising from the underlying lattice structure of the crystal and invariance to rigid-body rotations. (There is an earlier 'linearized' version of the theory in [10,11], due to Khachaturyan [29,30], Khachaturyan & Shatalov [31] and Roitburd [35,36], which is still nonlinear; for discussions of the relationship between the two theories see Bhattacharya [15], Kohn [32] and [11].)

Configurations of the crystal are described by invertible mappings $y : \Omega \to \mathbf{R}^3$, where $\Omega \subset \mathbf{R}^3$ is the region occupied by the crystal in a reference configuration. We suppose that $\Omega$ is bounded and open with sufficiently smooth boundary $\partial\Omega$. The total elastic energy is given by

(3.25) $$I(y) = \int_\Omega W(Dy(x))\,dx,$$

where $Dy$ denotes the gradient of $y$, and $W$ is the free-energy function of the crystal. In order to describe phase transformations it is important to allow $W$ to depend on temperature, but here we assume that the temperature i s held constant, and so we suppress this dependence. We temporarily ignore all other contributions to the energy, in particular any energy associated with interfaces across which $Dy$ jumps. For simplicity, consider a problem in which the deformed position of every point of $\partial\Omega$ is specified, so that

(3.26) $$y\,|_{\partial\Omega} = \bar{y}(\cdot),$$

for a given mapping $\bar{y} : \partial\Omega \to \mathbf{R}^3$. We now consider the variational principle

(3.27) $$\text{Minimize } I(y) \text{ subject to (3.26)}.$$

The key result of the theory is then that for $W$ appropriate for crystals the minimum is in general *not attained*, finer and finer microstructure being needed to get closer and closer to the infimum of $I$. In this way the static theory explains the occurence of microstructure as being a natural consequence of the behaviour of minimizing sequences for a problem of the calculus of variations with a non-attained minimum.

To give some insight into why the minimum of (3.27) need not be attained, consider the case of a cubic to tetragonal phase tranformation, such as occurs for InTl or NiMn. We suppose that the temperature is less than the transformation temperature, but take for the reference configuration the undistorted high-temperature cubic phase. Then it is natural to assume that the set $M$ of $3 \times 3$ matrices $A$ which minimize $W = W(A)$ is given by

$$(3.28) \qquad M = SO(3)U_1 \cup SO(3)U_2 \cup SO(3)U_3,$$

where

$$(3.29) \quad U_1 = \text{diag}\,(\eta_2, \eta_1, \eta_1), \quad U_2 = \text{diag}\,(\eta_1, \eta_2, \eta_1), \quad U_3 = \text{diag}\,(\eta_1, \eta_1, \eta_2),$$

and $\eta_1 > 0$, $\eta_2 > 0$ are the lattice parameters of the transformation. The matrices $U_i$ represent linear transformations of a cube into tetragons with sides parallel to the cubic axes, while the factors $SO(3)$ reflect the invariance of $W$ to rigid-body rotations. We suppose without loss of generality that $W(A) = 0$ for $A \in M$. An elementary (but slightly tricky) calculation shows that given any matrix $A \in SO(3)U_1$, say, there are precisely 2 matrices $B \in SO(3)U_2$ and 2 matrices $B \in SO(3)U_3$ which differ from $A$ by a matrix of rank one. Picking one of these 4 matrices we have

$$(3.30) \qquad\qquad B - A = a \otimes n$$

for some non-zero vectors $a, n \in \mathbf{R}^3$. Thus we can construct a sequence $y^\varepsilon$ of deformations whose gradients $Dy^\varepsilon$ take the values $A$ and $B$ alternately in layers normal to $n$ of thicknesses $\lambda\varepsilon$ and $(1-\lambda)\varepsilon$ respectively, where $0 < \lambda < 1$. Clearly $W(Dy^\varepsilon(x)) = 0$ for a.e. $x \in \Omega$. Let $0 \in \Omega$. By adding a suitable constant vector to each $y^\varepsilon$ we can assume without loss of generality that $y^\varepsilon(0) = 0$. Then as $\varepsilon \to 0$, $y^\varepsilon \to (\lambda A + (1-\lambda)B)x$ uniformly in $\Omega$, but $Dy^\varepsilon$ oscillates, converging only weak* in $L^\infty$ to $\lambda A + (1-\lambda)B$. Suppose now that the boundary data is given by $\bar{y}(x) = (\lambda A + (1-\lambda)B)x$. Of course $y^\varepsilon$ does not satisfy (3.26), but by modifying $y^\varepsilon$ in a thin boundary layer near $\partial\Omega$ it is easy to arrange that the modified sequence $\tilde{y}^\varepsilon$ satisfies (3.26) and that the energy in the boundary-layer tends to zero as $\varepsilon \to 0$, so that $\lim_{\varepsilon \to 0} I(\tilde{y}^\varepsilon) = 0$ and $\tilde{y}^\varepsilon$ is a minimizing sequence. Hence for these boundary conditions $\inf I = 0$. The nontrivial part of the analysis is to show that *any* minimizing sequence must behave like $\tilde{y}^\varepsilon$. In fact, it is proved in [11], using the weak continuity properties of Jacobians, that the gradient $Dy^\varepsilon$ of every minimizing sequence $y^\varepsilon$ has the same *Young measure* $(\nu_x)_{x \in \Omega}$, namely the constant measure

$$(3.31) \qquad\qquad \nu_x = \lambda \delta_A + (1-\lambda)\delta_B.$$

Said differently, the values of $Dy^\varepsilon(x)$ converge in measure to the set $\{A, B\}$ in such a way that the local volume fractions of points having gradient near $A$ (resp. near $B$) tends to $\lambda$ (resp. $1 - \lambda$).

For general boundary conditions $\bar{y}$ more complicated microstructure than simple layering seems to be necessary in order to get a minimizing sequence. The constructions in [11] use 'layers within layers'. For example, a minimizing sequence with Young measure

$$(3.32) \qquad \nu_x = \lambda[\nu \delta_A + (1-\nu)\delta_B] + (1-\lambda)[\mu \delta_C + (1-\mu)\delta_D]$$

can be constructed by having alternate layers of thicknesses $\lambda \varepsilon$ and $(1-\lambda)\varepsilon$ respectively, the layers of thickness $\lambda \varepsilon$ consisting of sublayers of thicknesses $\nu \varepsilon^2$, $(1-\nu)\varepsilon^2$ in which $Dy^\varepsilon(x)$ takes the values $A$ and $B$, and the layers of thickness $(1-\lambda)\varepsilon$ consisting of sublayers of thicknesses $\mu \varepsilon^2$, $(1-\mu)\varepsilon^2$ in which $Dy^\varepsilon(x)$ takes the values $C$ and $D$. In order to achieve compatibility, the matrices $A, B, C, D \in M$ must satisfy the relations

$$(3.33) \qquad B - A = a \otimes n, \quad D - C = b \otimes l,$$

$$(3.34) \qquad \nu A + (1-\nu)B - [\mu C + (1-\mu)D] = c \otimes m$$

for nonzero vectors $a, b, c, l, m, n$. Boundary layers are needed near the interfaces between the larger layers. Such microstructures are observed; see, for example, Arlt [1].

The above theory is quite successful, predicting the correct interface orientations in a variety of observed microstructures. It also illuminates some central questions concerning lower-semicontinuity and regularity in the calculus of variations. The theory predicts *infinitely fine* microstructures. This is both a good and a bad feature, good because extremely fine microstructures are commonly observed, and bad because these microstructures are of course not infinitely fine. The conventional explanation for limited fineness is that there is a small amount of interfacial energy which contributes significantly to the total energy of a fine microstructure. Model 1 in the next section suggests the interesting possibility that there could be additional dynamic effects that limit fineness.

**4. Dynamics and crystal microstructure.** In this section I explore the question of whether appropriate dynamical equations for crystals have the property that solutions produce finer and finer microstructure as time $t$ increases, thus providing a dynamical justification of the variational principle (3.27). This is a difficult question both from the point of view of modelling and of analysis. As regards modelling, it is not clear what dynamics should be assumed (and in particular what dissipation should be associated with the propagation of interfaces), while the analysis of even the simplest models seems to be beyond the scope of existing methods. For example, a simple viscoelastic model, ignoring thermal effects and the influence of anisotropy on dissipation, is given by the system

$$(4.35) \qquad \rho_R y_{tt} = \operatorname{div} D_A W(Dy) + \beta \Delta y_t,$$
$$(4.36) \qquad y\,|_{\partial \Omega} = \bar{y},$$

with initial data

$$(4.37) \qquad y(x,0) = y_0(x), \; y_t(x,0) = y_1(x),$$

where $\rho_R > 0$ is the density in the reference configuration, $\beta > 0$ is constant, and $y_0, y_1$ are given functions with $y_0 |_{\partial\Omega} = \bar{y}$. Under hypotheses on $W$ appropriate for crystals, Rybka [37] has proved the existence of solutions for the corresponding problem with zero-traction boundary conditions, and his techniques can probably be adapted for (4.36). However, I am not aware of any results concerning the asymptotic behaviour of solutions. A very interesting video of numerical computations of Swart & Holmes [40] for the case when $\Omega = (0,1)^2 \subset \mathbf{R}^2$, $y$ is a scalar, and $W$ has only a finite number of minimizers, shows in a striking way how microstructure is rapidly generated by solutions.

With a view to gaining insight into models such as (4.35)-(4.37), a study of some prototype one-dimensional models was carried out by Ball, Holmes, James, Pego & Swart [9]. A surprising difference was found between the asymptotic behaviour of solutions in two closely related models. In both models the unknown is a scalar $u = u(x,t)$ defined for $0 \leq x \leq \pi$, $t \geq 0$.

*Model 1* consists of the system

$$(4.38) \qquad u_{tt} = (u_x^3 - u_x)_x - \alpha u + \beta u_{xxt},$$
$$(4.39) \qquad u(0,t) = u(\pi,t) = 0$$

where $\alpha > 0, \beta > 0$ are constants, with initial conditions

$$(4.40) \qquad u(x,0) = u_0(x), \ u_t(x,0) = u_1(x).$$

Writing $w = u_t$, and

$$(4.41) \qquad V_1(u,w) = \int_0^\pi \left[\frac{1}{2}w^2 + \frac{1}{4}(u_x^2 - 1)^2 + \frac{\alpha}{2}u^2\right] dx,$$

a formal calculation shows that solutions to (4.38),(4.39) satisfy

$$(4.42) \qquad \frac{d}{dt}V_1(u,u_t) = -\beta \int_0^\pi |u_{xt}|^2 \, dx \leq 0,$$

so that $V_1$ is a Lyapunov function. A phase-plane analysis shows that there are uncountably many rest points for (4.38),(4.39). Uncountably many of these are weak relative minimizers of $V_1$ in $X = W_0^{1,\infty}(0,\pi) \times L^2(0,\pi)$, but none of them are strong relative minimizers, i.e. local minimizers in the energy space $W_0^{1,4}(0,\pi) \times L^2(0,\pi)$. Furthermore, $\inf_{\{u,w\}\in X} V_1(u,w) = 0$ but the minimum *is not attained*. The following theorem is proved in [9] (together with related existence, uniqueness and regularity theorems for (4.38),(4.39)):

THEOREM 4.1. *There is no solution of* (4.38),(4.39) *with initial data* $\{u_0, u_1\} \in X$ *such that*

$$(4.43) \qquad \lim_{t\to\infty} V_1(u,u_t) = 0.$$

Thus there is no solution of (4.38),(4.39) which realizes an absolute minimizing sequence for $V_1$! It is conjectured in [9] that in fact every solution converges to a rest point as $t \to \infty$. Theorem 4.1 and the conjecture are not inconsistent with the possibility that almost every solution converges to a weak relative minimizer of $V_1$; thus a result of the type discussed in the introduction could hold.

In the simpler case when $\alpha = 0$, Pego [33] proved that every solution to (4.38) converges to a rest-point as $t \to \infty$ when the boundary condition at $x = \pi$ is changed to that of zero traction, i.e. $\sigma(u_x) + \beta u_{xt} = 0$ at $x = \pi$, where $\sigma(z) = z^3 - z$. However, for this $\sigma$ and the zero displacement boundary conditions (4.39) it is not known whether every solution converges to a rest-point. If $\alpha = 0$ and we drop the $u_{tt}$ term in (4.38) then (4.38),(4.39) are seen to be equivalent to the equation

$$(4.44) \qquad \beta z_t = -\sigma(z) + \frac{1}{\pi} \int_0^\pi \sigma(z)\, dx$$

for $z = u_x$. Recently, Friesecke [27], following earlier work on a finite-dimensional version of (4.44) by Pego [34], has shown that each solution of (4.44) that is bounded in $L^\infty(\Omega)$ converges boundedly a.e. to a rest-point, provided that $\sigma$ is $C^1$ and is not constant on any interval. It remains to be seen whether these methods can be adapted to handle the case of (4.38),(4.39) with $\alpha > 0$.

*Model 2* consists of the system

$$(4.45) \qquad u_{tt} = (\|u_x\|^2 u_x - u_x)_x - \alpha u + \beta u_{xxt},$$
$$(4.46) \qquad u(0,t) = u(\pi,t) = 0$$

where $\alpha > 0, \beta > 0$ are constants, with initial conditions

$$(4.47) \qquad u(x,0) = u_0(x),\ u_t(x,o) = u_1(x),$$

which is obtained from (4.38)-(4.40) by replacing the term $u_x^3$ with the nonlocal expression $\|u_x\|^2 u_x$, where $\|u_x\| = \left(\int_0^\pi u_x^2\, dx\right)^{\frac{1}{2}}$. Defining

$$(4.48) \qquad V_2(u,w) = \int_0^\pi \left[\frac{1}{2} w^2 + \frac{\alpha}{2} u^2\right] dx + \frac{1}{4}(\|u_x\|^2 - 1)^2,$$

we find that solutions of (4.45),(4.46) satisfy

$$(4.49) \qquad \frac{d}{dt} V_2(u, u_t) = -\beta \int_0^\pi |u_{xt}|^2\, dx \leq 0.$$

An elementary Fourier analysis shows that there are a countable number of rest-points $\{\psi_k, 0\}$ for (4.45),(4.46). Also $\inf_{\{u,w\} \in X} V_2(u,w) = 0$, where $X = W_0^{1,2}(0,\pi) \times L^2(0,\pi)$ but the minimum *is not attained*. The following theorem is proved in [9]:

THEOREM 4.2. *$X$ can be written as the disjoint union of two dense sets $A_1, A_2$ of first and second category respectively. For initial data $\{u_0, u_1\} \in A_1$, $\{u, u_t\}$ converges in $X$ to a rest-point $\{\psi_k, 0\}$ as $t \to \infty$. For initial data $\{u_0, u_1\} \in A_2$,*

$$(4.50) \qquad \lim_{t \to \infty} V_2(u, u_t) = 0.$$

Thus in contrast to Model 1, the dynamics of Model 2 generically realize an absolute minimizing sequence for $V_2$.

**Acknowledgement.** The work reported here was supported by SERC grants GR/D73096, GR/E69690, GR/F86427 and the EEC grant EECST2J-0216-C.

## REFERENCES

[1] G. ARLT, *Twinning in ferroelectric and ferroelastic ceramics: stress relief*, J. Materials Science, 22 (1990), pp. 2655-2666.

[2] J. AUSLANDER AND P. SEIBERT, *Prolongations and stability in dynamical systems*, Ann. Inst. Fourier, Grenoble, 14 (1964), pp. 237–268.
[3] I. BAELE, G. V. TENDELOO, AND S. AMELINCKX, *Microtwinning in Ni-Mn resulting from the $\beta - \theta$ martensitic transformation*, Acta Metallurgica, 35 (1987), pp. 401–412.
[4] J. M. BALL, *Loss of the constraint in convex variational problems*, in Analyse Mathématique et Applications; Contributions en l'Honneur de J.-L.Lions, Gauthier-Villars, 1988, pp. 39–53.
[5] ———, *Dynamics and minimizing sequences*, in Problems involving change of type, K. Kirchgässner, ed., Springer Lecture Notes in Physics no. 359, Springer-Verlag, New York; Heidelberg, Berlin, 1990, pp. 3–16.
[6] J. M. BALL AND J. CARR, *Asymptotic behaviour of solutions of the Becker-Döring equations for arbitrary initial data*, Proc. Royal Soc. Edinburgh A, 108 (1988), pp. 109–116.
[7] ———, *The discrete coagulation-fragmentation equations; existence, uniqueness, and density conservation*, J. Statistical Physics, 61 (1990), pp. 203–234.
[8] J. M. BALL, J. CARR, AND O. PENROSE, *The Becker-Döring cluster equations; basic properties and asymptotic behaviour of solutions*, Comm. Math. Phys., 104 (1986), pp. 657–692.
[9] J. M. BALL, P. J. HOLMES, R. D. JAMES, R. L. PEGO, AND P. J. SWART, *On the dynamics of fine structure*, J. Nonlinear Sci., 1 (1991), pp. 17–90.
[10] J. M. BALL AND R. D. JAMES, *Fine phase mixtures as minimizers of energy*, Arch. Rat. Mech. Anal., 100 (1987), pp. 13–52.
[11] ———, *Proposed experimental tests of a theory of fine microstructure, and the two-well problem*, Phil. Trans. Roy. Soc. London A, (to appear).
[12] J. M. BALL AND G. KNOWLES, *Lyapunov functions for thermoelasticity with spatially varying boundary temperatures*, Arch. Rat. Mech. Anal., 92 (1986), pp. 193–204.
[13] Z. S. BASINSKI AND M. A. CHRISTIAN, *Experiments on the martensitic transformation in single crystals of indium-thallium alloys*, Acta Metallurgica, 2 (1954), pp. 148–166.
[14] N. P. BHATIA AND G. P. SZEGÖ, *Stability theory of dynamical systems*, vol. 161 of Grundlehren der mathematischen Wissenschaften, Springer-Verlag, Berlin, Heidelberg, New York, 1970.
[15] K. BHATTACHARYA, *Linear and nonlinear thermoelasticity theory for crystalline solids*, (in preparation).
[16] M. W. BURKART AND T. A. READ, *Diffusionless phase change in the indium-thallium system*, Trans. AIME J, Metals, 197 (1953), pp. 1516–1524.
[17] J. CARR, *Asymptotic behaviour of solutions to the coagulation-fragmentation equations, 1. the strong fragmentation case*, (to appear).
[18] M. CHIPOT AND D. KINDERLEHRER, *Equilibrium configurations of crystals*, Arch. Rat. Mech. Anal., 103 (1988), pp. 237–277.
[19] B. D. COLEMAN AND E. H. DILL, *On thermodynamics and the stability of motion of materials with memory*, Arch. Rat. Mech. Anal., 51 (1973), pp. 1–53.
[20] C. COLLINS AND M. LUSKIN, *The computation of the austenitic-martensitic phase transition*, in Partial Differential Equations and Continuum Models of Phase Transitions, M. Rascle, D. Serre, and M. Slemrod, eds., Lecture Notes in Physics 344, Springer-Verlag, 1990, pp. 34–50.
[21] P. DUHEM, *Traité d'Énergetique ou de Thermodynamique Générale*, Gauthier-Villars, Paris, 1911.
[22] J. L. ERICKSEN, *Thermoelastic stability*, in Proc $5^{th}$ National Cong. Appl. Mech., 1966, pp. 187–193.
[23] ———, *Special topics in elastostatics*, in Advances in Applied Mechanics, C.-S. Yih, ed., vol. 17, Academic Press, 1977, pp. 189–244.
[24] ———, *Constitutive theory for some constrained elastic crystals*, Int. J. Solids and Structures, 22 (1986), pp. 951–964.
[25] ———, *Introduction to the thermodynamics of solids*, Applied Mathematics and Mathematical Computation 1, Chapman and Hall, London, 1991.
[26] M. I. FREIDLIN AND A. D. WENTZELL, *Random perturbations of dynamical systems*, vol. 260 of Grundlehren der mathematischen Wissenschaften, Springer-Verlag, Berlin, Heidelberg, New York, Tokyo, 1984.
[27] G. FRIESECKE. to appear.
[28] M. Z. GUO, G. C. PAPANICOLAOU, AND S. R. S. VARADHAN, *Nonlinear diffusion limit for a system with nearest neighbour interactions*, Comm. Math. Phys., 118 (1988), pp. 31–59.
[29] A. G. KHACHATURYAN, *Some questions concerning the theory of phase transformations in solids*, Soviet Physics - Solid State, 8 (1967), pp. 2163–2168.
[30] ———, *Theory of Structural Transformations in Solids*, John Wiley, 1983.

[31] A. G. KHACHATURYAN AND G. A. SHATALOV, *Theory of macroscopic periodicity for a phase transition in the solid state*, Soviet Physics JETP, 29 (1969), pp. 557–561.
[32] R. V. KOHN, *The relaxation of a double-well energy*, Continuum Mechanics and Thermodynamics, (to appear).
[33] R. L. PEGO, *Phase transitions in one-dimensional nonlinear viscoelasticity: admissibility and stability*, Arch. Rat. Mech. Anal., 97 (1987), pp. 353–394.
[34] ———, *Stabilization in a gradient system with a conservation law*, (preprint).
[35] A. L. ROITBURD, Kristallografiya, (1967), p. 567 ff. in Russian.
[36] ———, *Martensitic transformation as a typical phase transformation in solids*, Solid State Physics, 33 (1978), pp. 317–390.
[37] P. RYBKA, *Dynamical modeling of phase transitions by means of viscoelasticity in many dimensions*, Proc. Royal Soc. Edinburgh, (to appear).
[38] M. SLEMROD, *Trend to equilibrium in the Becker-Döring cluster equations*, Nonlinearity, 2 (1989), pp. 429–443.
[39] I. W. STEWART, *A global existence theorem for the general coagulation-fragmentation equation with unbounded kernels*, Math. Meth. Appl. Sci., 11 (1989), pp. 627–648.
[40] P. J. SWART AND P. J. HOLMES, *Dynamics of phase transitions in nonlinear viscoelasticity*. Video, Cornell University, 1991.
[41] T. URA, *Sur les courbes définies par les équations différentielles dans l'espace à m dimensions*, Ann. Sci. Ecole Norm. Sup., 70 (1953), pp. 287–360.
[42] ———, *Sur le courant extérieur à une région invariante*, Funk. Ekv., 2 (1959), pp. 143–200.
[43] S. R. S. VARADHAN, *Scaling limits for interacting diffusions*, Comm. Math. Phys., 135 (1991), pp. 313–353.

# Chapter 2
# INTERMEDIATE ASYMPTOTICS IN MICROMECHANICS

**G. I. BARENBLATT**
Academy of Sciences
Russia

## 1 Introduction

The construction of continuum mechanics, general model of motions and/or equilibrium of real deformable bodies, proceeds in the following way. After proper definition of the basic concepts such as observer, continuous medium, etc., the general covariance principle and the conservation laws are introduced. According to covariance principle the laws of motions of the continuous media should be expressed by the equations equally valid for all observers. This principle of equality of all observers claims the invariant nature of the properties of the medium and its motions and therefore greatly simplifies the investigation. Conservation laws express the general fact that matter, momentum, angular momentum, energy, etc., do not arise from nothing and do not disappear by themselves. In combination with the covariance principle, conservation laws lead to such concepts as flux vector, stress tensor, couple stresses, etc.

Conservation laws and covariance principle taken alone are insufficient for designing mathematical models of motions of real bodies. To achieve this goal it is necessary to provide the continuum by physical properties. To provide means here to propose a certain a priori model of continuum, adequate for the motions and bodies under consideration but in general valid for restricted classes of real bodies motions only. Such models are represented by certain relations between the properties of motion (or state) of continuous medium and internal forces acting in it, so-called constitutive equations.

Simple examples are the classical models of continuous media, such as Hooke's elastic solid, or Newtonian viscous fluid. The model of an ideally linear elastic body is characterized by three constants: density $\varrho$, and two elastic constants, Young modulus $E$ and Poisson ratio $\nu$. More recent extension of the model of an elastic solid for the bodies with cracks assumes that outside of infinitely thin cracks, the body is perfectly linear elastic, but near the edges of the cracks cohesion forces are acting on the crack surfaces whose distribution over crack surface is autonomous, i.e., independent of form and position of crack outside the edge regions. For this extended model an additional constant is needed: fracture thoroughness, or cohesion modulus, an integral characteristic of cohesion forces. In the model of Newtonian viscous fluids, two (assumed to be constant) viscosity coefficients are introduced along with density $\varrho$.

It is provided in classical models that the constants entering constitutive equations being determined from some experiments preserve their values in arbitrary motions.

Utilization of an approach similar to the classic one causes essential difficulties when applied to many new materials entering modern research and high technology, like synthetic and especially composite materials.* The fluids with non-Newtonian properties attract the attention of engineers and (especially in what concerns biological fluids) nature explorers.

Researchers attempted at first to use modified classical models for new materials. So, four-, and even eight-constant models appeared. This way, however, is unreliable. It is not only the growing complexity of the determination of new constants that pricks up, and not even the loss of clear physical sense of these constants. Much worse is that these constants "cease to be constants," i.e., the range of universality where the constants of the model can be considered truly universal ones becomes so narrow that the models can lose their predictive capacity.

The reason is the following. Every material has its own relaxation time $\tau$, i.e., the characteristic time when the shear stresses are preserved under an imposed fixed shear strain. In its turn every process introduces its own characteristic time $T$. Therefore for each process in each material a governing dimensionless parameter appears, called the Deborah number

$$De = \frac{\tau}{T}. \qquad (1.1)$$

For "fluid-like" behaviour of a material in a process $De \ll 1$, for "solid-like" behaviour $De \gg 1$.

We note now that normal durability of a human experiment encloses about 18 decimal orders of magnitude—from $1ns = 10^{-9}s$ to $10^9 s \sim 30$ years, so $10^{-9}s < T < 10^9 s$ is the range of human experiments. For water at normal temperature $\tau \cong 10^{-12}s$, for steel $\tau \cong 10^{12}s$. Thus for the whole range of human experiment at normal temperature $De \ll 1$ for water and $De \gg 1$ for steel. But not for the well-known synthetic material, Silly Putty, for which $\tau \sim 1s$! This material demonstrates all types of behaviour, from perfectly elastic and brittle when shocked or torn quickly to perfectly viscous when it takes the form of a cup where it is put for an hour or less.

An approach was advanced for describing such mixed behaviour which seemed at first glance attractive: spatial and temporal nonlocality was introduced to the models. Practically it meant that the functionals were introduced to constitutive equations, e.g., so-called "simple body"

$$\underline{\sigma} = \underline{\Phi}(\underline{F}(\theta)). \qquad (1.2)$$

Here, $\underline{\sigma}$ is the stress in a given particle, $\underline{F}(\theta)$ the strain gradient in it, $\Phi$ constitutive functional, $\theta$ is the time in the whole range from the beginning of deformation to actual time moment $t: -\infty < \theta < t$.

Without going into details note that the appearance of functionals in constitutive equations means always that the approach is insufficient. Indeed, consider the simple example: gas flow in a tube with heat conducting walls. Imagine an obvious nonsense: we do not want to introduce temperature to our consideration. Evidently the density of a certain particle will become a pressure functional.

It is not necessary to explain that in gas dynamics another way is used: we introduce into consideration the temperature and energy equation added to the mathematical model.

---

*The legend of recent origin of composite materials is obviously not true. In fact all the alloys used since the very beginning of human civilization were perfect composites.

The temperature, however, is nothing but mean energy of microscopic molecular motion, and the energy equation can be interpreted as a kinetic equation for this directly observable characteristic of microstructure. Introduction into consideration of temperature simplified everything, at least the functionals disappear from consideration. It is curious to note that in mechanics of nonclassic bodies, especially in mechanics of polymers, do not consider as a rule the kinetics of microstructural transformations but introduce the functionals to constitutive equations, and, what is even more strange, believe in their universality.

## 2 Micromechanics

In last decades, and most intensively in last years, an alternative approach appeared more and more frequently in the practice of research. According to this approach, the properties of material microstructure, directly or indirectly observable, are introduced into consideration. The equations of macroscopic motions and those of kinetics of microstructural transformations are considered simultaneously.

In our opinion that is the very subject and approach of micromechanics. Thus, micromechanics in our understanding is the branch of continuum mechanics studying the motions for which the variations of microstructure are of governing value for macroscopic behaviour of bodies.

If the time scales of processes under consideration are such that the variations of microstructure can be considered either as instantaneous or as negligible, a classical approach, neglecting microstructural variations and using time-independent constitutive equations, can be applied.

Concerning the term "micromechanics," often it is used in a more narrow sense denoting the technique of obtaining macroscopic properties of bodies on the basis of analysis of their microstructure. From our viewpoint this is only a certain part of the subject of micromechanics, which is much wider.

We have here no time and place for entering a comprehensive historical analysis of the origin and formation of micromechanics. It seems necessary, however, to mention here several milestone works doubtless related to micromechanics in the sense just mentioned. At the same time they belong to the researchers whose reputation in mechanics is out of discussion—they are generally accepted as outstanding representatives of the mechanical community. The last point is especially important: the mechanical community in general is rather conservative in what concerns the subject of mechanical research. Therefore the example of outstanding people within the mechanical community is important: to be modern in mechanics you should know micromechanics.

The first to be mentioned is Th. von Kármán. In a series of papers summarized by his remarkable Maryland lecture entitled "Aerothermochemistry" [1] von Kármán formulated the general problem of aerothermochemistry as the problem of fluid dynamics with chemical transformations accompanied by heat generation. For the transformation rate $W$ of an active component of mixture he used semiempirical formulae of Arrhenius type:

$$W \sim \alpha^n \exp(-E/kT). \tag{2.1}$$

Here $\alpha$ is the mass concentration of active component under consideration, $E$ a constant called activation energy, $k$ the Boltzmann constant, $n$ the "reaction order," another constant. Heat generation is assumed to be proportional to reaction rate.

It is instructive that von Kármán did not intend to invent a new branch of science being motivated by the internal needs of science itself. Just the opposite: he was one of the first who understood that, for instance, the problem of space vehicle reentry to the atmosphere cannot be solved without the proper consideration of the chemical transformation and dissociation of air. More clear for him was the necessity to create this new branch of science in his attempt to design a rational theory of liquid fuel combustion in jet engines. He referenced widely and promoted the pioneering papers of Ya. B. Zeldovich, N. N. Semenov, D. A. Frank-Kamenetsky, and V. G. Levich (see [12]) concerning flame propagation and flame ignition in combustible gases and hydrodynamics of reacting fluids.

In the general lecture on the XIV Congress of IUTAM [2] devoted to "microhydrodynamics," G. G. Batchelor summarized a series of works concerning micromotions of small bodies in fluids. Subsequent papers by himself as well as his students, especially E. J. Hynch, applied the results and approaches of microhydrodynamical studies to create the hydrodynamics of suspensions.

In his papers [3], [4] entitled "Micromechanics" B. Budiansky emphasized the growing attention to "mechanics of very small things" in the mechanical community in the last years. He considered specialized problems, which were very important from the general viewpoint of micromechanics of solids: transformation (toughening due to phase changes) and strongly localized buckling of fibers in composites (a peculiar and very important form of their damage).

It is instructive to look at the photographs of microstructural pictures from mentioned papers of Batchelor [2] and Budiansky [3] to represent better the objects that recently entered the field of interest of the mechanical community.

The last but not least point to be discussed here: whether the coining of a new term is important, legitimizing a new branch of continuum mechanics and so giving it a status of certain independence?

I suppose, yes, it is. The examples of the theory of oscillations, considering the oscillations in a way irrelevant to their physical nature, functional analysis, cybernetics, and to a lesser extent cynergetics, confirm it. A general approach appears, as well as a unified style of analysis of new phenomena, to a certain extent new general ideology. Seemingly uncoordinated results appear in unified form. The transfer of results from one subject to another becomes possible as well as the prediction of results based on the previous experience. Therefore the legitimization of micromechanics to a unified new branch of continuum mechanics seems to be expedient.

## 3  Intermediate asymptotics

The set of equations of micromechanical models is usually very complicated, so analytical solutions of such sets may be obtained only as a rare exception, and even a numerical solution of such sets under given initial and boundary conditions is difficult to obtain. We have, however, to understand better what kind of result is really needed in micromechanical studies. In fact, the researcher in micromechanics even more than in other branches of continuum mechanics actually is not so much interested in a table of values of the function $f$ describing the way in which the property $a$ being studied depends on the governing parameters

$$a_1, \cdots, a_k; \quad b_1, \cdots, b_m$$
$$a = f(a_1, \cdots, a_k; b_1, \cdots, b_m)^* \qquad (3.1)$$

In micromechanical studies researchers are interested mostly in physical laws that determine the major features of the phenomenon under consideration. As a rule huge pads of tables of values of the function $f$ do not increase our understanding of these laws; moreover, often they kill it: in the cooperation of man and computer it is the man who appears to be a weak point!

In fact, the physical laws investigators are really interested in reveal mostly in the development of the phenomenon at times and distances away from initial moment and boundaries such that the effects of random features of initial and/or boundary distributions and/or fine details in the spatial structure of boundaries have already disappeared; however, the system is still far from its ultimate state. Such development is described by a special kind of asymptotics of the solutions, which was called intermediate asymptotics.

Dimensional analysis allows one to represent (3.1) in dimensionless form (see, e.g., [5], [6]):

$$\Pi = \Phi(\Pi_1, \cdots, \Pi_m), \tag{3.2}$$

where dimensionless parameters $\Pi, \Pi_i$ are

$$\Pi = \frac{1}{a_1^p \cdots a_k^r}, \qquad \Pi_i = \frac{b_i}{a_1^{p_i} \cdots a_k^{r_i}}.$$

Intermediate asymptotics always correspond to the case when at least one of the parameters, say $\Pi_1$, is small (large times, large distances, etc.).

It is assumed often that under such conditions we can simply drop the parameter $\Pi_1$ from consideration, and consequently drop the corresponding dimensional parameter $b_1$, so that (3.2) is replaced by a relation with an argument less

$$\Pi = \Phi_1(\Pi_2, \cdots, \Pi_m). \tag{3.3}$$

Such replacement, however, is correct only if there exists a finite and nonzero limit of the function $\Phi$ at $\Pi_1 \to 0$. If it is so, we can replace at small $\Pi_1$ function $\Phi$ by its limiting value $\Phi(0, \Pi_2, \cdots, \Pi_m) = \Phi_1(\Pi_2, \cdots, \Pi_m)$ and representation (3.3) is possible. Remarkable examples are known when this assumption is true, strong concentrated explosion among them (G. I. Taylor, 1941; J. von Newmann, 1941; L. I. Sedov, 1946).

In general, however, such finite limit does not exist and the parameter $\Pi_1$ cannot be dropped, however small it could be, so the number of arguments in (3.2) in general cannot be reduced.

There exists, however, a very important exceptional case when such reduction again becomes possible. In fact, we remind ourselves that what we really are interested in is not the limit at $\Pi_1 \to 0$, but intermediate asymptotics corresponding to small, however finite, values of the parameter $\Pi_1$. There is the special case when the function $\Phi$ has at small $\Pi_1$ a power-type asymptotics

$$\Phi \simeq \Pi_1^{\alpha_1} \Phi_1\left(\frac{\Pi_2}{\Pi_1^{\alpha_2}}, \cdots, \frac{\Pi_m}{\Pi_1^{\alpha_m}}\right). \tag{3.4}$$

---

*We suppose that governing parameters $a_1, \cdots, a_k$ have independent dimensions, whereas the dimensions of quantities $a, b_1, \cdots, b_m$ can be represented as power monomials in the dimensions of $a_1, \cdots, a_k$:

$$[a] = [a_1]^p \cdots [a_k]^r; \qquad [b_i] = [a_1]^{p_i} \cdots [a_k]^{r_i}; \qquad i = 1, \cdots, m$$

$[a]$ denotes the dimensions of quantity $a$.

Substituting (3.4) into (3.2) we again obtain the relation similar to (3.3) with an argument less:

$$\Pi^* = \Phi_1(\Pi_2^*, \cdots, \Pi_m^*),$$

$$\Pi^* = \frac{\Pi}{\Pi_1^{\alpha_1}} = \frac{a}{a_1^{p-\alpha_1 p_1} \cdots a_k^{r-\alpha_1 r_1} b_1^{\alpha_1}}, \qquad (3.5)$$

$$\Pi_i^* = \frac{\Pi_i}{\Pi_1^{\alpha_i}} = \frac{b_i}{a_1^{p_i-\alpha_i p_1} \cdots a_k^{r_i-\alpha_i r_1} b_1^{\alpha_i}}.$$

The difference between (3.3) and (3.5) is essential. The exponents $p - \alpha_1 p_1, \cdots, r_m - \alpha_m r_1$ cannot be obtained by dimensional analysis alone, because the numbers $\alpha_i$ are unknown: asymptotics (3.4) is a fine property of the special problem under consideration and does not follow from dimensional analysis and/or some other general considerations. Moreover, the parameter $b_1$ remains in the resulting equation (3.5) although in power-type combinations with other parameters only.

It is worthwhile to emphasize that this special case occurs in many physical problems. A well-known example is the success of L. Kadanoff, V. Pokrovsky, A. Patashinsky, and K. Wilson in statistical physics. Several terms are coined for it: self-similarity of the second kind, incomplete self-similarity with respect to a parameter; the physicists prefer the short term "scaling." The previous case when a finite limit of function $\Phi$ does exist is called self-similarity of the first kind or complete self-similarity with respect to a parameter.

Scaling is closely related to the intermediate asymptotics of traveling wave type:

$$a = F(x - \lambda t + c). \qquad (3.6)$$

Indeed, if we put $x = \ln \xi$, $t = \ln \tau$, $c = -\ln A$, we obtain

$$a = F\left(\ln \frac{\xi}{A\tau^\lambda}\right) = \Phi\left(\frac{\xi}{A\tau^\lambda}\right), \qquad (3.7)$$

intermediate asymptotics of the scaling type.

In the case of scaling, the dimensional parameter $b_1$ cannot simply be dropped opposite to the case of complete self-similarity. Let us reduce, however, parameter $b_1$ arbitrarily, say multiply it by a certain arbitrary small number $\lambda$, and renormalize properly the quantities $a, b_2, \cdots, b_m$. Then the one-parametric transformation group

$$a' = \lambda^{\alpha_1} a, \qquad b_1' = \lambda b_1, \qquad b_i' = \lambda^{\alpha_i} b_i \qquad (3.8)$$

will leave the intermediate asymptotics (3.5) invariant. The group (3.8) can be called naturally the renormalization group.

In some cases we meet the following situation. Let the function $\Phi_1$ depend on a parameter $\epsilon$, so that for $\epsilon = 0$ all quantities $\alpha_i$ are equal to zero and complete similarity holds. It is possible to solve the problem for small $\epsilon$ using the renormalization group and the perturbation technique, generally speaking the singular one [7], [8]. Important results were obtained by this approach by N. Goldenfeld and his colleagues Y. Oono, O. Martin, and F. Liu [9].

We note that special cases of complete and incomplete self-similarity (scaling) by no means exhaust the possibilities that may occur for small $\Pi_1$. In general the self-similarity (complete and incomplete) for small values of parameter $\Pi_1$ does not exist and the number of arguments in (3.2) cannot be reduced in intermediate asymptotics of micromechanical problems. The advantages of both types of self-similarities are, however, so important that the researchers should never lose their chance leaving complete similarity or scaling unnoticed.

## 4  Example: Neck Propagation (Cold Drawing) in Stretched Polymeric Bars

This example is especially important: for the micromechanics of deformable solids it seems to have the same fundamental value as the problem of flame propagation for aerothermochemistry. In stretching polymeric bars of such materials as nylon with constant grip extension rate, the straining at first proceeds uniformly over the specimen under monotonously increasing load. Suddenly a constriction, the "neck," appears in a certain place of specimen; the load drops, however, opposite to metallic specimens the neck does not proceed to fracture, but starts to propagate along the bar with constant velocity preserving its shape like a solid body (Figure 1).

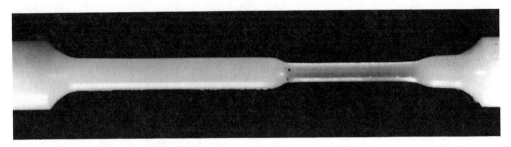

Figure 1: Neck propagation in polymeric bar.

During the neck propagation the load remains constant, until the neck covers the whole specimen.*

Whole straining of the specimen proceeds in the neck region; outside the neck the strain rate is negligibly small, so the straining at this stage proceeds nonuniformly over the specimen length.

The inelastic orientation deformation and strain-hardening of the material take place in the neck. Microscopic and x-ray investigations showed that supramolecular microstructural elements of the polymer undergo a transformation: they transfer from a rather isotropic state to one oriented in the direction of extension. Thus the microstructural elements are elongated, strain-hardened, and their cross-sections are reduced. An average relative cross-section reduction of a microstructural element will be denoted by $\alpha : 0 < \alpha < 1$. It is important that the density of the material, as shown in the experiments, remains practically invariable in the process of orientational transformation. The phenomenon of neck propagation in polymers was discovered by W. H. Carothers and J. W. Hill in 1932 [10] (they called it "cold drawing"; its principal value for mechanics of solids was practically immediately emphasized by A. Nadai [11].

For the mathematical model of the phenomenon it is essential that due to microinhomogeneity the transformation of microstructural elements takes place not simultaneously over the specimen cross-section. So, in a certain intermediate cross-section within the neck, a part of microstructural elements is in transformed oriented state, whereas the remaining part of elements preserves its untransformed state.

Let us introduce the concentration $n(x,t)$—the relative volume of transformed oriented

---

*If the specimen is sufficiently long and there is no special cooling, the neck after some time starts to propagate by jumps and the load starts to oscillate.

microstructural elements; $x$ is the coordinate along the specimen axis; $t$ is the time. The explicit taking into account of the material transformation into oriented state requires the supplementation of the macroscopic mechanical equations by an equation of kinetics of transformation of the microstructure.

We outline here the derivation of a basic model equation for the simpler isothermic case. For a spatially homogeneous transformation process the kinetic equation of microstructural transformation takes the form

$$\partial_t n = \frac{1}{\tau} q(n, \sigma, T), \tag{4.1}$$

where $q(n, \sigma, T)$ is a dimensionless function of concentration $n$, stress $\sigma$, and temperature $T$, $\tau$ is a constant of time dimensions—a characteristic scale of transformation process. The function $q(n, \sigma, T)$ will be taken in the Arrhenius form, analogous to (2.1):

$$q = (1 - n) \exp[-(U - \gamma\sigma/kT], \tag{4.2}$$

with the linearly stress-dependent activation energy (in fact linearly depending on the first invariant of the stress tensor). Here $U$ is the activation energy at zero stress, $\gamma$ is a kinetic constant. The basic feature of the relation (4.2) is that the transformation rate depends strongly on stress and temperature, due to large values of $U/kT$ and $\gamma\sigma/kT$ in the exponent.

Averaging over the cross-section we can write the kinetic equation in the form

$$\partial_t Sn + \partial_x Snu = \frac{1}{\tau} \int_\Omega q(n, \sigma(\omega), T) d\omega. \tag{4.3}$$

Here $S(x, t)$ is the cross-section area, $u(x, t)$ is the displacement velocity, and the integration is carried out over the specimen cross-section $\Omega$. The dependence of the transformation rate on the concentration is rather weak, the process under consideration is an isothermic one, so the quantities $n$ and $T$ in (4.3) can be taken corresponding to the mean values over the cross-section.

The most essential part is that the variation of the transformation rate over the cross-section due to microinhomogeneity is very strong, so by no means the integral in the right-hand side of (4.3) can be replaced by the cross-section area $S$ times the function $q$ corresponding to mean stress. We use here the following approach to calculate this integral. We introduce here the function $\varrho(x - \xi)$, which gives the contribution to bulk reaction rate over the section with coordinate $x$ of the elements where the stress is equal to average stress at some cross-section with coordinate $\xi$. Then the integral in the left-hand side of (4.3) can be transformed to

$$\frac{S}{\tau} \int_{-\infty}^{\infty} q[n, \sigma(\xi, t), T] \varrho(x - \xi) d\xi. \tag{4.4}$$

The following relations hold

$$\int_{-\infty}^{\infty} \varrho(x - \xi) d\xi = 1, \qquad \int_{-\infty}^{\infty} (x - \xi) \varrho(x - \xi) d\xi = 0. \tag{4.5}$$

The first of these relations can be simply obtained from a comparison of (4.1) and (4.3)–(4.4) for the case of spatially homogeneous specimen deformation. The second relation (4.5) follows from the assumption of symmetry which we accept for the sake of simplification only. This assumption seems to be natural because the function $\varrho$ decreases strongly with the growth of its argument modulus. Introduce the correlation length $L$ by a relation

$$L^2 = \frac{1}{2} \int_{-\infty}^{\infty} (x - \xi)^2 \varrho(x - \xi) d\xi. \tag{4.6}$$

The order of magnitude of this length scale $L$ depends on the mechanism of transformation of supramolecular structural elements and varies from several structural element length scales to larger values comparable with specimen width. Expanding the integrand in (4.4) in a Taylor series near $x$ and bearing in mind rapid decreasing of the correlation function $\varrho(x - \xi)$ as well as the relations (4.5), (4.6) we reduce the kinetic equation (4.3) to the instructive form

$$\partial_t n + u \partial_x n = \frac{q}{\tau} + \partial_x \left[ \left( \frac{L^2}{\tau} \partial_\sigma q \right) \partial_x \sigma \right]. \tag{4.7}$$

It is seen that the transformation microinhomogeneity and nonlocality following from it lead to an interesting phenomenon: quasi-tensodiffusional flux of transformed material with a strongly stress and temperature dependent tensodiffusivity coefficient

$$D = \frac{L^2}{\tau} \partial_\sigma q, \tag{4.8}$$

which may (see (4.2)) reach large values under tension being negligible when the applied stress is absent.

The kinetic equation (4.7) closes the system consisting of macroscopic mechanical equations

$$\partial_t S + \partial_x S u = 0, \tag{4.9}$$

the mass balance equation under constant density condition;

$$\sigma S = \sigma_0 S_0 = \text{const}, \tag{4.10}$$

quasistatic equilibrium equation, as well as an obvious relation between concentration $n(x,t)$ and cross-section area $S(x,t)$:

$$S/S_0 = 1 - (1 - \alpha)n. \tag{4.11}$$

Here $\sigma_0$, $S_0$ are correspondingly stress and cross-section area in the nondeformed part of the specimen far from the neck.

The uniform neck propagation—an inelastic deformation wave—is described by a traveling wave solution to the obtained set of equations for which all unknown functions depend on a single variable $\zeta = x + Vt$. We note that the grips extension rate $V_1$ and wave velocity $V$ are related by an obvious equation

$$V_1 = (1 - \alpha)V/\alpha. \tag{4.12}$$

In the isothermic case under consideration the set of the equations can be reduced to a single ordinary differential equation for the stress $\sigma(\zeta)$:

$$V \frac{d\sigma}{d\zeta} = \frac{q}{\tau} \sigma(1 - \alpha) + \sigma(1 - \alpha) \frac{d}{d\zeta} \left[ \left( \frac{L^2}{\tau} \partial_\sigma q \right) \frac{d\sigma}{d\zeta} \right], \tag{4.13}$$

under the boundary conditions

$$\sigma = \sigma_0, \quad \frac{d\sigma}{d\zeta} = 0 (\zeta \to -\infty); \quad \sigma = \frac{\sigma_0}{\alpha}, \quad \frac{d\sigma}{d\zeta} = 0 (\zeta \to \infty). \tag{4.14}$$

(The neck, we remember, propagates to the left.) The problem is similar to that of determination of the flame propagation velocity in the combustion theory [12]. Indeed, putting

$$z = \frac{L^2}{\tau} \partial_\sigma q \frac{d\sigma}{d\zeta}, \tag{4.15}$$

we reduce (4.13) to the equation of the first order

$$\frac{dz}{d\sigma} = \frac{V}{\sigma(1-\alpha)} - \frac{L^2}{\tau^2} \frac{q \partial_\sigma q}{z}, \tag{4.16}$$

under two boundary conditions

$$z = 0, \quad \sigma = \sigma_0; \quad z = 0, \quad \sigma = \sigma_0/\alpha. \tag{4.17}$$

The solution to this problem does not exist for an arbitrary couple of values of the parameters $V$, $\sigma_0$. We have here a typical situation of a nonlinear eigenvalue problem: an asymptotic integration gives a relation between $V$ and critical stress $\sigma_0$ necessary and sufficient for the existence of a single solution to the eigenvalue problem (4.16), (4.17). This relation occurs to be

$$\ln \frac{V}{V_*} = -\frac{U}{kT} + \frac{\gamma \sigma_0}{kT}, \quad V_* = \frac{\alpha L}{\tau \sqrt{2} \ln\left(\frac{1}{\alpha}\right)}. \tag{4.18}$$

The specially performed experiments confirm this linear relation between the critical stress and logarithm of the neck propagation velocity and give reasonable estimates for the kinetic constants which agree with the values of these constants obtained from different independent experiments (more detailed presentation of theoretical model and relevant experiments can be found in [13]–[16]).

## 5  Example: Scaling Laws for Developed Turbulent Shear Flows in Cylindrical Tubes

Since the early 1930s it has been known that the average velocity distribution $u(y)$ within an intermediate interval of distances $y$ from the wall (outside the tiny viscous layer near the wall and close vicinity of the tube axis) with equal accuracy can be represented in two different forms.

(1) Power law, depending on Reynolds number

$$\varphi = C\eta^\alpha; \quad \varphi = \frac{u}{u_*}, \quad \eta = \frac{u_* y}{\nu}. \tag{5.1}$$

(Here $u_* = (\tau/\varrho)^{1/2}$, $\tau$ is wall shear stress; $\varrho$ is fluid density; $\nu$ is fluid kinematic viscosity; $C$, $\alpha$ are dimensionless constants known to be weakly dependent on the global Reynolds number of the flow Re $== \bar{u}d/\nu$; $\bar{u}$ is mean fluid velocity; $d$ is tube diameter).

(2) Universal, independent of Re, logarithmic law

$$\varphi = \frac{1}{\kappa} \ln \eta + C_1, \tag{5.2}$$

where $\kappa = 0.417$ is called the von Kármán constant, and $C_1 = 5.5$.

Until recently it was commonly believed that logarithmic law (5.2) has some advantages—it can be derived from the seemingly plausible assumption that at large Re average velocity gradient should be independent of molecular viscosity. On the contrary, power law (5.1) was considered merely as a convenient representation of empirical data deprived from any theoretical basis.

Rather recently it was shown [17], however, that the power law (5.1) can be obtained from a different assumption (scaling, incomplete self-similarity) not less rigorously than the logarithmic law is usually derived from the simpler assumption of complete self-similarity, i.e., independence of the Reynolds number.

Mean velocity distribution is governed by flow microstructure, vortex dissipative structure of turbulent flow. Recent experimental results showed that this structure is most irregular, fractal, miltifractal, or even more complicated. Therefore it seems natural that the influence of molecular viscosity does not appear even at very large flow Reynolds numbers.

Therefore the recent result [18] seems natural—instructive coincidence with classical experimental data of Nikuradze [19] of power law (5.1) with the following parameters

$$\alpha = \frac{3}{2\ln \text{Re}}, \qquad C = \frac{\sqrt{3} + S\alpha}{2\alpha}, \tag{5.3}$$

so that the power law (5.1) can be represented in the following quasi-universal form

$$\psi = \frac{1}{\alpha} \ln \frac{2\alpha\varphi}{\sqrt{3}+S\alpha} = \ln \eta, \qquad \alpha = \frac{3}{2\ln \text{Re}}. \tag{5.4}$$

All 16 series, 256 experimental points represented in the tables of the paper [19] are presented in Figure 2 in coordinates $\psi$, $\ln \eta$. Experimental points appeared to be situated mainly close to the bisectrix in accordance with (5.4). The hit of experimental points on the bisectrix is an indication in favor of scaling law (5.1), i.e., in favor of incomplete self-similarity.

In fact, the difference between the two laws (5.1) and (5.2) has apparently very deep nature: from the power law (5.1) follows the fractality of microstructure—vortex dissipative structure—of developed turbulent shear flow.

## 6 Example: Microstructure of Oceanic Turbulence

Under the upper oceanic layer where the density is nearly homogeneous there exists a layer of sharp density increase (upper pycnocline) where the observers discovered using low-inertia gauges (Figure 3a) a peculiar microstructure of temperature and, consequently, density field: the thin layers with sharp temperature variation alternating with layers of homogeneous temperature [20], [21]. Similar peculiarities were observed in the velocity field. In the proposed model [23], [22] it was assumed that this peculiar microstructure is due to spatial intermittency of turbulence [24]: turbulence under the conditions of strong stable stratification (density variation in a strong gravity field) is concentrated in turbulent spots, squeezed by ambient nonturbulent fluid. These spots are formed (Figure 3b) due to breaking of internal waves in a shear flow of stably stratified fluid.

A micromechanical approach to this problem is natural requiring simultaneous consideration of hydrodynamic equations together with the equations of dynamics of spots and

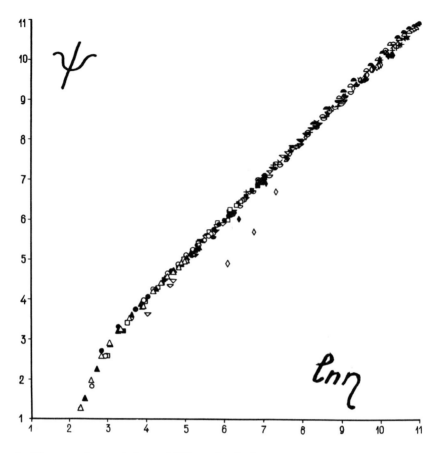

Figure 2: The experimental data of Nikuradze [19] are in agreement with quasi-universal power law (5.4).

| | | | | |
|---|---|---|---|---|
| △ | – | $\mathrm{Re} = 4 \cdot 10^3$, | ▲ – | $\mathrm{Re} = 6.1 \cdot 10^3$, |
| ○ | – | $\mathrm{Re} = 9.2 \cdot 10^3$, | ● – | $\mathrm{Re} = 1.67 \cdot 10^4$, |
| □ | – | $\mathrm{Re} = 2.33 \cdot 10^4$, | ■ – | $\mathrm{Re} = 4.34 \cdot 10^4$, |
| ▽ | – | $\mathrm{Re} = 1.05 \cdot 10^5$, | ▼ – | $\mathrm{Re} = 2.05 \cdot 10^5$, |
| ▽ | – | $\mathrm{Re} = 3.96 \cdot 10^5$, | ▼ – | $\mathrm{Re} = 7.25 \cdot 10^5$, |
| ◇ | – | $\mathrm{Re} = 1.11 \cdot 10^6$, | ◆ – | $\mathrm{Re} = 1.536 \cdot 10^6$, |
| + | – | $\mathrm{Re} = 1.959 \cdot 10^6$, | × – | $\mathrm{Re} = 2.35 \cdot 10^6$, |
| ◠ | – | $\mathrm{Re} = 2.79 \cdot 10^6$, | ◗ – | $\mathrm{Re} = 3.24 \cdot 10^6$ |

of the turbulence dynamics within the spots [25], which play the role of the equations of microstructure kinetics.

The equation for the spot thickness $h$ has the form of the nonlinear heat conduction equation [22], [25]

$$\partial_t h = \kappa \Delta h^5, \qquad (6.1)$$

where $t$ is the time, $\Delta$ is the horizontal Laplace operator, and $\kappa$ is a constant, defined by the intensity of stratification. Naturally the details of the initial form of the spot are unknown and immaterial, so intermediate asymptotics play here a decisive role. So, the

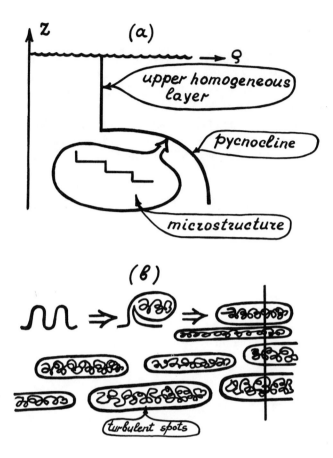

Figure 3: (a) Oceanic microstructure in the upper pycnocline, (b) Turbulent spots.

spot thickness at the intermediate asymptotic stage can be represented in self-similar form

$$h = h_0(t) F(r/r_0(t)), \qquad (6.2)$$

where $r$ is the radial polar coordinate, $h_0(t)$ is the maximal spot thickness, and $r_0(t)$ is the spot radius. Analysis showed that the function $F$ is nearly constant until the very contour of the spot going abruptly to zero near the spot edge, so the turbulent spot on the intermediate-asymptotic stage has the form of a disc. The radius of the spot $r_0(t)$ increases very slowly, and its thickness is slowly decreasing:

$$h_0(t) \sim (t - t_0)^{-1/5}, \qquad r_0(t) \sim (t - t_0)^{1/10}. \qquad (6.3)$$

The dynamics of turbulence within the spot was also considered [25]. It has some peculiarities: after the formation of the spot it decays as a rule and this stage of decay can be rather long. Later the turbulent energy starts to increase under the combined action of shear and reducing due to homogeneity of water inside the spot the work against buoyancy force—the basic sink of turbulent energy in stratified fluid.

## 7 Conclusions

Micromechanics, a new branch of continuum mechanics studying the motions where the variations of microstructure are of governing value for macroscopic behaviour of bodies, entered in the last two decades the podium of mechanical science. In his remarkable paper [4] Budiansky stressed that further proceeding of micromechanics requires sophisticated microscopic observations and microtesting as well as supercomputing. Indeed, microscopic observations tell us what things really look like and what really happens at microscopic scales. Supercomputing is the only way to make the mathematical models of cooperation of microstructural elements. It is very important to be able to extract from the results of microscopic experiments and supercomputing proper images of micromechanical phenomena. Intermediate asymptotics can be of great help in creating such images.

## Acknowledgement

The author wishes to acknowledge the hospitality and generosity of the Mathematics Department of the Rensselaer Polytechnic Institute where this paper was prepared. I want to express my special thanks to Professor Julian D. Cole, a great man in applied mathematics and mechanics whose bright creativity and superb human qualities are the reason of my affection and admiration during many years.

## References

[1] Th. von Kármán, *From Low Speed Aerodynamics to Astronautics*, Pergamon Press, Oxford, 1963, pp. 29–46.

[2] G. K. Batchelor, *Development in microhydrodynamics*, in Theoretical and Applied Mechanics, W. T. Koiter, ed., North-Holland, The Netherlands, 1976, pp. 33–55.

[3] B. Budiansky, *Micromechanics*, Computers and Structures, vol. 16, Nos. 1–4 (1981), pp. 3–12.

[4] ———, *Micromechanics* II, Proceedings of the 10th US National Congress of Applied Mechanics, Austin, Texas, 1986, pp. 1–8.

[5] P. W. Bridgman, *Dimensional Analysis*, Yale University Press, New Haven, 1922.

[6] G. I. Barenblatt, *Dimensional Analysis*, Gordon and Breach, New York, 1987.

[7] M. D. Van Dyke, *Perturbation Methods in Fluid Mechanics*, The Parabolic Press, Stanford, 1975.

[8] J. D. Cole, *Perturbation Methods in Applied Mathematics*, Blaisdell, Waltham, Toronto, London, 1968.

[9] N. Goldenfeld, O. Martin, Y. Oono, and F. Liu, *Anomalous dimensions and renormalization group in a nonlinear diffusion process*, Phys. Rev. Letters, vol. 64, No. 2 (1990), pp. 1361–1364.

[10] W. H. Carothers and J. W. Hill, *Studies of polymerization and ring formation. XV. Artificial fibers from synthetic linear condensation superpolymers*, J. Am. Chem. Soc., vol. 54 (1932), pp. 1579–1587.

[11] A. Nadai, *Theory of Flow and Fracture of Solids*, vol. 1, McGraw-Hill, New York, 2nd edition, 1950.

[12] Ya. B. Zeldovich, G. I. Barenblatt, V. B. Librovich, and G. M. Makhviladze, *The Mathematical Theory of Combustion and Explosions*, Consultants Bureau, New York and London, 1985.

[13] G. I. Barenblatt, *Methods of combustion theory in the mechanics of deformation, flow and fracture of polymers*, in Deformation and Fracture of High Polymers, H. H. Kausch, J. A. Hassell, and R. I. Jaffee, eds., Plenum Press, New York, London, 1973, pp. 91–111.

[14] ———, *Neck propagation in polymers*, Rheologica Acta, vol. 13 (1974), pp. 924–933.

[15] G. I. Barenblatt, V. A. Volodchenkov, and D. Ya. Pavlov, *Isothermic neck propagation in polymers, Experimental study*, Bull. USSR Ac. Sci., Ser. Mech. of Solids, No. 4 (1973), pp. 172–177. (In Russian.)

[16] G. I. Barenblatt, V. A. Volodchenkov, I. M. Kershtein, and D. Ya. Pavlov, *Isothermic neck propagation in polymers. Comparison with creep process*, Bull. USSR Ac. Sci., Ser. Mech. of Solids, No. 5 (1974), pp. 144–156. (In Russian.)

[17] G. I. Barenblatt and A. S. Monin, *Similarity laws for turbulent stratified shear flows*, Archive for Rat. Mech. and Analysis, vol. 70 (1979), pp. 307–317.

[18] G. I. Barenblatt, *On the scaling laws (incomplete self-similarity with respect to Reynolds number) for the developed turbulent flows in cylindrical tubes*, CR Acad. Sci., Paris, 1991.

[19] J. Nikuradze, *Gesetzmässigkeiten der turbulent Strömung in glatten Rohren*, VDI-Forschungsheft, No. 356, 1932.

[20] K. N. Fedorov, *Thermohaline Fine Structure of Oceanic Waters*, Gidrometeoizdat, Leningrad, 1976. (In Russian.)

[21] A. S. Monin and R. V. Ozmidov, *Oceanic Turbulence*, Gidrometeoizdat, Leningrad, 1981. (In Russian.)

[22] G. I. Barenblatt, *Dynamics of turbulent spots and intrusions in stably stratified fluid*, Bull. USSR Ac. Sci., Ser. Atm. and Oceanic Physics, vol. 14, No. 2 (1978), pp. 195–205. (In Russian.)

[23] G. I. Barenblatt and A. S. Monin, *The origin of oceanic microstructure*, in Proceedings of XII Symposium on Naval Hydrodynamics, US Nat. Ac. Sci., Washington, D.C., 1979, pp. 574–581.

[24] O. M. Phillips, *The generation of clear-air turbulence by the degradation of internal waves*, in The Atmospheric Turbulence and Propagation of Radio Waves, Nauka, Moscow, 1967, pp. 130–138.

[25] G. I. Barenblatt, *Dynamics of turbulent spots in stably stratified fluid*, in Mathematical Approaches in Hydrodynamics, T. Miloh, ed., SIAM, Philadelphia, 1991, pp. 373–381.

# Chapter 3
# COMPUTER VISION: MATHEMATICS AND COMPUTING

**MICHAEL BRADY**
University of Oxford
United Kingdom

## 1 Introduction

This paper reports on the work of the author and his colleagues at the University of Oxford on the development of implemented sensor-based systems, especially their mathematical underpinnings and the parallel computer architectures that support real-time implementation. The systems we describe are:

- a four-camera system developed to track several targets, typically people, as they move independently around in a room. The system degrades gracefully in the sense that if one or more of the cameras is unplugged it continues to track, albeit with increased uncertainty in the positional estimates of the targets;
- a free-ranging robot vehicle that uses sonar sensors to detect obstacles encountered along its nominal trajectory, and which can make appropriate detours to avoid them, even if this entails re-planning the entire trajectory;
- a real-time implementation of a structure-from-motion vision algorithm that can compute the three-dimensional layout of the scene it is driving through, and can determine which parts are driveable. The system works by tracking a set of features of the image surface ("corners") that need to be extracted from the image sequence accurately and reliably;
- the combination of stereo vision and shape-from-shading in a system that helps opthalmologists monitor the development of glaucoma by measuring the swelling of the optic disk as a result of heightened interocular pressure.

The systems are representative of recent work in robotics that combines mathematical analyses of signals and images, parallel architectures and algorithms, and developments in electronics to make complete working systems.

## 2 Intensity changes

An image is a discretized, quantised sampling of an intensity distribution. Being an energy distribution, intensity values are positive and finite. Modern charge-coupled device cameras give low noise, high resolution images. As the imaging surface is etched onto a fixed set of sites (as opposed to earlier steerable electronic beam cameras), CCD images have sufficient geometric accuracy to support three-dimensional vision processes such as stereo or structure-from-motion. The computer representation of an image is as an array of brightness values $I(x,y,t)$ (we consider only monochrome images in this paper, otherwise a wavelength parameter would be added), consisting typically, for fixed time $t$, up to $512 \times 512$ pixels. Current work mostly aims to achieve video rate temporal sampling.

Not all pixels carry the same amount of information, even as a constant signal carries no information. Our ability as humans to readily interpret line drawings made from images, suggests that an important early processing step is to extract intensity changes, a suggestion reinforced by physiological studies of simple cells in visual cortex. Indeed, under certain weak conditions, an image can be reconstructed from its intensity changes. A (static) image can be regarded as a surface, in which case the extraction of intensity changes amounts to developing numerical approximations to the gradient $\nabla I$ and image (local) surface curvatures. One popular approach is first to smooth the image with a Gaussian filter $G_\sigma$, then find zero-crossings of the Laplacian of the smoothed surface [Marr and Hildreth 1980]. By the derivative theorem of convolutions,

$$\Delta(G_\sigma * I) = (\Delta G_\sigma) * I,$$

(where $*$ denotes convolution). The operator $\Delta G_\sigma$ resembles a Mexican hat and approximates the response of retinal ganglion cells. In practice, isotropic operators such as $\Delta G_\sigma$ or directional derivatives such as $\partial I/\partial \vec{n}$, where $\vec{n} = \nabla I/\|\nabla I\|$ are themselves ineffective for producing useful edge maps. In 1983, Canny found that a succession of *non*-linear postprocessing steps greatly improved the set of edge points produced. Of these, non-maximum suppression was particularly important. Canny's algorithm is as follows:

1. *smooth image*, with a Gaussian convolution;

2. *estimate gradient*, with the $3 \times 3$ finite difference kernel whose rows for $I_x$ are $(-1,0,1)$, and which are the columns for $I_y$. Estimate the gradient magnitude, $M(x,y) = I_x^2 + I_y^2$;

3. *directional non-maximum suppression* Erect a local coordinate frame at the point $(x,y) = (0,0)$, and without loss of generality, assume that the direction of the gradient is between $(1,0)$ and $(1,1)$. Estimate the gradient at $x = 1$ by sub-pixel interpolation from

$$M^+ = M_{11} I_y(0,0) + M_{10}(I_x(0,0) - I_y(0,0))$$

Similarly, estimate $M^-$ between (-1,0) and (-1,-1). Finally, mark $M_{00}$ as an edge point if (i) it exceeds a threshold and (ii) it is greater than both $M^+$ and $M^-$. Clearly, this technique for suppressing potential edge points whose gradients are smaller than those interpolated nearby can be generalised.

4. *contour extension* to fill in small gaps by estimating the local tangent to the intensity change contour;

5. *thin edge contours* to be one bit wide.

The Canny edge detector performed very well, for its time, but mostly serves to emphasise the importance of non-linear filtering. Over the past five years three novel non-linear filters have been developed: Fleck's application of finite-resolution cell complexes to compute high frequency texture edges; Owens and Morrone's local energy model based on Hilbert quadrature pairs and local phase; and morphological filtering.

Mathematical morphology has been developed by Serra, Matheron, and their colleagues over the past twenty years [Serra, 1982]. Originally designed for binary images, the basic operators *dilation* and *erosion* have been extended to grey-scale morphology. The basic operators can be combined in a variety of ways, of which the *open* and *close* are the most important. As their names suggest, the open filter removes thin neck-like structures on the boundary of a region, leaving low curvature portions intact. The close operation fills in narrow inlets. The difference between the dilation (which swells a shape) and its erosion locates intensity changes. Until recently, applications of morphological filters had an *ad hoc* flavour governed by the choice of the shape and size of the structuring element. Recent work by Maragos, Arce and Stephenson, Noble, and others have analysed the filter characteristics of morphological filters, compared them to the median filter, while Noble has analysed morphological analogues of the image surface local differential geometry. Noble's [1990] thesis provides a good introduction to morphology and edge and corner detection. Unlike edge detectors such as Canny's algorithm, in which the local assumption is that there is an edge of low curvature, the corners of images of polyhedral objects are not suppressed. Indeed, even high curvature points, such as the tines of a fork can be detected as continuous curves.

To summarise so far, in many cases the significant intensity changes can be extracted automatically from images; non-linear filtering is a crucial requirement.

Corners carry even more information than edges, though they are correspondingly more difficult to extract accurately and reliably. Later, we describe a system that computes structure from motion by tracking corners over a sequence of images. How can corners be modeled and extracted? As before, let $\overline{n}$ be the unit vector in the direction of the intensity gradient. The edge curvature $\kappa_{\overline{n}\perp}$ can be measured by the image surface normal curvature in the tangential direction orthogonal to $\overline{n}$.

Elementary differential geometry shows that:

$$\kappa_{\vec{n}\perp} = \frac{1}{1+I_x^2+I_y^2} \frac{I_y^2 I_{xx} + I_x^2 I_{yy} - 2I_x I_y I_{xy}}{I_x^2 + I_y^2}$$

This formula was discretised and used in an early corner finder by Kitchen and Rosenfeld. A later implementation by Zuniga and Haralick worked out the expression analytically for the coefficients of a bicubic polynomial which was locally fit to the image surface. Both implementations were highly sensitive to noise. An alternative is to measure the local image autocorrelation function; but this tends to place corners at some remove from the ideal location, and while this does not affect tracking, it does affect the estimation of three-dimensional depth.

Recently, Wang and Brady [1991] developed a novel corner finder that works well in practice. It estimates the tangential curvature $\kappa_{\vec{n}\perp}$ using the linear interpolation and non-maximum suppression technique described above for Canny's edge finding algorithm. Suppose that the image gradient $|\nabla I|$ is above a threshold at the image point $O$, and let the edge tangent be $\vec{t} = \vec{n}^\perp$, not necessarily aligned with pixel sites. The intensity function $I$ is interpolated at locations $a, b, c, d$, at one and two pixel distances removed from $O$, and the second directional derivative $D_{\vec{t}}^2 = \kappa_{\vec{n}\perp}$ is estimated from the finite difference approximation:

$$D_{\vec{t}}^2 I = \frac{1}{r}(-2a - b + 6O + c + 2d),$$

where $r$ varies spatially, eg $r = \nabla I / I_x$ when the tangential direction is less than $\pi/4$. Corner points are marked where (i) $D_{\vec{t}}^2 I > \omega |\nabla I|$, and (ii) non-maxima are suppressed.

The Wang-Brady corner finder has been implemented on a parallel architecture *PARADOX* that consists of a commercial image processing device, a network of Transputers, and an interface board that interconnects the two. Corners can be located at 10Hz.

## 3 Decentralised Kalman filtering

The Kalman filter maintains an estimate of the state $x$ of a system and a covariance estimate $P$ of its uncertainty. Bar Shalom and Fortmann [1988] provide a lucid introduction to the theory of Kalman filtering. The application of the Kalman filter to sensor-guided control is used to reduce the discrepancy between the planned and actual states of a vehicle increase, as does the state uncertainty, when no sensor measurements are made (see John Leonard's [1991] recent thesis for more detail and illustrations. Suppose the vehicle senses a planar surface. The uncertainty orthogonal to the wall decreases sharply; but continues to increase in the direction tangential to the wall. If a second wall is sensed, the state uncertainty is sharply reduced.

More formally, assume that the linearised system dynamics at the $k$th time step are given by:

$$\vec{x}(k+1) = \mathbf{F}(k)\vec{x}(k) + \mathbf{G}(k)\vec{u}(k) + \vec{v}(k)$$

where $\vec{u}(k)$ is the (known) control signal, and $\vec{v}(k)$ is a zero-mean, white, Gaussian noise process with covariance matrix $\mathbf{Q}(k)$. There are ways of weakening the linearity assumption but they are omitted from this paper. Usually, the state itself is not measured directly, rather there is a measurement equation:

$$\vec{z}(k) = \mathbf{H}(k)\vec{x}(k) + \vec{w}(k),$$

where, again, $\vec{w}(k)$ is a zero-mean, white, Gaussian noise process with covariance $\mathbf{R}(k)$. Let the estimated state after $k$ steps; but taking account of sensor readings up to the $k-1$st step, be denoted by $\hat{x}(k|k-1)$ and let its covariance matrix be denoted $\mathbf{P}(k|k-1)$. In its simplest form, the Kalman filter algorithm proceeds as follows (see, for example, [Bar Shalom and Fortmann 1988, page 61]):

1. *state estimation and uncertainty*: maintain $\hat{x}(k|k)$ and $\mathbf{P}(k|k)$; initially $\hat{x}(0|0)$ and $\mathbf{P}(0|0)$ are estimated from "knowledge" of the system.

2. *prediction*

$$\begin{aligned}\hat{x}(k+1|k) &= \mathbf{F}(k)\hat{x}(k|k) + \mathbf{G}(k)\vec{u}(k) \\ \mathbf{P}(k+1|k) &= \mathbf{F}(k)\mathbf{P}(k|k)\mathbf{F}(k)^T + \mathbf{Q}(k)\end{aligned}$$

3. *measurement prediction*: $\hat{z}(k+1|k) = \mathbf{H}(k+1)\hat{x}(k+1|k)$, and *innovation* (difference between predicted and actual measurement): $\vec{z}(k+1) - \hat{z}(k+1|k)$

4. *update of gain matrices*:

$$\begin{aligned}\mathbf{S}(k+1) &= \mathbf{H}(k+1)\mathbf{P}(k+1|k)\mathbf{H}^T(k+1) + \mathbf{R}(k+1) \\ \mathbf{W}(k+1) &= \mathbf{P}(k+1|k)\mathbf{H}^T(k+1)\mathbf{S}^{-1}(k+1)\end{aligned}$$

5. *update of state and covariance estimate*:

$$\begin{aligned}\hat{x}(k+1|k+1) &= \hat{x}(k+1|k) + \mathbf{W}(k+1)(\vec{z}(k+1) - \hat{z}(k+1|k)) \\ \mathbf{P}(k+1|k+1) &= \mathbf{P}(k+1|k) - \mathbf{W}(k+1)\mathbf{S}(k+1)\mathbf{W}^T(k+1)\end{aligned}$$

The Kalman filter has the property that, under certain reasonable assumptions, it is the optimal state estimator. Not that it is without problems in practice. Among the more severe of these are (i) the difficulty of computing a good *initial state estimate* $\hat{x}(0|0)$; (ii) the difficulty of determining appropriate gain matrices;

and (iii) the difficulty of identifying and approximation real plants in the simple form shown.

Our group at Oxford has not contributed to the theory of the Kalman filter so much as to the way in which it may be computed. In a typical fielded system, a set of sensors make independent measurements of components of the state and report them, at each step, to a central processor that runs the Kalman filter. If one of the sensors ceases to operate, the system continues to run, albeit it with increased state uncertainty. If, however, the central processor ceases to operate, the whole system fails. An obvious alternative design is to distribute the Kalman filter amongst the sensors (in the current parlance this makes them "smart" sensors) and enable them to communicate their state estimates amongst themselves. Rao, Durrant-Whyte, and Sheen [1991] have shown that the equations of the Kalman filter can be partitioned so that such a fully decentralised system converges to the same global optimum as the centralised system. The system degrades gracefully as "smart" sensors cease to operate, and upgrades gracefully as new "smart" sensors are introduced into the system.

Our first application of these ideas was in collaboration with the British Aerospace (BAe) Sowerby Research Centre, as part of an Esprit collaboration *SKIDS*. Four cameras are placed in the corners of a room that is 10 metres by 6 metres. They are multiplexed through a single image processing system (on grounds of cost, ideally there would be four such), and then through an interface card, to four Transputers that are fully interconnected. One Transputer is dedicated to each camera, and receives from the image processing system, at each time step (currently 1/7th second), a difference image that gives a (poor) representation of image motion as seen from that camera. At each iteration, a Transputer (i) performs a connected component analysis to find moving blobs in its difference image; (ii) finds the point at the centre of the bottom of the blob (hopefully this corresponds to the feet on the floor); (iii) knowing the camera transformation from the world frame, de-projects the ray through this point to intersect it with the ground plane; (iv) transmits this estimate to the other three cameras and receives theirs; and (v) computes the updated Kalman estimate of the positions of the moving objects. In his forthcoming thesis, Rao [1991] shows how the system can be extended to track several moving targets. The graceful degradation refered to above is confirmed in practice as the system continues to track targets as one, then two, cameras are unplugged and then plugged back in.

## 4 Autonomous Guided Vehicle

Our second application has centred on an Autonomous Guided Vehicle (AGV). Our research aimed to equip autonomous guided vehicles, in particular a state-of-the-art British product developed by GEC Fast (in collaboration with the Caterpillar company) designed for transporting material around factories, with sensory capabilities, and with planning capabilities that rely upon sensed data. The AGV is controlled using a GEM-80 industrial controller that integrates information from an infrared ranging system that reads reflecting bar-coded targets, and from odometry. The fielded industrial system uses a *landbase* software system that models the pathways and places in a factory and is used for route planning. For safety, the vehicle is equipped with a plastic bumper that stops the vehicle on contact with an unexpected obstacle. The vehicle follows cubic spline trajectories planned between set points established in the landbase. Implicitly, it assumes that the real factory environment in which the AGV operates matches exactly the software model that is enshrined in the landbase.

The particular applications that we have worked on require the vehicle to acquire objects whose positions are substantially uncertain and which therefore needed to be sensed: (i) acquiring objects from an input bay, such as pallets dumped by a fork-lift truck; and (ii) cleaning an area that is well-known but which may contain unexpected obstacles. We have equipped the vehicle with a number (12) of sonar sensors. These are multiplexed into a Transputer system that interfaces to the vehicle's industrial controller. The vehicle sets out to follow a planned trajectory, but may encounter obstacles located by the sonar sensors. An algorithm determines if the obstacle can be by-passed by a simple detour around it, in which case a cubic spline patch is inserted into the planned trajectory. If it is considered that the object cannot be by-passed in this manner, a search algorithm that contains a model of the environment (without unexpected obstacles) is called to formulate a new trajectory to the goal. Recent work on the sonar guided vehicle aims to integrate decision theory with the Kalman filter and a probabilistic model of possibly moving obstacles to determine dynamically the optimal policy (trajectory). Do we, for example, cut our losses if the navigable gap around an obstacle seems too narrow for comfort, instead planning an entirely different route?

In another Esprit project *VOILA*, we have been collaborating with Plessey's Roke Manor Research Laboratory on a real-time implementation of a structure from motion algorithm *DROID* originally developed by C. J. Harris and J. M. Pike[1987]. *DROID* determines the three-dimensional structure of a scene by tracking image features over a sequence of frames. It works as follows:

1. The state of the system at time $t$ consists of a set of feature points, each of which has an associated vector $[x_i(t)\ y_i(t)\ z_i(t)\ \epsilon_x^i(t)\ \epsilon_y^i(t)\ \epsilon_z^i(t)\ \lambda_i]^\mathsf{T}$. The first three components are the estimated three-dimensional position of the feature. The second three components are the semi-axes of an ellipsoidal representation of uncertainty. Initially, the uncertainty is low in the image coordinates, but large in depth. As the vehicle circumnavigates a feature, the uncertainty in depth reduces by intersecting ellipsoids. The final component $\lambda$ is a flag that says whether the feature is temporarily invisible (eg because it is occluded).

2. Given an estimate of the motion between the imaging position at time $t$ and time $t + 1$, the state prediction at time $t + 1$ can be made, and, after perspective projection using a pinhole model, the positions of the feature

points in the $t + $ 1st image can be predicted.

3. Features can be extracted from the $t + $ 1st image. Originally, features defined by image autocorrelation were used. Recently, we developed the Wang-Brady corner finder described in Section 2, and those feature points are used instead. This is by far the most computationally intensive step of the computation.

4. The *correspondence* is computed between the predicted locations of points in step 2 and the feature points found in step 3. Ambiguous matches are decided by matching descriptions of the predicted and found points that involve the intensity value and gradients.

5. The state estimate is updated.

6. The image locations of the feature points are triangulated, using the Delaunay triangulation that minimises long thin triangles. The triangles are deprojected into the scene (recall that the $z$ component of each feature point has been estimated. The surface normal of each deprojected triangular facet is computed. Driveable regions are computed as the transitive closure of the facets lying close to the horizontal plane, starting just in front of the vehicle.

Wang and Brady report a number of careful experiments with the *DROID* algorithm, using a parallel implementation of their corner finder. Current work aims at a parallel implementation of the rest of the *DROID* system.

# 5 Monitoring Glaucoma

The basic anatomy of glaucoma, and a survey of previous computational techniques to monitors its development, presented in [Lee and Brady 1990, 1991; Lee 1991]. The abnormally high inter-ocular pressure that is characteristic of glaucoma results in the swelling, or extrusion, of a tiny structure called the optic disk, which marks the blind spot, where the optic nerve leaves the retina for the brain. The optic disk has a diameter approximately $1.8mm \pm 0.2mm$, and within its rim is concave to a depth of about 200 microns. The optic disk is about $24mm$ from the cornea. Images of the optic disk, taken with a fundus camera after dilating the patient's pupil occupy about 200 pixels. We have developed a system that reconstructs the three-dimensional surface of the optic disk.

The algorithm that we have implemented works as follows:

1. *Capture images* $I_l^1, I_l^2$ and $I_r$. As will become apparent, $I_l^1$ and $I_r$ form a stereo pair of images. Images $I_l^1$ and $I_l^2$ are taken from the same vantage point; but with two different illumination directions. In our experiments to date, the camera has been moved physically from $I_l^1$ to $I_r$, but this will soon be replaced by a specially designed camera that we have patented.

2. *Locate the optic disk boundary* using the process sketched below. This is a key step in rectifying the images for subsequent stereo processing.

3a. *Perform stereo processing* in two steps:

   3a(i) *rectify the images* after blanking out the image regions that correspond to the optic disks located at step 2.

   3a(ii) *stereo match* using the PMF disparity gradient algorithm [Pollard, Mayhew, and Frisby 1991] This yields a set of disparities, that can be converted knowing the calibration of the system to depths, but only at the sparse set of locations matched by PMF. In our current system, the sparse set marks the edges of the blood vessels that are easily detectable in the images as "tree-like" structures. The result is a sparse depth map.

3b. *Photometric stereo* is performed as step 3b in parallel with step 3a. The process is described below in more detail. It yields a dense set of estimates for the *gradient* of the optic disk surface.

4. *Integrate stereo and photometric stereo.* The sparse depth map computed at step 3a is integrated with the dense gradient map computed at step 3b using a finite element technique.

5. *display the computed surface* or measure *clinical parameters* of the optic disk, such as its area, depth, etc.

The process of finding the boundary of the optic disk is described in more detail in [Lee and Brady 1991]. A key requirement of subsequent rectification is that the boundary be located accurately. Simple analytical models such as the boundary being approximated by an ellipse do not work well in practice as many optic disks have significant protrusions in their boundaries. An alternative is to try to fit a "snake" energy minimising curve to the boundary. In this approach, the curve is given by $\vec{v}s$ and is chosen to minimize a functional such as:

$$\oint \left(\frac{d\vec{v}}{ds}\right)^2 ds + \lambda \oint \left(\frac{\partial I}{\partial \vec{n}}\right)^2 ds + \mu \sum_i (\vec{v}_i - \vec{d}_i)^2$$

where $\vec{n}$ is the unit normal to the disk boundary and $\vec{d}_i$ are given values from the image data. To be effective, blood vessels first need to be removed as they form deep "ravines" in the image surface of the optic disk that distort the above energy function. [Lee and Brady 1991] have invented a novel, application specific, morphological filter that removes the blood vessels. It exploits the observation that the blood vessels cross the disk boundary approximately orthogonally. The morphological filter is defined most simply in a polar coordinate transform of the image as a rectangular filter. It works remarkably well and reliably. The optic disk boundary is used in the rectification process: its interior is blanked out for the left and right images and the least squares image transform to minimize the disparity of the blood vessels in the fundus outside the optic disk.

Photometric stereo is a process invented by Bob Woodham [1978] that fixes the viewpoint, and varies the direction of the lighting in order to estimate the gradient of the viewed surface. The intensity (grey level) at a given image location depends on three things: (i) the colour or albedo $\rho$ of the corresponding scene location; (ii) the power of the light source $E_s$; and (iii) the reflectance function that describes how light is reflected off the surface locally. In general, reflectance functions are complex functions of the local surface gradient, the view direction, and the light source direction. Lee and Brady [1990] argue that Lambert's cosine law provides a good approximation for the reflectance function of fundus tissue in the optic disk. That is, for each image location $(x, y)$ the intensity is given by:

$$I = \rho \ E_s \ \frac{1 + p \cdot p_s + q \cdot q_s}{(1 + p^2 + q^2)^{1/2}(1 + p_s^2 + q_s^2)^{1/2}}$$

where $[p \ q]^\mathsf{T} = [\partial z/\partial x \ \partial z/\partial y]^\mathsf{T}$ is the local surface gradient, and where $[p_s \ q_s]^\mathsf{T}$ represents the light source direction. Unfortunately, the albedo $\rho(x, y)$ varies spatially in the optic disk images. If however images are taken from *two* directions $[p_{s_1} \ q_{s_1}]^\mathsf{T}$ and $[p_{s_2} \ q_{s_2}]^\mathsf{T}$ and the intensity values divided $I_1(x, y)/I_2(x, y)$, the spatially varying albedo can be eliminated. The resulting formula for $I_1/I_2$ can be simplified considerably by careful choice of light source directions. We choose: $q_{s_1} = q_{s_2} = 0; p_{s_1} = -p_{s_2} = p_s; E_{s_1} = E_{s_2}$, in which case the expression simplifies to:

$$\frac{I_1(x,y)}{I_2(x,y)} = \frac{1 + p(x,y)p_s}{1 - p(x,y)p_s},$$

which can be solved for $p(x, y) = \partial z(x, y)/\partial x$. This gives good results for synthetic data; but the gradient estimates need to be filtered for real images. To do this, Lee [1991] proposes a least squares filter that takes account of the approximately known geometry of the optic disk and shows sample results.

We have designed a stereo camera with two light source directions that enable us to implement the process sketched in this section accurately and quickly. The design is currently under construction prior to clinical trials.

## Acknowledgements

The distributed Kalman filter and its application to surveillance is due to my colleague Hugh Durrant-Whyte and his students Bobby Rao and John Leonard. It is a continuing pleasure to work with Hugh on sensor-guided control. The AGV work was joint with my colleagues Huosheng Hu and Han Wang. The glaucoma work was joint with my student Simon Lee, and we acknowledge the use of the PMF software developed at Sheffield University. Chris Harris of Plessey and his colleagues invented the DROID algorithm whose parallel implementation is reported in this paper. I also acknowledge my former students Alison Noble and

John Canny for the morphological work and edge finder. Work on this paper was supported by the Science and Engineering Research Council ACME Directorate, by the EC on the VOILA and SKIDS contracts, Alison Noble was supported by IBM UK, and Simon Lee by a Croucher Foundation award.

# 6 References

Bar Shalom, Y., and Fortmann, Thomas E. *Tracking and Data Association*, Academic Press, 1988

Rao, B.S.Y., Durrant-Whyte, H.F., and Sheen, J.A., "A Fully Decentralized Multi-Sensor System For Tracking and Surveillance", Oxford University Engineering Science Department report no. OUEL 1886/91, (Int. Jnl. Rob. Res. to appear)

Harris, C. G., and Pike, J. M., "3D positional integration from image sequences", Proc. 3rd Alvey Vision Conference, p233-236, 1987

Lee, S., "Visual monitoring of glaucoma", D. Phil. thesis, Oxford University Engineering Science, 1991

Lee, S., and Brady, Michael, "Integrating stereo and photometric stereo to monitor the development of glaucoma", Image and Vision Computing, 9:39-44, 1990

Lee, S., and Brady, Michael, "Optic nerve head boundary detection", Proc. British Machine Vision Conference, Glasgow, Sept., 1991. To be published by Springer Verlag

Leonard, John J. "Directed sonar sensing for mobile robot navigation", D. Phil. thesis, Oxford University Engineering Science, 1991

Marr, D., and Hildreth, E. C., "Theory of edge detection", Proc. Roy. Soc. London, 207, 301-328, 1980

Noble, J. A. "Descriptions of image surfaces", D. Phil. thesis, Oxford University Engineering Science, 1990

Pollard, S. B., Mayhew, J. E. W. and Frisby, J. P., "Implementation details of the PMF stereo algorithm", pages 33-39, in *3D model recognition from stereoscopic cues* Mayhew and Frisby (eds.), MIT Press 1991.

Rao, B. Y. S., *Data Fusion Methods in Decentralized Sensing Systems*, D. Phil. thesis, Oxford University Engineering Science, 1991

Serra, J., *Image analysis and mathematical morphology*, Academic Press, 1982

Wang, Han, and Brady, Michael, "Corner detection for 3D vision using array processors", Proc. BARNAIMAGE 91, Barcelona, Spain, to be published by Springer Verlag, 1991

Woodham, R. J., "Photometric stereo: a reflectance map technique for determining surface orientation from image intensity", SPIE Image Understanding Systems and Industrial Applications, 155: 136 - 143, 1978

# Chapter 4
# ADAPTED WAVE FORM ANALYSIS, WAVELET-PACKETS AND APPLICATIONS

| R. COIFMAN | Y. MEYER | V. WICKERHAUSER |
|---|---|---|
| Yale University | Université Paris-Dauphine | Washington University |
| New Haven, Connecticut | France | St. Louis, Missouri |
| (pictured) | | |

**Introduction.**
Adapted wave form analysis, refers to a collection of FFT like adapted transform algorithms.

Given a function or an operator these methods provide a special orthonormal basis relative to which the function is well represented, and the operator is described by a sparse matrix. The selected basis functions are chosen inside predefined libraries of oscillatory localized functions (waveforms). These algorithms are of complexity $N \log N$ opening the door for a large range of applications in signal and image processing, as well as in numerical analysis.

Our goal is to describe and relate traditional windowed Fourier Transform methods to wavelet, wavelet-packet base algorithms by making explicit their dual nature and relative role in analysis and computation.

Starting with a recent refinement of the windowed sine and cosine transforms we will derive an adapted local sine transform show it's relation to wavelet and wavelet-packet analysis and describe an analysis tool-kit illustrating the merits of different adaptive and non-adaptive schemes.

We end with sample applications to signal and image compression statistical factor analysis, and numerical analysis and P.D.E.

## §1. Windowed FFT and Adapted Window Selection.
We start with a description of an algorithm to compute the Fourier expansion of a function on a union of two adjacent intervals of the same size, in terms of the Fourier expansion on each interval.

Let $f$ be defined on $[0, 2]$

$$f = f^0 + f^1 \quad \text{where} f^0 = \begin{cases} f & x \in [0,1] \\ 0 & x \notin [0,1] \end{cases}$$

we want to compute

$$\hat{f}_m = \frac{1}{\sqrt{2}} \int_0^2 f(t) e^{-2\pi i m \frac{t}{2}} dt$$

in terms of $\hat{f}_n^0 = \int_0^1 f(t)e^{-2\pi int}dt$ and $\hat{f}_n^1 = \int_1^2 f(t)e^{-2\pi int}dt$. Clearly, when $m = 2n$ we have
$$\hat{f}_{2n} = \frac{1}{\sqrt{2}}\{\hat{f}_n^0 + \hat{f}_n^1\}.$$

For $m = 2n + 1$ we define
$$d_n = \frac{1}{\sqrt{2}}\{\hat{f}_n^0 - \hat{f}_{n+1}^1\}$$

and find
$$\boxed{\hat{f}_{2n+1} = \frac{1}{\pi i}\sum \frac{d_k}{(n-k+\frac{1}{2})}}.$$

In fact,
$$\hat{f}_{2n+1} = \frac{1}{\sqrt{2}}\int_0^1 [f(t) - f(t+1)]e^{-i\pi t}e^{-2\pi int}dt.$$

Since $d_n$ are the Fourier coefficients on $[0,1]$ of $f(t)-f(t+1)$, and $\frac{1}{\pi i(n+\frac{1}{2})}$ are the coefficients of $e^{-i\pi t}$, we obtain the coefficients of $\hat{f}_{2n+1}$ by convolving these sequences.

A fast way to compute $\hat{f}_{2n+1}$ is to compute the inverse transform on $(0,1)$ of $d_n$, multiply by $e^{-it/2}$ and recompute the transform on $(0,1)$. This procedure, when discretized, leads to a Fourier transform algorithm of complexity $\leq N(log_2 N)^2$ (in complex arithmetic). (A faster algorithm can be obtained by implementing an order N computation for the convolution)

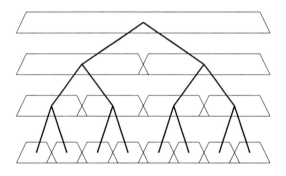

FIGURE 1. SCHEMATIC DESCRIPTION.

We see that in order to compute the transform on the large interval, we can start with adjacent pairs of small intervals, combine coefficients to obtain the expansion on their union, and continue until we reach the top level. As a result we have obtained all dyadic windowed Fourier transform as intermediate computations.

Clearly every disjoint collection of intervals equipped with an orthogonal basis on each provides us with an orthogonal basis for the union. A natural question that arises in connection with the windowed Fourier transform is how to place the windows (see Figures 2,3 where the effect of the window selection on the number of large coefficients is visible).

The signal is a combination of three linear chirps. By choosing small windows we find three main frequencies per window (the vertical axis is the frequency axis while the horizontal is the time axis).

In Figure 3 the windows are larger leading to a less efficient intertwined representation of the signal.

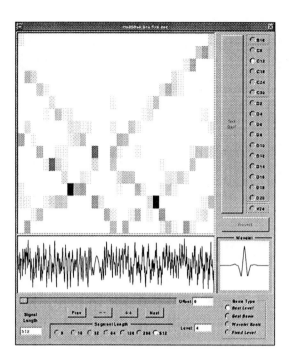

FIGURE 2. OPTIMAL WINDOW SELECTION.

For the moment let us consider the question of optimizing the windows to obtain an efficient representation of a function.

We can proceed as follows:

We start with the adjacent small intervals and consider expansion coefficient in each separately. We then compute the expansion coefficients on their union. We can now choose that expansion for which the number of coefficients needed to capture 99% of the energy is smallest (or that expansion whose "cost" is smallest; information cost, coding cost, error cost).

We compare the cost of the chosen expansions on two adjacent unions of pairs to the expansion on their union and again pick the best.

We continue until we reach an optimal distribution of windows (see Figure 4 where the windows were adapted to the voice recording).

The procedure described above, although natural, is not very useful if we take the windowed Fourier transform with discontinuous windows, since the discontinuity introduces "large" expansion coefficients, (a cosine basis on each interval is somewhat better). On the other hand, it is well known that we cannot find a smooth window function $\omega(x)$ supported on $(-\frac{1}{2}, \frac{3}{2})$ such that $\omega(x - k)e^{2i\pi nx}$ are orthogonal. (This would imply $\int \omega(x)\omega(x - 1)e^{-2\pi imx}dx = 0$ for all $m$ i.e. $\omega(x)\omega(x - 1) = 0$).

Recently Daubechies ,Jaffard,and Journe as well as Malvar observed that by taking equal windows and sines or cosines orthogonality can be maintained. Coifman and Meyer [3] observed that the windows can be chosen to different sizes enabling adaptive constructions as above. (See Figures 5,6)

We start by defining this library of trigonometric waveforms. These are localized sine transforms associated to covering by intervals of **R** (more generally, of a manifold).

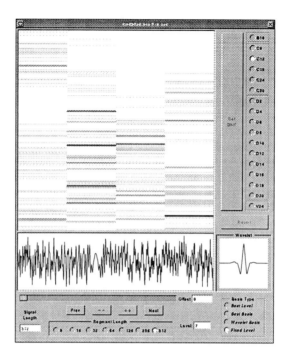

FIGURE 3. LARGE WINDOW SELECTION.

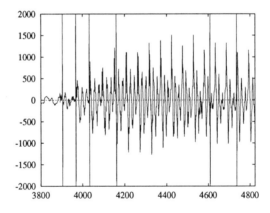

FIGURE 4. OPTIMAL WINDOW SELECTION.

We consider a cover $\mathbf{R} = \bigcup_{-\infty}^{\infty} I_i$ $I = [\alpha_i \alpha_{i+1})$ $\alpha_i < \alpha_{i+1}$, write $\ell_i = \alpha_{i+1} - \alpha_i = |I_i|$ and let $p_i(x)$ be a window function supported in $[\alpha_i - \ell_{i-1}/2, \alpha_{i+1} + \ell_{i+1}/2]$ such that

$$\sum_{-\infty}^{\infty} p_i^2(x) = 1$$

and

$$p_i^2(x) = 1 - p_i^2(2\alpha_{i+1} - x) \quad \text{for} \quad x \quad \text{near} \quad \alpha_{i+1}$$

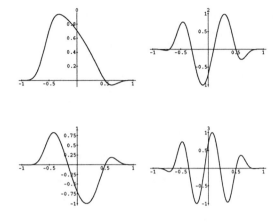

FIGURE 5. LOCAL TRIGONOMETRIC WAVEFORMS.

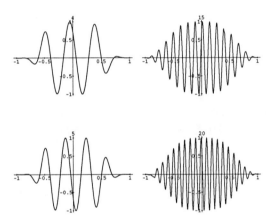

FIGURE 6. LOCAL TRIGONOMETRIC WAVEFORMS.

then the functions

$$S_{i,k}(x) = \frac{2}{\sqrt{2\ell i}} p_i(x) \sin[(2k+1)\frac{\pi}{2\ell_i}(x - \alpha_i)]$$

form an orghonormal basis of $L^2(\mathbf{R})$ subordinate to the partition $p_i$. The collection of such bases forms a library of orthonormal bases.

It is easy to check that if $H_{I_i}$ denotes the space of functions spanned by $S_{i,k}$ $k = 0, 1, 2, ...$ then $H_{I_i} + H_{I_{i+1}}$ is spanned by the functions

$$P(x)\frac{2}{\sqrt{2(\ell_i + \ell_{i+1})}} \sin[(2k+1)\frac{\pi}{2(\ell_i + ell_{i+1})}(x - \alpha_i)]$$

where

$$P^2 = p_i^2(x) + p_{i+1}^2(x)$$

is a "window" function covering the interval $I_i \cup I_{i+1}$. This fundamental identity permits the useful implementation of the adapted window algorihm described in Figure 1. (Other possible libraries can be constructed. The space of frequencies can be decomposed into pairs of symmetric windows around the origin, on which a smooth partition of unity is constructed. This and other constructions were obtained by one of our students E. Laeng [L].

Higher dimensional libraries can also be easily constructed,(as well as libraries on manifolds) leading to new and direct analysis methods for linear transformations.)

**Relation to Wavelets - Wavelet Packets.**

We consider the frequency line $\mathbf{R}$ split as $\mathbf{R}^+ = (0, \infty)$ union $\mathbf{R}^- = (-\infty, 0)$. On $L^2(0, \infty)$ we introduce a window function $p(\xi)$ such that $\sum_{k=-\infty}^{\infty} p^2(2^{-k}\xi) = 1$ and $p(\xi)$ is supported in $(3/4, 3)$ clearly we can view $p(2^{-k}\xi)$ as a window function above the interval $(2^k, 2^{k+1})$ and observe that

$$\sin\left[(j + \frac{1}{2})\pi \left(\frac{\xi - 2^k}{2^k}\right)\right] p(2^{-k}\xi) = s_{k,j}$$

form an orthonormal basis of $L^2(\mathbf{R}^+)$. Similarly $c_{k,j} = \cos\left[(j + \frac{1}{2})\pi \left(\frac{\xi-2^k}{2^k}\right)\right] p(2^{-k}\xi)$ gives another basis. If we define $S_{k,j}$ as an odd extension to $\mathbf{R}$ of $s_{k,j}$ and $C_{k,j}$ as an even extension, we find $S_{k,j} \perp C_{k',j'}$ permitting us to write $C_{k,j} \pm iS_{k,j} = e^{\pm ij\pi\xi/2^k} \hat{\psi}(\xi/2^j)$ where $\hat{\psi}(\xi) = e^{i\pi/2\xi} p(\xi)$ is the Fourier transform of the base wavelet $\Psi$ (see Meyer).

We therefore see that wavelet analysis corresponds to windowing frequency space in "octave" windows $(2^k, 2^{k+1})$.

A natural extension therefore is provided by allowing all dyadic windows in frequency space and adapted window choice. This sort of analysis is "equivalent" to wavelet packet analysis.

The wavelet packet analysis algorithms permit us to perform an adapted Fourier windowing directly in time domain by successive filtering of a function into different regions in frequency. The dual version of the window selection provides an adapted subband coding algorithm.

This new library of orthonormal bases constructed in time domain is called the Wavelet packet library. This library contains the wavelet basis, Walsh functions, and smooth versions of Walsh functions called wavelet packets.See Figure 7

We'll use the notation and terminology of [4], whose results we shall assume.

We are given an exact quadrature mirror filter $h(n)$ satisfying the conditions of Theorem (3.6) in [4], p. 964, i.e.

$$\sum_n h(n-2k)h(n-2\ell) = \delta_{k,\ell}, \quad \sum_n h(n) = \sqrt{2}.$$

We let $g_k = h_{l-k}(-1)^k$ and define the operations $F_i$ on $\ell^2(\mathbf{Z})$ into "$\ell^2(2\mathbf{Z})$"

(1.0)
$$F_0\{s_k\}(i) = 2\sum s_k h_{k-2i}$$
$$F_1\{s_k\}(i) = 2\sum s_k g_{k-2i}.$$

The map $\mathbf{F}(s_k) = F_0(s_k) \oplus F_1(s_k) \in \ell^2(2\mathbf{Z}) \oplus \ell^2(2\mathbf{Z})$ is orthogonal and

(1.1)
$$F_0^* F_0 + F_1^* F_1 = I$$

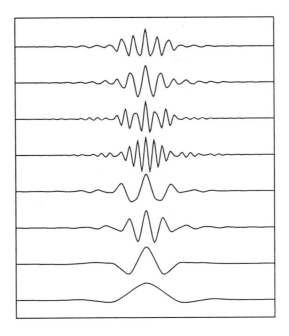

FIGURE 7. WAVELET PACKET LIBRARY.

We now define the following sequence of functions.

(1.2) $$\begin{cases} W_{2n}(x) = \sqrt{2} \sum h_k W_n(2x - k) \\ W_{2n+1}(x) = \sqrt{2} \sum g_k W_n(2x - k). \end{cases}$$

Clearly the function $W_0(x)$ can be identified with the scaling function $\varphi$ in [D] and $W_1$ with the basic wavelet $\psi$.

Let us define $m_0(\xi) = \frac{1}{\sqrt{2}} \sum h_k e^{-ik\xi}$ and

$$m_1(\xi) = -e^{i\xi} \bar{m}_0(\xi + \pi) = \frac{1}{\sqrt{2}} \sum g_k e^{ik\xi}$$

**Remark.** The quadrature mirror condition on the operation $\mathbf{F} = (F_0, F_1)$ is equivalent to the unitarity of the matrix

$$\mathcal{M} = \begin{bmatrix} m_0(\xi) & m_1(\xi) \\ m_0(\xi + \pi) & m_1(\xi + \pi) \end{bmatrix}$$

Taking the Fourier transform of (1.2) when $n = 0$ we get

$$\hat{W}_0(\xi) = m_0(\xi/2) \hat{W}_0(\xi/2)$$

i.e.,

$$\hat{W}_0(\xi) = \prod_{j=1}^{\infty} m_0(\xi/2^j)$$

and

$$\hat{W}_1(\xi) = m_1(\xi/2) \hat{W}_0(\xi/2) = m_1(\xi/2) m_0(\xi/4) m_0(\xi/2^3) \cdots$$

More generally, the relations (1.2) are equivalent to

$$\hat{W}_n(\xi) = \prod_{j=1}^{\infty} m_{\varepsilon_j}(\xi/2j) \tag{1.3}$$

and $n = \sum_{j=1}^{\infty} \varepsilon_j 2^{j-1} (\varepsilon_j = 0 \text{ or } 1)$.

The functions $W_n(x - k)$ form an orthonormal basis of $L^2(\mathbf{R}^1)$. a We define a <u>library</u> of wavelet packets to be the collection of functions of the form $W_n(2^\ell x - k)$ where $\ell, k \in \mathbf{Z}, n \in N$. Here, each element of the library is determined by a scaling parameter $\ell$, a localization parameter $k$ and an oscillation parameter $n$. (The function $W_n(2^\ell x - k)$ is roughly centered at $2^{-\ell}k$, has support of size $\approx 2^{-\ell}$ and oscillates $\approx n$ times).

We have the following simple characterization of subsets forming orthonormal bases.

**Proposition.** *Any collection of indices $(\ell, n)$ such that the intervals $[2^\ell n, 2^\ell n + 1)$ form a disjoint cover of $[0, \infty)$ gives rise to an orthonormal basis of $L^2$*[1].

(These intervals correspond to the partition of frequency space alluded to in §1.)

Motivated by ideas from signal processing and communication theory we were led to measure the "distance" between a basis and a function in terms of the Shannon entropy of the expansion. More generally, let $H$ be a Hilbert space.

Let $v \in H$, $\|v\| = 1$ and assume

$$H = \oplus \sum H_i$$

an orthogonal direct sum. We define

$$\varepsilon^2(v, \{H_i\}) = -\sum \|v_i\|^2 \ln \|v_i\|^2$$

as a measure of distance between $v$ and the orthogonal decomposition.

$\varepsilon^2$ is characterized by the Shannon equation which is a version of Pythagoras' theorem. Let

$$H = \oplus (\sum H^i) \oplus (\sum H_j)$$
$$= H_+ \oplus H_-$$

$H^i$ and $H_j$ give orthogonal decompositions $H_+ = \sum H^i$, $H_- = \sum H_j$. Then

$$\varepsilon^2(v; \{H^i, H_j\}) = \varepsilon^2(v, \{H_+, H_-\}$$
$$+ \|v_+\|^2 \varepsilon^2 \left(\frac{v_+}{\|v_+\|}, \{H^i\}\right)$$
$$+ \|v_-\|^2 \varepsilon^2 \left(\frac{v_-}{\|v_-\|}, \{H_j\}\right)$$

This is Shannon's equation for entropy (if we interpret as in quantum mechanics $\|P_{H_+} v\|^2$ as the "probability" of $v$ to be in the subspace $H_+$).

This equation enables us to search for a smallest entropy space decomposition of a given vector.

---

[1] We can think of this cover as an even covering of frequency space by windows roughly localized over the corresponding intervals.

In fact, for the example of the first library restricted to covering by dyadic intervals we can start by calculating the entropy of an expansion relative to a local trigonometric basis for intervals of length one, then compare the entropy of an adjacent pair of intervals to the entropy of an expansion on their union. Pick the expansion of minimal entropy and continue until a minimum entropy expansion is achieved (see Figure 1).

In practice, discrete versions of this scheme can be implemented in $CN \log N$ computations (where $N$ is the number of discrete samples $N = 2^L$.)

Of course, while entropy is a good measure of concentration or efficiency of an expansion, various other information cost functions are possible, permitting discrimination and choice between various special function expansion.

## §2. Wavelet Packet and Adapted Waveform Analysis.

We would like to summarize some obvious implications of the preceding discussion. Wavelet packet analysis consists of a versatile collection of tools for the analysis and manipulation of signals such as sound and images, as well as more general digital data sets. The user is provided with a collection of standard libraries of waveforms, which can be chosen to fit specific classes of signals. These libraries come equipped with fast numerical algorithms, enabling real time implementation of a variety of signal processing tasks such as compression, feature extraction for recognition and diagnostics, data transformation, and manipulation.

The process of analysis of data usually starts by comparing acquired segments of data with stored "known" samples. As a model, consider how a real sampled signal is analyzed with such libraries. The example of voice or music can be used for illustration.

Voice signals consist of modulated oscillations as can be seen in Figure 4, representing a segment of a recording of the word "armadillo." Such a general signal is a superposition of different structures occurring on different time scales at different times. One purpose of analysis is to separate and sort these structures. The oscillations are analogous to musical notes, and the analysis is equivalent to choosing the best instrument to match to the voice, then finding the musical score to describe the word.

A musical note can be described by four basic parameters: intensity (or amplitude), frequency, time duration, and time position. Wavelet packets or localized sinusoids are indexed by the same parameters. In addition, there are other parameters corresponding to choice of library, i.e., the instrument or recipe used to generate all the waveforms. For wavelet packets these extra numbers are the quadrature mirror filter coefficients; for local cosines, they are the smooth window coefficients.

The process of analysis compares a sound or other signal with all elements of a given library and picks up large correlations, notes which are good fits to segments of the signal. A most-concentrated orthogonal subset of these good notes can then be chosen. This "best basis" realization provides an economical transcription, an efficient superposition of oscillatory modes on different time scales. When ordered by decreasing intensity, this transcription sorts the main features out in order of importance. It permits rebuilding the signal to a specified accuracy with the fewest waveforms. It can be used to compress signals for digital transmission and storage.

Of more practical value is the ability to compute and manipulate data in compressed parameters. This ability is particularly important for recognition and diagnostic purposes. As an illustration, consider a hypothetical diagnostic device for heartbeats, in which fifty consecutive beats are recorded. We would like to use this data as a statistical foundation for detection of significant changes in the next batch of beats. Theoretically this can be done by factor analysis, or the Karhunen–Loéve transformation; unfortunately, the computation involving raw data is too large to be practical. But when the recorded data is efficiently compressed to a few parameters in the single statistical best basis, the factor analysis (if needed) can be performed in real time. The deviation of the next few heartbeats from their predecessors can be computed "on the fly," and significant changes can be flagged immediately.

In another example, consider a very large three dimensional atmospheric pressure map, and the problem of calculating the evolution of the pressure. In this case it is natural to break up the computation as a sum of interactions within different scales, with some limited interaction between adjacent scales, Such a breakup is automatic if the pressure map is expressed in the wavelet basis, which in this case is also the natural choice for compression of the data.

Such algorithms, which first compress a large set of measurements in order to compute with fewer parameters, can dramatically reduce the time needed to transform and manipulate data. They generalize the classical transform methods, like FFT, by custom building a fast transform for each specific application, merging beautifully the technologies of data compression and numerical analysis.

### References

1. R. Coifman, *Adapted multiresolution analysis, computation, signal processing and operator theory*, ICM 90 (Kyoto).

2. R. Coifman, Y. Meyer, and V. Wickerhauser, *Wavelet Analysis and Signal Processing Proceeding conference on Wavelets Lowell Mass 1991*.

3. R. Coifman and Y. Meyer, *Remarques sur l'analyse de Fourier à fenêtre*, série I, C. R. Acad. Sci. Paris **312** (1991), 259–261.

4. I. Daubechies, *Orthonormal bases of compactly supported wavelets*, Communications on Pure and Applied Mathematics **XLI** (1988), 909–996.

5. E. Laeng, *Une base orthonormale de $L^2(R)$, dont les éléments sont bien localisés dans l'espace de phase et leurs supports adaptés à toute partition symétrique de l'espace des fréquences*, série I, C. R. Acad. Sci. Paris **311** (1990), 677–680.

6. Available by anonymous ftp, *ceres.math.yale.edu*, InterNet address 130.132.23.22.

# Chapter 5
# LARGE-SCALE NONLINEAR CONSTRAINED OPTIMIZATION*

| A. R. CONN | NICK GOULD | PH. L. TOINT |
|---|---|---|
| IBM T.J. Watson Research Center Yorktown Heights, NY (pictured) | Rutherford Appleton Laboratory England | Facultés Universitaires ND de la Paix Belgium |

**1. Introduction.** Our purpose in this paper is to present an overview of the state-of-the art in large-scale nonlinear optimization. This article is a personal response to the questions, why we are interested in large-scale nonlinear optimization, what are the difficulties and what kind of progress has been made. Although we have made some effort to be complete in our references and thus hope to provide a useful bibliography, we have not attempted to be as complete in our overview. Rather, we have tried to include enough details of the general issues to indicate the nature and reasons for some of the current research in the field. Moreover, the length of treatment is frequently an indicator that the work is new and less well-known, and a brief mention does not mean that the work is relatively less important.

It seems appropriate to first state the most general form of the problem that we are addressing, namely

$$(1.1) \qquad \underset{x \in \Re^n}{\text{minimize}} \ f(x)$$

subject to the general (possibly nonlinear) inequality constraints

$$(1.2) \qquad c_j(x) \leq 0, \quad 1 \leq j \leq l,$$

to the (possibly nonlinear) equality constraints

$$(1.3) \qquad c_j(x) = 0, \quad l+1 \leq j \leq m,$$

and the simple bounds

$$(1.4) \qquad l_i \leq x_i \leq u_i, \quad 1 \leq i \leq n.$$

---

* This research was supported in part by the Advanced Research Projects Agency of the Department of Defense and was monitored by the Air Force Office of Scientific Research under Contract No F49620-91-C-0079. The United States Government is authorized to reproduce and distribute reprints for governmental purposes notwithstanding any copyright notation hereon.

Here, $f$ and the $c_j$ are all assumed to be twice-continuously differentiable and any of the bounds in (1.4) may be infinite.

At the outset we should make it clear that we only expect to obtain local minimizers. This is in marked contrast to combinatorial optimizers who, typically, are only interested in global solutions. This presents no problems in convex programming, where all local minima are indeed global (for example, in linear programming), but even for small, general nonlinear programming problems it is usually extremely difficult to verify globality. For large problems, it is practically impossible. Fortunately, in many situations, an algorithm that determines local optima suffices.

Our primary interest here is in problems that involve a large number of variables and/or constraints. Consequently, it seems worthwhile to elaborate as to what we mean by large.

Firstly, this notion is clearly *computer dependent*. What is large on an Apple Macintosh is significantly different from what is large on an IBM 3090 or a Cray 2. The first machine has a substantially smaller memory and storage than the other two, and therefore has more difficulty handling problems involving a large amount of data. Secondly, a highly nonlinear problem in one hundred variables could be considered large, whereas in linear programming it is possible to solve problems in five million variables. The notion of size is thus *problem dependent*. It also depends upon the *structure of the problem*. Many large-scale nonlinear problems arise from the modelling of very complicated systems that may be subdivided into loosely connected subsystems. This structure may often be reflected in the mathematical formulation of the problem and exploiting it is often crucial if one wants to obtain an answer efficiently. The complexity of the structure is often a key factor in assessing the size of a problem. Lastly, the notion of a large problem depends upon the *frequency* with which one expects to solve a particular instance or closely related problem. When one anticipates solving the same class of problems many times, one can afford to expend a significant amount of energy analyzing and exploiting the underlying structure. Thus, although it is not possible to say categorically that a problem in say seven hundred variables is large, suffice it to say that, today, a problem in fifty variables is small and a generally nonlinear problem in five thousand variables and one thousand nonlinear constraints is large.

One might suppose that intellectual curiosity alone is sufficient reason to be interested in large-scale nonlinear optimization. However, although we readily admit to the fact that this is an important element of our interest (and indeed if our research had been confined to the publication of theoretical articles, arguably the main one), much of our joint effort has been devoted to the time consuming and often tedious task of writing software, preparing input and testing. The salient point is that there is a need for algorithms to solve large-scale nonlinear optimization problems. The accurate modelling of physical and scientific phenomena frequently leads to such problems. Nature loves to optimize: minimum energy, minimum potential difference, shortest paths. Moreover, the universe certainly is not linear. If the model is to be accurate (for example, if it is derived from a discretization of a continuous process), the number of variables is necessarily large. Another area where large nonlinear problem arise naturally is in economics, where one often wishes to maximize profit (or minimize losses) in complex situations involving many parameters. The proliferation of large linear models, rather than nonlinear ones, is sometimes a consequence of our lack of knowledge concerning the phenomena being modelled, in which case assuming linearity is about the simplest assumption one can make. As our knowledge improves, often the models are refined and nonlinearity should be introduced. In our opinion, the frequent use of linear models is not an indication that nonlinear problems do not abound. Rather, it is a statement of the desire to use an algorithm (the simplex method) that is readily understood and is well-known to be suitable for large problems. In particular, it is one of our tasks to convince you that you should consider solving nonlinear programs, when they are more appropriate. As a necessary corollary, it should be emphasized that solutions to large nonlinear problems on moderate workstations in a reasonable amount of time are currently quite possible. Furthermore, in practice one is often only seeking marked improvement rather than assured optimality (another reason why globality is not necessarily an issue). This fact makes even problems that at first sight seem impossible (for example, control problems that one wishes to solve in something like real-time), tractable.

Without a doubt, the ubiquity of powerful workstations and the availability of supermachines (both parallel and sequential) have encouraged research in algorithms for large-scale problems. However, we concur with a remark that Martin Beale once made that he would 'much rather work with today's algorithms on yesterday's computers than with yesterday's algorithms on today's computers' [128].

**2. Examples of Applications.** As we already stated, our interest in developing algorithms for large-scale optimization was created out of necessity. There is an increasing demand for such software as the size and nonlinearity of the problems that practitioners are interested in solving steadily grows. The same evolution that leads to larger and larger nonlinear optimization models for physical phenomena is observed in data fitting, econometrics and operations research models. In particular, it is perhaps worth listing some examples:

- Discretizations of variational calculations and optimal control problems involving both state and control variables.

    These arise, for example, in quantum physics, tidal flow analysis, design (of aircraft, journal bearings and other mechanical devices), structural optimization, ceramics manufacturing, chemical process control and satellite piloting.

- Nonlinear equations arising both in their own right and in the solution of ordinary and partial differential equations.

    These occur, for instance, in elasticity, semiconductor simulation, chemical reaction modelling and radiative transfer.

- Nonlinear least squares or regression.

    Some examples include fluid dynamic calculations, tomography (both seismic and medical), combustion, isomerization and metal coating thickness assessment.

- Nonlinear approximation.

    These include antenna design, power transmission, maximum likelihood and robust regression.

- Nonlinear networks.

    Examples occur in traffic modelling, energy and water distribution/management systems, and neural networks. These problems can have hundreds of thousands of variables but they are tractable because of their very special structure.

- Other interesting problems occur in macro- and micro-economics, equilibrium calculations, production planning, energy scenario appraisal and portfolio analysis. These problems often give rise to quadratic programs, particularly in portfolio analysis.

The increasing interest in the solution of large-scale nonlinear optimization problems is also related to the realisation by users that today's advances in computer technology are making the solution of such problems possible.

Recent articles and books devoted primarily to large-scale optimization include [19], [20], [24], [33], [35], and [133]. Some examples of applications are given in [59], [87], [95], [96], [123] and [131]. Background material on nonlinear optimization is given in [54], [60],

[61], [70] and [108]. An overview of what is involved in a mathematical programming system, especially with respect to linear problems, is given in [129].

**3. What are the difficulties?** Efficient algorithms for small-scale problems do not necessarily translate into efficient algorithms for large-scale problems. This is unfortunate, since in the past twenty years rather sophisticated and reliable techniques for small-scale problems have been developed (see [108] for good surveys). As a consequence, it is just not adequate to take existing optimization software for small problems and apply it to large ones, hoping that the increased capacity in computing will take care of the growth in problem size. By contrast, we could expect that an efficient method for large-scale problems be at least moderately efficient for small-scale problems.

Perhaps the most important difficulty in the large-scale context is that of *exploiting structure*. The fact that we are able to solve large problems at all is because they are structured. Even in linear programming, a problem in one thousand variables devoid of structure (happily, usually an indication of a bad formulation) severely taxes codes. Thus, it is absolutely essential for efficient algorithms to exploit structure. Moreover, this means exploiting more than just sparsity. Unfortunately, this exploitation often complicates the question of stability, that is, the ability of an algorithm to guarantee that small perturbations in the data will only result in small perturbations to the solution for 'satisfactorily conditioned' problems. By contrast, algorithms for small problems have the possibility to ignore structure.

Another significant difficulty is that of *scaling*. Perhaps the main reason algorithms for small-scale problems do not necessarily translate into efficient algorithms for large-scale problems is that in order to be able to handle large problems the algorithms have to necessarily be as simple as possible. Consequently, relative to many of the more successful algorithms for small, dense problems, the amount of information available at any given iteration may be severely restricted. This makes designing algorithms that are scale invariant (in the sense that, assuming infinite precision arithmetic, quasi-Newton methods for unconstrained optimization are invariant under linear transformations) more difficult for large-scale problems.

One consequence of the necessity to keep the algorithms for large-scale nonlinear optimization '*simple*' is that typically one has rather incomplete information available at every iteration. Thus it becomes difficult to successfully merging two distinct (and in many ways, conflicting) aspects of any nonlinear programming algorithm. The first aspect is that which guarantees global convergence. By global convergence (not to be confused with convergence to global optima), we mean convergence to a stationary point from any starting point. Essentially, this is a weak requirement (the method of steepest descent with a suitable line search condition will suffice) which combined with the desire to stress simplicity encourages us to use a steepest-descent-like method. On the other hand, ultimately we want faster convergence. This means using Newton's method or quasi-Newton methods, although in the case of large problems a fast linear rate might suffice.

From a more mundane point of view there are very real difficulties *inputting* large problems. In particular, the amount of information present in the structure of a large-scale optimization problem, although crucial for the acceptable performance of algorithms, is also very difficult to specify in a complete and understandable format. A standard input that is a simple formal language in which these structural concepts could be expressed unambiguously, has been rather well-established and successful in the more restricted domain of linear programming [47]. We have extended this input format to the nonlinear case (see [36]). However, particularly because of the necessity of exploiting structure in a rather general sense, the resulting standard is not as simple as it was in the linear case.

Other approaches are based upon using a high-level modelling language. The high-level aspect makes these rather more user friendly but they require an interpreter and are thus not normally in the public domain. Moreover, they do not, currently, exploit structure as generally as we would like. Well-known examples of this approach are GAMS ([14]) and AMPL ([65]).

Obviously, a primary requirement in evaluating the quality of an algorithm is to have a good set of *test problems*. In the first instance, it is by no means obvious what constitutes such a set for large-scale optimization, in the main because of the complexity of the problems and the lack of experience in solving them. Nevertheless, important starts have been made, and although we do not yet have collections as readily available as those for linear programming and sparse linear algebra (see [57] and [66]), we currently have over nine hundred problem instances in (our) 'standard input format' [40]. We are also asking for more test problems from the community ([125]). Moreover, as a part of the MINPACK-2 project 'a collection of significant optimization problems' is being made, see [4]. Earlier collections for unconstrained problems include [15] and [102].

Just as important is the ability to *evaluate results*. Given the number of variables (in both senses of the word) at hand, it is a complex task to interpret the results of testing. It is fair to say that, at present, we require more established test problems and a broader experience of the behaviour with various algorithms.

**4. Current Approaches.** Having considered the difficulties, we now examine how they are addressed. It is convenient to consider three broad classes, namely, approaches based upon classical large linear programming, approaches based upon small-scale nonlinear programming and approaches based upon a mixed linear programming/interior point method, even though the ideas in each approach are not mutually exclusive.

**4.1. Approach based upon classical large linear programming.** We will begin with a terse and somewhat eccentric summary of the simplex method. For those who need further details, an excellent recent survey article, that includes the interior point method that is relevant to the third approach below, is given in [73]. Consider the linear programming problem in the form

$$(4.5) \qquad \underset{x \in \Re^n}{\text{minimize}} \quad c^T x$$

subject to the $m$ ($\leq n$) general linear constraints

$$(4.6) \qquad Ax = b$$

and to the simple bound constraints

$$(4.7) \qquad x \geq 0.$$

The solution to the above problem normally occurs at a vertex of the feasible region, that is a point defined by the equations (4.6) and $n - m$ of the variables lying on their bounds (4.7). Without loss of generality, we can assume that the last $n - m$ components of $x$ are at their bounds. We call such variables *non-basic* — the remaining $m$ variables are termed *basic*. We may thus consider the activities (i.e. those constraints satisfied as equalities) to be given by

(4.8) $$Cx = \begin{pmatrix} B & N \\ & I \end{pmatrix} x = \begin{pmatrix} b \\ 0 \end{pmatrix}.$$

Now

(4.9) $$C^{-1} = \begin{pmatrix} B^{-1} & -B^{-1}N \\ & I \end{pmatrix}.$$

The fact that the $k^{th}$ column of $C^{-1}$ is orthogonal to the other $n-1$ rows of C, along with the fact that we start at a vertex of the feasible polytope and insist on following a path of objective-improving feasible vertices to optimality is really the heart of the simplex method. The first statement means that by moving along this $k^{th}$ column the remaining equations corresponding to the other $n-1$ rows of $C$ stay active. Being at a vertex ensures that we can refer to $C^{-1}$. Objective-improving is just a matter of sign and maintaining feasibility requires that we move to an adjacent vertex.

More importantly, from the point of view of this article, the method is efficient because it exploits heavily the structure of $B$, making use of techniques such as the Markowitz strategy, [93], and sparse Bartels-Golub, [6], updating of $LU$ factors (see [117] for further details and the first implementation in Fortran). Another important feature of linear programming software is the ability to have crash starts, i.e. a relatively simple method for finding a good starting basis (see, for example [76]).

The highly successful package MINOS, [106] can be viewed as an extension of the simplex method as a reduced gradient technique. Its origins come from [118] and [119]. One should also note that for practitioners who are used to linear programming approaches, MINOS serves as an *extremely* useful bridge to nonlinear programming. In particular, MINOS replaces

(4.10)
$$\begin{aligned}
& \underset{x \in R^n, y \in R^m}{\text{minimize}} && F(x) + c^T x + d^T y \\
& \text{subject to} && f(x) + A_1 y = b_1 \\
& && A_2 x + A_3 y = b_2 \\
& \text{and} && \\
& && l_x \leq x \leq u_x \\
& && l_y \leq y \leq u_y
\end{aligned}$$

with

(4.11)
$$\begin{aligned}
& \underset{x \in R^n, y \in R^m}{\text{minimize}} && F(x) + c^T x + d^T y + \lambda_k^T(f(x) - \tilde{f}(x)) + \tfrac{1}{2}\rho(f(x) - \tilde{f}(x))^T(f(x) - \tilde{f}(x)) \\
& \text{subject to} && \\
& && \tilde{f}(x) + A_1 y = b_1 \\
& && A_2 x + A_3 y = b_2 \\
& \text{and} && \\
& && l_x \leq x \leq u_x \\
& && l_y \leq y \leq u_y,
\end{aligned}$$

where

$$\tilde{f}(x) = f(x_k) + J_k(x - x_k),$$

and $J_k$ denotes the Jacobian of $f$ evaluated at $x_k$. In other words, the nonlinear contribution to the constraints is linearized so that we can exploit linear programming technology. It then formulates a quadratic model for the artficial objective function. A reduced gradient technique is used, that is one determines a search direction that maintains the current activities to first-order (i.e. the linearized approximations that were active stay active). Writing the activities that are determined by the general linear constraints as

$$\hat{A}x = \begin{pmatrix} B & S & N \end{pmatrix} x = b, \tag{4.12}$$

this means that our search direction is given by

$$h = Zd, \tag{4.13}$$

where

$$Z^T = \begin{pmatrix} -[B^{-1}S]^T & I & 0 \end{pmatrix}. \tag{4.14}$$

This follows directly from the fact that

$$\hat{A}Z = 0 \text{ and } (0 \ 0 \ I)Z = 0. \tag{4.15}$$

Analogously to the simplex method, the columns of $B$ correspond to basic variables and the columns of $N$ correspond to non-basic variables. However, because of the nonlinearity of the objective function, we are no longer able to ensure that optima lie at vertices (the number of columns of $B$ and $N$ may not add up to the dimension of the space). The ensuing deficiences are made up by the columns of $S$, the so-called superbasic columns. Because of the similarities in the resulting linear algebra the exploitation of structure is much the same as that in the simplex method. It is worth pointing out that exploitation of the structure of $Z$ and the simple bounds is especially attractive in the context of network problems (see, for example, [51], [82], [127] and [126]).

It should be clear that the fewer superbasic columns, the closer the problem is to a linear programming problem. MINOS works particularly well when there are relatively few superbasics.

A related approach that was one of the earliest successful pieces of software that could handle large nonlinear problems was an implementation of the generalised reduced gradient method of Abadie ([1]) by Lasdon ([90]). A quadratic programming algorithm that uses similar ideas to MINOS and is for large-scale problems is given in [75].

Not surprisingly, the earliest approach to large-scale nonlinear optimization was a successive linear programming technique (see [81]). A more recent successive linear programming technique that uses an exact $l_1$ penalty function and incorporates trust region constraints is given in [64], although numerical results are given for small problems only.

**4.2. Approaches based upon small-scale nonlinear programming.**

**4.2.1. Sequential quadratic programming.** One of the best known techniques for nonlinear programming is the so-called sequential quadratic programming approach (see, for example, [61], Chapter 12 and [70], Chapter 6). Recent work by Eldersveld [58] and colleagues uses the augmented Lagrangian. The vector $s$ represents slack or surplus variables (see below for some motivation for this function and the introduction of slack or surplus variables)

$$f(x) - \lambda^T [c(x) - s] + [c(x) - s]^T [c(x) - s]/\mu, \tag{4.16}$$

with the quadratic programming search-direction subproblem

(4.17)
$$\begin{aligned} \underset{p,q}{\text{minimize}} \quad & \tfrac{1}{2}p^T H p + g^T p \\ Ap - q &= -[c(x_k) - s_k] \\ \hat{l} \leq \begin{bmatrix} p \\ q \end{bmatrix} &\leq \hat{u}, \end{aligned}$$

using a suitable symmetric matrix $H$, to solve (1.1) to (1.4). A protoype implementation has been developed that uses a modification of the MINOS code. They use for the active set

(4.18)
$$\hat{A} = \begin{pmatrix} B & S & N \\ 0 & 0 & I \end{pmatrix}$$

and solve

$$Z^T H Z y_z = -Z^T (g + Hp),$$

$$y = Z y_z,$$

where

$$Z^T = \begin{pmatrix} -[B^{-1}S]^T & I & 0 \end{pmatrix}$$

and $Z^T H Z$ is small. Noting that one needs $H$ to evaluate the gradient of the quadratic objective function, they make use of the fact that if we define $Q = [Z\ Y]$, choosing $Z$ and $Y$ so that $Z^T HY = 0$ and $Y^T HY = I$, then we can write $H = Q^{-T}(Q^T HQ)Q^{-1}$, where

$$Q^T H Q = \begin{bmatrix} Z^T H Z & 0 \\ 0 & I \end{bmatrix}.$$

For example, taking

$$Y = \begin{bmatrix} B^{-1} & 0 \\ 0 & 0 \\ 0 & I \end{bmatrix}$$

then

$$Q^{-1} = \begin{bmatrix} 0 & I & 0 \\ B & S & 0 \\ 0 & 0 & I \end{bmatrix}.$$

Details are given in [58]. An approach that also uses sequential quadratic programming, but 'solves' the quadratic program using an interior point method (see below) is given by [13]. Although these methods hold promise, computational experience to date has been insufficient to make definitive statements as to their effectiveness. Other sequential quadratic programming based methods include [92] and [105].

**4.2.2. The LANCELOT project.** The approach to which we wish to devote much of the rest of this article is based upon the adaptation of trust region methods to the problem with simple bounds. The method is extended to general constraints by using an augmented Lagrangian function and the bounds are handled directly via projections that are easy to compute. We use group partial separability (a generalisation of sparsity, introduced in [79]) to allow efficient storage and updating of matrices in matrix-vector product form. This approach has the further advantage that accurate approximations to the second derivatives of the element functions, normally being of low rank, are easier to obtain than for the assembled matrices. This structure is extremely general. Indeed, any sufficiently differentiable function with a sparse Hessian matrix may be written in this form. An introduction to group partial separability is given by [34]. The entire project has resulted in a substantial amount of software that is available at nominal cost for research purposes. There is also a book ([41]) to accompany the software. Returning to the underlying concepts of **LANCELOT**, we will now give some details.

Trust region methods in the context of unconstrained optimization have been able to combine a rather intuitive framework and robust numerical implementations with a powerful and elegant theoretical foundation. An excellent reference is [101]. The basic idea is to model the objective function (by a quadratic given by the first three terms of a Taylor's series expansion about the current point $x^k$, for example). One then 'trusts' this model in a neighbourhood (called the trust region) of $x^k$. The next step is to approximately minimize the model in the trust region, thereby obtaining a point $x^k + s^k$, say. One now determines how well the model actually predicted the change in the true objective function. If good descent is obtained, the next iterate, $x^{k+1}$, is set to $x^k + s^k$ and the trust region is expanded. If moderate descent is obtained, the next iterate, $x^{k+1}$ is set to $x^k + s^k$ and the trust region remains unchanged. Otherwise, $x^{k+1}$ is set to $x^k$ and the trust region is contracted. The beauty of such an approach is that, when the trust region is small enough and the problem smooth, the approximation is good, provided the model gradient is sufficiently accurate. Moreover, assuming one does at least as well as the minimum along the steepest descent direction of the model within the trust region (that determines the so-called Cauchy point), one can ensure convergence to a stationary point ([18]). In addition, eventually the trust region is expanded sufficiently that it does not interfere with the subsequent iterates, and thus, assuming that in this situation the underlying algorithm is sufficiently sophisticated, one can ensure fast asymptotic convergence. Details are given in [101].

The algorithm that is at the heart of **LANCELOT** is a method for which all the constraints are just simple bounds. The extension of the above ideas are relatively straightforward in this case. Essentially, one generalizes the Cauchy point to the minimum along the *projected* gradient path within the trust region, where the projection is with respect to the simple bounds. It is important to note that it is trivial computationally to compute such a projection: components of $x$ that hit a bound just remain fixed. This approach was first carried out by McCormick in [94], and independently by Bertsekas [10] and Levitin and Polyak [91]. More recently it has been exploited extensively in the context of large-scale optimization by many authors, see for example [32], [52], [103], and [104]. As in the unconstrained case, global convergence can be guaranteed, provided one does at least as well as the generalized Cauchy point. One obtains better convergence, and ultimately a satisfactory asymptotic convergence rate, by further reducing the model function. In the context of **LANCELOT**, this is achieved by fixing the activities determined by the generalized Cauchy point and further reducing the model within the feasible region and trust region using just the remaining free variables. Updating of the trust region size is handled in exactly the same way as it is in the unconstrained case. The basic algorithm can be

summarised as follows:
- Find the generalized Cauchy point based upon a local (quadratic) model.
- Fix activities to those at the generalized Cauchy point.
- Using the free variables further reduce the model within the feasible region and the trust region. Of course this may, and typically does, introduce new activities in addition to those determined by the generalized Cauchy point.
- Determine whether the current point is acceptable and update the trust region radius accordingly.

Provided the quadratic model is reasonable, we are able to prove that we converge to a Kuhn-Tucker point. Moreover, we identify the correct active constraints (activities) after a finite number of iterations assuming that strict complementarity is satisfied and the activities determined by the generalised Cauchy point are kept active when the model is further reduced. Details are given in [31].

The extension to general constraints is carried out by means of an augmented Lagrangian function. In order to understand this extension we need to motivate this function. We have known for nearly two hundred years ([89], part 1, section 4, article 2), that a solution to

$$(4.19) \quad \begin{array}{c} \text{minimize} \quad f(x) \\ x \in R^n \\ \text{subject to} \quad c(x) = 0 \end{array}$$

is a feasible stationary point of the Lagrangian

$$(4.20) \quad f(x) - \lambda^T c(x).$$

It is only since the Second World War, that we have recognised ([48]) that we can solve (4.19) using the quadratic penalty function

$$(4.21) \quad \begin{array}{c} \text{minimize} \quad f(x) + c^2(x)/\mu \\ x \in R^n \end{array}$$

as $\mu$ tends to zero from above. The idea here is that as $\mu$ becomes small the 'penalty term' $c^2(x)/\mu$ forces one to become feasible. This intuitive idea was not accorded a sound theoretical basis until the work of Fiacco and McCormick (see for example [60]). Augmented Lagrangians combine both ideas, thereby convexifying the Lagrangian and circumventing the necessity of requiring small $\mu$ by, instead, approximating the Lagrange multipliers, $\lambda$. Thus we use the problem

$$(4.22) \quad \begin{array}{c} \text{minimize} \quad f(x) - \lambda^T c(x) + c^2(x)/\mu. \\ x \in R^n \end{array}$$

This approach was first suggested by K. J. Arrow and R. M. Solow in [3] but is better known through the work of Hestenes and Powell in [85] and [115].

In **LANCELOT** one thus solves the general problem by first introducing slack or surplus variables, if necessary, to change inequalities to equalities. Subsequently one minimizes the augmented Lagrangian

$$(4.23) \quad \Phi(x, \lambda, S, \mu) = f(x) + \sum_{i=1}^{m} \lambda_i c_i(x) + \frac{1}{2\mu} \sum_{i=1}^{m} s_{ii} c_i(x)^2$$

(where the diagonal matrix $S$ is introduced to incorporate scalings) subject to the explicit bounds, using the earlier algorithm[1]. This approach can be summarised as follows:

1. Test for convergence using the two following conditions.
   Sufficient stationarity — the projected gradient of the augmented Lagrangian with respect to the simple bounds is sufficiently small;
   Sufficient feasibility — the norm of the constraint violations is sufficiently small.
2. Use the simple bounds algorithm to find a sufficiently stationary approximate minimizer of $\Phi$ (considered as a function of $x$ only) subject to simple bounds.
3. If sufficiently feasible (both the 'local convergence' values here and in 2) are greater than the test for convergence, in general) update the multipliers and decrease the tolerances for stationarity and feasibility.
4. Otherwise, decrease the penalty parameter and reset tolerances for stationarity and feasibility.

We are able to show, under suitable conditions, that we converge to a first-order stationary point for the nonlinear programming problem. Furthermore, if we have a single limit point, we eventually stop reducing the penalty parameter, $\mu$. Under somewhat stronger conditions we are able to show that one requires only a single iteration of the simple bounds algorithm to satisfy the conditions of the third item above. Details of these important properties are given in [38] and [42].

As we have already seen, a significant (and often dominant) cost in optimization is solving a linear system. Typically these arise from the necessity to determine an approximate stationary point for a quadratic function — equivalently, the necessity to solve a linear system whose coefficient matrix is a symmetric matrix. If the system is large there are two possible approaches. The first is to use direct methods based upon multifrontal techniques. These use partial assembly and dense matrix technology on sparse matrices. General details are given in [56] and an application in the context of **LANCELOT** is given in [43]. Our experience to date, however, has been that an iterative approach is more robust. The most popular such approach is preconditioned conjugate gradients. For ease of motivation we first consider conjugate gradients without a preconditioner. Directions, $d_i$, are called conjugate with respect to a positive definite matrix $A$ if $d_1^T A d_2 = \langle d_1, d_2 \rangle_A = 0$. In other words, they are orthogonal in the $A$-metric. The best known conjugate set of vectors are the set of orthonormal eigenvectors. If one considers minimizing a strictly convex quadratic form $\frac{1}{2}x^T A x - b^T x + c$, it is easy to see from the geometry that if one minimizes along the eigenvectors of $A$, then at each stage one determines the minimum of the quadratic on the space spanned by the eigenvectors used, and thus, after at most $n$ steps, if $A$ is $n$ by $n$, the quadratic function's (unique) minimum is determined.

The appeal of conjugate-gradient methods is that this finite termination result for quadratics is true for general conjugate directions. Moreover, the attraction for large-scale optimization is that such an orthogonalisation can be determined via a three term recurrence and thus the method is particularly simple and only requires that we store three vectors.

However, if $n$ is large, performing $n$ steps may be prohibitively expensive. Moreover, a quadratic is only being used to model a nonlinear problem and so what we have is really a 'moving quadratic'. What makes this technique remain attractive is the use of preconditioners. The essential result is that whenever one has multiple eigenvalues, the conjugate gradient method *minimizes the quadratic* in the space spanned by the corresponding eigen-

---

[1] It is worth noting that MINOS uses for its objective function an augmented Lagrangian function with corresponding constraints $f - \tilde{f} = 0$, whose relaxation can be considered as *trusting* the linear approximation to $f$.

vectors. If we can cluster eigenvectors (i.e. approximately have multiple eigenvectors) we can reduce the number of iterations for good approximations to minimizers, from $n$ to the number of clusters. The perfect way to do this in the quadratic case is to precondition with $A^{-1}$ — but then this is equivalent to doing Newton's method. Surprisingly one can often do very well by using very crude approximations to $A^{-1}$ (diagonal matrices, for instance). A good reference is [67].

It is fortunate that most optimization problems in thousands of variables are structured; fortunate but demanding of respect. If one considers the arrow-head matrix ($a_{i,1}, a_{1,i}, a_{i,i}$ non-zero, $i = 1, 2, \ldots n$, all other entries zero), it is clear that one neither wants to input or nor wants to store this matrix as a dense matrix, for large $n$. A little less obvious is the fact that if one does Gaussian elimination without pivoting, after the first column is updated the remaining $n - 1$ by $n - 1$ block will be full — in other words, fill-in is disastrous. On the other hand, if we first reverse the ordering of the rows and columns (which amounts to changing the orderings of the equations and the labellings of the variables) there is no fill-in at all. Since optimal numerical stability typically dictates row and column orderings and limitations on storage motivate one to minimize fill-in, we are immediately aware of a major conflict in numerical linear algebra when one wants to account for structure. One form of compromise is known as threshold pivoting (see, for example, [56]).

In **LANCELOT** we take the point of view that invariant subspaces are more important than sparsity. For example, consider $f(x) = x_{50}^4$, and $F(x) = \left(\sum_{i=1}^{5,000,000} x_i\right)^4$, $x \in R^{5,000,000}$. In the first case the Hessian is sparse, while in the second case the Hessian is dense. But they both have an invariant subspace of dimension $n - 1$. In the first case, it is the orthogonal complement of $e_{50}$ (the vector with a single non-zero entry, one, in the fiftieth component), and in the second case it is the orthogonal complement of $e$ (the vector of all ones). We exploit invariant subspaces by writing our functions as $f(x) = \sum_{i=1}^m g_i \left(l_i(x) + \sum_{j \in I_i} w_j f_j(x)\right)$, where $g_i$ is a scalar function, the $l_i$ are linear functions and the $w_j$ are weights for the nonlinear functions $f_j(x)$. The essential point is that the rank of $\nabla_{xx} f_j$) is much smaller than $n$ and the null space of $\nabla_{xx} f_j$ is fixed. This, and the use of linear transformations to consider expressions like $e^{(x+y)}$ as $e^u$, $u = x + y$, enables us to use very compact representations of the problems we are optimizing. In particular we can store these as *dense* matrices. We note, however, that it is no longer reasonable to expect these matrices to be positive definite. This has led to a revival of interest in rank one secant methods. For an introduction to these considerations, with many more details, the reader is urged to read [34]. A related means of exploiting structure is given in [96]. For recent work on rank one updating see [37], [55], [88], [111], and [132].

Although we are unaware of the details, an augmented Lagrangian approach that is designed for large-scale optimization has been developed by Contesse, [46]. There are some similarities between the approach of [38], [45] and [64]. Fletcher and Sainz de la Maza use a piecewise linear model and an $l_1$ penalty function for the merit function but improvements over the Cauchy point involve projected Hessian approximations.

Current work in **LANCELOT** has been to consider the special case of convex constraints. The motivation is that, in particular, linear constraints are very common, are too simple to handle effectively in the same manner as we handle general constraints, but are nevertheless too complicated to handle the projections easily in the context of large problems. We use a combination of a simple piecewise linear line search (with only two pieces) and the trust region approach. The projections are handled approximately, but the approximation has to be 'good enough'. Details are given in [45]. In particular, a special case gives a convergence proof for sequential linear programming in the case of convex constraints. In the case of nonlinear networks, Sartenaer, in [122], has obtained some very encouraging numerical

| Problem | $n$ | $m$ |
|---|---|---|
| 1-d nonlinear boundary value problem | 5002 | 0 |
| Economic model from Thailand | 2230 | 1112 |
| 1-d variational problem from ODEs | 1001 | 0 |
| 2-d variational problem from PDEs | 5184 | 0 |
| 3-d variational problem from PDEs | 4913 | 0 |
| Nonlinear network gas flow problem | 2734 | 2727 |
| Chemical reaction problem | 5000 | 5000 |
| Oscillation problem in structural mechanics | 5041 | 0 |
| Nonlinear optimal control problem | 9006 | 7000 |
| Maximum pivot growth in Gaussian Elimination | 3946 | 3690 |
| Nonlinear network problem on a square grid | 13284 | 6724 |
| Nonlinear optimal control problem | 10001 | 5000 |
| Hydro-electric reservoir management | 2017 | 1008 |
| Minimum surface problem with nonlinear boundary conditions | 15625 | 0 |
| Economic equilibrium | 1825 | 730 |
| Nonlinear optimal control problem | 7011 | 5005 |
| Orthogonal regression problem | 8197 | 4096 |
| Analysis of semiconductors | 1002 | 1000 |
| Elastic-plastic torsion problem | 14884 | 0 |

FIG. 1. *Some typical examples.*

results.

In addition we have a test-bed of around nine hundred problems written in our standard data format. We have solved almost all these examples, many of which are of a substantial size. We expect to have them available electronically via netlib or something similar some time this year (1992). Some typical examples of applications we have solved using **LANCELOT** are tabulated below, in Figure 1. Detailed analysis of these results and our interpretation will be reported separately.

What about other current work on **LANCELOT** and the future? As an alternative to the augmented Lagrangian we are implementing a Lagrangian barrier method [44], that is closely related to the modified and shifted barrier function method (see, for example, [68], and [114]). We use the Lagrangian barrier function

$$\Psi(x, \lambda, s) = f(x) - \sum_{i=1}^{m} \lambda_i s_i \log(s_i - c_i(x)),$$

which we can then optimize with respect to the simple bounds. A possible choice is $s_i = \mu \lambda_i^{\alpha_\lambda}$, where $\mu$ is a penalty parameter and $0 < \alpha_\lambda \leq 1$.

We also would like to exploit group partial separability at a more fundamental level in our trust region algorithm. We have been able to use the structure of the problem explicitly in the definition of the trust region and have suitable convergence properties [39] and we are currently investigating the computational implications. In order to exploit linear constraints more successfully we are investigating interior point methods. Finally, we always need to perform more testing.

**4.3. Approach based upon a mixed linear programming/interior point method.**
Although not strictly speaking interior point methods, we have already mentioned modified

barrier methods above. More generally, interior point methods, although originally developed for nonlinear programming, have had a spectacular success in the context of linear programming and have thereby generated new interest in their use in nonlinear optimization. See [134], for example, for background material. Encouraging results have already been obtained in the context of linear-like problems (see [23], [25], [26], [100] and [120]) and quadratic programming ([2], [9], [12], [22] [72], [83], [84], [99], [112], and [137]). Nash and Sofer have computational experience with a barrier method applied to a thousand variable nonlinear problem with bound constraints, [107]. Work in convex programming includes [5], [53], [86], [97], [98], [135] and [136]. It seems reasonable to expect further developments within nonlinear programming, especially since these techniques appear to be especially appropriate for large-scale problems.

**4.4. Other issues.** There is obviously a number of issues that are relevant to our subject, but that fall slightly out of the context of the current paper. We now briefly mention some of the most important ones.

Both primal and dual degeneracy are intrinsic difficulties in that they are often a manifestation of the problem that can be troublesome to the method of solution. By primal degeneracy, we mean that the dual variables are not uniquely defined. By dual degeneracy, we mean that some of the dual variables are zero. Moreover degeneracy, especially primal degeneracy, is not a rare occurrence. Revelent work includes [17], [50], [62], [63], [69] and [121]. This is an important area in which there is a need for further research.

The primary level of formulation is clearly important. Augmented Lagrangians are rather different from barrier functions. If one exploits the structure using group partial separability this has a profound effect on the design of the software and the algorithmic techniques used. Trust region methods typically make different demands from line search approaches. Small scale sequential quadratic programming techniques do not have much in common with the approach taken in **LANCELOT**.

Parallelism is starting to have its effect on optimization. Firstly, numerical linear algebra plays a fundamental role in optimization. Recently there has been much work implementing such methods on advanced architectures. An overview of many of the major issues is given in [74]. A lucid description, illustrated by considering the Cholesky factorization, is given in [130]. Secondly, parallelism can be exploited at the basic level of the optimization algorithm. Simple examples are the computation of 'extra' quantities in parallel (for example, speculative steps), or independent runs with different choices for some of the algorithm parameters. Often the most effective algorithms for very large problems are those which are very simple but not efficient if implemented in a sequential environment. However, with the possibility of intelligently running what amount to several instances in parallel, a much more effective algorithm results.

We think that although they are not scale invariant, truncated Newton methods are especially important in the context of large-scale nonlinear optimization. When incorporated with an iterative technique such as preconditioned conjugate directions, they enable us to handle the difficulty of deciding whether to solve inner iterations accurately or inaccurately, with the limited information available, whilst still being able to ensure a satisfactory asymptotic convergence rate and global convergence. For example, it is not easy to see how to achieve the same ends in the context of sequential quadratic programming using active set strategies.

Anyone who has tried inputting significant problems appreciates the potential of automatic differentiation. Recently, this has become a very active area of research (see, for example, [77] and [78]), and efforts are underway to provide automatic differentiation tools that promise to make this technology readily usable to the optimization community ([11]

and [80]).

After the great success of quasi-Newton methods there was much hope that such techniques could be adapted to the manipulation of large Hessian approximations. Unfortunately, this turned out not to be the case and the results have been disappointing (see [124]). On the other hand, sparse finite difference schemes have been successful. Since the pioneering work of Curtis, Powell and Reid [49] there has been significant progress, some of which exploits parallelism (see [21], [27], [28], [29], [30], [71] [113], and [116]).

An alternative approach is to use limited memory quasi-Newton updating. This uses the information of only a few, most recent, steps to define a variable metric approximation to the Hessian (see, for example [16] and [110]). These methods have proved to be very useful for solving certain large unstructured problems and Nocedal, [109], claims it is competitive with the partitioned quasi-Newton method on partially separable problems in which the number of element variables exceeds five or six, at least when the cost of evaluating the objective function is relatively low. Recently Bartholomew-Biggs and Hernandez ([8]) have been able to solve large problems using limited memory approximations to the inverse of the Lagrangian in the sequential quadratic programming framework of [7].

We should also mention that the current familiarity of users with sophisticated computer environments has created a high expectation for a user friendly interface to software. Unfortunately, developers have been too busy, as yet, coping with the algorithmic complexities to have devoted much time to the important practicalities of a first rate interface.

**4.5. In conclusion.** We would like to emphasize that it is possible to solve large nonlinear constrained problems in thousands of variables in acceptable time on reasonable workstations. At least two software packages, **LANCELOT** and MINOS, are available. Input is important, and significant progress in both modelling languages and a standard input format, have been made.

This is a vibrant, challenging and useful research area. Our hope is that, in the not too distant future, practitioners will be solving nonlinear models rather than linear ones, when the former is the most appropriate one to consider. Prefering to solve linear models, only because we understand how to solve large linear programs, should no longer be the normal practice.

## REFERENCES

[1] J. ABADIE AND J. CARPENTIER, *Généralisation de la méthode du gradient réduit de Wolfe au cas des contraintes non-linéaires*, in Proceedings IFORS Conference, D. Hertz and J. Melese, eds., Amsterdam, 1966, John Wiley, pp. 1041–1053.

[2] K. M. ANSTREICHER, D. DEN HERTOG, C. ROOS, AND T. TERLAKY, *A long step barrier method for convex quadratic programming*, Tech. Rep. 90-53, Faculty of Mathematics and Computer Science, Delft University of Technology, Delft, Netherlands, 1990.

[3] K. J. ARROW AND R. M. SOLOW, *Gradient methods for constrained maxima with weakened assumptions*, in Studies in Linear and Nonlinear Programming, L. H. Arrow and H. Uzawa, eds., Stanford, CA, 1958, Stanford University Press.

[4] B. M. AVERICK, R. G. CARTER, AND J. J. MORÉ, *The MINPACK-2 test problem collection*, Tech. Rep. ANL/MCS-TM-150, Applied Mathematics Division, Argonne National Labs, Argonne, IL, 1991.

[5] O. BAHN, J. L. GOFFIN, J. P. VIAL, AND O. D. MERLE, *Implementation and behaviour of an interior point cutting plane algorithm for convex programming: an application to geometric programming*, tech. rep., GERARD, Faculty of Management, McGill University, McGill University, Montreal, 1991.

[6] R. H. BARTELS AND G. H. GOLUB, *The simplex method of linear programming using the LU decomposition*, Communications of the ACM, 12 (1969), pp. 266–268.

[7] M. BARTHOLOMEW-BIGGS, *Recursive quadratic programming methods based on the augmented Lagrangian function*, Mathematical Programming Study, 31 (1987), pp. 21–42.

[8] M. C. BARTHOLOMEW-BIGGS AND M. DE F. G. HERNANDEZ, *Some improvements to the subroutine OPALQP for dealing with large problems*, tech. rep., Numerical Optimization Center, Hatfield Polytechnic, Hatfield, UK, 1992.

[9] M. BEN DAYA AND C. M. SHETTY, *Polynomial barrier function algorithm for convex quadratic programming*, Report J85-5, School of ISE, Georgia Institute of Technology, Atlanta, Georgia, 1988.

[10] D. P. BERTSEKAS, *Projected Newton methods for optimization problems with simple constraints*, SIAM J. Control Optim., 20 (1982), pp. 221–246.

[11] C. BISCHOF, A. CARLE, G. CORLISS, P. HOVLAND, AND A. O. GRIEWANK, *ADIFOR: Generating derivative codes from Fortran programs*, Tech. Rep. MCS-P263-0991, Argonne National Labs, Argonne, IL, 1991.

[12] P. T. BOGGS, P. D. DOMICH, J. E. ROGERS, AND C. WITZGALL, *An interior point method for linear and quadratic programming problems*. Mathematical Programming Society COAL Newsletter, August 1991.

[13] P. T. BOGGS AND J. W. TOLLE, *A truncated SQP algorithm for large scale nonlinear programming problems*, January 1992. Presented at the sixth IIMAS-UNAM workshop on numerical analysis and optimization.

[14] A. BROOKE, D. KENDRICK, AND A. MEERAUS, *GAMS: a User's Guide*, The Scientific Press, Redwood City, USA, 1988.

[15] A. G. BUCKLEY, *Test functions for unconstrained minimization*, Tech. Rep. CS-3, Computing Science Division, Dalhousie University, Dalhousie, Canada, 1989.

[16] A. G. BUCKLEY AND A. LENIR, *QN-like variable storage conjugate gradients*, Math. Prog., 27 (1983), pp. 155–175.

[17] S. BUSOVAČA, *Handling degeneracy in a nonlinear $l_1$ algorithm*, Tech. Rep. Tech. Rept. CS-85-34, Univ. of Waterloo, Dept. of Computer Science, Univ. of Waterloo, Waterloo, Ontario N2L 3G1, 1985.

[18] A. CAUCHY, *Méthode générale pour la résolution des systèmes d'équations simultannées*, Comptes Rendus de l'Académie des Sciences, (1847), pp. 536–538.

[19] T. F. COLEMAN, *Large Sparse Numerical Optimization*, in Lecture Notes in Computer Science #165, Springer-Verlag, Berlin, 1984.

[20] ———, *Large Scale Numerical Optimization: Introduction and overview*, in Encyclopedia of Computer Science and Technology, Marcel Dekker, Inc., New York, 1992 (to appear).

[21] T. F. COLEMAN AND J.-Y. CAI, *The cyclic coloring problem and estimation of sparse Hessian matrices*, SIAM J. Appl. Math., 20 (1983), pp. 187–209.

[22] T. F. COLEMAN AND L. HULBERT, *A globally and superlinearly convergent algorithm for convex quadratic programs with simple bounds*, Tech. Rep. TR 90-1092, Department of Computer Science, Cornell University, Ithaca, NY, 1990. To appear in SIAM Journal on Optimization.

[23] T. F. COLEMAN AND Y. LI, *A global and quadratic affine scaling method for (augmented) linear $l_1$ problems*, in Proceedings of the 13th Biennial Numerical Analysis Conference Dundee 1989, G. Watson, ed., Longmans, 1989.

[24] ———, eds., *Large-Scale Numerical Optimization*, SIAM, Philadelphia, 1990.

[25] ———, *A global and quadratic affine scaling method for linear $l_1$ problems*, Mathematical Programming, (1992 (to appear)).

[26] ———, *A global and quadratically convergent method for linear $l_\infty$ problems*, SIAM Journal on Optimization, (1992 (to appear)).

[27] T. F. COLEMAN AND J. J. MORÉ, *Estimation of sparse Jacobian matrices and graph coloring problems*, SIAM J. Numer. Anal., 20 (1983), pp. 187–209.

[28] ———, *Estimation of sparse Hessian matrices and graph coloring problems*, Mathematical Programming, 28 (1984), pp. 243–270.

[29] ———, *Software for estimating sparse Jacobian matrices*, TOMS, 10 (1984), pp. 329–345.

[30] T. F. COLEMAN AND P. E. PLASSMAN, *Solution of nonlinear least-squares problems on a multiprocessor*, SIAM Journal on Scientific and Statistical Computing, 13 (1992).

[31] A. R. CONN, N. I. M. GOULD, AND PH. L. TOINT, *Global convergence of a class of trust region algorithms for optimization with simple bounds*, SIAM Journal on Numerical Analysis, 25 (1988), pp. 433–460. See also *SIAM Journal on Numerical Analysis*, 26:764-767, 1989.

[32] ———, *Testing a class of methods for solving minimization problems with simple bounds on the variables*, Mathematics of Computation, 50 (1988), pp. 399–430.

[33] ———, eds., *Large-Scale Optimization*, vol. 45, no. 3 of Mathematical Programming, North-Holland, 1989.

[34] ———, *An introduction to the structure of large scale nonlinear optimization problems and the lancelot project*, in Computing Methods in Applied Sciences and Engineering, R. Glowinski and A. Lich-

newsky, eds., Philadelphia, 1990, SIAM, pp. 42–51.
[35] ———, eds., *Large-Scale Optimization — Applications*, vol. 48, no. 1 of Mathematical Programming, North-Holland, 1990.
[36] ———, *A proposal for a standard data input format for large-scale nonlinear programming problems*, Report CS-89-61, Department of Computer Science, University of Waterloo, Waterloo, Ontario, Canada N2L 3G1, 1990.
[37] ———, *Convergence of quasi-Newton matrices generated by the symmetric rank one update*, Mathematical Programming, 50 (1991), pp. 177–195.
[38] ———, *A globally convergent augmented Lagrangian algorithm for optimization with general constraints and simple bounds*, SIAM Journal on Numerical Analysis, 28 (1991), pp. 545–572.
[39] ———, *Convergent properties of minimization algorithms for convex constraints using a structured trust region*, tech. rep., Department of Mathematics, FUNDP, Namur, Belgium, 1992.
[40] ———, CUTE: *a collection of constrained and unconstrained test examples for nonlinear programming*, 1992.
[41] ———, LANCELOT: *a Fortran package for large-scale nonlinear optimization*, Springer-Verlag, Berlin, Heidelberg and New York, 1992.
[42] ———, *On the number of inner iterations per outer iteration of a globally convergent algorithm for optimization with general nonlinear constraints and simple bounds*, in Proceedings of the 14th Biennial Numerical Analysis Conference Dundee 1991, D. F. Griffiths and G. Watson, eds., Longmans, 1992 (to appear).
[43] A. R. CONN, N. I. M. GOULD, M. LESCRENIER, AND PH. L. TOINT, *Performance of a multifrontal scheme for partially separable optimization*, in Proceedings of the Sixth Mexico-United States Numerical Analysis Workshop, S. Gómez, J. P. Hennart, and R. Tapia, eds., 1992 (to appear).
[44] A. R. CONN, N. I. M. GOULD, R. POLYAK, AND PH. L. TOINT, *A globally convergent Lagrangian barrier algorithm for optimization with general inequality constraints and simple bounds*, tech. rep., IBM Thomas J. Watson Research Center, P.O.Box 218, Yorktown Heights, NY 10598, U.S.A., 1992.
[45] A. R. CONN, N. I. M. GOULD, A. SARTENAER, AND PH. L. TOINT, *Global convergence of a class of trust region algorithms for optimization using inexact projections on convex constraints*, SIAM Journal on Optimization, (1992 (to appear)).
[46] L. CONTESSE AND J. VILLAVICENCIO, *Resolución de un modelo económico de despacho de carga eléctrica mediante el método de penalización Lagrangeana con cotas*, Revista del Instituto Chileno de Investigación Operativa, (1982), pp. 80–112.
[47] IBM CORPORATION, *Mathematical programming system/360 version 2, linear and separable programming-user's manual*, Tech. Rep. H20-0476-2, IBM Corporation, 1969.
[48] R. COURANT, *Variational methods for the solution of problems of equilibrium and vibrations*, Bull. Amer. Math. Soc., 49 (1943), pp. 1–23.
[49] A. R. CURTIS, M. J. D. POWELL, AND J. K. REID, *On the estimation of sparse Jacobian matrices*, IMA J. Appl. Math., 13 (1974), pp. 117–120.
[50] A. DAX, *A note on optimality conditions for the Euclidean multifacility location problem*, Math. Prog., 36 (1986), pp. 72–80.
[51] R. S. DEMBO AND J. G. KLINCEWICZ, *A scaled reduced gradient algorithm for network flow problems with convex separable costs*, Mathematical Programming, 15 (1981), pp. 125–147.
[52] R. S. DEMBO AND U. TULOWITSKI, *On the minimization of quadratic functions subject to box constraints*, Tech. Rep. B71, Yale School of Management, Yale University, New Haven, USA, 1983.
[53] D. DEN HERTOG, C. ROOS, AND T. TERLAKY, *A potential reduction method for a class of smooth convex programming problems*, Tech. Rep. 90-01, Faculty of Mathematics and Computer Science, Delft University of Technology, Delft, Netherlands, 1990.
[54] J. E. DENNIS JR. AND R. B. SCHNABEL, *Numerical Methods for Unconstrained Optimization and Nonlinear Equations*, Prentice-Hall, Englewood Cliffs, NJ, 1983. Russian edition, Mir Publishing Office, Moscow, 1988, O. Burdakov, translator.
[55] J. E. DENNIS JR. AND H. WOLKOWICZ, *Sizing and least-change secant methods*, SIAM J. Numer. Anal., (1992 (to appear)).
[56] I. S. DUFF, A. M. ERISMAN, AND J. K. REID, *Direct methods for sparse matrices*, Clarendon Press, Oxford, UK, 1986.
[57] I. S. DUFF, R. G. GRIMES, AND J. G. LEWIS, *Sparse matrix test problems*, ACM Transactions on Mathematical Software, 15 (1989), pp. 1–14.
[58] S. K. ELDERSVELD, *Large-scale Sequential Quadratic Programming Algorithms*, PhD thesis, Department of Operations Research, Stanford University, Stanford, CA, 1991. Also available as a technical report.

[59] J. E. FALK AND G. P. MCCORMICK, *Computational aspects of the international coal trade model*, in Spacial price equilibrium: Advances in theory, computation and application, Lecture Notes in Economics and Mathematical Systems #249, P. Harker, ed., Springer-Verlag, Berlin, Heidelberg and New York, 1986, pp. 73–117.

[60] A. V. FIACCO AND G. P. MCCORMICK, *Nonlinear Programming: Sequential Unconstrained Minimization Techniques*, John Wiley & Sons, New York, NY, 1968. Reprinted as *Classics in Applied Mathematics 4*, SIAM, 1990.

[61] R. FLETCHER, *Practical Methods of Optimization*, J. Wiley and Sons, Chichester, second ed., 1987.

[62] ———, *Degeneracy in the presence of roundoff errors*, Linear Algebra and its Applications, 106 (1988), pp. 149–183.

[63] ———, *Resolving degeneracy in quadratic programming*, Tech. Rep. NA/135, Department of Mathematical Sciences, University of Dundee, 1991.

[64] R. FLETCHER AND E. SAINZ DE LA MAZA, *Nonlinear programming and non-smooth optimization by successive linear programming*, Math. Prog., 43 (1989), pp. 235–256.

[65] R. FOURER, D. M. GAY, AND B. W. KERNIGHAN, *AMPL: A mathematical programming language*, computer science technical report, AT&T Bell Laboratories, Murray Hill, USA, 1987.

[66] D. M. GAY, *Electronic mail distribution of linear programming test problems*. Mathematical Programming Society COAL Newsletter, December 1985.

[67] P. E. GILL AND W. MURRAY, *Conjugate-gradient methods for large-scale nonlinear optimization*, Tech. Rep. SOL79-15, Operations Research Department, Stanford University, Stanford, USA, 1979.

[68] P. E. GILL, W. MURRAY, M. A. SAUNDERS, AND M. H. WRIGHT, *Shifted barrier methods for linear programming*, Tech. Rep. SOL88-9, Operations Research Department, Stanford University, Stanford, USA, 1988.

[69] ———, *A practical anti-cycling procedure for linearly constrained optimization*, Mathematical Programming, 45 (1989), pp. 437–474.

[70] P. E. GILL, W. MURRAY, AND M. H. WRIGHT, *Practical Optimization*, Academic Press, New York, London, Toronto, Sydney and San Francisco, 1981.

[71] D. GOLDFARB AND PH. L. TOINT, *Optimal estimation of Jacobian and Hessian matrices that arise in finite difference calculations*, Math. Comp., 43 (1984), pp. 69–88.

[72] D. GOLDFARB AND S. LIU, *An $O(n^3 L)$ primal interior point algorithm for convex quadratic programming*, Math. Prog., 49 (1991), pp. 325–343.

[73] D. GOLDFARB AND M. J. TODD, *Linear programming*, in Optimization, G. Nemhauser, A. R. Kan, and M. Todd, eds., vol. 1 of Handbooks in Operations Research and Management Science, North-Holland, Amsterdam, 1989, ch. 2.

[74] G. H. GOLUB AND C. F. VAN LOAN, *Matrix Computations*, Johns Hopkins University Press, Baltimore, Maryland, second ed., 1989.

[75] N. I. M. GOULD, *An algorithm for large-scale quadratic programming*, IMA Journal of Numerical Analysis, 11 (1991), pp. 299–324.

[76] N. I. M. GOULD AND J. K. REID, *New crash procedures for large systems of linear constraints*, Mathematical Programming, 45 (1989), pp. 475–502.

[77] A. O. GRIEWANK, *Direct calculation of Newton steps without accumulating Jacobians*, in Large-Scale Numerical Optimization, T. F. Coleman and Y. Li, eds., SIAM, Philadelphia, 1990, pp. 115–137.

[78] A. O. GRIEWANK AND G. F. CORLISS, *Automatic Differentiation of Algorithms: Theory, Implementation, and Application*, SIAM, Philadelphia, 1991.

[79] A. O. GRIEWANK AND PH. L. TOINT, *On the unconstrained optimization of partially separable functions*, in Nonlinear Optimization, M. Powell, ed., Academic Press, London, 1982.

[80] A. O. GRIEWANK, D. JUEDES, J. SRINIVASAN, AND C. TYNER, *ADOL-C, a package for the automatic differentiation of algorithms written in C/C++*, ACM Trans. Math. Software, (to appear). Also appeared as Preprint MCS-P180-1190, Mathematics and Computer Science Division, Argonne National Laboratory, 9700 S. Cass Ave., Argonne, IL 60439, 1990.

[81] R. E. GRIFFITH AND R. A. STEWART, *A nonlinear programming technique for the optimization of continuous processing systems*, Manage. Sci., 7 (1961), pp. 379–392.

[82] M. D. GRIGORIADIS, *An efficient implementation of the network simplex method*, Mathematical Programming, 26 (1986), pp. 83–111.

[83] C.-G. HAN, P. M. PARDALOS, AND Y. YE, *Computational aspects of an interior point algorithm for quadratic programming problems with box constraints*, in Large-Scale Numerical Optimization, T. Coleman and Y. Li, eds., Philadelphia, 1990, SIAM, pp. 92–102.

[84] ———, *Solving some engineering problems using an interior-point algorithm*, Tech. Rep. CS-91-04, Computer Science Department, The Pennsylvania State University, University Park, PA, 1991.

[85] M. R. HESTENES, *Multiplier and gradient methods*, J. Optim. Theory Appl., 4 (1969), pp. 303–320.

[86] F. JARRE AND M. A. SAUNDERS, *Practical aspects of an interior-point method for convex programming*, Tech. Rep. SOL91-9, Department of Operations Research, Stanford University, Stanford, USA, 1991.

[87] A. P. JONES, *The chemical equilibrium problem: An application of SUMT*, Tech. Rep. RAC-TP-272, Reserach Analysis Corporation, McLean, Virginia, 1967.

[88] H. KHALFAN, R. H. BYRD, AND R. B. SCHNABEL, *A theoretical and experimental study of the symmetric rank one update*, SIAM J. on Optimization, (1992 (to appear)).

[89] J. L. LAGRANGE, *Théorie des Fonctions Analytiques*, Impr. de la République, Paris, 1797.

[90] L. S. LASDON, A. D. WAREN, A. JAIN, AND M. RATNER, *Design and testing of a generalized reduced gradient code for nonlinear programming*, TOMS, 4 (1978), pp. 34–50.

[91] E. LEVITIN AND B. POLYAK, *Constrained minimization problems*, USSR Comput. Math. and Math. Phys., 6 (1966), pp. 1–50.

[92] D. MAHIDHARA AND L. LASDON, *An SQP algorithm for large sparse nonlinear programs*, working paper, MSIS Department, School of Business, University of Texas, Austin, TX 78712, 1990.

[93] H. M. MARKOWITZ, *The elimination form of the inverse and its application to linear programming*, Management Sci., 3 (1957), pp. 255–269.

[94] G. P. MCCORMICK, *Anti-zig-zagging by bending*, Manage. Sci., 15 (1969), pp. 315–320.

[95] ———, *Computational aspects of nonlinear programming solutions to large-scale inventory problems*, Tech. Rep. Technical Memorandum Serial T-63488, George Washington University, Institute of Management Science and Engineering, Washington,DC, 1972.

[96] G. P. MCCORMICK AND A. SOFER, *Optimization with unary functions*, Mathematical Programming, 52 (1991), pp. 167–179.

[97] S. MEHROTRA AND J. SUN, *An interior point algorithm for solving smooth convex programs based on Newton's method*, in Contemporary Mathematics, J. Lagarias and M. Todd, eds., AMS, Providence, Rhode Island, 1990, pp. 265–284.

[98] R. C. MONTEIRO, *The global convergence of a class of primal potential reduction algorithms for convex programming*, report, Systems and Industrial Engineering Department, University of Arizona, Tucson, Arizona, 1991.

[99] R. C. MONTEIRO AND I. ADLER, *Interior path following primal-dual algorithms, part ii: convex quadratic programming*, report, Department of IEOR, University of California, Berkeley, California, 1987.

[100] R. C. MONTEIRO, I. ADLER, AND M. G. RESENDE, *A polynomial-time primal-dual affine scaling algorithm for linear and convex quadratic programming and its power series extension*, Report ESRC 88-8, Department of IEOR, University of California, Berkeley, California, 1987.

[101] J. J. MORÉ, *Recent developments in algorithms and software for trust region methods*, in Mathematical Programming: The State of the Art, A. Bachem, M. Grötschel, and B. Korte, eds., Berlin, 1983, Springer Verlag, pp. 258–287.

[102] J. J. MORÉ, B. S. GARBOW, AND K. E. HILLSTROM, *Testing unconstrained optimization software*, ACM Transactions on Mathematical Software, 7 (1981), pp. 17–41.

[103] J. J. MORÉ AND G. TORALDO, *Algorithms for bound constrained quadratic programming problems*, Numerische Mathematik, 14 (1979), pp. 14–21.

[104] ———, *On the solution of large quadratic programming problems with bound constraints*, SIAM Journal on Optimization, 1 (1991), pp. 93–113.

[105] W. MURRAY AND F. J. PRIETO, *A sequential quadratic programming algorithm using an incomplete solution of the subproblem*, Tech. Rep. SOL90-12, Operations Research Department, Stanford University, Stanford, USA, 1990.

[106] B. A. MURTAGH AND M. A. SAUNDERS, *MINOS 5.1 user's guide*, Tech. Rep. SOL83-20R, Department of Operations Research, Stanford University, Stanford, USA, 1987.

[107] S. G. NASH AND A. SOFER, *A barrier method for large-scale constrained optimization*, Tech. Rep. 91-10, Department of Operations Research and Applied Statistics, George Mason University, Fairfax, Virginia 22030, 1991.

[108] G. L. NEMHAUSER, A. H. G. R. KAN, AND M. J. TODD, eds., *Optimization*, vol. 1 of Handbooks in Operations Research and Management Science, North-Holland, Amsterdam, 1989.

[109] J. NOCEDAL, *Theory of algorithms for uncontrained optimization*, Acta Numerica, 1 (1992 (to appear)).

[110] J. NOCEDAL AND D. C. LIU, *On the limited memory BFGS method for large scale optimization*, Mathematical Programming, Series B, 45 (1989), pp. 503–528.

[111] M R. OSBORNE AND L. P. SUN, *A new approach to the symmetric rank-one updating algorithm*, Math. Prog., (1992 (to appear)).

[112] P. M. PARDALOS, C.-G. HAN, AND Y. YE, *Interior point algorithms for solving nonlinear optimization problems*. Mathematical Programming Society COAL Newsletter, August 1991.

[113] P.E. PLASSMAN, *Sparse Jacobian estimation and factorization on a multiprocessor*, in Large-Scale Numerical Optimization, T. Coleman and Y. Li, eds., Philadelphia, 1990, SIAM, pp. 152–179.

[114] R. POLYAK, *Modified barrier functions (theory and methods)*, Research report RC 15886 (#70630), IBM T. J. Watson Research Center, P.O.Box 218, Yorktown Heights, NY 10598, U.S.A., 1990.

[115] M. J. D. POWELL, *A method for nonlinear constraints in minimization problems*, in Optimization, R. Fletcher, ed., Academic Press, New York, NY, 1969, pp. 283–298.

[116] M. J. D. POWELL AND PH. L. TOINT, *On the estimation of sparse Hessian matrices*, SIAM J. Numer. Anal., 16 (1979), pp. 1060–1074.

[117] J. K. REID, *Fortran subroutines for handling sparse linear programming base*, Tech. Rep. AER-R-8269, AERE Harwell Laboratory, Harwell, UK, 1976.

[118] S. M. ROBINSON, *A quadratically convergent algorithm for general nonlinear programming problems*, Math. Prog., 3 (1972), pp. 145–156.

[119] J. B. ROSEN AND J. KREUSER, *A gradient projection algorithm for nonlinear constraints*, in Numerical Methods for Nonlinear Optimization, F. Lootsma, ed., Academic Press, New York, NY, 1972, pp. 297–300.

[120] A. S. RUZINSKY AND E. T. OLSON, $l_1$ *and* $l_\infty$ *minimization via a variant of Karmarkar's algorithm*, IEEE Trans. Acoustics, Speech and Signal Processing, 37 (1989), pp. 245–253.

[121] D. M. RYAN AND M. R. OSBORNE, *On the solution of highly degenerate linear programs*, Mathematical Programming, 41 (1988), pp. 385–392.

[122] A. SARTENAER, *On some strategies for handling constraints in nonlinear optimization*, PhD thesis, Department of Mathematics, FUNDP, Namur, Belgium, 1991.

[123] D. A. SCHRADY AND U. C. CHOE, *Models for multi-item continuous review inventory policies subject to constraints*, Naval Research Logistics Quarterly, 18 (1971), pp. 451–463.

[124] D. C. SORENSEN, *An example concerning quasi-Newton estimation of a sparse Hessian*, SIGNUM Newsletter, 16 (1981), pp. 8–10.

[125] PH. L. TOINT, *Call for test problems in large scale nonlinear optimization*, COAL Newsletter, 16 (1987), pp. 5–10.

[126] PH. L. TOINT AND D. TUYTTENS, *On large scale nonlinear network optimization*, Mathematical Programming, Series B, 48 (1990), pp. 125–159.

[127] ——, *LSNNO: a Fortran subroutine for solving large scale nonlinear network optimization problems*, ACM Transactions on Mathematical Software, (1990 (to appear)).

[128] J. A. TOMLIN, *The influences of algorithmic and hardware developments on computational mathematical programming*, in Mathematical Programming— Recent Developments and applications, M. Iri and K. Tanabe, eds., Tokyo, 1988, KTK Scientific Publishers, pp. 159–175.

[129] J. A. TOMLIN AND J. S. WELCH, *Mathematical programming systems*, Tech. Rep. RJ 7400 (69202), IBM Thomas J. Watson Research Center, P.O.Box 218, Yorktown Heights, NY 10598, U.S.A., 1990.

[130] C. F. VAN LOAN, *A survey of matrix computations*, Tech. Rep. CTC90TR26, Cornell Theory Center, Cornell University, Ithaca, NY, 1990.

[131] P. WERBOS, *Backpropagation: past and future*, in Proceedings of the 2nd International Conference on Neural Networks, New York, 1988, IEEE.

[132] H. WOLKOWICZ, *Measures for symmetric rank-one updates*, Math. Prog., (1992 (submitted)).

[133] M. H. WRIGHT, *Optimization and large scale computation*, in Very Large Scale Computation in the $21^{st}$ Century, J. Mesirov, ed., Philadelphia, 1991, SIAM.

[134] ——, *Interior methods for constrained optimization*, Acta Numerica, 1 (1992 (to appear)).

[135] S. J. WRIGHT, *An interior-point algorithm for linearly constrained optimization*, Tech. Rep. MCS-P162-0790, Argonne National Labs, Argonne, IL, 1990.

[136] Y. YE, *Interior point algorithms for linear, quadratic and linearly constrained convex programming*, PhD thesis, Engineering and Economic Systems Department, Stanford University, Stanford, CA, 1987.

[137] ——, *Interior point algorithms for quadratic programming*, in Recent Developments in Mathematical Programming, S. K. (to appear), ed., Gordan and Breach Scientific Publishers, London, 1992.

## Chapter 6
# TIME-SPLITTING METHODS FOR ADVECTION-DIFFUSION-REACTION EQUATIONS ARISING IN CONTAMINANT TRANSPORT*

| CLINT N. DAWSON | MARY F. WHEELER |
|---|---|
| Rice University | Rice University |
| Houston, Texas | Houston, Texas |
|  | (pictured) |

**1. Introduction.** In this paper we consider two time-splitting algorithms for solving systems of nonlinear advection-diffusion-reaction problems of the form:

(1.1) $$\phi_i \frac{\partial c_i}{\partial t} - \nabla \cdot D(u)\nabla c_i + u \cdot \nabla c_i = (\tilde{c}_i - c_i)\tilde{q} + \phi R_i \mathcal{R}_i(c_1, c_2, \ldots, c_M),$$
$$x \in \Omega, \ t \in J,$$

(1.2) $$u \cdot n = D(u)\nabla c_i \cdot n = 0, \quad x \in \partial\Omega, \ t \in J,$$

and

(1.3) $$c_i(x, 0) = c_i^0(x), \quad x \in \Omega,$$

where $i = 1, 2, \ldots, M$, $\Omega$ is a bounded domain in $\mathbf{R}^d$, $J = [0, T]$. These types of equations arise in the modeling of contaminant transport in groundwater. In this context, $c_i$ is the concentration of component $i$ (e.g., contaminants, other nutrients, microorganisms, etc.); $\tilde{q} = \max(q, 0)$ is nonzero at source points only; $u$ is the Darcy velocity; $D(u)$ is the hydrodynamic diffusion/dispersion tensor; $\phi_i = \phi R_i$; where $\phi$ is porosity and $R_i$ is the retardation factor due to adsorption; and $\mathcal{R}_i$ are kinetic terms which account for biodegradation of contaminants, utilization of nutrients, and growth and decay of microorganisms. For $d = 2$, $D$ generally has the form

(1.4) $$D(u) = D_m I + \frac{\alpha_l}{|u|} \begin{bmatrix} u_x^2 & u_x u_y \\ u_x u_y & u_y^2 \end{bmatrix} + \frac{\alpha_t}{|u|} \begin{bmatrix} u_y^2 & -u_x u_y \\ -u_x u_y & u_x^2 \end{bmatrix},$$

where $D_m$ is the molecular diffusivity and $\alpha_l$ and $\alpha_t$ are the longitudinal and transverse dispersivities, respectively. Furthermore, in (1.1) $\tilde{c}$ is specified at injection wells and $\tilde{c} = c$ at production wells.

For convenience, we assume (1.1) is $\Omega$-periodic; i.e., we assume all functions in (1.1) are spatially $\Omega$-periodic. This is physically reasonable, since no-flow boundaries are generally treated by reflection, and because in general interior flow patterns are much more important than boundary effects. Thus, the no-flow boundary conditions above can be dropped.

---

*This research was supported by Department of Energy.

This paper is divided into six additional sections. In the second section we briefly describe the importance of modeling the fate and transport of contaminants and the possible potential of biorestoration in cleanup strategies. In the third section we establish mathematical notation. The modified method of characteristics (MMOC) for linear advection problems with $u$ assumed to be given is described in Section 4. In Section 5 a time-splitting method is formulated in which the advection-diffusion equations are treated by the MMOC and subsequently, reactions are computed using many small time steps. Theoretical error estimates for this algorithm are given, including a new result in which the dispersion tensor is assumed to be only postive semi-definite. This time-splitting algorithm has been used extensively to model contaminant transport and biodegradation in two and three spatial dimensions [2, 3, 4, 15, 16, 17]. Some numerical experiments of *in-situ* biorestoration are presented in Section 6. A new time-splitting in which the reactions are computed along characteristics is formulated and analyzed in Section 7.

**2. Contaminant Transport and Biodegration.** Groundwater contamination is of growing concern to the United States. Of the fifty largest cities in the United States, over thirty-four obtain their drinking water from groundwater. In addition groundwater supplies the needs of over 50% of the U. S. population. At the same time the growth of industrial wastes has risen from approximately 57 million metric tons in 1980 (with as little as 10% disposed over properly) to 265 million metic tons in 1989. These contaminants include inorganic chemicals such as arsenic, nitrate, fluoride, radium, and lead and organic chemicals such as gasoline, DDT, PCB, and detergents, and volatile organics such as chlorinated solvents (TCE), carbon tetrachloride, and vinyl chloride, biological matter, and radioactive compounds. Sources of groundwater contamination include land disposal of waste materials, water wells, sewage and waste water disposal systems, leaks and spills, oil, gas and mining activities, agricultural practices, chemical injection wells, and stormwater runoff.

One of the most promising restoration techniques is bioremediation. Here indigenous microflora are used to remove subsurface pollutants by biodegradation. Numerous laboratory and field studies have shown microbes degrade various compounds in both aerobic and anaerobic conditions and control contaminant movement; see [10, 11, 13, 14] and the references therein. *In-situ* biodegradation involves enhancement of these natural processes by introducing nutrients and dissolved oxygen into the system.

The biorestoration process involves transport of substrates, nutrients, and microorganisms, and interaction of components in the aqueous phase with the solid phase through adsorption and biodegradation. Many factors must be considered, such as the characteristics of the organisms, the primary substrate and nutrients, cometabolic substrates, contaminant concentration and location, available electron acceptors, and environmental conditions, including adsorption, ion exchange, the effect of colloidal particles, soil pH and mineral composition, temperature, rock heterogeneity, fractures, effective dispersivity, and changes in rock properties due to microbial activity. Much of the field and laboratory data used to estimate these effects is incomplete and/or inexact, and is scale-dependent.

Mathematical modeling serves as a valuable link between the experimental studies and scientific theory. It is especially useful in understanding the mechanisms of interactive transport, sorption, and biodegradation, analyzing the sensitivity and significance of various parameters in a model and providing insight into field data collection activities, and efficiently investigating various strategies for remediation and mitigation.

In Section 6 we briefly describe the governing equation of flow and transport with biodegradation in a saturated porous medium. For simplicity we assume linear sorption and aerobic conditions. More general conditions such as the Michaelis-Menten kinetics can be treated with the numerical techniques discussed in this paper.

The coupled nonlinear advection-diffusion-reaction system consists of $M_s$ electron donors or substrates (contaminants) and $M_n$ electron acceptors or nutrients and a system of $M_x$ ordinary differential equations involving microbial mass. The flow is given by Darcy's Law and the continuity equation. Transport of microbes can also be treated if one assumes a system of advection-diffusion-reaction equations rather than a system of ordinary differential equations.

**3. Notation and definitions.** On $\Omega$, let $L^2(\Omega)$ and $L^\infty(\Omega)$ denote the standard Banach spaces, with norms $\|\cdot\|$ and $\|\cdot\|_\infty$, respectively. Let $H^m(\Omega)$ and $W_\infty^m(\Omega)$ denote the standard Sobolev spaces, with norms $\|\cdot\|_m$ and $\|\cdot\|_{W_\infty^m}$, respectively. Let $(\cdot,\cdot)$ denote the $L^2$ inner product on $\Omega$.

Let $[a,b] \subset [0,T]$ denote a time interval, $X = X(\Omega)$ a Banach or Sobolev space. To incorporate time dependence, we use the notation $L^p(a,b;X)$ and $\|\cdot\|_{L^p(a,b;X)}$ to denote the space and norm, respectively, of $X$-valued functions $f$ with the map $t \to \|f(\cdot,t)\|_X$ belonging to $L^p(a,b)$. If $[a,b] = [0,T]$, we simplify our notation and write $L^p(X)$ for $L^p(0,T;X)$.

Let $\Delta t = T/N^*$ for some positive integer $N^*$, $t^n = n\Delta t$, $n = 0,\ldots,N^*$, and $f^n = f(t^n)$. In our time-splitting procedure we will also use a "small" time step $\Delta t_s$, with $\Delta t_s = \Delta t/N$, where $N$ is another positive integer. Let $f^{j,n} = f(t^n + j\Delta t_s)$, $j = 0,\ldots,N$.

Finally, let $\mathcal{M}_h$ denote a finite dimensional subspace of $H^1(\Omega)$ consisting of continuous piecewise polynomials of degree $\leq k$ on a quasi-uniform mesh of diameter $\leq h$.

**4. Description of the method.** Before defining our time-splitting procedure we comment on the MMOC applied to the single advection-diffusion equation

(4.5) $$\phi_i \frac{\partial c_i}{\partial t} + u \cdot \nabla c_i - \nabla \cdot D \nabla c_i = 0,$$

where $\phi_i = \phi R_i$. We write

$$\phi_i \frac{\partial c_i}{\partial t} + u \cdot \nabla c_i$$

as a directional derivative. Let $\tau_i$ denote the unit vector in the direction $(u, \phi_i)$ in $\Omega \times J$ and set $\psi_i = \sqrt{|u|^2 + \phi_i^2}$. Then one obtains

(4.6) $$\psi_i \frac{\partial c}{\partial \tau_i} - \nabla \cdot D \nabla c_i = 0.$$

Let

$$\check{x} = x - u_i \Delta t,$$

and

$$\check{c}(x) = c(\check{x}),$$

where $u_i = u/\phi_i$. Then

$$\psi_i \frac{\partial c_i}{\partial \tau_i} \approx \phi_i \frac{c_i^n - \check{c}_i^{n-1}}{\Delta t}.$$

For analysis of the MMOC procedure and application to problems in porous media the reader is referred to [7, 8, 9, 12].

In our error analysis, we will compare the approximate solution to a projection in $\mathcal{M}_h$. Let $\tilde{C}_i^n \in \mathcal{M}_h$ satisfy

(4.7) $\left(\phi_i \dfrac{\tilde{C}_i^n - \tilde{C}_i^{\check{z}\,n-1}}{\Delta t}, \chi\right) + (D\nabla \tilde{C}_i^n, \nabla \chi) = (q(\tilde{c}_i^n - \tilde{C}_i^n), \chi) + (\mathcal{R}_i(c^n), \chi),$

$$\chi \in \mathcal{M}_h, \quad n \geq 1, \quad i = 1, \ldots, M,$$

and

(4.8) $$\tilde{C}_i^0(x) = C_i^e(x, 0).$$

Here, if $D$ is uniformly positive definite, that is, $0 < D_* \leq D(u)$, $C_i^e(x,t)$ is the "elliptic projection" of $c(x,t)$ into the space $\mathcal{M}_h$, given by

$$(D\nabla C_i^e, \nabla \chi) + (C_i^e, \chi) + (\tilde{q} C_i^e, \chi) = (D\nabla c_i, \nabla \chi) + (c_i, \chi) + (\tilde{q} c_i, \chi), \quad \chi \in \mathcal{M}_h.$$

When $D$ is positive semidefinite, $C_i^e$ is the $L^2$ projection, given by

(4.9) $$(C_i^e(\cdot, t) - c_i(\cdot, t), \chi) = 0, \quad \chi \in \mathcal{M}_h.$$

**5. The first time-splitting approach.** The time-split procedure for solving (1.1) is a collection of maps $C_i : \{t^0, t^1, \ldots, t^{N^*}\} \to \mathcal{M}_h$ defined by

(5.10) $\left(\phi_i \dfrac{C_i^n - \bar{C}_i^{\check{z}\,N,n-1}}{\Delta t}, \chi\right) + (D\nabla C_i^n, \nabla \chi) = (q(\tilde{c}_i^n - C_i^n), \chi),$

$$\chi \in \mathcal{M}_h, \quad n \geq 1, \quad i = 1, \ldots, M,$$

where

(5.11) $$\bar{C}_i^0(x) = \tilde{C}_i^0(x).$$

We will discuss two different choices for the function $\bar{C}_i(x)$. The first choice, presented here and analyzed in [15], is based on solving the system of ordinary differential equations

(5.12) $$c_t = \mathcal{R}(c), \quad c = (c_1, \ldots, c_M), \quad \mathcal{R} = (\mathcal{R}_1, \ldots, \mathcal{R}_M)$$

along lines of constant $x$. For simplicity, in this paper, we will assume the system (5.12) is approximated by explicit Euler; however, our analysis can be generalized to higher-order explicit or implicit techniques [15]. In this case, $\bar{C}^{N,n-1}$ is defined recursively by

(5.13) $$\bar{C}_i^{0,n-1} = C_i^{n-1},$$

and for $j = 1, \ldots, N$,

(5.14) $$\bar{C}_i^{j,n-1} = \bar{C}_i^{j-1,n-1} + \Delta t_s \mathcal{R}_i(\bar{C}^{j-1,n-1}).$$

Note that (5.10) and (5.13)-(5.14) represent a time-splitting approach, where at the beginning of each time step, we first solve (5.12) using explicit Euler, and use the results of this step as the "initial condition" for the transport step, where advection and diffusion are approximated by the MMOC.

In [15], the following error estimate is derived for the method outlined above.

THEOREM 5.1. *Let $c_i$ satisfy (1.1) and assume the data and $c_i$ are sufficiently smooth, $0 \leq \phi_* \leq \phi_i$, $D(u)$ is uniformly positive definite, and $\mathcal{R}_i$ is Lipschitz continuous, $i =$*

$1,\ldots,M$. Let $C_i \in \mathcal{M}_h$ be the approximation given by (5.10), (5.13)-(5.14). Then there exists a positive constant $K$, such that

$$\text{(5.15)} \qquad \max_n \|\phi_i^{1/2} C_i^n - c_i^n\| \leq K(\Delta t + h^{(r+1)}), \quad i = 1, \ldots, M,$$

where $r$ is the maximum degree of polynomials in $\mathcal{M}_h$.

The smoothness assumptions on coefficients and data needed to prove Theorem 5.1 are given in [5, 15].

Theorem 5.1 can be modified to treat the case where $D(u)$ is positive semidefinite using the techniques in [5]; in this case the exponent on $h$ is $r$. This rate is optimal for the case $r = 2$ [6].

**6. Numerical results for first time-splitting approach.** In this section, we present some two-dimensional results for a three-component system. Let $c_1 = S$ denote the concentration of contaminant, $c_2 = O$ the concentration of dissolved oxygen, and $c_3 = M$ the concentration of microorganisms. We will assume microorganisms are immobile, and that biodegradation is described by the Monod kinetic equations. Thus, the system of equations to be solved is

$$\text{(6.16)} \qquad \phi R_1 \frac{\partial S}{\partial t} - \nabla \cdot D(u)\nabla S + u \cdot \nabla S = (\tilde{c}_1 - S)\tilde{q} + \phi R_1 \mathcal{R}_1(S, O, M),$$

$$\text{(6.17)} \qquad \phi \frac{\partial O}{\partial t} - \nabla \cdot D(u)\nabla O + u \cdot \nabla O = (\tilde{c}_2 - O)\tilde{q} + \phi \mathcal{R}_2(S, O, M),$$

and

$$\text{(6.18)} \qquad \frac{\partial M}{\partial t} = \mathcal{R}_3(S, O, M),$$

where

$$\text{(6.19)} \qquad \mathcal{R}_1(S, O, M) = -M \cdot k \cdot \left(\frac{S}{K_S + S}\right) \cdot \left(\frac{O}{K_O + O}\right)$$

$$\text{(6.20)} \qquad \mathcal{R}_2(S, O, M) = -M \cdot k \cdot f \cdot \left(\frac{S}{K_S + S}\right) \cdot \left(\frac{O}{K_O + O}\right)$$

and

$$\text{(6.21)} \quad \mathcal{R}_3(S, O, M) = M \cdot k \cdot Y \cdot \left(\frac{S}{K_S + S}\right) \cdot \left(\frac{O}{K_O + O}\right) + k_c \cdot Y \cdot OC - b \cdot M.$$

In these equations, $k$ is the maximum substrate utilization rate per unit mass microorganisms, $Y$ is the microbial yield coefficient, $K_S$ is the substrate half saturation constant, $K_O$ is the oxygen half saturation constant, $b$ is the microbial decay rate, $f$ is the ratio of oxygen to substrate consumed, $k_c$ is the first order decay rate of natural organic carbon, and $OC$ is the natural organic carbon concentration. Further details of this formulation and parameter selection can be found in [1].

In the simulations described below, we assumed a rectangular, heterogeneous aquifer of 2000 by 800 feet, with no-flow boundary conditions on the horizontal boundaries and pressure (inflow and outflow) boundary conditions on the vertical boundaries, with a pressure drop of 28 feet across the region. Figure 1 shows a contour plot of an initial substrate plume, which we generated numerically by introducing sources of contamination throughout the region, and assuming constant, nonzero background concentrations of dissolved oxygen and microorganisms. Figure 2 shows the amount of oxygen left in the system after five years of contaminant injection and biodegradation.

Next, we place a series of injection and production wells throughout the system which inject dissolved oxygen and produce a mixture of dissolved oxygen and substrate. The intent is to stimulate biodegradation and contain the contaminant plume. One problem of interest in determining biorestoration strategies is to determine "optimal" flow rates and well patterns. Here we have adopted a strategy based on graphically monitoring the contaminant plume movement, and changing well locations and flow rates at one year intervals. In Figure 3-6, we present the contaminant solution at one, three, five, and seven years. The well locations are superimposed on the plots, a cross denotes an injection well and a circle a production well. Note that after seven years, most of the contaminant has been removed.

The physical parameters used in the simulations were $\phi = 0.25$, $\alpha_l = 10$ feet, and $\alpha_t = 1$ foot. The flow rates varied between 20 and 40 square feet / day. The biodegradation parameters were $k = 0.17/\text{day}$, $k_c = 2.7 * 10^{-6}$, $OC = 750$, $F = 3.0$, $Y = 0.13$, $K_S = 0.13$ mg/l, $K_O = 0.1$ mg/l, and $b = .01/\text{day}$.

In the numerical simulator, different non-uniform grids were used with each change in the well pattern, with grids concentrated around the wells. A constant $\Delta t$ of 4 days was assumed, and $\Delta t_s = .004$ days.

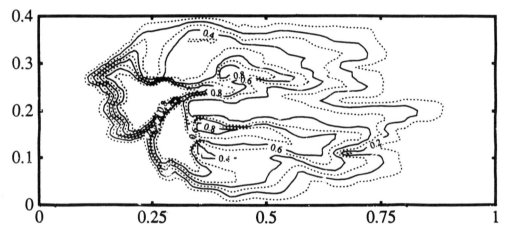

FIG. 1. *Initial substrate plume*

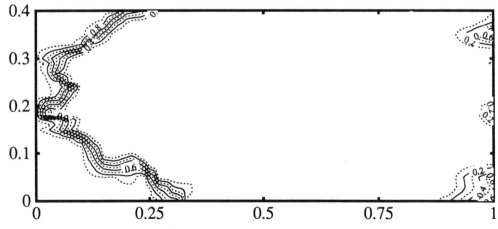

FIG. 2. *Initial oxygen plume*

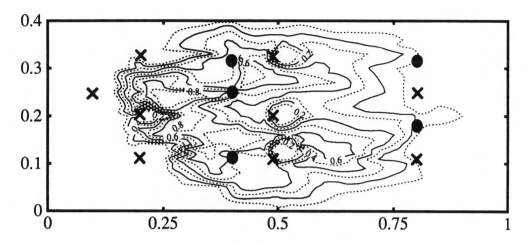

FIG. 3. *Substrate plume after 1 year of biorestoration*

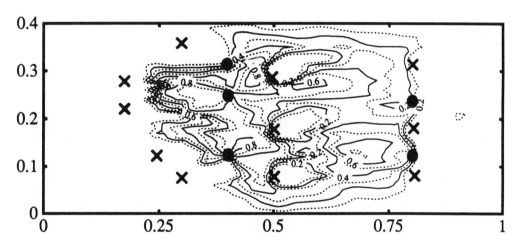

FIG. 4. *Substrate plume after 3 years of biorestoration*

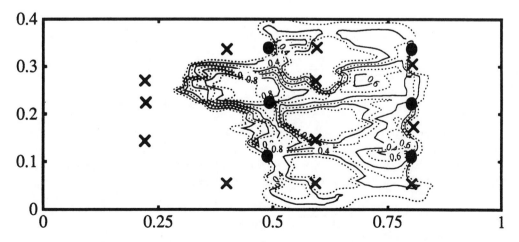

FIG. 5. *Substrate plume after 5 years of biorestoration*

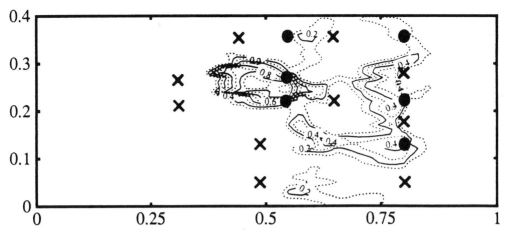

FIG. 6. *Substrate plume after 7 years of biorestoration*

**7. The second time-splitting approach.** We now present another time-splitting approach. At each time step, we advect and react by solving

$$(7.22) \qquad \psi_i \frac{\partial c_i^R}{\partial \tau_i} = \phi_i \mathcal{R}_i(c_1, c_2, \ldots, c_{i-1}, c_i^R, c_{i+1}, \ldots, c_M),$$

and the solution generated from this step is used as initial data for the equation

$$(7.23) \qquad \phi_i \frac{\partial c_i^D}{\partial t} - \nabla \cdot D \nabla c_i^D = \tilde{q}(\tilde{c}_i - c_i^D).$$

That is, we advect and react along the characteristic direction $\tau_i$. For problems with small diffusion, this approach should give a more accurate representation of the physics of the problem, since the reactions "move with the flow."

Let $\tau_{i,j}^{n-1}(x)$ denote the point $(x - u_i(N-j)\Delta t_s, t^{n-1})$, and $\tau_{i,j}^{j,n-1}(x) = (x - u_i(N-j)\Delta t_s, t^{n-1} + j\Delta t_s)$. Assume $C_1^{n-1}, \ldots, C_M^{n-1} \in \mathcal{M}_h$ are known. Define

$$(7.24) \qquad \bar{C}_i^{0,n-1}(x) = C_i^{n-1}(\check{x}).$$

For $j = 0, \ldots, N-1$, set

$$(7.25) \quad \bar{C}_i^{j+1,n-1} = \bar{C}_i^{j,n-1} + \Delta t_s \mathcal{R}_i(C_1(\tau_{i,j}^{n-1}), C_2(\tau_{i,j}^{n-1}), \ldots, C_{i-1}(\tau_{i,j}^{n-1}),$$
$$\bar{C}_i^{j,n-1}, C_{i+1}(\tau_{i,j}^{n-1}), \ldots, C_M(\tau_{i,j}^{n-1})).$$

Note that (7.24)-(7.26) is an explicit Euler approximation to (7.22). Next, we solve

$$(7.26) \quad \left(\phi_i \frac{C_i^n - \bar{C}_i^{N,n-1}}{\Delta t}, \chi\right) + (D\nabla C_i^n, \nabla \chi) = (\tilde{q}(\tilde{c}_i^n - C_i^n), \chi),$$
$$\chi \in \mathcal{M}_h, \quad n \geq 1, \quad i = 1, \ldots, M.$$

In the error analysis below, $K$ will represent a generic constant, independent of discretization parameters.

In order to analyze (7.24)-(7.26), let $\zeta_i = C_i - \tilde{C}_i$, and $\xi_i = c_i - \tilde{C}_i$, where $\tilde{C}_i$ is given by (4.7) and (4.8). Then, subtract (4.7) from (7.26) and use (7.24) and (7.26) to obtain

$$(7.27) \quad \left(\phi_i \frac{\zeta_i^n - \check{\zeta}_i^{n-1}}{\Delta t}, \chi\right) + (D\nabla \zeta_i^n, \nabla \chi) + (\tilde{q}\zeta_i^n, \chi)$$
$$= (\phi_i(\bar{\mathcal{R}}_i^n - \mathcal{R}_i(c^n)), \chi), \quad \chi \in \mathcal{M}_h, \quad n \geq 1, \quad i = 1, \ldots, M,$$

where

$$\bar{\mathcal{R}}_i^n = \sum_{j=0}^{N-1} \frac{\Delta t_s}{\Delta t} \mathcal{R}_i(C_1(\tau_{i,j}^{n-1}), C_2(\tau_{i,j}^{n-1}), \ldots, C_{i-1}(\tau_{i,j}^{n-1}), \tag{7.28}$$
$$\bar{C}_i^{j,n-1}, C_{i+1}(\tau_{i,j}^{n-1}), \ldots, C_M(\tau_{i,j}^{n-1})).$$

Set $\chi = \zeta_i^n$ to obtain

$$\frac{\|\phi_i^{1/2}\zeta_i^n\|^2 - \|\phi_i^{1/2}\zeta_i^{n-1}\|^2}{\Delta t} + \|D^{1/2}\nabla\zeta_i^n\|^2 + \|q^{1/2}\zeta_i^n\|^2 \tag{7.29}$$
$$\leq |(\phi_i(\bar{\mathcal{R}}_i^n - \mathcal{R}_i(c^n)), \zeta_i^n)|.$$

In order to analyze the term on the right side of (7.29), assume a two component system ($M = 2$). Assume $i = 1$ and drop the subscript on $\mathcal{R}$ momentarily, and consider

$$\left| \phi_i \sum_{j=0}^{N-1} \frac{\Delta t_s}{\Delta t} \mathcal{R}(\bar{C}_1^{j,n-1}, C_2(\tau_{1,j}^{n-1})) - \mathcal{R}(c_1^n, c_2^n) \right| \tag{7.30}$$

$$\leq \phi_1 \sum_{j=0}^{N-1} \frac{\Delta t_s}{\Delta t} \left\{ |\mathcal{R}(\bar{C}_1^{j,n-1}, C_2(\tau_{1,j}^{n-1})) - \mathcal{R}(\bar{c}_1^{j,n-1}, C_2(\tau_{1,j}^{n-1}))| \right.$$
$$+ |\mathcal{R}(\bar{c}_1^{j,n-1}, C_2(\tau_{1,j}^{n-1})) - \mathcal{R}(\bar{c}_1^{j,n-1}, c_2(\tau_{1,j}^{n-1}))|$$
$$+ |\mathcal{R}(\bar{c}_1^{j,n-1}, c_2(\tau_{1,j}^{n-1})) - \mathcal{R}(\bar{c}_1^{j,n-1}, c_2(\tau_{1,j}^{j,n-1}))|$$
$$+ |\mathcal{R}(\bar{c}_1^{j,n-1}, c_2(\tau_{1,j}^{j,n-1})) - \mathcal{R}(c_1(\tau_{1,j}^{j,n-1}), c_2(\tau_{1,j}^{j,n-1}))|$$
$$\left. + |\mathcal{R}(c_1(\tau_{1,j}^{j,n-1}), c_2(\tau_{1,j}^{j,n-1})) - \mathcal{R}(c_1^n, c_2^n)| \right\}$$
$$\equiv |T_1| + |T_2| + |T_3| + |T_4| + |T_5|.$$

Here $\bar{c}_1^{0,n-1}(x) = c_1^{n-1}(\check{x})$, and for $j = 0, \ldots, N-1$,

$$\bar{c}_1^{j,n-1} = \bar{c}_1^{j-1,n-1} + \Delta t_s \mathcal{R}(\bar{c}_1^{j-1,n-1}, c_2(\tau_{1,j}^{j,n-1})). \tag{7.31}$$

Let $L$ be the Lipschitz constant for $\mathcal{R}$. By (7.24), (7.26), and (7.31), we find

$$|\bar{C}_1^{j,n-1} - \bar{c}_1^{j,n-1}| \leq |\bar{C}_1^{0,n-1} - \bar{c}_1^{0,n-1}| + \tag{7.32}$$
$$L \sum_{k=0}^{j-1} \frac{\Delta t_s}{\Delta t} \left[ |\bar{C}_1^{k,n-1} - \bar{c}_1^{k,n-1}| + |C_2(\tau_{1,k}^{n-1}) - c_2(\tau_{1,k}^{k,n-1})| \right].$$

For the last term on the right side of (7.32) we note that

$$|C_2(\tau_{1,k}^{n-1}) - c_2(\tau_{1,k}^{k,n-1})| \leq |C_2(\tau_{1,k}^{n-1}) - c_2(\tau_{1,k}^{n-1})| + |c_2(\tau_{1,k}^{n-1}) - c_2(\tau_{1,k}^{k,n-1})| \tag{7.33}$$
$$\leq |C_2(\tau_{1,k}^{n-1}) - c_2(\tau_{1,k}^{n-1})| + Kj\Delta t_s$$

by the definition of $\tau_{i,j}^{n-1}$ and $\tau_{i,j}^{j,n-1}$. Thus

$$|\bar{C}_1^{j,n-1} - \bar{c}_1^{j,n-1}| \leq |\bar{C}_1^{0,n-1} - \bar{c}_1^{0,n-1}| + K\Delta t + \tag{7.34}$$
$$L \sum_{k=0}^{j-1} \frac{\Delta t_s}{\Delta t} \left[ |\bar{C}_1^{k,n-1} - \bar{c}_1^{k,n-1}| + |C_2(\tau_{1,k}^{n-1}) - c_2(\tau_{1,k}^{k,n-1})| \right].$$

Applying this recursion relationship to each entry in the sum we find

(7.35) $|\bar{C}_1^{j,n-1} - \bar{c}_1^{j,n-1}|$
$$\leq K\left[|\bar{C}_1^{0,n-1} - \bar{c}_1^{0,n-1}| + \Delta t + \sum_{k=0}^{N-1} \frac{\Delta t_s}{\Delta t}|C_2(\tau_{1,k}^{n-1}) - c_2(\tau_{1,k}^{n-1})|\right],$$

where $K$ is a sufficiently large constant which grows exponentially with $N$. Multiplying by $\frac{\Delta t_s}{\Delta t}$ and summing on $j$, $j = 0, \ldots, N-1$, we find

$$(7.36)\quad |T_1| \leq K\phi_1\left[|C^{n-1}(\check{x}) - c^{n-1}(\check{x})| + \Delta t + \sum_{k=0}^{N-1} \frac{\Delta t_s}{\Delta t}|C_2(\tau_{1,k}^{n-1}) - c_2(\tau_{1,k}^{n-1})|\right].$$

Similarly

$$(7.37)\quad |T_2| \leq K\phi_1 \sum_{k=0}^{N-1} \frac{\Delta t_s}{\Delta t}|C_2(\tau_{1,k}^{n-1}) - c_2(\tau_{1,k}^{n-1})|.$$

By the definitions of $\tau_{1,j}^{n-1}$ and $\tau_{1,j}^{j,n-1}$,

$$(7.38)\quad |T_3| \leq L\phi_1 \sum_{j=0}^{N-1} \frac{\Delta t_s}{\Delta t}|c_2(\tau_{1,j}^{n-1}) - c_2(\tau_{1,j}^{j,n-1})|$$
$$\leq K \sum_{j=0}^{N-1} \frac{\Delta t_s}{\Delta t} j\Delta t_s$$
$$\leq K\Delta t.$$

For $T_4$, we have

$$(7.39)\quad |T_4| \leq L\phi_1 \sum_{j=0}^{N-1} \frac{\Delta t_s}{\Delta t}|\bar{c}_1^{j,n-1} - c(\tau_{1,j}^{j,n-1})|.$$

Let $c_1^R$ denote the solution to the initial value problem (7.22), with initial condition $c_1^R(0) = c_1^{n-1}(\check{x})$, then

$$(7.40)\quad |\bar{c}_1^{j,n-1} - c(\tau_{1,j}^{j,n-1})| \leq |\bar{c}_1^{j,n-1} - c_1^R(\tau_{1,j}^{j,n-1})| + |c_1^R(\tau_{1,j}^{j,n-1}) - c_1(\tau_{1,j}^{j,n-1})|.$$

The first term represents the error in applying Euler's method to (7.22), thus

$$(7.41)\quad |\bar{c}_1^{j,n-1} - c_1^R(\tau_{1,j}^{j,n-1})| \leq K\Delta t_s \|(c_1^R)_{\tau\tau}\|_\infty.$$

Writing the differential equation (1.1) as

$$(7.42)\quad \psi_1 \frac{\partial c_1}{\partial \tau_1} - \nabla \cdot D\nabla c_1 = \phi_1 \mathcal{R}(c_1, c_2),$$

and integrating (7.42) and (7.22) from $\tau_* = \tau_{1,0}^{n-1}$ to $\tau^* = \tau_{1,j}^{j,n-1}$, we find

$$(7.43)\quad |c_1^R(\tau_{1,j}^{j,n-1}) - c_1(\tau_{1,j}^{j,n-1})|$$
$$\leq \int_{\tau_*}^{\tau^*} |\psi_1^{-1}(\nabla \cdot D\nabla c_1)|d\tau + \int_{\tau_*}^{\tau^*} |\psi_1^{-1}\phi_1(\mathcal{R}(c_1^R, c_2) - \mathcal{R}(c_1, c_2))|d\tau$$
$$\leq K\left[\Delta t + \int_{\tau_*}^{\tau^*} |c_1^R - c_1|d\tau\right].$$

Thus, by Gronwall's Lemma applied to (7.43), and (7.41),

(7.44) $$|T_4| \leq K\Delta t.$$

Finally,

(7.45) $$|T_5| \leq L\left[|c_1(\tau_{1,j}^{j,n-1}) - c_1^n| + |c_2(\tau_{1,j}^{j,n-1}) - c_2^n|\right]$$
$$\leq K\Delta t.$$

Combining the estimates for $T_1 - T_5$, we find

$$\left|\phi_1 \sum_{j=0}^{N-1} \frac{\Delta t_s}{\Delta t} \mathcal{R}(\bar{C}_1^{j-1,n-1}(x), C_2(\tau_{1,j}^{n-1}(x)) - \mathcal{R}(c_1^n(x), c_2^n(x))\right|$$
$$\leq K\left\{\Delta t + |\phi_1(C_1^{n-1}(\check{x}) - c_1^{n-1}(\check{x}))| + \phi_1 \sum_{j=0}^{N-1} \frac{\Delta t_s}{\Delta t}\left|C_2(\tau_{i,j}^{n-1}(x)) - c_2(\tau_{i,j}^{n-1}(x))\right|\right\}$$
$$\leq K\left\{\Delta t + |\phi_1 \check{\zeta}_1^{n-1}(x)| + |\phi_1 \check{\xi}_1^{n-1}(x)|\right.$$
$$\left. + \phi_1 \sum_{j=0}^{N-1} \frac{\Delta t_s}{\Delta t}\left[|\zeta_2^{n-1}(\tau_{i,j}^{n-1}(x))| + |\xi_2^{n-1}(\tau_{i,j}^{n-1}(x))|\right]\right\}.$$

Under the assumption of no-flow boundary conditions, one can show [5]

$$\|\check{f}\| \leq \|f\|$$

for any $L^2(\Omega)$ function $f$. Thus, assuming $\phi_2/\phi_1 \leq K$,

$$|(\phi_1(\bar{\mathcal{R}}_1^n - \mathcal{R}_1(c^n)), \zeta_1^n)| \leq K\|\phi_1^{1/2}\bar{\mathcal{R}}_1^n - \mathcal{R}_1(c^n)\|^2 + K\|\phi_1^{1/2}\zeta_1^n\|^2$$
$$\leq K\left[\Delta t^2 + \|\phi_1^{1/2}\zeta_1^{n-1}\|^2 + \|\phi_1^{1/2}\xi_1^{n-1}\|^2\right.$$
$$\left. + \|\phi_2^{1/2}\zeta_2^{n-1}\|^2 + \|\phi_2^{1/2}\xi_2^{n-1}\|^2\right]$$
$$+ K\|\phi_1^{1/2}\zeta_1^n\|^2.$$

Substituting into (7.29) and rearranging some terms we find

(7.46) $$\frac{\|\phi_1^{1/2}\zeta_1^n\|^2 - \|\phi_1^{1/2}\zeta_1^{n-1}\|^2}{\Delta t} + \|D^{1/2}\nabla\zeta_1^n\|^2 + \|q^{1/2}\zeta_1^n\|^2$$
$$\leq \frac{\|\phi_1^{1/2}\check{\zeta}_1^{n-1}\|^2 - \|\phi_1^{1/2}\zeta_1^{n-1}\|^2}{\Delta t} + K\|\phi_1^{1/2}\zeta_1^n\|^2$$
$$+ K\left[\Delta t^2 + \|\phi_1^{1/2}\zeta_1^{n-1}\|^2 + \|\phi_1^{1/2}\xi_1^{n-1}\|^2 + \|\phi_2^{1/2}\zeta_2^{n-1}\|^2 + \|\phi_2^{1/2}\xi_2^{n-1}\|^2\right].$$

By Lemma 3.1 in [5], we have

(7.47) $$\frac{\|\phi_1^{1/2}\check{\zeta}_1^{n-1}\|^2 - \|\phi_1^{1/2}\zeta_1^{n-1}\|^2}{2\Delta t} \leq K\|\phi_1^{1/2}\zeta_1^{n-1}\|^2.$$

Applying this inequality to (7.46) for each component $i = 1, \ldots, M$, adding these equations together, multiplying the result by $\Delta t$ and summing on $n$, and applying Gronwall's Lemma we find

(7.48) $$\max_n \|\phi_i^{1/2}\zeta_i^n\| \leq K\Delta t, \quad i = 1, \ldots, M.$$

Using arguments found in, e.g., [5], one can show

(7.49) $$\max_n \|\phi_i^{1/2} \xi_i^n\| \leq K(\Delta t + h^{r+1}).$$

Applying the triangle inequality we obtain the following result.

THEOREM 7.1. *Let $c_i$, $i = 1, \ldots, M$ satisfy (1.1) and assume the data and $c_i$ are sufficiently smooth. Let $C_i \in \mathcal{M}_h$ be the approximation given by (7.24)-(7.26). Then there exists a positive constant $K$, such that*

(7.50) $$\max_n \|C_i^n - c_i^n\| \leq K(\Delta t + h^{r+l}), \quad i = 1, \ldots, M,$$

*where $r$ is the maximum degree of polynomials in $\mathcal{M}_h$, and $l = 1$ if $D(u)$ is uniformly positive definite, $l = 0$ if $D(u)$ is positive semi-definite.*

## REFERENCES

[1] R. C. Borden and P. B. Bedient, *Transport of dissolved hydrocarbons influenced by oxygen-limited biodegradation 1. Theoretical development*, Water Resour. Res. 22, pp. 1973-1982, 1986.

[2] C. Y. Chiang, C. N. Dawson, and M. F. Wheeler, *Modeling of in-situ biorestoration of organic compounds in groundwater*, to appear in Transport in Porous Media.

[3] C. N. Dawson, M. F. Wheeler, T. M. Nguyen, and S. W. Poole, *Simulation of hydrocarbon biodegradation in groundwater*, CRAY Channels 8, 3, pp. 14-19, 1986.

[4] C. N. Dawson, M. F. Wheeler, T. M. Nguyen, and S. W. Poole, *Simulation of subsurface contaminant transport with biodegradation kinetics*, Proceedings of Third International Symposium on Science and Engineering on CRAY Supercomputers, Mendota Heights, MN, pp. 75-86, 1987.

[5] C. N. Dawson, T. F. Russell, and M. F. Wheeler, *Some improved error estimates for the modified method of characteristics*, SIAM J. Numer. Anal. 26, pp. 1487-1512, 1989.

[6] C. N. Dawson, T. F. Dupont, and M. F. Wheeler, *The rate of convergence of the modified method of characteristics for linear advection equations in one dimension*, Mathematics for Large Scale Computing, ed. J. C. Diaz, pp. 115-126, 1989.

[7] J. Douglas and T. F. Russell, *Numerical methods for convection-dominated diffusion problems based on combining the method of characteristics with finite element or finite difference procedures*, SIAM J. Numer. Anal. 19, pp. 871-885, 1982.

[8] R. E. Ewing, T. F. Russell, and M. F. Wheeler, *Simulation of miscible displacement using a mixed method and a modified method of characteristics*, SPE 12241, Proceedings of Seventh SPE Symposium on Reservoir Simulation, Society of Petroleum Engineers, Dallas, pp. 71-81, 1983.

[9] R. E. Ewing, T. F. Russell, and M. F. Wheeler, *Convergence analysis of an approximation of miscible displacement in porous media by mixed finite elements and a modified method of characteristics*, Comp. Meth. Appl. Mech. Eng. 47, pp. 73-92, 1984.

[10] J. S. Kindred and M. A. Celia, *Contaminant transport and biodegradation 2. Conceptual model and test simulations*, Water Resour. Res. 25, pp. 1149-1159, 1989.

[11] H. S. Rifai *Numerical techniques for modeling in situ biorestoration and biodegradation of organic contaminants in ground water*, Ph. D. Thesis, Rice University, 1989.

[12] T. F. Russell, M. F. Wheeler, and C. Y. Chiang, *Large-scale simulation of miscible displacement*, Proceedings of SEG/SIAM/SPE Conference on Mathematical and Computational Methods in Seismic Exploration and Reservoir Modeling, ed. W. E. Fitzgibbon, Society for Industrial and Applied Mathematics, Philadelphia, pp. 85-107, 1986.

[13] J. M. Thomas, M. D. Lee, P. B. Bedient, R. C. Borden, L. W. Canter, and C. H. Ward, *Leaking underground storate tanks: remediation with emphasis on in situ biorestoration*, Environmental Protection Agency, 600/2-87,008, January, 1987.

[14] United States Department of Energy, *Site-directed subsurface environmental initiative, five year summary and plan for fundamental research in subsoils and in groundwater*, FY1989-FY1993, DOE/ER 034411, Office of Energy Research, April 1988.

[15] M. F. Wheeler and C. N. Dawson, *An operator-splitting method for advection-diffusion-reaction problems*, MAFELAP Proceedings VI (J. A. Whiteman, ed.), Academic Press, London, pp. 463-482, 1988.

[16] M. F. Wheeler, C. N. Dawson, P. B. Bedient, C. Y. Chiang, R. C. Borden, and H. S. Rifai, *Numerical simulation of microbial biodegradation of hydrocarbons in groundwater*, in Proceedings of AGWSE/IGWMCH Conference on Solving Ground Water Problems with Models, National Water Wells Association, pp. 92-108, 1987.

[17] B. Wood and C. N. Dawson, *Effects of lag and maximum growth in contaminant transport and biodegradation modeling*, to appear.

# Chapter 7
# ON MODULATION EQUATIONS OF THE GINZBURG-LANDAU TYPE

**WIKTOR ECKHAUS**
University of Utrecht
The Netherlands

**1 Introduction.** Modulation equations describe (approximately) the evolution of wave-like phenomena in nonlinear systems. They arise in many domains of science, such as fluid dynamics, reaction-diffusion processes, nonlinear optics, electric forcing of liquid crystals, etc. In spite of the large variety of sources only few equations play a predominant role and occur "generically." The most important are: the *Nonlinear Schrödinger equation* (NLS) and the *Ginzburg-Landau equation* (G.L.).

The nonlinear Schrödinger equation arises when one studies dispersive waves in neutral media, that is in problems where linear theory produces waves which neither grow nor decay in time and which have a group velocity that is different from the phase velocity. In canonical form the equation looks as follows:

$$i\frac{\partial \psi}{\partial \tau} + \frac{\partial^2 \psi}{\partial \xi^2} + s|\psi|^2\psi = 0; \quad s = \pm 1 \tag{1.1}$$

where $\psi(\xi, \tau)$ is a complex valued function, $\xi \in (-\infty, \infty)$, $\tau \geq 0$. The equation is famous for the occurence of solitons. (For a study of the generic nature of the NLS-equation one can consult Calogero & Eckhaus (1987), (1988) and Calogero (1991)).

The Ginzburg-Landau equation describes (at near critical conditions) the evolution of patterns through instabilities and bifurcation. It is given by:

$$\frac{\partial \Phi}{\partial \tau} = \left(\bar{\mu}_0 + \beta |\Phi|^2\right) \Phi - (\mu_2 + i\nu_2)\frac{\partial^2 \Phi}{\partial \xi^2} \tag{1.2}$$

$\Phi(\xi, \sigma)$ is again complex-valued, $\bar{\mu}_0, \mu_2, \nu_2 \in \mathbf{R}$, $\beta$ is in general complex.

In a mathematical sense, by taking $\bar{\mu}_0 = \mu_2 = 0$ and $\text{Re}\beta = 0$, one finds (after cosmetic rescaling) that the G.-L. equation (1.2) contains the NLS equation (1.1). However, the exercise is academic: in the derivation of (1.2) one essentially assumes that the real parts of the coefficients on the r.h.s. are of order unity.

In this paper we discuss various formal methods for deriving the G.-L. equation and describe recent rigorous results on the validity of that equation as a "universal model equation". We shall also discuss the role that the G.-L. equation plays in the study of dynamics of patterns and finally consider another generic modulation equation of the Ginzburg-Landau type which is applicable when (1.2) ceases to be valid. As a preparation we summarize, in section 2, some classical results on linear stability analysis and the dynamics of bifurcations, which are needed as a background for the subsequent discussion.

*Historical remarks.* In connection with certain particular problems in fluid dynamics the equation (1.2) was first derived by Segel (1969) and Newel & Whitehead (1969), while in DiPrima, Eckhaus & Segel (1971) the equation was derived for a large class of problems. The name "Amplitude equation" or "Envelope equation" was favoured for a while. More recently the name "Ginzburg-Landau" has become firmly established. The reason is to be found in Landau's collected papers (1965), which contain a paper with Ginzburg on superconductivity. However, the equation to be found there is a *stationary version of* (1.2), and is hence of little relevance to our discussion.

## 2 The classical linear stability analysis and the dynamics of bifurcations.

Let us consider, to fix the ideas, classical stability problems of fluid dynamics, such as the Taylor-Couette problem of flow between rotating cylinders, Bénard's experiment on a layer of fluid heated from below, or Poiseuille flow between parallel walls. (As a general reference one can use Drazin & Reid (1981)). The mathematics of these problems (and many other problems to which our discussion will apply) is as follows:

Consider $V(x, Y, t)$, a vector-function, $x \in (-\infty, \infty)$, $Y \in D \subset \mathbf{R}^n$, $n = 0, 1, 2$, which satisfies the "full" problem

$$L(V) = 0, \qquad (2.1)$$

with boundary conditions on the boundary $\partial D$ of $D$. The operator $L$ symbolizes for instance the Navier-Stokes equations. A basic solution $u_0(Y)$ is given (laminar flow, or fluid at rest), which satisfies

$$L(u_0) = 0 \text{ and } B.C. \text{ on } \partial D \qquad (2.2)$$

To investigate the stability of $u_0(Y)$ one writes

$$V = u_0(Y) + U(x, Y, t). \qquad (2.3)$$

For the pertubation $U$ one gets

$$\mathcal{L}_R(u) + N(u) = 0 \qquad (2.4)$$

where $\mathcal{L}_R$ is the linearization of $L$ at $u_0$, i.e.

$$\mathcal{L}_R(U) = \lim_{\lambda \to 0} \frac{1}{\lambda} \{L(u_0 + \lambda U) - L(u_0)\} \qquad (2.5)$$

$N(U)$ are the nonlinear terms. $\mathcal{L}_R$ contains a control parameter $R$ (Reynolds-number, Taylor's-number, Rayleigh's-number...) which can be varied at will by manipulating the experimental machinery.

In *linearized stability analysis* one studies the solutions of

$$\mathcal{L}_R(U) = 0 \qquad (2.6)$$

with homogeneous boundary conditions on $\partial D$. The equation (2.4) and hence also (2.6) usually is invariant with respect to translation in the $t$ and the $x$ directions. One can therefore reduce the problem, by Fourier decomposition, to the study of linear waves of the structure

$$U(x, Y, t) = e^{ikx} e^{\mu t} \varphi_k(Y) + c.c. \qquad (2.7)$$

where $k$ is the wave number, arbitrary but fixed. $\mu(k, R)$ follows from the eigenvalue problem in the $Y$-variables, to which (2.6) reduces after substitution of (2.7). If the transversal dimension is zero (i.e. $V$ is a function of $x$ and $t$ only), then $\mu$ simply follows from the dispersion relation. Otherwise, in all that follows, $\mu(k, R)$ will mean the "lowest eigenvalue", that is: for any fixed $k, R$ all other eigenvalues $\mu^{(n)}(k, R)$ satisfy

$$\operatorname{Re}\mu_n \leq \operatorname{Re}\mu. \tag{2.8}$$

$\varphi_k(Y)$ is the corresponding eigenfunction. Let us write

$$\mu(k, R) = \mu^{(r)}(k, R) + i\nu(k, R). \tag{2.9}$$

The stability question now becomes the question of the sign of $\mu^{(r)}$. In problems mentioned at the beginning of this section (and in a great many other problems) one gets a result summarized in fig.1.

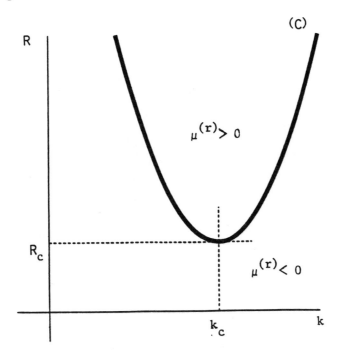

Fig. 1: Generic stability results.

In the $R - k$ plane there is a parabola-like critical curve $(C)$ on which $\mu^{(r)} = 0$. If $R > R_c$, then for an interval of values of $k$ within this critical curve the corresponding linear waves (2.7) grow with time. This means that nonlinear effects must be taken into account.

Let us consider an arbitrary but fixed value $k = k_0$. The nonlinear terms in (2.4) will generate harmonics of the simple linear wave (2.7). One therefore looks for space-periodic solutions of (2.4):

$$U(x, Y, t) = \sum_{n=-\infty}^{\infty} e^{ink_0 x} \Phi_n(Y, t) \tag{2.10}$$

with $\Phi_n = \Phi_n^*$, the complex conjugate. Bifurcation analysis produces the following results.

For $(k_0, R)$ near the critical curve $(C)$ there exist small solutions of (2.4) given by

$$U(x,Y,t) = \varepsilon A_{k_0}(t)\varphi_{k_0}(Y)e^{ik_0 x} +$$
$$+ \varepsilon^2 |A_{k_0}(t)|^2 G_{k_0}^{(0)}(Y) + \varepsilon^2 [A_{k_0}(t)]^2 G_{k_0}^{(2)}(Y)e^{2ik_0 x} \qquad (2.11)$$
$$+ \text{c.c.} + o(\varepsilon^2)$$

where $\varepsilon$ is of the order of magnitude of $\sqrt{|\mu^{(r)}|}$. The amplitude-function $A_{k_0}(t)$ satisfies the equation

$$\frac{dA_{k_0}}{dt} = \mu(k_0, R)A_{k_0} + \varepsilon^2 \beta(k_0, R)|A_{k_0}|^2 A_{k_0} + o(\varepsilon^2). \qquad (2.12)$$

The nonlinear coefficient $\beta$ can be computed explicitly for any explicitly given equation (2.4), and in particular for problems (2.4) which are of the structure

$$\frac{\partial}{\partial t}S(U) = \hat{\mathcal{L}}_R(U) + \sum_{l=1}^{N} P^{(l)}(U)Q^{(l)}(U) \qquad (2.13)$$

where $S$, $\mathcal{L}_R$, $P^{(l)}$, $Q^{(l)}$ are linear differential operators, independant of $t$ and invariant with respect to translation in $x$, (see for example Eckhaus (1965)).

The *amplitude equation* (2.12) is sometimes called *Landau's equation*, because of the conjuncture formulated by Landau that nonlinear effects could be represented by an equation of this structure (see Landau and Lifschitz (1959)). However, Landau gave no indication how the crucial coefficient $\beta$ should be computed.

If, as a result of computations, one finds $\text{Re}(\beta) < 0$, then for $\mu^{(r)} > 0$ supercritical bifurcation occurs and it follows from (2.12) that

$$\lim_{t \to \infty} |A_k(t)|^2 = -\frac{\mu^{(r)}}{\varepsilon^2 \text{Re}(\beta)}. \qquad (2.14)$$

In what follows we shall assume that this is the case. The bifurcation is called *pitchfork* if $\text{Im}(\mu) = 0$, and is a *Hopf-bifurcation* if $\text{Im}(\mu) \neq 0$.

**3 Modulations by an algorithmic derivation.** Let us consider values of $R$ slightly higher than $R_c$ and introduce

$$\varepsilon = \sqrt{(R - R_c)} \qquad (3.1)$$

then for values of $k$ in an interval of which the extent is proportional to $\varepsilon$, space-periodic solutions exist. The situation is sketched in fig 2.

In order to magnify the $k$-interval of interest we write

$$k - k_c = \varepsilon \sigma. \qquad (3.2)$$

Let us now make a nearly trivial observation: *any periodic solution with $k$ in the interval (3.2) is a slow modulation of the space-periodic solution for $k = k_c$*. This follows from the identity

$$e^{ikx} = e^{ik_c x} \cdot e^{i\sigma \bar{\xi}}, \quad \bar{\xi} = \varepsilon x \qquad (3.3)$$

We introduce next, instead of (2.11), a more general representation of solutions

$$U(x, Y, t) = \varepsilon \bar{\Phi}(\bar{\xi}, t)e^{ik_c x}\varphi_{k_c}(Y) + \text{c.c.} + O(\varepsilon^2) \qquad (3.4)$$

and pose the problem: *to formulate an equation for $\bar{\Phi}(\bar{\xi}, t)$ such that when substituting*

# MODULATION EQUATIONS OF THE GINZBURG-LANDAU TYPE

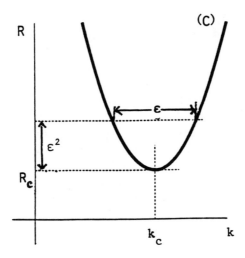

Fig. 2: Near-critical conditions

$$\bar{\Phi}(\bar{\xi}, t) = e^{i\sigma\bar{\xi}} A_k(t) \tag{3.5}$$

*one recovers the amplitude equation* (2.12). To solve the problem we need a representation of the coefficients of (2.12) at near critical conditions. Using Taylor series one gets

$$\mu^{(r)} = \mu_0 (R - R_0) + \varepsilon^2 \mu_2 \sigma^2 + o(\varepsilon^2) \quad \mu_0 > 0, \quad \mu_2 < 0 \tag{3.6}$$

$$\nu = \nu_0 + \nu_1 \varepsilon \sigma + \nu_2 \varepsilon^2 \sigma^2 + o(\varepsilon^2) \tag{3.7}$$

$$\beta = \beta_0 + O(\varepsilon); \quad \beta_0 = \beta(k_c, R_c) \tag{3.8}$$

The equation (2.12) takes the form

$$\begin{aligned}\frac{dA_k}{dt} &= i(\nu_0 + \varepsilon\sigma\nu_1 + \varepsilon^2\sigma^2\nu_2)A_k + \\ &+ \varepsilon^2(\mu_0 + \mu_2\sigma^2)A_k + \varepsilon^2\beta_0|A_k|^2 A_k + o(\varepsilon^2)\end{aligned} \tag{3.9}$$

The problem formulated above is now easily solved: just replace $A_k$ by $\bar{\Phi}$ and $\sigma$ by $\frac{\partial}{\partial \bar{\xi}}$. We get in this way the equation

$$\begin{aligned}\frac{\partial \bar{\Phi}}{\partial t} &= i\nu_0\bar{\Phi} + \varepsilon\nu_1 \frac{\partial \bar{\Phi}}{\partial \bar{\xi}} + \varepsilon^2 \left[\mu_0 + \beta_0|\bar{\Phi}|^2\right] \bar{\Phi} \\ &- \varepsilon^2 (\mu_2 + i\nu_2) \frac{\partial^2}{\partial \bar{\xi}^2} + o(\varepsilon^2)\end{aligned} \tag{3.10}$$

It is easily verified that by substitution of (3.5) one indeed recovers (3.9). The final step consists of the transformations

$$\bar{\Phi} = e^{i\nu_0 t}\Phi \tag{3.11}$$

$$\xi = \bar{\xi} + \varepsilon\nu_1 t = \varepsilon(x + \nu_1 t) \tag{3.12}$$

$$\tau = \varepsilon^2 t \tag{3.13}$$

The result is the Ginzburg-Landau equation

$$\frac{\partial \Phi}{\partial \tau} = \left[\mu_0 + \beta_0|\Phi|^2\right]\Phi - (\mu_2 + i\nu_2)\frac{\partial^2 \Phi}{\partial \xi^2} \tag{3.14}$$

**Remarks** (I) Note that the variables $\xi$ and $\tau$, occurring in the G.-L. equation are *slow* variables. In particular the space-like variable $\xi$ is attached to coordinate moving with *group velocity* (3.12)

(II) It is amusing to note that the algorithmic derivation ("replace $\sigma$ by $\frac{\partial}{\partial \xi}$") is closely analogous to the "derivation" of Schrödinger equation of quantum mechanics from Hamilton formulation of classical mechanics by the "Heisenberg trick".

**4 Ginzburg-Landau equation is a universal model equation.** The only link between the modulation equation (3.14) and the original nonlinear problem (2.4) or (2.13) is the assumption of generic stability results and the numerical values of the lowest eigenvalue $\mu$ and the nonlinear coefficient $\beta$. No other details of the original problem are needed. It follows that the Ginzburg-Landau equation should be expected to be a universal approximate equation, applicable to large classes of problems. However, the algorithmic derivation is only suggestive in this direction: it does not show, even on a formal level, that (3.4) with (3.14) indeed produces approximations of solutions of (2.13). This is why more serious derivations of modulation equations are important. Such derivations have been produced, by various procedures. We discuss them in this section, and describe also recent rigorous results on the validity of the G.-L. equation.

**4.1 The multiple-scaling ansatz.** Consider the class of equations (2.13), i.e.

$$\frac{\partial}{\partial t} S(U) = \hat{\mathcal{L}}_R(U) + \sum_{l=1}^{N} P^{(l)}(U) Q^{(l)}(U). \tag{4.1}$$

On the basis of the results of section 3 the following *ansatz* should not be surprising:

$$U(x, Y, t) = \sum_{n=-\infty}^{\infty} \delta_n(\varepsilon) \psi_n(\xi, Y, \tau, t) e^{i k_c n x} \tag{4.2}$$

with

$$\xi = \varepsilon(x + \nu_1 t); \quad \tau = \varepsilon^2 t \tag{4.3}$$

$$\psi_{-n} = \psi_n^*; \quad \delta_{-n}(\varepsilon) = \delta_n(\varepsilon) \tag{4.4}$$

The orders of magnitude $\delta_n(\varepsilon)$ can easily be established by an educated guess: eg. (4.2) should hold in particular for space-periodic solutions. One then finds (and part of the result is given explicitly in (2.11)):

$$\delta_1 = \varepsilon, \quad \delta_0 = \varepsilon^2, \quad \delta_n = \varepsilon^n \text{ for } n \geq 2. \tag{4.5}$$

In this setting one can start the formal machinery of multiple-scale expansions, which involves solvability conditions of Fredholm-type. This has been done systematically in Doelman (1990) where it has been established that the Ginzburg-Landau equation indeed is a formal approximation for solutions of (4.1) at near critical conditions (and under the assumption of a generic stability result). Doelman (1990) also considers the derivation of modulation equations which allow for slow modulations in a second (unbounded) transversal variable. This involves a Squire-type transformation and a carefull analysis of the boundary conditions of the problem.

**4.2 Clustered Fourier-modes by a discrete analysis.** By the *ansatz* (4.2) it is not excluded that (4.1) may admit small solutions which are not slow modulations of the basic linear waves. This is why one of the oldest derivations of the G.-L. equation for systems (4.1), given in DiPrima, Eckhaus & Segel (1971), is not only of historical interest. In that paper, slow modulations are not introduced at the outset, but arise as a result of the analysis. This is done as follows:
Consider a very fine subdivision of the $k$-axis given by

$$k = m\Delta k, \quad \Delta k = O(\varepsilon), \quad m = 0, \mp 1, \mp 2, \ldots \quad (4.6)$$

and a corresponding discrete Fourier-decomposition

$$U(x, Y, t) = \sum_k e^{ikx} \delta_k(\varepsilon) \psi_n(Y, t). \quad (4.7)$$

Inspired by the order-of-magnitude distribution for space-periodic solution, we introduce a distribution of $\delta_n$'s as sketched in fig. 3.

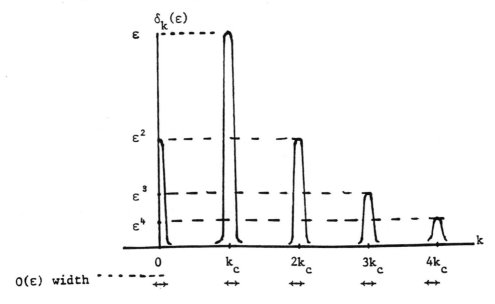

Fig. 3: Clustered mode-distribution.

The order functions $\delta_k(\varepsilon)$ have magnitudes indicated in fig.3 in intervals $|k - nk_c| = O(\varepsilon)$ and tail off very rapidly outside these intervals (in the next subsection we shall describe precisely how this happens).

In DiPrima, Eckhaus, Segel (1971) it was shown that for solutions of (4.1) the clustered mode-distribution persists in time, i.e. when initial conditions have Fourier-components as in fig. 3, this remains true for the corresponding solutions. With this as a starting point, the Ginzburg-Landau equation follows in due course of the (formal) analysis.

**4.3 Clustered modes by Fourier-transformation and the validity of the G.-L. equation.** Recently, Van Harten (1990) developed a rigorous theory of the Ginzburg-Landau equation for a prototype subclass of (4.1). He considers $U(x,t)$, $x \in (-\infty, \infty)$, $t \geq 0$ which satisfy

$$\frac{\partial U}{\partial t} = \hat{\mathcal{L}}_R(U) + N(U^2) \quad (4.8)$$

with $\hat{\mathcal{L}}$, $N$ linear operators of the structure

$$\hat{\mathcal{L}}_R = \mu\left(\frac{1}{i}\frac{\partial}{\partial x}, R\right) \tag{4.9}$$

$$N = 2\pi\rho\left(\frac{1}{i}\frac{\partial}{\partial x}, R\right). \tag{4.10}$$

Introducing the Fourier-transform

$$\Phi(k,t) = \int_{-\infty}^{\infty} U(x,t)e^{-ikx}\,dx \tag{4.11}$$

one gets

$$\frac{\partial \Phi}{\partial t} = \mu(k)\Phi + \rho(k)\Phi * \Phi \tag{4.12}$$

where $\mu(k)$ and $\rho(k)$ are symbols of the operators defined in (4.9), (4.10) and $\Phi * \Phi$ is the convolution, i.e.

$$\Phi * \Phi = \int_{-\infty}^{\infty} \Phi(k')\Phi(k-k')\,dk' \tag{4.13}$$

Inspired by results described in section 4.2 Van Harten considers clustered mode-distributions as in fig. 3, with sufficiently fast decay outside the intervals $|k - nk_c| = O(\varepsilon)$. This distribution is invariant under convolution. It is easy to derive formally the Fourier-transformed equivalent of the G.-L. equation. The main body of Van Harten's (1990) paper is the proof of validity, which is necessarily rather technical and requires a subtle choice of suitable Banach-spaces to accomodate all types of interesting solutions of the G.-L. equation and yet be able to use a contraction mapping argument.

The main result can be described as follows: consider solutions of (4.8) with initial conditions $U(x,0) = U_0(x)$ such that the Fourier transform $\Phi^0(k) = \delta_k(\varepsilon)\psi^0(k)$ with $\delta_k(\varepsilon)$ the clustered mode-distribution of fig. 3. Then the corresponding solution of the G.-L. equation is an approximation of $U(x,t)$ with an $O(\varepsilon^2)$ error on any compact interval of the intrinsic $\frac{1}{\varepsilon^2}$ time-scale, on which the solution of G.-L. is bounded.

Very recently (in fact after the ICIAM 91 meeting) this author (Eckhaus (1991)) has supplemented Van Harten's results as follows:
Consider solutions of (4.8) on long time-scales but short as compared to the G.-L. time-scale, explicitly

$$0 < t < \frac{T}{\varepsilon^\nu}, \quad 0 < \nu < 2 \tag{4.14}$$

and let the initial conditions for the corresponding Fourier-transform satisfy

$$\Phi(k,0) = O(\varepsilon), \quad \forall k \in \mathbf{R}. \tag{4.15}$$

and some mild technical decay-condition for $|k| \to \infty$. Then on time-scales (4.14) the corresponding solutions of (4.12) settle to

$$\Phi(k,t) = \delta_k(\varepsilon)\psi_k(k,t), \quad \psi_k = O(1) \tag{4.16}$$

with $\delta_k(\varepsilon)$ the clustered mode-distribution of fig. 3. In fact $\delta_k(\varepsilon)$ can be described analytically with aid of the function $f(k,k_0)$, given by

$$f(k,k_0) = \frac{\varepsilon^2}{(k-k_0)^2 + \varepsilon^2}. \tag{4.17}$$

Clearly, $f(\varepsilon, k_0) = O(1)$ when $|k - k_0| = O(\varepsilon)$, and decays to $O(\varepsilon^2)$ outside that region. In Eckhaus (1991) it is shown rigorously that

$$\delta_k(\varepsilon) = \text{Max}\left\{\varepsilon \sum_{n=0}^{N} \varepsilon^{|n-1|} [f(k, nk_c)]^N, \varepsilon^{N+1}\right\} \tag{4.18}$$

where $N$ is an arbitrarily large positive number. The maximum should be considered in the sense of orders of magnitude. It follows that outside the intervals $|k - nk_c| = O(\varepsilon)$ the clustered mode-distribution is smaller than any power of $\varepsilon$. This result combined with Van Harten's (1990) proof, shows in essence that the Ginzburg-Landau equation is an attractor for all $O(\varepsilon)$ solutions of equations of the type (4.8), and hence is truly a universal model equation.

An extension of the analysis described in this section to more general classes of problems (4.1) is a subject of current research.

**4.4 Other validity results.** In the recent years there has been a considerable interest in the problem of validity of the G.-L. equation. Let us consider first the work of Collet & Eckmann (1990) on the so called *Swift-Hohenberg equation*

$$\frac{\partial U}{\partial t} = -\left(1 + \frac{\partial^2}{\partial x^2}\right)^2 U + \lambda U - U^3 \tag{4.19}$$

The essential difference with (4.8) is the fact that (4.18) has a cubic nonlinearity. Collet & Eckmann (1990) give a proof of validity of the G.-L. equation for a rather large class of initial conditions, working directly in the $x$-space, hence without Fourier-transformation. However, it is illuminating to reflect on the difference between quadratic and cubic non-linearities using Fourier-decomposition.

The cubic nonlinearity $|\Phi|^2\Phi$ in the G.-L. equation arises, in the case of the Swift-Hohenberg equation, through direct self-interaction of the basic mode cluster at $|k - k_c| = O(\varepsilon)$. Other mode clusters are not involved in the first approximation. In the case of quadratic non-linearity as in (4.8), and more in general (4.1), the cubic term $|\Phi|^2\Phi$ arises through interaction of mode clusters at $|k - nk_c| = O(\varepsilon)$, $n = 0, 1, 2$. This makes the analysis and the proof of validity much more complicated.

We also mention a simplified proof for (4.18) given in Kirrmann, Schneider & Mielke (1991).

On the other hand, definite progress in validity results for some specific problems of the more general class (4.1) has been achieved by certain special methods. The interested reader should consult Iooss, Mielke & Demay (1989), Iooss & Mielke (1991) and Mielke (1991).

**5 Instability of periodic solutions.** For $R > R_c$ there is a non-empty interval of values of the wave-number $k$ such that for each $k$ within that interval a space-periodic solution exists. The problem of stability of these solutions, subject to perturbations containing all wave-numbers was first studied in Eckhaus (1963), (1965), by a direct perturbation analysis on the class of problems (2.13). In later years, when the Ginzburg-Landau equation became available, these results were rederived and extended, taking as a starting point the G.-L. equation. This was done notably by Stuart & DiPrima (1978). The short-cut in analysis, introduced in this way, is considerable and is justified if one assumes that the G.-L. equation indeed governs *all* small solutions of the "full problems" (2.13). Let us summarize these results.

We have introduced
$$k - k_c = \varepsilon \sigma. \tag{5.1}$$
Using this variable, the main results are depicted in fig. 4.

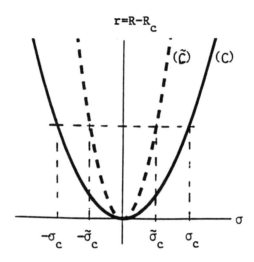

Fig. 4: Instability of space-periodic solutions.

Within the parabola $(C)$, containing possible periodic solutions there is an inner parabola $(\tilde{C})$ containing stable periodic solutions. Periodic solutions between $(C)$ and $(\tilde{C})$ are unstable. Particularly simple results follow if the lowest eigenvalue $\mu$ and the nonlinear coefficient $\beta$ are real. In that case
$$\tilde{\sigma}_c = \frac{1}{\sqrt{3}} \sigma_c. \tag{5.2}$$
These results are true under the condition
$$S := 1 + \frac{\nu_2 \mathrm{Im}(\beta_0)}{\mu_2 \mathrm{Re}(\beta_0)} > 0 \tag{5.3}$$
(For the definition of $\nu_2, \mu_2$ and $\beta_0$ see (3.6), (3.7), (3.8)). *If $S < 0$ then there are no stable periodic solutions.*

**6 Life on the Ginzburg-Landau manifold.** In the case of supercritical bifurcations the G.-L. equation can be tranformed into a canonical form given by
$$\frac{\partial \Phi}{\partial \tau} = \left\{ 1 - (1+ib)|\Phi|^2 \right\} \Phi + (1+ia) \frac{\partial^2 \Phi}{\partial \xi^2} \tag{6.1}$$
with $a, b \in \mathbf{R}$.

The search for solutions of (6.1), others than space-periodic solutions, has been an object of active research in recent years. As an example let us describe with a bit of detail *stationary quasi-periodic solutions*, which exist when $a = b = 0$ (Kramer & Zimmerman (1985)).

We write
$$\Phi(\xi, \tau) = \rho(\xi) e^{i\theta(\xi)} \tag{6.2}$$

with $\rho(\xi)$, $\theta(\xi)$ real. Equation (6.1) decomposes into

$$\rho\theta_{\xi\xi} + 2\rho_\xi\theta_\xi = 0 \qquad (6.3)$$

$$\rho_{\xi\xi} - \rho\theta_\xi^2 + \rho - \rho^3 = 0 \qquad (6.4)$$

Equation (6.3) can be integrated and one finds

$$\Omega = \rho^2\theta_\xi \qquad (6.5)$$

with $\Omega$ an arbitrary constant (at this stage). Substitution into (6.4) leads to a second integral

$$K = \rho_\xi^2 + \rho^2 - \frac{1}{2}\rho^4 + \frac{\Omega^2}{\rho^2} \qquad (6.6)$$

Hence we find a two-parameter family of solutions. From (6.5)

$$\theta(\xi) = \Omega \int^\xi \frac{d\xi'}{\rho^2(\xi')} \qquad (6.7)$$

For each $\Omega$ in the range

$$0 < \Omega < \sqrt{\frac{4}{27}}$$

the phase-portrait for $\rho(\xi)$ is sketched in fig. 5.

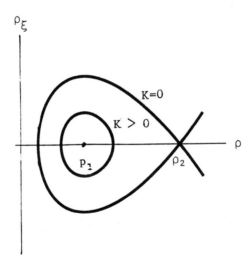

Fig. 5: Phase-protrait for $\rho(\xi)$.

For $K > 0$ the solutions (6.2) are (in general) quasi periodic. Notice that there are two stationary points $\rho_1$ and $\rho_2$ for $\rho(\xi)$ to which there correspond periodic solutions

$$\Phi_{1,2}(\xi) = \rho_{1.2}\exp\{ip_{1.2}\xi\} \qquad (6.8)$$

$$p_{1.2} = \Omega/\rho_{1.2}^2 \qquad (6.9)$$

It turns out that $p_1$ (corresponding to the center in fig. 4) is in the unstable region of the periodic solution, while $p_2$ (corresponding to the saddle) is in the stable region. As $\Omega$ increases toward $\sqrt{\frac{4}{27}}$ the two values $p_1$ and $p_2$ get closer to each other and coalesce at the stability boundary $p_1 = p_2 = \frac{1}{\sqrt{3}}$. Kramer & Zimmerman (1985) argue that the quasi-periodic solutions are all unstable. A rigorous demonstration of this conjecture has not been given yet.

Other interesting solutions of (6.1) in the case $a = b = 0$ are *fronts* studied extensively in Collet & Eckmann (1990). These are solutions of the structure

$$\Phi(\xi, \tau) = \nu(\xi - c\tau) \qquad (6.10)$$

which are such that $\nu(z)$ tends to zero as $z \to +\infty$, and to a nonzero value as $z \to -\infty$. It turns out that such solutions exist for all values of $c$.

Other analytical results on quasi-periodic, homoclinic and heteroclinic solutions, in particular in the complex case $a, b \neq 0$, can be found in Bernhoff (1988), Doelman (1989), Holmes (1986). A particularly intriguing case arises when

$$1 + ab < 0 \qquad (6.11)$$

so that all space-periodic solutions of (6.1) are unstable. This case has been studied by finite-dimensional projections

$$\Phi(\xi, t) = \sum_{n=-N}^{N} Z_n(\tau) e^{inq\xi} \qquad (6.12)$$

with $q$ a (bifurcation-) parameter. This leads to the system of equations

$$\frac{dZ_n}{d\tau} = \left[1 - n^2 q^2 (1 + ia)\right] Z_n - (1 + ib) \sum Z_k Z_l Z_m^* \qquad (6.13)$$

with the summation over all $k, l, m$ such that

$$k + l + m = n \qquad (6.14)$$

$$|k|, |l|, |m| \leq N \qquad (6.15)$$

Ghidaglia & Heron (1987) derived upper- and lower- bounds for the global attractor of (6.13) (see also Doering *et al* (1988)). Other interesting estimates on $Z_n$'s are in Doelman (1990), (1991). Numerical studies of (6.13) are reported in Moon, Huerre & Redekopp (1983), Keefe (1985), Doelman (1990), (1991) and others. They all display chaotic behavior.

**7 Degenerate bifurcations: strong selection or rejection of space-periodic patterns.** All of our preceding discussion is based on the (implicit) assumption that $\text{Re}\beta(k, R)$ is not small at $(k, R)$ near critical conditions. It is by this assumption that the scaling of the solutions is consistent (see (2.11), (2.14)). It does happen however in various applications that, as a result of lengthy calculations, one finds that $\text{Re}\beta(k, R)$ is zero near $(k_c, R_c)$. This case has been analysed in Eckhaus & Iooss (1989), where one can also find a list of references to physical problems exhibiting such behavior (convection in rotating fluids, Jeffery-Hamel flow, Taylor-experiment with counter- rotating cylinders, Blasius boundary-layer, convection in binary fluids).

The first step of the analysis is to take the amplitude equation to higher order:

$$\frac{dA_k}{dt} = \mu(k, R) + \hat{\varepsilon}\beta(k, R)|A_k|^2 A_k + \hat{\varepsilon}^4 \gamma(k, R)|A_k|^4 A_k \qquad (7.1)$$

where $\hat{\varepsilon}$ is an as yet unknown scaling of the solutions.

In the situation which we consider here

$$\text{Re}\beta(k,R) = \delta\beta_0^{(r)} + (k-k_c)\beta_1^{(r)} + \ldots \quad (7.2)$$

with $\delta$ a second small parameter, which arises in the results of computations. Let us assume that

$$\text{Re}\gamma = -\gamma_0^{(r)} + O(k-k_c), \quad \gamma_0^{(r)} > 0 \quad (7.3)$$

Then it turns out that

$$\hat{\varepsilon} = \sqrt{\varepsilon}, \quad \varepsilon = \sqrt{|R-R_c|}. \quad (7.4)$$

Hence, the bifurcations in the degenerate case are larger then the classical ones.

Studying (7.1) one finds a rich variety of bifurcating solutions, including supercritical *and* supcritical bifurcations. In supercritical conditions the region of existence of space-periodic solutions may include parts of the $(R,k)$-plane in which solutions decay by linear theory (for details see publication quoted above).

To study the stability of periodic solutions a modulation equation is needed. It turns out that for most purposes of the analysis (and for simplicity of presentation) one can take

$$\delta = \varepsilon \quad (7.5)$$

By the algorithmic derivation the following modulation equation is obtained

$$\begin{aligned}\frac{\partial\Phi}{\partial t} &= i\varepsilon\left\{\beta_0^{(i)}|\Phi|^2 + \varepsilon\gamma_0^{(i)}|\Phi|^4\right\} + \varepsilon^2\left\{\mu_0 + \beta_0^{(r)}|\Phi|^2 - \gamma_0^{(r)}|\Phi|^4\right\}\Phi \\ &\quad - i\varepsilon^2\left\{\hat{\beta}_1\frac{\partial\Phi}{\partial\xi}|\Phi|^2 - \check{\beta}_1\Phi^2\frac{\partial\Phi^*}{\partial\xi}\right\} - (\mu_2 + iv_2)\frac{\partial^2\Phi}{\partial\xi^2} + o(\varepsilon^2)\end{aligned} \quad (7.6)$$

where we have used the obvious notation

$$\beta_0^{(i)} = \text{Im}[\beta(R_c,k_c)] \quad (7.7)$$

$$\gamma_0^{(i)} = \text{Im}[\gamma(R_c,k_c)] \quad (7.8)$$

Notable differences with the classical case are as follows:

- If $\beta_0^{(i)} \neq 0$ (and is not small) then the oscillations of $\Phi$ are on a more rapid time-scale ($\tau' = \varepsilon t$) than the evolution of $|\Phi|$, for which the natural time is $\tau = \varepsilon^2 t$.

- The equation contains first order derivatives with respect to $\xi$.

- One finds that $\hat{\beta}_1 + \check{\beta}_1 = \beta_1^{(r)}$, however, $\hat{\beta}_1 - \check{\beta}_1$ does not follow from the algorithmic derivation and has to be computed from the "full" problem (See Doelman (1990)).

In the case that $\beta_0^{(i)} = O_s(1)$ one has, as condition for existence of stable periodic solutions

$$v_2\beta_0^{(i)} > 0 \quad (7.9)$$

which is similar to the condition $1 + ab > 0$ in the classical case. But further results are very much different. We write again

$$r = R - R_c, \quad k - k_c = \varepsilon\sigma. \quad (7.10)$$

The main result is as follows: *Stable space-periodic solutions are restricted to a $O(\varepsilon^{\frac{1}{2}})$ neighbourhood of a curve $\Gamma$ in the $r - \sigma$ plane.*

96  ECKHAUS

For the further discussion we introduce a crucial quantity $s$, through

$$s := \left[\frac{1}{2}\beta_1^{(r)}\right]^2 + \gamma_0^{(r)}\mu_2 \qquad (7.11)$$

For $s < 0$, two examples of disposition of the curve $\Gamma$ are given in fig. 6

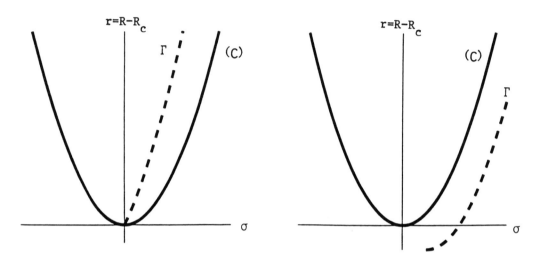

Fig. 6: Strong pattern selection for $s < 0$.

We see that for each value of $R$ in the interval of existence of the curve $\Gamma$ a wavelength of the stable space-periodic solution is selected, with only a very small incertitude. This is quite unlike the classical case, where a whole interval of stable solutions exists.

When $s > 0$, a new phenomena appears. The curve $\Gamma$ in that case is sketched in fig. 7.

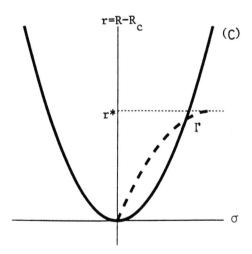

Fig. 7: Strong pattern selection and rejection for $s > 0$.

The curve $\Gamma$ stops at $R - R_c = r^*$, with $r^* = O(\varepsilon^2)$. In this situation one has a strong pattern selection for $R_c < R < R_c + r^*$, while for $R > R_c + r^*$ all space-periodic solutions are unstable.

The case occurs notably for the flow in Blasius boundary-layer (by quasi-parallel approximation). This problem has some further complications but has nevertheless been fully treated by a slight generalization of the analysis described here (see Eckhaus & Iooss (1989)).

*Concluding remarks.*

- Admitting modulations in two space-direction (and using Squires theorem) Doelman (1990) found that $\Gamma$ is a non-planar curve in the $(R - k - l)$ space ($l$ being the wave number in the transversal unbounded direction). This means that stable periodic solutions are 3-dimensional.

- In Doelman & Eckhaus (1991) the degenerate bifurcation problem was studied for the case in which the amplitude equation (7.1) has real coefficients, or coefficients with small imaginary parts. In the real case one finds again stationary space-periodic solutions *and* quasiperiodic solutions. As in classical cases, the disappearance of quasi-periodic solutions is strongly related to the stability boundaries of the periodic solutions. In the case of small imaginary parts a rich collection of solutions of the modulation equation (7.6) has been found.

# References

[1.] A.J. Bernhoff, *Slowly varying fully nonlinear wavetrains in the Ginzburg-Landau equation.* Phys. D. 30 (1988).

[2.] F. Calogero, *Why are certain PDE's both widely applicable and integrable?* In: What is Integrability, V.E. Zakharov Ed. Springer Verlag (1991).

[3.] F. Calogero and W. Eckhaus, *Nonlinear evolution equations, rescalings, model PDE's and their integrability.* I and II, Inverse Problems 3 (1987) and 4 (1988).

[4.] P. Collet & J.-P. Eckmann, *Instabilities and fronts in extended systems* Princeton University Press (1990).

[5.] P. Collet & J.-P. Eckmann, *The time dependent amplitude equation for the Swift-Hohenberg problem.* Comm. Math. Physics 132 (1990)

[6.] R.C. DiPrima, W. Eckhaus, L.A. Segel, *Nonlinear wave-number interactions at near-critical flows.* Journ. Fluid Mech. 49 (1971).

[7.] A. Doelman, *Slow time-periodic solutions of the Ginzburg-Landau equation* Phys. D. 40 (1989).

[8.] A. Doelman, *On the nonlinear evolution of patterns.* Thesis, The University of Utrecht, Mathematics Department (1990).

[9.] A. Doelman, *Finite dimensional models of the Ginzburg-Landau equation* Nonlinearity (1991).

[10.] A. Doelman & W. Eckhaus, *Periodic and quasiperiodic solutions of degenerate modulation equation.* To appear in Physica D (1991).

[11.] C.R. Doering, J.D. Gibbon, D.D. Holm & B. Nicolaenco, *Low dimensional behaviour in the complex Ginzburg-Landau equation* Nonlinearity (1988)

[12.] P.G. Drazin & W.H. Reid, *Hydrodynamics Stability* Cambridge University Press (1981).

[13.] W. Eckhaus, *Stabilité des solutions périodiques.* J. Mécanique II (1963).

[14.] W. Eckhaus, *Studies in nonlinear stability theory.* Springer Verlag (1965).

[15.] W. Eckhaus. Manuscript in preparation (1991).

[16.] W. Eckhaus & G. Iooss, *Strong selection or rejection of spatially periodic patterns in degenerate bifurcations.* Physica D. 39 (1989)

[17.] J.M. Ghidaglia & B. Heron, *Dimension of the attractors associated to the Ginzburg-Landau equation.* Phys. D. 28 (1987)

[18.] V.L. Ginzburg & L.D. Landau: Paper nr. 73 in Collected Papers of Landau, Pergamon Press (1965).

[19.] A. van Harten, *On the validity of Ginzburg-Landau's equation.* Dept. of Mathematics, Utrecht, Preprint 635 (1990). To appear in Journal of Nonlinear Science (1991).

[20.] P. Holmes, *Spatial structure of time-periodic solutions of the Ginzburg-Landau equation.* Phys. D. 23 (1986)

[21.] G. Iooss, A. Mielke & Y. Demay, *Theory of steady Ginzburg-Landau equation in hydrodynamic stability problems.* Europ. Journ. Mech. B/Fluids (1989).

[22.] G. Iooss & A. Mielke, *Bifurcating time-periodic solutions of Navier-Stokes equations in infinite cylinders.* To appear in Journ. of Nonlinear Sciences (1991).

[23.] L.R. Keefe, *Dynamics of perturbed wavetrain solutions to the Ginzburg-Landau equation.* Stud. Appl. Math 73 (1985).

[24.] P. Kirrmann, G. Schneider, A. Mielke. *The validity of modulation equations for extended systems with cubic nonlinearities.* Preprint, Mathematischer Institut A, Stuttgart (1991).

[25.] L. Kramer & W. Zimmerman, *On the Eckhaus Instability for spatially periodic patterns.* Physica D. 16 (1985).

[26.] L.D. Landau & E.M. Lifschitz, *Fluid mechanics* English edition by Pergamon Press (1959).

[27.] W. Mielke, *Reduction of PDE's with domains with several unbounded direction.* Mathematisches Institut A, stuttgart. Preprint (1991).

[28.] H.T. Moon, P. Huerre, L.G. Redekopp, *Transition to chaos in Ginzburg-Landau equation* Phys. D. 7 (1983)

[29.] A.C. Newel & J.A. Whitehead, *Finite bandwidth finite amplitude convection.* Journal Fluid Mech. 38 (1969).

[30.] L.A. Segel, *Distant side walls cause slow amplitude modulation in cellular convection.* Journ. Fluid Mech. (1969).

[31.] J.T. Stuart & R.C. DiPrima, *The Eckhaus and the Benjamin-Feir resonance mechanisms.* Proc. Roy. Soc. London A 362 (1978).

# Chapter 8
# MODELLING THE SOLIDIFICATION OF POLYMERS: AN EXAMPLE OF AN ECMI COOPERATION*

A. FASANO
Università di Firenze
Italy

**1. About ECMI.** First of all I wish to thank the organizers for their invitation to speak at ICIAM 91. This honor takes in my case a particular meaning. As the president of the European Consortion for Mathematics in Industry (ECMI) I like to believe that my presence here proves that ECMI is now recognized also as a research promoting institution.

ECMI is a very young association: its foundation was announced during the previous ICIAM conference in Paris. During these first few years ECMI has successfully accomplished some of its goals. For instance the educational activities (postgraduate program, continuing education, modelling weeks, etc.) are running at regime and I am very proud to say that they are stimulating considerable interest in the U.S. too.

ECMI is different from any other national or international mathematical society, since it addresses almost exclusively to Industrial Companies and Academic Institutions and its efforts are concentrated on promoting their cooperation in the area of the so-called Industrial Mathematics, operating on a European scale. In such a framework, our Educational Program (coordinated by an Educational Committee and mainly funded by EEC) is an important service provided to Industries and in general to the scientific community.

Needless to say, some of the tasks ECMI is facing are lifetime commitments. Everybody who has frequent contacts with Industrial Companies knows very well their attitude towards the idea of discussing their problems with academic mathematicians. There is indeed an entire spectrum of reactions, ranging from immediate enthusiasm (in very few cases) to radicated adversion. Of course the average varies in different countries and the overall average is still far from our wish.

However, the number and the quality of the research projects presently under investigation in the various ECMI centers are the best evidence of the fact that there is a great deal of interest in this direction and I am sure that ECMI is to play an important role in the research on mathematical problems coming from Industry.

I regret I cannot proceed any further in this sketchy illustration of ECMI, since my presentation is supposed to be about a specific technical problem, but before coming to that I like to remenber that from the very beginning ECMI received a great impulse as a catalyst for research by Professor Hansjörg Wacker, who was the president of ECMI during 1990 and who died untimely a few months ago. We miss not only his apassionate work, but also the good man and the dear friend he was. This talk is dedicated to his memory.

---

*Work partially supported by the Italian CNR.

## 2. The solidification of polymers.

A. <u>Participating Institutions.</u>   The core of this presentation deals with a complex phenomenon, namely the (partial) crystallization of polymers in a variable thermal field.

The research (still in progress) is being performed by various ECMI members and is also sponsored by the Italian National Research Council (CNR).

The industrial partner is HIMONT-ITALIA, a company based in Ferrara (Italy), whose laboratories have provided the experimental data. It was a nice experience working with Dr. S. Marzullo from Himont, who gave us continous assistance and advise about the physical background and about the modelling itself.

In Florence, at the Math. Department U. Dini, two groups worked jointly the first one (including D. Andreucci, M. Primicerio and myself) had the main responsibility of the mathematical model and of the related theoretical work, as well as of the general coordination of the research. The second one (composed by M. Bianchini and A. Pasquali) has produced much work about the identification of some key parameters.

At the I.A.N. in Pavia M. Paolini (CNR) and C. Verdi (Math. Department, Univ. of Milano) performed the numerical computation for the full model in 3-D geometry, developing a rather complex and flexible code.

At SASIAM (Bari) C. Capasso with L. Borrelli and Li Peng made a numerical simulation of the isothermal process, which proved to be a basic step in the construction of the mathematical model.

Prof. G.C. Alfonso (Univ. of Genova) was also participaing and gave valuable suggestions.

The remainder of this section is a brief illustration of the physical problem to be studied. The mathematical model will be introduced gradually. After some general remarks (Sect. 3) the isothermal case is considered first (Sect. 4 and 5). A short survey of the literature (Sect. 6) precedes the statement of the full model (Sect. 7). Numerical calculations are presented in Sect. 8. The presentation is closed by an overview of open problems (Sect. 9) and by some conclusive remarks (Sect. 10).

B. <u>The physical process.</u>   Let me recall that a polymer molecule is a long chain of identical elements (monomers). Such molecules are produced by means of a particular chemical process (polymerization). In a given polymer the number of monomers in a molecule is statistically distribuited around some average (typically $1500 \div 2000$) and there will also be considerably shorter molecules (oligomers).

Thus a first important remark is that polymerization always produces some degree of heterogeneity, which in part explains some of the peculiar features of polymer solidification. It is superfluous to recall that there is a great variety of polymers and that their applications are incredibly diversified, from objects used in the every day life to components destined to sophisticated technologies. There is a growing interest in their medical applications (e.g. for fixing or replacing parts of the skeleton). It is therefore clear that it is sometimes crucial to obtain products with well defined mechanical properties.

For a given polymer we can define a melting temperature $T_m$ such that for all temperatures $T > T_m$ the polymer is in the liquid phase (its molecules moving freely).

Change of phase begins when the temperature drops below $T_m$. The main characteristic of the solidification process is that crystals may be generated only in some temperature range $(T_g, T_m)$, $T_g$ being called the glassy transition temperature. Such an interval can be quite large: for instance $T_g = 333°K$, $T_m = 501°K$ for Nylon-6 at normal pressure.

Crystallization proceeds in two stages:
1) Nucleation,
2) Crystal growth.

Both nucleation rate and growth rate depend on temperature. They can be described as bell-shaped functions in the interval $(T_g, T_m)$ and vanishing outside. Typical curves are reported in Fig. 2.1.a, b.

For temperatures below $T_g$ no more crystals are formed, the existing ones stop growing and the residual liquid phase is converted into an amorphous solid. More precisely we should say that it acquires an enormously high viscosity, characteristic of the glassy state.

The number of crystals in the solidified product and their size distribution can have a great influence on the mechanical properties. For this reason it is necessary to develop a mathematical model that is able to predict also the microscopic structure after solidification.

Another important aspect of the process, that we are not going to discuss further, is the dependence of $T_g, T_m$ on the pressure. This becomes quite important in mould injection processes in which very high pressure pulses are applied in order to complete mould filling, thus shifting $T_g, T_m$ towards higher values.

Before we discuss our mathematical model, we need to describe crystallization in more detail.

I.<u>Nucleation</u>. A nucleus is formed when a segment of a molecule takes a regular pattern (it may also happen that more than one molecule enter the same nucleus). The size of a nucleus depends on the temperature at which it is generated even though there are no precise data over the entire range $(T_g, T_m)$, we can say that the typical diameter of a nucleus is around 100Å, so that its volume is frequently neglected in computing the total volume occupied by crystals.

In this talk only spatially homogeneous nucleation will be considered.

II. <u>The spherulites</u>. A "crystal" in a polymer is a much more complex structure than a set of molecules arranged in a specific lattice.

What grows out of a nucleus is instead an aggregate of "crystalline ribbons" (Fig. 2.2) proceeding in radial directions and separated by amorphous regions. The resulting crystalline unit is called a spherulite.

The presence of amorphous phase within the spherulites is the cause of the fact that complete crystallization is never achieved.

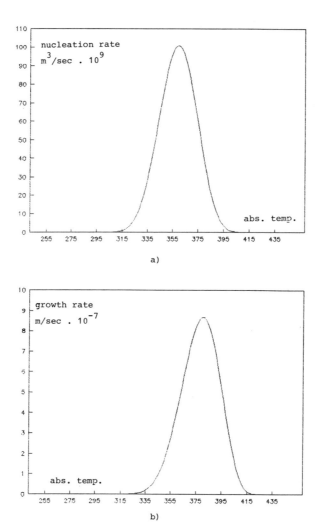

Fig. 2.1. Approximate nucleation rate (a) and crystal growth rate (b) for polypropilene as functions of temperature, extrapolated from available data.

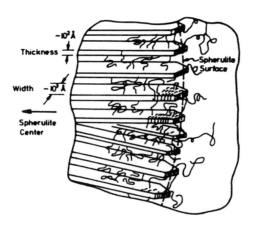

Fig. 2.2. Schematic representation of a spherulite.

## 3. The mathematical model: preliminary remarks.

The main quantities to be predicted as functions of space and time are
a) the volume fraction occupied by the crystalline phase (crystallinity),
b) the size distribution of the spherulites,
c) the temperature field.

Of course we are referring to quantities averaged over a representative elementary volume. Moreover, we shall not deal with the microscopic structure of the spherulites, which will be described in the model by spheres with a volume equivalent to their crystalline component.

We anticipate that a remarkable feature influencing crystal growth is the fact (well known in the literature) that the spherulites can impinge with each other.

Another relevant aspect is the fact that crystallization and temperature evolution are strongly coupled, since the local value of temperature determines the nuceation rate and the spherulites growth rate, while the formation of crystals acts as a distributed heat source, owing to the release of latent heat.

In order to formulate the model, it will be convenient to proceed step by step, dealing first with the simplest case, i.e. isothermal crystallization.

## 4. Isothermal crystallization.

Isothermal processes can be experimentally investigated by placing a small sample of the polymer in conctact with a thermostat (the time during which the sample is cooled down to the selected temperature should be as short as possible in order to reduce nucleation in the transient regime). The thermostat will remove the crystallization latent heat from the sample. Such a quantity can be measured and provides a calorimetric determination of the crystallinity as a function of time.

The fundamental advantage of isothermal crystallization is that the nucleation rate and the growth rate of the spheres equivalent to the crystalline component of the spherulites are constant. Such quantities will be denoted by $\dot{N}_0$ and $\dot{R}_0$, respectively, and for each polymer can be deduced from experimental curves like those in Fig. 2.1.

Of course the phenomenon is not as simple as it can appear from the above statement, because it is also influenced by other effects of different nature, namely:
a) geometrical effects, due to impingement, progressively reducing the increase of crystallinity,
b) chemical effects, to be attributed to the fact that as crystallization proceeds the residual liquid phase is enriched of the less crystallizable components (oligomers).

In our model we try to take both phenomena into account by introducing depressing factors for the nucleation and growth rate, wich tend to zero when the crystallinity $w$ approaches its maximum value $w_m$.

More precisely we write

$$(4.1) \qquad \dot{N} = \dot{N}_0 \left(1 - \frac{w}{w_m}\right)^p ,$$

$$(4.2) \qquad \dot{R} = \dot{R}_0 \left(1 - \frac{w}{w_m}\right)^q ,$$

where $p$ and $q$ are positive constants.

One first question is: *disregarding non-geometrical effects, can we choose $p$ and $q$ in some appropriate way so that (4.1), (4.2) provide a correct description of the phenomenon?*

A natural choice for $p$ appears to be $p = 1$. Indeed the factor $1 - \frac{w}{w_m}$ is nothing but the volume fraction available for nucleation. On the contrary, there is no apparent reason indicating that (4.2) makes sense for some value of $q$. The determination of such a value of $q$ was indeed a succesfull step of this research.

Given $\dot{N}_0$, $\dot{R}_0$, a rigorous determination of the function $w(t)$ is provided by Avrami's formula (with $w_m = 1$)

$$(4.3) \qquad w(t) = 1 - e^{-\tilde{w}(t)},$$

where $\tilde{w}(t)$ is the "virtual crystal volume", i.e. the volume calculated by letting the crystals grow indefinitely (overlapping whith each other) and counting also crystals generated in regions already occupied by other crystals.

Avrami's formula was first derived by Kolmogorov in 1937 [14] on the basis of a simple probabilistic argument, leading to the conclusion that

$$(4.4) \qquad \frac{dw}{dt} = \frac{d\tilde{w}}{dt}(1-w),$$

from which (4.3) follows by integration (with the initial condition $w(0) = 0$).

Avrami obtained the formula independently and applied it in a series of papers [6, 7, 8].

Most of the models developed in the literature take Avrami's formula (or (4.4)) as a basis and try to adapt it to the case of polymers (see Sect. 6 below).

Avrami's formula is particularly simple, since $\tilde{w}(t)$ can be calculated easily. Neglecting the volume of the nuclei we obtain

$$(4.5) \qquad \tilde{w}(t) = \frac{\pi}{3} \dot{N}_0 \dot{R}_0^3 t^4.$$

The approach based on (4.1), (4.2) is by far more flexible for two reasons:
(i) by varying the exponents $p$ and $q$ we can include non-geometric effect. For instance, taking $p > 1$ reduces the probability of nucleation, particularly towards the end of the process, and increasing $q$ has a similar effect on the growth rate,
(ii) by means of (4.1), (4.2) we can predict not only the evolution of crystallinity (as we shall see even in non-isothermal cases), but we can also calculate the size distribution of our equivalent spheres, which is roughly the size distribution of the spherulites.

With these objectives in mind we revert to our original question of finding $q$ such that (4.1) (with $p = 1$) and (4.2) match Avrami's formula (for $w_m = 1$).

Let us first write down the consequences of (4.1), (4.2) for the isothermal model.

If at the time $t$ the symbol $\rho(t, \tau)$ denotes the radius of a "crystal" (i.e. an equivalent sphere) born at the time $\tau$, namely

$$(4.6) \qquad \rho(t, \tau) = r_0 + \int_\tau^t \dot{R}(s)\, ds,$$

the crystallization rate is

$$(4.7) \qquad \frac{dw}{dt} = \frac{4}{3}\pi r_0^3 \dot{N}(t) + 4\pi \dot{R}(t) \int_0^t \dot{N}(\tau) \rho^2(t, \tau)\, d\tau.$$

Neglecting $r_0$ and using (4.1), (4.2), we arrive at the following integro-differential equation for $w$:

$$(4.8) \qquad \frac{dw}{dt} = 4\pi \dot{R}_0 \left(1 - \frac{w}{w_0}\right)^q \cdot \int_0^t \dot{N}_0 \left(1 - \frac{w}{w_m}\right)^p \left[\int_\tau^t \dot{R}_0 \left(1 - \frac{w}{w_m}\right)^q ds\right]^2 d\tau$$

with the inital condition $w(0) = 0$.

At SASIAM's laboratories a spectacular numerical simulation (for $w_m = 1$) was performed through the following steps [2]:

1) random generation of nuclei ($r_0 = 0$), at a constant rate and according to some statistics, discarding those nuclei that happen to be created at points already occupied by crystals (represented by spheres),
2) constant radial growth until two spheres (or circles or segments: the simulation was repeated for 2-D and 1-D) collide. The computer visualizes the stopping points, which lie on hyperboloids (hyperbolas in 2-D), as it can be shown easily,
3) computation of the crystal volume fraction until the process is completed (in practice the process is stopped when $1 - w$ is negligibly small).

The series of pictures in Fig. 4.1 illustrates a 2-D simulation with a small number of crystals generated (so that the impingement boundaries are clearly visible).

The resulting function $w(t)$, calculated for several runs and over a large spectrum of values for $\dot{N}_0$, $\dot{R}_0$, has been compared with the one deduced from Avrami's formula and with the solution of (4.8), taking $p = 1$ and trying to identify $q$.

Of course there is an excellent agreement between the results of the simulation and Avrami's formula (since they reproduce the same process). The agreement with (4.8) is surprisingly good if we take $q = 0.7$, as shown in the two cases of Fig. 4.2a, b (equally good results were obtained for all the other tested values of $\dot{N}_0$, $\dot{R}_0$).

Although we have no theoretical explanation, we can conclude that (4.1), (4.2) with $p = 1$ and $q = 0.7$ give the correct description of impingement.

It is well known, however, that real polymer crystallization processes are generally slower than it is predicted by Avrami's formula. Thus we have now to reconsider the parameters identification, taking into account the experimental data.

**5. Matching experimental data for isothermal crystallization.** As we have recalled in the previous section, there are effects of chemical and physical origin tending to slow down crystallization.

In the spirit of the approach based on (4.1), (4.2), we can try to match the experimental data available for polypropilene by redefining $p$ and $q$.

Here we found that there is some degree of uncertainty, since we can let $p$ and $q$ vary almost independently (with $q \geq p$) between 1 and 1.25 and adjust the value of $\dot{R}_0$ accordingly, always getting a satisfactory agreement with the data (see [1] where a least squares method was used). A typical result is shown in Fig. 5.1.

We concluded that in order to obtain a sharper information, a more detailed experimental description is needed of the early stage of crystallization, when impingement is not important and the phenomenon is dominated by crystal growth, thus allowing a reliable identification of $\dot{R}_0$.

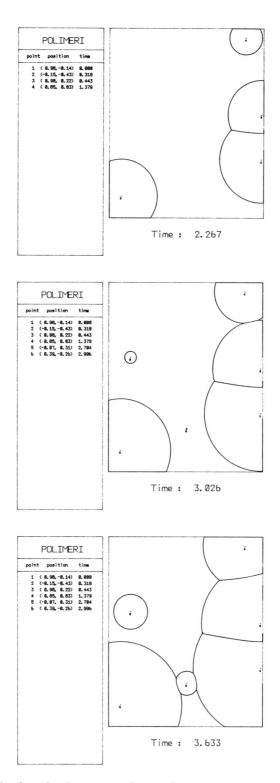

Fig. 4.1a-c. Growth of randomly generated crystals in a unit square
(2-D computer simulation). The sequence of pictures refers to the following times (sec.):
2.267 (a), 3.026 (b), 3.633 (c).

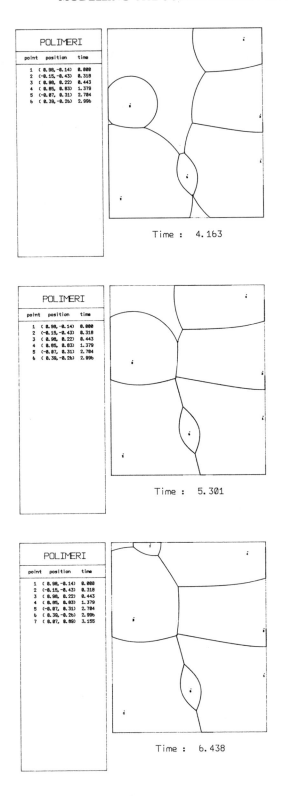

Fig. 4.1d-f. Continuing the sequence of pictures of Fig.4.1a-c for the times (sec.): 4.163 (d), 5.301 (e), 6.438 (f).

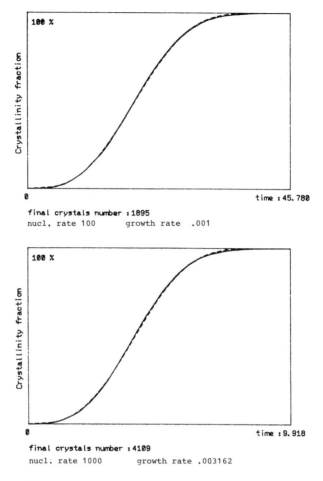

Fig. 4.2a, b. Crystallinity vs. time according to Avrami's formula (dotted lines), to the stochastic computer simulation (solid lines), and to formula (4.8) (broken lines) with $p = 1$, $q = 0.7$ for two different sets of values of nucleation rate and growth rate. Time in seconds. Simulations in a square whose side is taken as the unit length. Units for nucleation and growth rates are expressed accordingly.

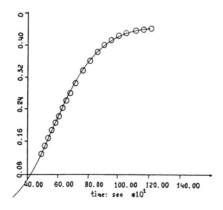

Fig. 5.1. Crystallinity vs. time as predicted by (4.8) with $p = q = 1$ (continuos curve), compared with experimental data for polypropilene for an isothermal process at $134°C$.

It could be remarked that after all such an incertainty is not a bad feature of the model. On the contrary we can say that the model does not appear to be affected in a very critical way by the choice of the parameters, at least as far as we are content with a prediction of crystallinity.

**6. A short review of other models.** Owing to the importance of the problem it is little surprise that a considerable amount of work has been done on this subject.

We have already recalled the papers by Kolmogorov and Avrami about the basic question of impingement.

A modification of Avrami's formula proposed by Tobin in [26, 27] consists in replacing (4.4) with

$$(6.1) \qquad \frac{dw}{dt} = \frac{\tilde{w}}{dt}(1-w)^2,$$

thus leading to

$$(6.2) \qquad w = (1-w)\tilde{w}$$

instead of (4.3).

Other models guess a differential equation for $w$, on a heuristic basis. In this class we find the models of Malkin [15, 16]

$$(6.3) \qquad \frac{dw}{dt} = (\frac{4}{3}\pi \dot{N}_0 r_0^3 + Cw)(1-w)$$

($C$ being a positive constant) and the model proposed by Berger and Schneider [9]

$$(6.4) \qquad \frac{dw}{dt} = Kw^{1-\beta}(1-w)^{1+\beta}$$

with $K > 0$ (depending on temperature) and $\beta \in (0,1)$ (other references for this kind of approach are [23] and [10]).

We will come back to (6.4) for an interesting comparison with our model.

The replacement of the factor $(1-w)$ with $\left(1 - \frac{w}{w_m}\right)$ was first suggested by Mandelkern [17].

Mazzullo [18], [20] has considered a different scheme with free boundaries.

Generalization of Kolmogorov's argument to age-dependent crystal growth are studied in [21], [22].

Non-isotropic growth has been considered in [29, 30].

An interesting comparison among different models can be found in [2].

It is worth mentioning the so-called isokinetic assumption (already formulated in Avrami's papers).

A crystallization process is said to be in the isokinetic range if the ratio $\dot{N}/\dot{R}$ is constant, i.e. $\dot{N}/\dot{R} = \dot{N}_0/\dot{R}_0$.

According to (4.1), (4.2) this means that we take $p = q$.

In such a case it can be shown that (4.8) yields

$$\text{(6.5)} \qquad \frac{dw}{dt} = K\left(\frac{w}{w_m}\right)^{\frac{3}{4}}\left(1 - \frac{w}{w_m}\right)^q,$$

which coincides with (6.4) if one takes $q = 1.25$ in (6.5) and $\beta = 0.25$ in (6.4). This remark is due to Mazzullo [19].

Although such coincidence is not in itself a proof that either model is undisputably correct, the compatibility between the two different approaches is nevertheless encouraging, since - as far as we know - the choice $p = q = 1.25$ is not in contrast with the experimental data.

We conclude this section by recalling that Avrami's rule has been used extensively in the study of solid-solid transitions (see e.g. [11], [12], [Vi 87]) with the help of Scheil's additivity rule [24] when dealing with non-isothermal processes.

One of the main advantages of our model is that we can pass to the non-isothermal case in a natural way, just by inserting in (4.1), (4.2) the temperature dependence of $\dot{N}_0$, $\dot{R}_0$, thus keeping in the model the information about the size distribution of the equivalent spheres modelling the spherulites.

## 7. Non-isothermal crystallization.

It is absolutely obvious that the crystallinity at the end of solidification depends in a crucial way on the cooling process. If the sample is quenched below the crystallization range the resulting product will be basically amorphous, while high values of $w$ can be achieved only if enough time is spent at temperatures corresponding to large values of $\dot{N}_0$, $\dot{R}_0$.

All we have to do now is to draw the consequences of (4.1), (4.2), taking into account the temperature dependence of $r_0$, $\dot{N}_0$, $\dot{R}_0$.

For a crystal nucleated at point $x$ at time $\tau$, the radius at time $t$ will be

$$\text{(7.1)} \qquad \rho(t,\tau,x) = r_0(T(x,\tau)) + \int_\tau^t \dot{R}_0(T(x,s))\left[1 - \frac{w(x,s)}{w_m}\right]^q ds.$$

For simplicity we disregard the fact that also $w_m$ can depend on $T$.

The rate of change of $w$ can be written as follows

$$\text{(7.2)} \qquad \frac{\partial w}{\partial t} = \frac{4}{3}\pi r_0^3(T(x,t))\dot{N}_0(T(x,t))\left[1 - \frac{w(x,t)}{w_m}\right]^p$$
$$+ 4\pi \dot{R}_0(T(x,t))\left[1 - \frac{w(x,t)}{w_m}\right]^q \int_0^t \dot{N}_0(T(x,\tau))\left[1 - \frac{w(x,\tau)}{w_m}\right]^p \rho^2(t,\tau,x)\,d\tau,$$

with $\rho(t,\tau,x)$ given by (7.1).

The second governing equation expresses the thermal energy balance

$$\text{(7.3)} \qquad c(x,t)\frac{\partial T}{\partial t} = \text{div}[k(x,t)\nabla T] + L\frac{\partial w}{\partial t}$$

where $c$ is the thermal capacity, $k$ the heat conductivity and $L$ denotes the latent heat of crystallization.

The system above must be supplemented by the initial and boundary conditions pertaining to the specific cooling process.

In [4] the resulting problem is proved to have a unique global classical solution. The solution is obtained as the fixed point of an appropriate mapping in a suitable Banach space.

However, such a result is limited to cooling processes (the temperature decreasing everywhere). Cases in which the temperature increases (at least locally) are also of great interest. Indeed crystallization can be restarted in a polymer by taking it back to the appropriate temperature range. A partial solution of the problem including non-monotonic time variation of temperature is given in [5], but the problem deserves a deeper analysis.

Another generalization (which however looks no difficult) is to allow $c$ and $k$ in (7.3) to depend on $w$ and on $T$.

## 8. Numerical results for the full non-isothermal model.

The results, which will appear in [3], are summarized below. The first step in [3] was to reformulate (7.1), (7.2) in terms of a system of o.d.e.'s. This transformation, already used in [25], appears more convenient for the numerical treatment and presents also some interesting physical interpretation. The study of [3] was based on a rigorous theoretical analysis of the convergence and of the stability of the method implemented.

The domain (simulating a mould) is obtained by reflecting the planar figure in Fig. 8.1 around the horizontal axis and rotating it around the vertical axis. The polymer considered is polypropilene.

Fig. 8.1. Quarter section of the axisymmetric domain used for the numerical computation of the full non-isothermal model. The domain is obtained by reflection accross the horizontal axis and by rotation around the vertical axis.
The mesh used in the f.e.m. is also shown.

The initial temperature is taken to be the melting temperature of polypropilene ($T_m = 180°C$) and at the boundary it is assumed a linear radiation law, the external temperature being fixed at $27°C$.

Figs. 8.2 a-c show the lines of constant temperature after $50, 100$ and $155$ seconds, respectively.

The corresponding lines $w = const.$ are shown in Figs. 8.3 a-c.

We remark the appearance of an internal maximum for $w$ in a region in which thermal conditions are most favourable for crystallization.

The same sort of behaviour can be observed for other quantities like the density of nuclei and the average radius of crystals (Figs. 8.4 a-c and 8.5 a-c).

## 9. Open question and generalization.

The model presented in the preceding sections refers to solidification of a polymer with a single species of crystals, with spatially homogeneous properties, and occuring at a constant pressure.

It would be not difficult to include effects such as heterogeneous nucleation and polymorphism (i.e. the presence of more than one crystal species with diversified kinetics).

What appears instead more complicated is to incorporate the effects of pressure variations, which could cause the temperature to cross several times the crystallization range in both directions. This problem appears to be naturally related to the question of modelling reheating processes (at constant pressure), which, as we have said, has not yet received a completely satisfactory solution.

Another effect not described in the model is the so-called secondary crystallization, i.e. the possible formation of crystalline structures in the amorphous phase within the spherulites.

On the experimental side it would be necessary to have a precise description of crystal growth far from impingement, as well as series of data concerning the evolution of all the main quantities during a cooling process.

Besides all that, perhaps the main open question is a central problem: how to control the cooling process in order to get a desired crystal size distribution in the final product. As we said, the importance of this problem comes from the necessity of designing the internal structure of the specimen so that it meets some specific requirements.

An entirely different class of problems arises in the mould injection phase. Here we have two types of difficulties: the presence of free boundaries (i.e. the boundary confinig the crystallization region) and the variable rheological regime of the flowing phase, which includes growing crystals at least near the boundary with the solid and has rheological parameters strongly dependent on temperature. The mould injection problem has been considered e.g. in [13], but the problem in its complete generality is, as far I know, open.

## 10. Conclusions.

We have developed a model for polymer crystallization, based on the idea of modelling crystalline aggregates (spherulites) by means of equivalent spheres, "tuning" nucleation and growth rate in order to match experimental data for isothermal problems.

The model is formulated in such a way that for a general cooling process it can describe in a natural way the time evolution of the crystalline component along with the temperature distribution, taking into account not only the classical geometrical aspect of mutual impingement of crystals, but also other features which are peculiar to polymers.

Moreover, the model is flexible enough to incorporate even more specific aspects.

The fact that the model is mathematically well posed and that numerical computation can be performed successfully producing good results, justifies the high degree of confidence we have acquired on it.

Due to the complexity of the problem and to its multiple aspects there are challenging questions that remain open and that are interesting and stimulating subjects for further research.

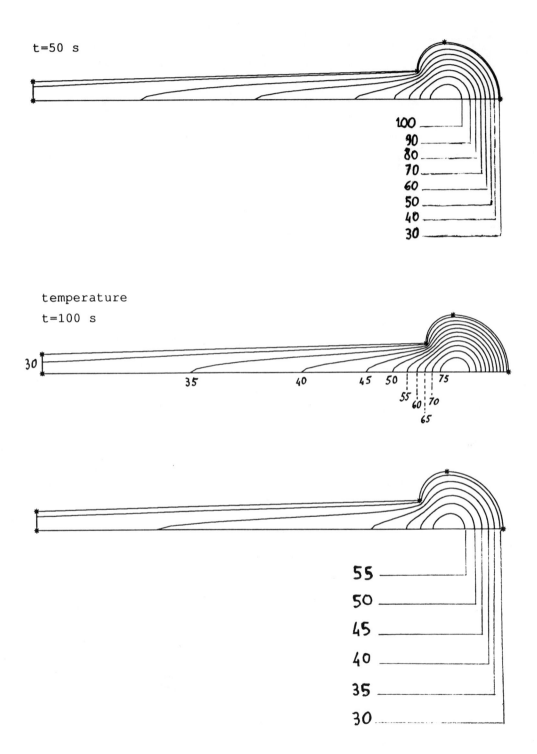

Fig. 8.2a-c. Isothermal lines after 50 sec. (a), 100 sec. (b), and 155 sec (c). External temperature $27°C$. Initial temperature $180°C$.

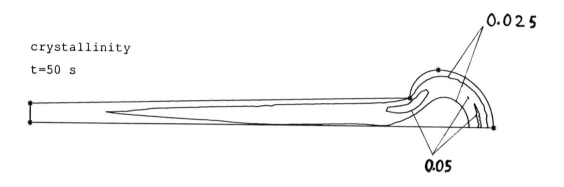

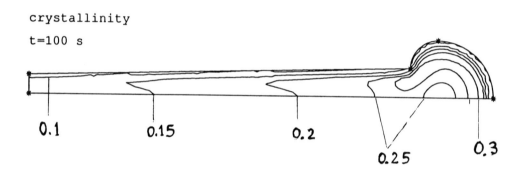

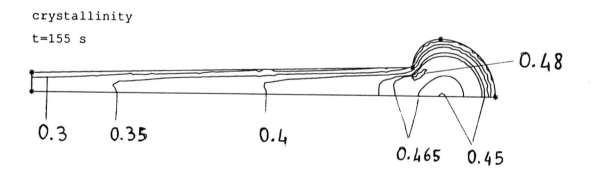

Fig. 8.3a-c. Lines of constant crystallinity after 50, 100, and 155 sec.

# MODELLING THE SOLIDIFICATION OF POLYMERS

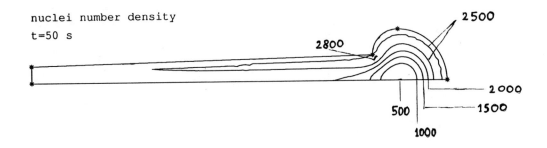

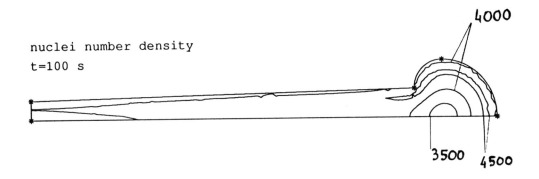

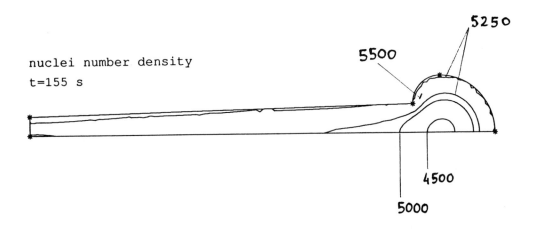

Fig. 8.4a-c. Lines of constant nuclei density after 50, 100, and 155 sec.

116  FASANO

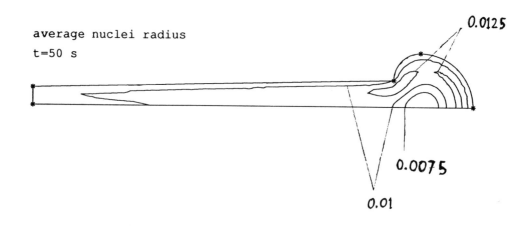

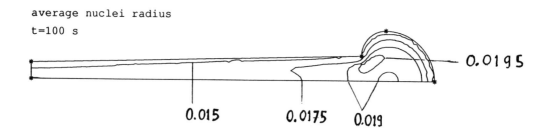

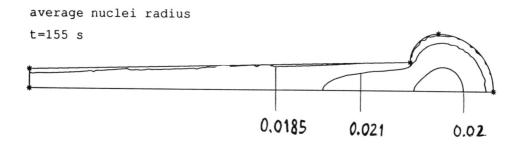

Fig. 8.5a-c. Lines of constant average crystal radii after 50, 100, and 155 sec.

# REFERENCES

1. D. ANDREUCCI, M. BIANCHINI, A. PASQUALI, *Identification of parameters in polymer crystallization*, in preparation.

2. D. ANDREUCCI, L. BORRELLI, V. CAPASSO, P. LI, M. PRIMICERIO, *Polymer crystallization kinetics: the effect of impingements*, SASIAM, sep. (1991)

3. D. ANDREUCCI, A. FASANO, S. MAZZULLO, M. PAOLINI, M. PRIMICERIO, C. VERDI, *Numerical simulation of polymer crystallization.* To appear.

4. D. ANDREUCCI, A. FASANO, M. PRIMICERIO, *On mathematical model for the crystallization of polymers*, Proc. ECMI 90, Hj. Wacker, W. Zulehner eds., Teubner-Kluwer, (1991) 3-16.

5. D. ANDREUCCI, A. FASANO, M. PRIMICERIO, *A mathematical model for non-isothermal crystallization*

6. M. AVRAMI, *Kinetics of phase change, Part. I*, J. Chem. Phys. 7, (1939) 1103-1112.

7. M. AVRAMI, *Id., Part. II*, Ibid. 8, (1940) 212-224.

8. M. AVRAMI, *Id., Part.III*, Ibid. 9, (1941) 117-184.

9. J. BERGER, W. SCHNEIDER, *A zone model of rate controlled solidification*, Plastic and Rubber Processing and Applications 6, (1986) 127-133.

10. G. BRANDEIS et al., *Nucleation, Crystal Growth and the thermal regime of cooling magmas*, J. Geophysical Res. 89, (1984) 10161-10177.

11. J.W. CHRISTIAN, *The Theory of Transformations in Metals and Alloys*, Pergamon Press, Oxford, (1965)

12. W.J. HAYES, *Mathematical models in material sciences*, M. Sc. Thesis, Oxford, (1985)

13. M.R. KAMAL, Polym. Eng. Sci. 26 , (1986) 92.

14. A.N. KOLMOGOROV, *Statistical theory of crystallization of metals*, Bull. Acad. Sci. USSR Mat. Ser. 1, (1937) 355-359.

15. A.Ya. MALKIN V.P. BEGHISEV, I.A. KEAPIN, S.A. BOLGOV, *General treatment of polymer crystallization kinetics, Part I*, Polym. Eng. Sci. 24, (1984) 1396-1401.

16. Id., *Part II*, Ibid. 24, (1984) 1402-1408.

17. L. MANDELKERN, *Crystallization of polymers*, Mac Grow Hill, N.Y., (1964)

18. S. MAZZULLO, *Crystallization of polymers processing and the associate free and moving boundary problems*, Himont Italia Rep., (1987)

19. S. MAZZULLO, *Modello di granulazione, accrescimento e arresto della cristallizzazione nei polimeri semicristallini*, Himont Italia Rep., (1990)

20. S. MAZZULLO, M. PAOLINI, C. VERDI *Polymer crystallization and processing: free boundary problems and their numerical approximations*, Proc. ECMI 88, S. McKee ed., Teubner-Kluwer, (1990) 437-443.

21. T. OHTA, Y. ENEMOTO, R. KATO, *Domain growth with time-dependent front velocity*, preprint (1990).

22. S. OHTA, T. OLITA, K. KAWASAKI, *Domain growth in systems with multiple-degenerate ground states*, Physica 140A, (1987) 478-505.

23. J. RABESIAKA, A.I. KOVACS *Isothermal crystallization kinetics of polyethilene III*, J. Appl. Phys. 32, (1961) 2314-2320.

24. E. SCHEIL, *Anlaufzeit den Austenitumwandung*, Ark.für Eisenhüttenwesen 8, (1935) 565-579.

25. W. SCHNEIDER, A. KO .. ppl, J. BERGER, *Non-isothermal crystallization. Crystallization of polymers*. International Polymer Procesing 2 (1988) 151-154.

26. M.C. TOBIN, *Theory of phase transition kinetics with growth site impingement*, Part I, J. Polym. Sci, Polym. Phys. Ed. 12, (1974) 394-406.

27. Id., *Part II*, Ibid. 14, (1976) 2253-2257.

28. A. VISINTIN, *Mathematical models of solid-solid phase transition in steel*, IMA J. Appl. Math. 39, (1987) 143-157.

29. A. ZIABICKI, *Theoretical analysis of oriented and non-isothermal crystallization*, Part I, Colloid & Polym. Sci. 252, (1974) 207-221.

30. Id., *Part II*, Ibid. 252, (1974) 433-447.

# Chapter 9
# DISCRETE MATHEMATICS IN MANUFACTURING

**MARTIN GRÖTSCHEL**
Konrad-Zuse-Zentrum
für Informationstechnik and
Technische Universität, Berlin
Germany

**1. Introduction.** Computer aided design, flexible manufacturing and computer integrated manufacturing have become technological buzzwords of our time. We are fascinated when we see driverless vehicles transport parts through a factory, watch robots executing complicated movements, or observe automated assembly lines producing goods at a speed, and with a quality, unimaginable with manual production methods. However, when our initial fascination is gone, and we examine the details, we quickly realize that enormous improvements are still possible. In fact, improvements can often be made without costly technical changes: by organizing the production flow in a different way, by designing the products better, by scheduling the jobs differently, or by controlling the machines in a more effective manner. Manufacturing, in general, provides rich opportunities for important mathematical contributions to significant real-world problems and, simultaneously, provides a virtually untapped source of mathematical problems, interesting in their own right.

Although the number of mathematically oriented journals, books and papers in the manufacturing field is rapidly increasing, there is still a huge gap between what could be done, and what actually is done. It is my opinion that there are at least two reasons for this phenomenon. In my experience, many of the talented engineers who build and operate complicated manufacturing systems so ingeniously, simply do not have the background in the rather new mathematical techniques that are necessary to handle the issues to be discussed in the sequel; and they sometimes do not believe that mathematics can help. Secondly, only a few mathematicians are willing to go through the laborious and occasionally painful process of understanding, analyzing and modeling complex manufacturing systems and then discussing their findings with the practitioners. Both parties suffer as a result. Companies in particular miss opportunities for more efficient and cost-effective production, and mathematicians opportunities to identify and solve challenging problems, problems that arise in connection with one of the most fascinating technical developments of our time.

It is not my intention here to survey the mathematical problems that arise in this area. Rather, I will concentrate on those aspects that involve the techniques of discrete mathematics. Many of the problems I am aware of are combinatorial optimization problems. Due to the richness of the field of manufacturing, it is impossible to list all the different problem types. Thus, I will concentrate on practical applications and their mathematical models, applications that the members of my research group have worked on in recent years in cooperation with industry. This work was begun at the University of Augsburg and is continuing in Berlin. In most of the cases reported here, the industry partners were, and often still are, branches of Siemens and Siemens Nixdorf.

I have organized this paper following the natural method of design and production in a typical electronics company. The design phase marks the beginning of a product. I will outline issues from this phase in Section 2. In the next phase, components are produced. The related issues of machine control and the like will be discussed in Section 3. In the final phase, the parts are assembled.

The complex management and scheduling problems of highly automated assembly systems will be described in Section 4. I will touch upon the various mathematical problems that arise in these phases only very briefly. A few remarks about the mathematical techniques involved can be found in Section 5. A glimpse of further important issues such as logistics, distribution and safety aspects is also given in Section 5.

An electronics company was a natural choice for my presentation since by far the largest fraction of our projects have been joint efforts with the Siemens corporation. However, other types of companies, such as automotive, chemical and machine construction companies, could just as well have served to demonstrate the use of mathematics in manufacturing applications.

**2. The Design Phase.** Companies with bad products and effective production will not survive. Good products, products that customers value, are what makes a company successful. But the financial success of a company depends to a large extent on how an idea is realized and on how the resulting product is manufactured. The transformation of an idea into a producible item is called the design phase. Disregarding the (very vital) aspects of style, or look-and-feel of a product, what is important from our perspective is that a product is designed in such a way that it can be easily and cheaply manufactured. This task is hard to quantify and tremendously complex. The usual approach to tackle it is to break the task (often hierarchically) into several subproblems that are more manageable and to hope that the overall solution is "reasonable".

We outline this general aspect here by describing a few tasks that arise in the design of electronic circuits.

**2.1. VLSI Design.** By looking at the computers on our desks and comparing them with machines 10 or 20 years old, we can observe the incredible improvements that very large scale integration (VLSI) has brought about. Hundreds of thousands or even millions of transistors integrated on a few square centimeters of silicon (a chip) perform an enormous number of operations at breathtaking speed. This large scale integration is one of the most significant technological revolutions of our time. Mathematics is used here in various stages of the chip design phase. A brief outline follows.

Once the full task of a circuit is described, the logic has to be determined that will perform all the desired operations. This logic is then cast in silicon. Today's approach for physically realizing the logic design is to begin with predefined small cells that perform certain simple logic operations, and then to connect these cells with wires (so-called nets) so that the combination of cells and nets realizes the abstract logic model physically. Depending on the chosen technology, the design rules and customer requirements, the cells have a certain shape (usually rectangular) and size, and the wires need a certain width and require a certain distance to other wires or units on the chip. Given cells and nets together with additional technological constraints, the task is to place the cells and route the nets so that either the size of the resulting chip is as small as possible, or the resulting chip fits into a given frame. There are many additional side constraints that have to be met. For instance, certain cells have to be close to each other, certain nets have to have a maximum or minimum length, etc.

Why do we want the chip to be small? Of course, smaller chips can usually be run at higher cycle speeds and are faster, but a major reason is that the yield of chip production decreases nonlinearly with increasing chip size. Modern chip production in futuristic, extremely clean and almost fully automated factories is basically a fixed cost operation, relatively independent of the number of wafers processed. Therefore, it makes a tremendous difference whether 20% or 80% of the chips produced per day are defective or not. Also, the smaller the chip size, the more likely is a higher, top quality output.

Let us now look at a specific technology, the so-called sea-of-cells technology, one that is currently in wide use. Here a rectangular "master chip" is given. Among the feasible master chips, a master is usually chosen that is as small as possible so that one can hope to realize the given circuit on it. This master is subdivided into, say, $m$ base cells. All logic cells are rectangular and have a size equal to some multiple of the base cell size. Suppose $n$ logic cells are given, connected by $z$ nets. The task roughly is to assign the $n$ logic cells to the base cells so that all cells fit, no two logic cells overlap, all nets can be routed, and such that the total net length (the sum of the lengths of all nets) is as small as possible.

It turns out that this problem is enormously complicated. At least at present, it seems impossible to handle the placement and routing problem simultaneously in a sound mathematical model for realistic problem instances.

Thus, the whole task is subdivided into several hierarchical problems. Depending on instance sizes and algorithmic approaches, the following phases are considered in general chip design: global placement, global routing, local placement, local routing, layer assignment and (possibly) compaction. These phases are processed in an iterative manner and may be repeated until satisfactory results are achieved. An excellent account of this area can be found in [L90].

**2.2. Placement in Sea-of-Cells Technology.** We will now look at the placement problem (without distinguishing between global and local placement) for chip design in sea-of-cells technology. Figure 1 shows a small master consisting of 26 × 18 base cells and the outer frame of pad cells; the nine dark rectangles are logic cells placed on this master.

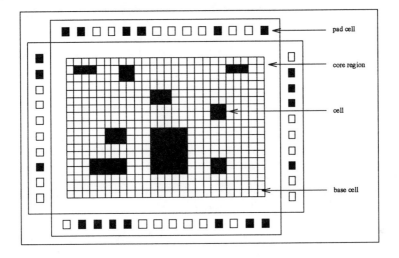

Figure 1: Example of a sea-of-cells master

In the last decade many algorithms have been developed for a solution of this placement problem. They can be categorized as follows.

The most popular and probably still most frequently used approach is based on the min-cut heuristic for bipartitioning graphs. We define a hypergraph $G = (V, E)$, where the node set represents the logic cells. For every net of the given circuit, we define a hyperedge consisting of all logic cells that the net connects. The idea is to recursively partition the node set of the hypergraph into two sets of roughly equal size such that the number of edges in the associated cut is as small as possible; [SK72], [Br77]. The main drawback when working in this scheme is that the global view of the problem gets lost. Moreover, the seemingly easy problem of finding a minimum cut with both sides of about equal size is an $\mathcal{NP}$-hard problem itself.

Another class of algorithms has been developed that model the placement problem mathematically in such a way that a relaxation of it can be solved to optimality. One such relaxation can be interpreted as a (nonlinear) energy model, where the objective function approximates the total wiring length and the side constraints assure that the trivial placement (all cells are assigned to the same base cell) is excluded [KlSJ88]. The solution of such a relaxation is usually not feasible for the placement problem (e.g., cells overlap) and thus the solution of the relaxation has to be modified heuristically in order to obtain a solution of the underlying discrete optimization problem.

We now address the placement problem by introducing boolean variables

$$x_{ik} = \begin{cases} 1, & \text{if cell } i \text{ is assigned to base cell } k, \\ 0, & \text{otherwise.} \end{cases}$$

Let $o(ik, jl)$ and $d(ik, jl)$ denote, respectively, the number of overlapping base cells and the distance if cells $i$ and $j$ are assigned to base cells $k$ and $l$. Then a corresponding quadratic 0/1-optimization problem can be stated as follows:

$$(1.1) \quad \min \sum_{k=1}^{m} \sum_{l=1}^{m} \sum_{i=1}^{n} \sum_{j=1}^{n} (c_{ij} d(ik, jl) + \lambda \cdot o(ik, jl)) x_{ik} x_{jl}$$

$$s.t. \quad \sum_{k=1}^{m} x_{ik} = 1 \quad \text{for all} \quad i = 1 \ldots n,$$

$$x_{ik} \in \{0, 1\} \quad \text{for all} \quad i = 1 \ldots n, \; k = 1 \ldots m,$$

where $\lambda$ is the penalty parameter for the overlaps, and the coefficients $c_{ij}$ denote the number of nets between cells $i$ and $j$. Problem (1.1) is a (slightly imprecise) model of the placement problem. The equations in (1.1) guarantee that all cells are placed, but the requirement that two cells may not overlap on a chip has been dropped since it leads to a tremendous number of additional inequalities. This requirement is taken into account indirectly by a suitable increase of the objective function value, when cell overlaps occur.

One can show that (1.1) is $\mathcal{NP}$-hard. The complexity of (1.1) and its sheer size for real problems suggest the idea of decomposing large instances into smaller ones. For example, the problem "soc4", mentioned later, has 2776 logic cells and 13440 base cells on the master chip. This leads to a quadratic 0/1-program with 37,309,440 variables. The decomposition is carried out in such a way that the global view does not get lost. For a solution of the decomposed problems, several heuristics have been developed; cf. [JMRW92], [W92] for a detailed outline of the procedure.

In Table 1 the quadratic 0/1-approach sketched above is compared to two state-of-the-art algorithms used in industry, namely the min-cut placement procedure and a method based on the energy model in [KlSJ88]. The three algorithms were applied to four electronic circuits, called soc1, ..., soc4, that consist of 602 to 2776 logic cells. Their performance was measured by the total wiring length estimated by an industrial routing algorithm for the respective placements. The resulting wiring lengths are reported in Table 1.

|      | min-cut | energy | 0/1 QP |
|------|---------|--------|--------|
| soc1 | 212258  | 180445 | 169892 |
| soc2 | 194732  | 189683 | 185592 |
| soc3 | 766622  | 652129 | 553575 |
| soc4 | 623159  | 506160 | 497285 |

Table 1

The running time of the quadratic 0/1-programming algorithm is about ten times as large as the roughly identical running times of the other two heuristics. Using certain clustering techniques for the quadratic approach in addition, see [W92], the running times can be made comparable with the running times of the other two heuristics without loss of the solution quality. The placement of chip "soc1" obtained with the quadratic 0/1-programming approach is shown in Figure 2.

**2.3. Routing.** Let us now focus on the next step in the hierarchical design process of electronic circuits, the routing problem. We assume that all cells (logic and I/O-cells) are placed, and a list of nets is given. Each net is viewed as a set of points, where each point corresponds to a terminal (also called pin) of a cell. The routing problem deals with connecting the pins of the nets by wires. In fact, the problem is quite complicated, because certain given design rules must be taken into account and an objective function, such as the total wire length, must be minimized. Since the routing problem is $\mathcal{NP}$-hard and usually of extremely large scale, the problem is often decomposed into special cases that can then be handled more easily. A large variety of such special cases is considered in practice and in the literature.

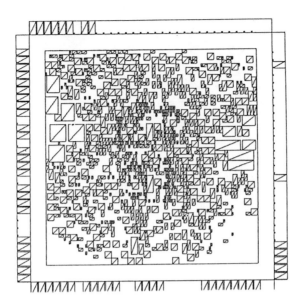

Figure 2: Cell placement in sea-of-cells technology

We want to model the routing problem as a Steiner tree packing problem in a graph. One way of introducing a graph $G = (V, E)$ here is to define nodes for subareas of the whole routing area, and to link nodes that represent adjacent subareas by an edge. In addition, we assign edge capacities and edge weights to each of the edges. The nets are represented in this graph by subsets of the node set. Let a graph $G = (V, E)$ and a node set $T \subseteq V$ be given. We call an edge set $S \subseteq E$ a Steiner tree in $G$ for $T$, if the graph $(V(S), S)$ (consisting of the edge set $S$ and all nodes $V(S)$ that appear as endnodes of edges in $S$) contains an $[s,t]$-path for each pair of nodes $s, t \in T$. With this definition, the routing problem can be stated formally as follows.

### (1.2) The Routing Problem

Given: A graph $G = (V, E)$ with edge capacities $c_e \in I\!N$ and edge weights $w_e \in I\!N$ for all $e \in E$ and a netlist $\mathcal{N} = \{T_1, \ldots, T_N\}$, $T_k \subseteq V, k = 1, \ldots, N$.

Problem: Find edge sets $S_1, \ldots, S_N \subseteq E$ such that
    (i)    $S_k$ is a tree in $G$ for $T_k$, $k = 1, \ldots, N$,
    (ii)   $|\{k | e \in S_k\}| \le c_e$ for all $e \in E$, and
    (iii)  $\sum_{k=1}^{N} \sum_{e \in S_k} w_e$ is minimal.

We call an $N$-tuple of edge sets $(S_1, \ldots, S_N)$ that satisfies (i) and (ii) of (1.2) a packing of Steiner trees. Problem (1.2) can then be designated as the Steiner tree packing problem. It is not surprising that problem (1.2) is $\mathcal{NP}$-hard. Indeed, it includes several $\mathcal{NP}$-hard problems as special cases; see, for example, [GaJ77], [Ka72], [KrL84], [KPS90].

Figure 3 indicates a $5 \times 3$ underlying grid graph (the grid is not drawn), in which each edge has capacity 1, and three nets, each consisting of three pins. The three Steiner trees forming the nets are drawn with solid, dashed and dotted lines, respectively.

Our approach to the Steiner tree packing problem is to consider it from a polyhedral point of view and to use linear programming techniques. We define, for a given instance $(G, \mathcal{N}, c, w)$, a polyhedron $P$ whose vertices correspond uniquely to the packings of Steiner trees for that instance. Thus, the Steiner tree packing problem reduces to the problem of minimizing the objective function $w^T x$ over the polyhedron $P$.

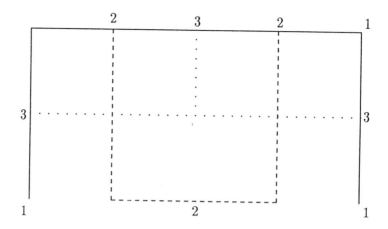

Figure 3: A packing of three Steiner trees

What we need for the application of linear programming techniques is a complete or at least "good" inequality description of the polyhedron $P$. In [M92] a large number of valid inequalities for $P$ is described. In fact, many of these inequalities also define facets of $P$ if the underlying graph $G$ is complete and the terminal sets are disjoint. The machinery involved in describing these inequalities is rather complicated and thus we omit detailed statements of these results. These inequalities form the basis for the development of cutting plane algorithms. We have implemented a cutting plane algorithm for special instances of the Steiner tree packing problem, when the graph $G$ is a rectangular grid graph and the terminals of the nets lie on the outer face of $G$. In VLSI design, these instances are known as switchbox routing problems. We have tested our algorithms on many benchmark problems from the literature. For a detailed documentation of these results see [M92].

At its current stage this approach is not yet able to handle instances as large as those that can be treated with the standard routing heuristics used presently in industry. The remarkable feature of this approach, however, is that very good (provable) lower bounds on the total wiring length can be computed, a result that none of the other currently used approaches can provide. In fact, for the (small) problem sizes considered so far, quality guarantees of less than 1% or even provable optimality have been achieved.

**2.4. Via Minimization.** We continue with our chip design example and assume that the placement and the routing phase have been completed successfully. Usually routing algorithms disregard the requirement that two different nets may not cross. Routings with such defects are referred to as transient. A transient routing of eigth nets is shown in Figure 4. The reason for ignoring net crossings is that the physical routing can be done on two or more layers of a chip. In case two nets cross in a transient routing, a wire segment of one of the nets can be assigned to a different layer so that no physical wire crossing occurs. The connections between the wire segments of a net on different layers are provided by so-called vias. In chip manufacturing the vias are produced by means of a delicate chemical process. Many vias on a chip increase the probability that, at the end of the production process, the chip will have a short or not work correctly. Moreover, vias use considerably more space than wires and thus a large number of vias may lead to an increase of the chip size.

Virtually the same problem occurs in the design of printed circuit boards. Here, the vias are produced by mechanically drilling a hole through the board (lasers are also used). A danger is that boards may crack in the drilling phase, another reason for minimizing the number of vias. Furthermore, the drilling process is quite time-consuming (see Section 3.2), and hence reducing the number of vias leads to a reduction in the overall production time.

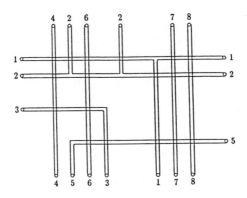

Figure 4: A transient routing

Due to these facts, it is desirable to assign the wires to layers of a chip or a printed circuit board in such a way that no two wires cross and the number of vias is minimum. We call this task the via minimization or layer assignment problem.

For the case of two layers (assuming that the transient routing contains no $k$-way junctions for $k \geq 4$) [Pi84] and [CKC83] have shown independently that this problem is solvable in polynomial time through reduction to a max-cut problem in a planar graph. However, these reductions do not cover certain side constraints required in practice. In many cases one of the two layers is preferred and pins are preassigned to some layer. [BaGJR88] pointed out that Pinter's reduction can be generalized to the via minimization problem subject to layer preference and pin preassignments. However, the max-cut problem that results from this reduction is $\mathcal{NP}$-hard. The transformation is elegant but technically rather involved and will not be explained here. The reader is invited to find the minimum number of vias for the problem shown in Figure 4 before looking at Figure 5 where an optimum solution needing four vias is displayed. The vias are indicated by the little open squares (where the nets change color, i. e., lay

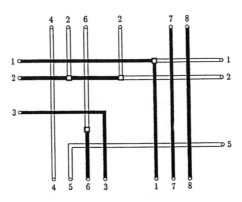

Figure 5: A via-minimal layer assignment

We have developed both simple and elaborate heuristics for the via minimization problem and an exact cutting plane algorithm that is based on polyhedral investigations of the max-cut problem. These algorithms are described in [BaGJR88] and [GJR89].

Tables 2 and 3 report the success of this approach. The symbols $C1,\ldots,C6$ denote six chips from industry. For each chip, the via minimization problem comes in two versions. In one case, no pin was preassigned, and in the other, certain pins were required to be on one of the layers of the chip. Of course, this latter requirement will generally result in more vias. Table 2 reports on the results of the via optimization without pin preassignments, Table 3 shows those with pin preassignments. The row labels have the following meaning:

| # nodes | $\widehat{=}$ number of nodes of the so-called reduced layout graph, |
| # edges | $\widehat{=}$ number of edges of the reduced layout graph. |

These two numbers indicate the sizes of the assoiated max-cut problems.

| # vias original | $\widehat{=}$ number of vias in the (original) industry design, |
| # vias heuristic | $\widehat{=}$ number of vias in our best heuristic solution, |
| # vias optimal | $\widehat{=}$ minimum number of vias for this chip found by the cutting plane algorithm, |
| improvement | $\widehat{=}$ improvement in percent ((# vias original $-$ # vias optimal) / # vias original). |

|                  | C 1    | C 2    | C 3    | C 4    | C 5    | C 6    |
|------------------|--------|--------|--------|--------|--------|--------|
| # nodes          | 827    | 979    | 1326   | 1201   | 1365   | 924    |
| # edges          | 1445   | 1775   | 2480   | 2606   | 2234   | 1740   |
| # vias original  | 421    | 434    | 683    | 650    | 782    | 630    |
| # vias heuristic | 272    | 347    | 513    | 475    | 610    | 525    |
| # vias optimal   | 264    | 334    | 500    | 467    | 608    | 516    |
| improvement      | 37.29% | 23.04% | 26.79% | 28.15% | 22.25% | 18.10% |

Table 2: Via minimization without pin preassignment

|                  | C 1    | C 2    | C 3    | C 4    | C 5    | C 6    |
|------------------|--------|--------|--------|--------|--------|--------|
| # nodes          | 828    | 980    | 1327   | 1202   | 1366   | 925    |
| # edges          | 1749   | 2102   | 2844   | 2915   | 2557   | 2008   |
| # vias original  | 421    | 434    | 683    | 650    | 782    | 630    |
| # vias heuristic | 302    | 376    | 563    | 504    | 645    | 585    |
| # vias optimal   | 302    | 376    | 561    | 482    | 643    | 585    |
| improvement      | 28.27% | 13.36% | 17.86% | 25.85% | 17.77% | 7.14%  |

Table 3: Via minimization with pin preassignment

These results were achieved with algorithms designed and implemented by M. Jünger and G. Reinelt. The heuristics approximately solve these via minimization problems within a few seconds, while the cutting plane algorithm (yielding a provable optimum solution) only requires a few minutes on a workstation. The results displayed in these tables show that this optimization approach results in a considerable reduction of the number of vias compared to the industry solutions and thus leads to more favorable chip designs. Moreover, optimum solutions could easily be computed and the heuristics (these are special purpose methods taking the special features of this problem into account) did extremely well.

This discussion of some features of the design phase of a manufacturing process was meant to show that even seemingly minor details such as via minimization lead to very interesting mathematical problems, and that their solution may have a significant impact on other aspects of production. The mathematical theory on topics of this type, whether combinatorial or continuous, is not very well-developed. Only few isolated cases have been analyzed. There is still considerable room for further exciting developments.

**3. Control of Machines.** Let us now suppose that the design phase has been completed. The product designers have determined how to subdivide the production process into various tasks and how the different parts of a product must be manufactured.

In this section we will consider the manufacturing process on a single machine. The additional problems that arise when taking transport and sequencing into account will be discussed in the next section. In the typical situation, an object enters a machine, and the machine, controlled by a computer program, performs various operations on that object. The questions that arise are: how to do the jobs as fast as possible, and how to schedule the jobs such that idle times, say those induced by retooling or positioning moves, are as short as possible.

We have noticed in our work, dealing with situations like this, that slightly different technical devices, a few details in machine capabilities, and decisions of the production managers may lead to quite different mathematical models. In fact, whole new ranges of problems do arise in this area. We will exemplify these statements by considering two problems in printed circuit board production. We follow the paper [GJR91] in our presentation.

**3.1. The Plotting Problem.** Complex printed circuit boards are usually produced by a photochemical process. For each layer of the board, the pattern of wires and contacts is produced by covering the board with light sensitive material, exposing this material to light, etching, cleaning, etc. The process is similar to the usual production of photographs. The structures that later should appear on the board are first "drawn" on a mask (a negative) that is placed between the board and the light source, so that certain parts of the board are not exposed to light. These unexposed areas are to form the conductors, pads and contacts of the layer. The question we address here is the generation of the masks. The masks are made of glass and the patterns on the glass are generated optically either using ultraviolet light or laser beams. In our case, a photo plotter is used for the mask production. Figure 6 shows an example of one layer of a printed circuit board. It is one of our test cases.

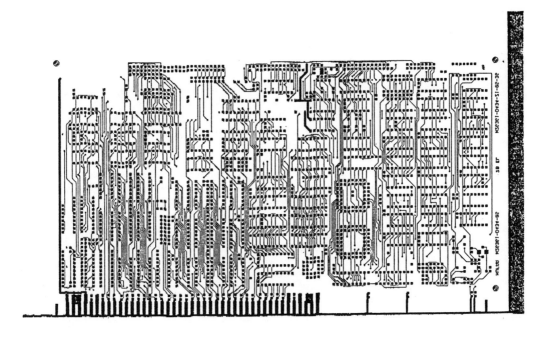

Figure 6: A layer of a printed circuit board

The photo plotter works as follows. It has two modes, a "drawing mode" for plotting lines and a "flashing mode" for plotting points. As one can see from Figure 6, points may be of various sizes and shapes, and lines may be of various widths. Before plotting, an aperture is chosen that produces the required shape or width.

Points are plotted by moving the light source to certain coordinates on the board, choosing the aperture, and flashing the light. Lines are plotted by moving the head to one end of the line, choosing the aperture, opening the shutter, moving along the line with the open shutter and closing the shutter at the end of the (not necessarily straight) line.

The above is nothing but a basic description of the principle of the plotting process. Plotting machines are offered in various mechanical forms. In some cases, only the head of the light source is moved, in some cases only the compound table, and in others the head moves in one direction and the table in the other. For our purposes, the actual mechanics are irrelevant. We only need to know the time it takes to move from one point to another. For the sake of exposition we will assume a model in which only the head moves.

There is, given a pattern, nothing to be done about the time needed for drawing and flashing. This process requires a certain fixed time depending on the plotter characteristics. What can be optimized is the time needed for positioning head moves, i. e., moves of the head without drawing. We will now describe the mathematical modelling of the plotting process in some detail. Depending on technological side constraints, there are several options of modelling this problem mathematically. We discuss two examples that came up in our application.

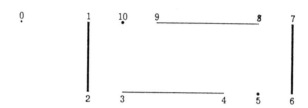

Figure 7: First plotting problem

Consider the line and point plotting problem shown in Figure 7. It seems obvious how to proceed. We begin at the starting position 0, move to point 1, choose the "drawing mode", and switch the aperture to "drawing thick lines". Then, while moving, we draw the line from 1 to 2, move to point 3 and change the aperture to "thin drawing", draw the line from 3 to 4, move to point 5 and change to the "point flashing mode", etc. The trouble here is that during every positioning move the aperture has to be changed. The head is (in many cases) a mechanically delicate device that — after a certain number of aperture changes — has to be readjusted or substituted. This is a costly procedure. Therefore it may be wise to proceed as follows. One first chooses an aperture, plots everything that can be plotted with this aperture, changes the aperture, etc. This approach decomposes the problem into various plotting subproblems and adds a new problem, namely that of an optimal sequence of apertures. For Figure 7, an optimum solution in this case would be as follows. We first choose the thick drawing aperture, go from 0 to 1, draw the line from 1 to 2, move to point 6, draw the line from 6 to 7, change to the thin drawing aperture and move to 8, draw the line from 8 to 9, move to 3, draw the line from 3 to 4, change to the flashing mode and move to 5, flash, move to 10, flash, and return to 0.

Clearly, the second choice produces longer (and sometimes substantially longer) positioning head moves and — provided that aperture changes do not require more time than the moves — longer machine running times.

One has three choices in practice. Either one ignores aperture changes and uses the first option, or one decides to decompose the plotting problem into subproblems (one for each mode and each aperture), or one mixes the two by penalizing those positioning moves where also aperture changes occur. In each case, the person responsible for mask plotting has to decide — based on his knowledge of the technical characteristics of the machine — whether or not aperture changes are considered crucial operations that have to be kept at a minimum and which of the three options should be used.

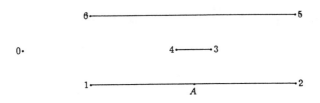

Figure 8: Second plotting problem

There is another option that may produce shorter positioning head moves: preemption. Consider Figure 8, where all three lines are drawn with the same aperture. An optimum solution here is to move from 0 to 1, draw the line from 1 to 2, move to 3, draw the line from 3 to 4, move to 5, draw the line from 5 to 6 and return to 0. We have so far made the assumption, without explicitly stating it, that whenever the plotter has started drawing a line it continues drawing until the other endpoint of the line is reached. If we allow interruptions (technically often called preemptions) we could do better. For instance, in Figure 8 we could draw the line from 1 to 2 only until point $A$ is reached then we move to point 3, draw the line from 3 to 4, return to point $A$ and continue drawing from $A$ to 2, etc. This choice would produce shorter positioning head moves than the "optimum solution" sketched above. So the question is whether preemptions should be allowed or not. There is no general answer. In each individual case one has to make a decision based on the particular conditions.

In our case the decision was to execute as few aperture changes as possible and not to allow preemptions. Therefore, the following combinatorial optimization problems arose.

### (3.1) Point Flashing Subproblem.
*Given an aperture, select all points that are flashed with this aperture and determine a shortest Hamiltonian path through these points.*

### (3.2) Line Drawing Subproblem.
*Given an aperture, select all lines that have to be plotted with this aperture and determine a sequence of these lines such that the total distance travelled by positioning moves is as short as possible.*

### (3.3) Aperture Sequencing Subproblem.
*Determine a sequence of apertures and, for each aperture, a starting point and a terminal point for the point flashing or line drawing process such that the total time for positioning moves (with and without aperture changes) is as short as possible.*

The aperture sequencing problem is by far the most complicated and we do not have any idea of how to solve it. Fortunately, in our practical problems, it was of almost no importance. We treated it by using a simple heuristic.

The point flashing problem is, after an easy transformation, equivalent to a symmetric travelling salesman problem (TSP). The line drawing problem turns out be a so-called rural postman problem. A rich mathematical theory and a substantial algorithmic toolbox exist for the TSP, see [LLRS85], whereas the rural postman problem has basically been neglected in the literature up to now.

To solve the plotting problem in practice we had to meet additional computational side constraints. The organization of the production process required the problem to be solved almost in real time, i. e., for each instance, five minutes of computation time were available on a 3 MIPS machine for the solution of the point flashing, the line drawing and the aperture sequencing problem. Within this time limit good solutions had to be produced. [GJR91] and [R92] describe some of the heuristics that were designed and implemented to satisfy these constraints.

The methodology developed also involved the fast solution of problems from computational geometry, such as computing Voronoi diagrams, Delaunay triangulations or convex hulls. These techniques were utilized to reduce problem sizes in order to satisfy the running time requirements.

Our codes were tested on real-world masks by the engineers at Siemens. The solutions of the industry heuristics were compared with our solutions by measuring the running times on the real plotting machines. The savings turned out to be tremendous. We had to solve line drawing problems with up to 38,621 lines and flashing problems with up to 2,496 flashes. The running time reductions for the positioning moves using our fastest heuristics ranged from 14% in the worst case up to 83% in the best case. Further improvements of an approximately additional 10% could be achieved by more elaborate and time-consuming heuristics. These savings resulted in capacity increases of the plotting machine in the range of 5% to 35%, see [GJR91] for more detail.

**3.2. The Drilling Problem.** In Section 1.4 we introduced the via minimization problem for chips. Its analogue for printed circuit boards (PCBs) has direct consequences for the printed circuit board production process: the fewer the vias, the shorter the production time.

The practical problem arising in this application is the following. To connect a conductor on one layer with a conductor on another layer or to position (in a later stage of the PCB production) the pins of IC's, holes have to be drilled through the board. The holes may be of different diameters. To drill two holes of different diameters consecutively, the head of the machine must move to a tool box and change the drilling equipment. This is quite time-consuming. Thus, it is clear at the outset that one has to choose some diameter, drill all holes of the same diameter, change the drill, drill the holes of the next diameter, etc.

There is no tool changeover problem here since, in any case, after loading of the boards the machine head is at the initial position (where the tool box is) and after having drilled all holes of one diameter it has to return to the initial position to pick up the new drill. Thus, our drilling problem can be viewed as a sequence of symmetric travelling salesman problems, one for each diameter resp. drill, where the "cities" are the initial position and the set of all holes that can be drilled with one and the same drill. The "distance" between two cities is the time it takes to move the head from one position to the other. The aim here again is to minimize the travel time for the head of the machine. The (quite substantial) time needed to drill a hole cannot be influenced at all. This is a fixed production time. As for the plotting application described before, severe bounds on the running time were required for our heuristic.

The problem sizes that came up in this particular application ranged from about 500 to about 2500 "cities". For the real test cases we obtained from Siemens Nixdorf the following was achieved. All instances were solved in less than 2 minutes on a 3 MIPS computer by our heuristics. The time needed for positioning moves by our solutions was up to 55% lower than the respective time of the solution used in industry. Since the drilling time is quite substantial this does not mean a halfing of the production time. The reduction of the positioning moves resulted in a decrease of the total production time of 5% to 20% on the average. The results were considered quite remarkable by the engineers responsible for the CNC-machines and it was decided to incorporate our codes into the software controlling the machines in order to speed up production.

The dots on Figures 9 and 10 indicate the holes of a typical drilling problem for printed circuit boards. Figure 9 shows the positioning moves of the original industry solution, Figure 10 the positioning moves of our solution. Our tour is 51.83% shorter.

A detailed account of these results is given in [GJR91]. A thorough description of the design and analysis of the heuristics is contained in [R92]. A side effect of this investigation was the compilation of a large number of real-world travelling salesman problems. A library, called TSPLIB, that includes the instances of this study and many other TSP instances was set up. It can be accessed by e-mail. A description of this library and its use can be found in [R91].

DISCRETE MATHEMATICS IN MANUFACTURING  **131**

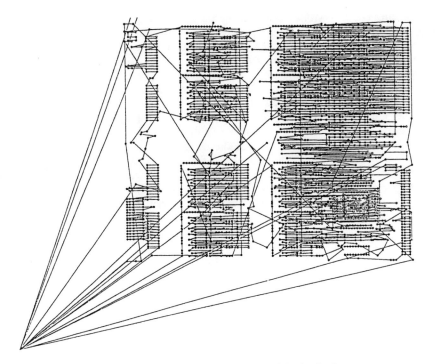

Figure 9: Positioning moves, original solution

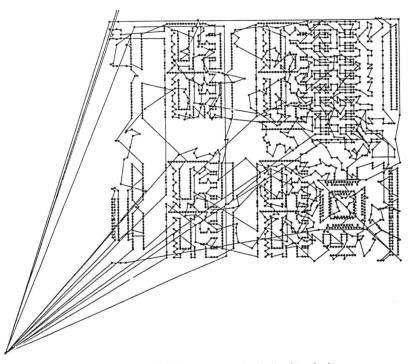

Figure 10: Positioning moves, "optimized" solution

**4. Assembling the Parts.** Having fine tuned the individual machines that produce the parts of a product and having improved their performance by using optimization methods as described in the previous section, we now focus on the task of putting the parts together to obtain the final product. A main problem here is to design the assembly process and to schedule the jobs to be executed in such a way that the production cost is small and the output per time unit high. These global goals are often replaced by (occasionally) more operational or measurable aims such as: all machines are utilized to the full, the work loads of the factory branches are balanced, stocks are kept at a minimum level under the condition that quick changes in the production volume and pattern are still possible, etc. No doubt, this is a tremendously difficult task.

Mathematics is, at present, rarely used in the design phase of the assembly process, i. e., when the machines are chosen, the factory floor is layed out, the transport systems are selected and dimensioned, etc. However, once these basic decisions are made, simulation systems are often employed that are based on mathematical tools. These simulation systems are used to get a feeling for the flow of goods, the throughput, the possible bottlenecks of the system, etc. Simulation methods have become indispensable and valued tools for the design of large production systems. Nevertheless, there is considerable room for the use of further and more sophisticated mathematical tools in this design process.

Mathematicians are, in general, only asked for their expertise in cases where the system does not work as expected, or when the work load of the system has changed considerably and ways are sought to run the system efficiently without making significant technical changes, or when certain components of the system turn out to be severe bottlenecks and reorganization of work seems feasible for curing this disease.

I will explain these aspects by means of two projects that were jointly carried out with Siemens Nixdorf, Augsburg.

**4.1. Flexible Printed Circuit Board Production.** The Siemens Nixdorf, Werk für Systeme in Augsburg produces, among other items, all main frame computers of the company. Important parts of computers and other electronic products are printed circuit boards (PCBs). Printed circuit board production is a division that manufactures PCBs for Siemens Nixdorf itself and other outside customers.

The whole PCB production process consists of various stages. We have already met two such phases in the previous section (plotting and drilling). We now focus on the final assembly. The PCB frame is finished and the task is to place various electronic components on the PCBs. There are components of different types and mechanical properties. For instance, some components are glued onto the board, some have "little legs" that have to be positioned into certain holes (drilled in a previous stage) and that are soldered in a later phase of the PCB production, etc.

In general terms, the flexible PCB assembly line (called FALKE at Siemens Nixdorf) has the following properties. At the head of the system, the so-called HEAD-cell, the boards are put onto carriers and fed into the system. There are automatic conveyor belts for the transport of the carriers. The carriers may enter some or all of six "cells", where each cell consists of a series of CNC-machines each carrying out certain feeding or other operations. There are buffers in front of each machine that can hold up to 25 carriers. Except for the initial system feeding and certain manual operations for the insertion of special parts, the entire process including transport, component feeding, etc. is controlled by a computer system and is fully automated. The whole assembly line is an impressive, automatic flexible manufacturing system.

The problem with the FALKE line was the following. Due to changes in the design of PCBs over time the distribution of work among the machines of the system was somewhat unbalanced. Certain machines were constantly working, while others were frequently idle. Although some machines had a higher total workload than others, it was not always the same machines that appeared to be the production bottleneck. So the question was: Is it possible to increase the throughput of the system without making any changes in the general control software or any technical changes? We were not allowed, for this type of investigation, to retool the machines, to change the order of jobs to be executed on individual PCB's, or to modify the control software for the conveyor system, the feeders or buffers. Such changes would simply have incurred costs and dangers too high for the reliability of the assembly line. The only control parameter left to be influenced by optimization was the feeding of the system.

At the time of analysis of the FALKE system about 1500 to 2000 circuit boards were produced per day consisting of about 20 to 40 different types. In total, about 500 different PCB types could be automatically assembled on the system, without changing the machine settings.

The managers of the FALKE system learn about the planned production volume for a day about one to two days ahead of time. This gives enough lead time to arrange for the shipment of required boards and components. Having no other option, we proposed using this time for also finding a sequence of the carriers such that the completion time of the whole set of PCBs is minimized.

An immediate question arises. Given a sequence of PCBs, how can one compute the time it needs to finish the work? Of course, we assume here that no artificial idling occurs, i.e., that all machines work and that each job is done at the earliest possible time, etc.

In fact, we were not able to come up with an analytic formula for the completion time. The difficulty is mainly due to long distance interdependencies that are hard to model analytically. For example, some carriers do not enter certain cells or machines (an average PCB is processed on about two thirds of the machines), carriers may overtake each other, a carrier that blocks another at some machine may be blocked in return by this other one at a later stage.

So we decided to design, very carefully, a special purpose simulation model of the FALKE line by measuring all transport times, feeding times and obeying all rules and side constraints of the system. Programming and validating this simulation tool took about 9 month of two persons' work. In the end, we had a simulation tool that could simulate an eight hour shift of the FALKE line in about 5 minutes on a PC. Two weeks of real production were recorded and the parameters of our simulation tool were adapted in such a way that it faithfully reproduced the real production in our computer model of the production line.

We then invented a number of heuristics to improve the sequencing of the jobs. A typical iterative improvement heuristic starts with some sequence of jobs and changes this sequence using some myopic optimization rule. Then it computes the completion time of the new sequence. If it is better (or only slightly worse), it takes the new sequence and applies the same rule. If after a number of applications of this rule, no improvements have been made, other sequence changing techniques are applied until no significant progress is visible. The trouble here is that evaluating the new sequence takes some minutes of running time, and thus, not too many sequences can be tested. To achieve acceptable running times for the heuristics, a fast lower bounding procedure for the completion time was invented. This lower bound was used as the objective function value for the heuristic. The real completion times were only occasionally computed by means of the simulation model.

After having tuned these heuristics, we ran them on the production data of the two weeks available to us. We improved upon the completion time, compared to the runs of the real system, in the range of 3.3% to 12.7%. On the average the completion time could be reduced by 6.7%.

This improvement was significantly less than the management had expected. But we could also show that the expectations were much too high. By means of our lower bounding procedure we could prove that there was not as much room for improvement as thought. We could show that, for the days considered, a maximum total speed up of about 20% might be possible. (We actually believe that this is an overestimation.) So, on the average, about 13% of further completion time reduction might be possible at most.

What turned out to be the most significant drawback of our heuristic solutions was the following property. Since we know that the HEAD-cell is manually operated we expected the person in charge to make a few sequencing errors. Thus we randomly perturbed our "optimized" carrier sequences a little and noticed that a few changes could result in the loss of all the gains that we had worked for so hard, i. e., our solutions turned out to be quite unstable. Due to this fact and the not so significant improvements, the company decided not to change the present system.

In some sense, our efforts were in vain. Techniques of combinatorial optimization did not lead here to significant improvements. What we learned is that we do not have a good understanding of such complex and complicatedly interlinked systems like the FALKE line. We do not seem to have the right tools yet to control these systems efficiently. In fact, what is really needed is an on-line control system that makes adjustments of the job schedule whenever interrupts, machine breakdowns, conveyor stops, etc. occur. I think we are still far away from a good mathematical understanding of such systems. In fact, it may be reasonable to consider designing less complex production systems, production systems that are efficiently controllable. After all, flexibility does not help if it can't be used.

This work has — to a large extent — been carried out by Petra Bauer together with Siemens Nixdorf engineers and is documented in [Bau90]. A paper describing in detail the findings surveyed above will appear in the near future.

**4.2. Optimizing a PC Factory.** The Werk für Arbeitsplatzsysteme, Augsburg is the plant of Siemens Nixdorf where all PCs (and some other related products) are manufactured. In the fall of 1989, we started a joint project aimed at looking for possibilities for production improvements. The goal, in the long run, was to design and implement a software package that optimizes the material flow through the factory and allows for the production of the required PCs, monitors, keyboards, etc. in shortest possible time with high flexibility. This is obviously a very ambitious goal, and we are aware of the fact that we are unlikely to achieve it. The point, though, is to keep the real goal in mind whenever optimization and simulation tools are designed for particular components and branches of the factory. Clearly, the output and efficiency of the whole factory is what really matters and not the speed of a few machines. Figure 11 shows

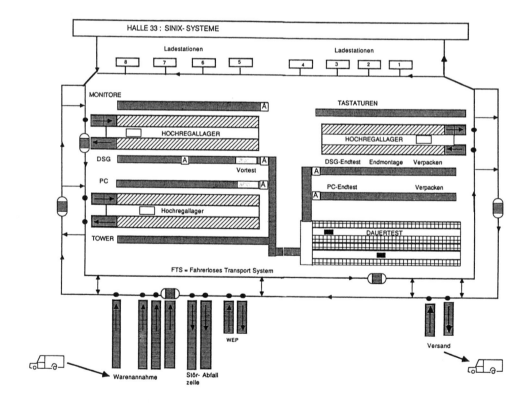

Figure 11: Sketch of the factory floor

a symbolic sketch of one hall of the factory. We give now a rough description of the operations involved in the assembly.

There is a receiving area, where all parts necessary for the production are supplied by trucks. Among these are, for instance, the printed circuit boards mentioned in the previous section. These are manufactured in a factory about one km away. The parts arrive in or are loaded into three types of boxes and given a bar code identifying them. These boxes are moved by a conveyor to several loading points where automatic driverless vehicles pick up the boxes and transport them to their destination. The vehicles are controlled by a software system that decides when which vehicle executes which transportation job.

A typical destination of such a vehicle is the Input/Output buffer of one of the many automatic storage systems. When the driverless vehicle arrives, a mechanical device moves the box on the I/O-buffer and signals to the control of the storage system that the box has to be transported by the stacker crane to its storage location.

The PC and other assembly lines are located on each side of the storage systems such that all parts needed at a certain point on the assembly line are delivered upon request automatically to the desired location by the stacker crane. The control system of the stacker crane decides where to put incoming boxes, when to remove empty boxes and in what sequence all current orders are processed. All this is done on-line.

One of the assembly lines is fully automated, robots do all of the jobs. At the other lines, manual work is involved and the speed of production can be adjusted by varying the number of workers at the lines.

Once a product like a PC is finished, a conveyor moves it to a "hot line" where it is tested by means of software for about twenty four hours. There is a scheduling and a capacity problem here. The control system of the hot line has to determine where to locate a PC, how to perform the tests and whether it pays to relocate the PCs at night in order to be in a position to do all the necessary transportation work fast in case of high traffic volume.

The PCs that have failed the test are moved to a repair area and may reenter the hot line later. The other PCs will run through further manual tests before they are packed and moved to the distribution center. Storing all the products (including manuals, software, etc.), combining these parts to existing orders, and combining orders to reasonable truck loads is another complicated and highly automized process that we do not want to describe here.

In a project that is mainly executed by Norbert Ascheuer and Atef Abdel Hamid with the help of some mathematics students and a support team of Siemens Nixdorf, we have designed and implemented a simulation model that covers, at present, the receiving area, the transportation system (i.e., the driverless vehicles), and the automatic storage systems. This simulation model has been validated by comparing its performance with the data of several weeks of real production. The various models have been validated individually and in joint operation.

The role of the simulation system is twofold. Its use as part of a planning system where decisions about the work load of the day and the distribution of the jobs are made is currently being tested. The simulation system is employed here to check not only whether all parts needed for the planned production volume are available but also to check whether the work force is sufficient or whether bottlenecks in the transportation system and the parts supply will arise.

The main objective of designing the simulation models, however, was to analyse the performance of optimization approaches and compare them with the existing system. To convince the engineers that our analysis is sound and that our system speedup predictions will show up in practice, validation was indispensable. Since outlining the whole work would be too space-consuming, I will concentrate here on the automatic storage system and the stacker crane operation.

The storage systems that are in use at the Siemens Nixdorf Werk für Arbeitsplatzsysteme are huge frames that hold all the material needed at the assembly lines next to them. The initial planning question is how to choose the size of the storage frame to make sure that it will be large enough to hold all the items required in the production. At this point in time, it is also decided which type of transportation and which storage boxes are going to be utilized. In our case, there are three types of boxes of different sizes. Each box type needs differently shaped storage locations.

We have mathematically modelled various optimization questions associated with the design and management of this storage system. A first problem is how to subdivide the frame into storage units such that the expected storage volume of different boxes is sufficient and all boxes are as close as possible to their output buffer, i.e., the problem is to determine locations such that all needed boxes fit and the expected moves of the stacker crane for box transportation are as short as possible. This leads to general integer programming approaches or set packing models. We are not sure yet whether or not our models achieve what is needed in practice. Further modelling iterations may be necessary.

Once the locations are fixed and once the system is in operation we have to decide, every time a new box arrives, where to locate it among the available empty storage locations such that the stacker crane moves are as short as possible. This requires on-line decisions that should take also into account which further boxes are already in the material flow pipeline and are likely to arrive soon. It turns out that, for two of the three box types, this problem can be formulated as an (ordinary) rectangular assignment problem. Typical sizes of such assignment problems are $20 \times 800$ and can be handled with the assignment code of Kleinschmidt, see [AKP89], in a few seconds on a PC. For one type of small boxes the associated model is a special case of the generalized assignment problem. Unfortunately, it is $\mathcal{NP}$-hard. We have developed fast heuristics, e.g., based on matching techniques, and a cutting plane algorithm for its solution. The heuristics provide satisfactory approximate solutions very quickly.

For a detailed description of all these aspects of the storage system and the theoretical and algorithmical investigations of the generalized assignment problem, see [AH92].

Let us now turn to the moves of the stacker crane. By analysing all the technical details, measuring running times of the crane, reaction times of the systems, times for moving boxes, etc. Norbert Ascheuer has implemented a simulation system that models the behaviour of the stacker crane and has compared it with the real moves. Table 4 shows the validation results for all operations of one week. One job consists of the positioning move, the pick up of a box, the move with the box, and the delivery of the box. The time needed for moving and handling one box on day 1, for instance, was 76.27 sec. on the average, see column ØRL of Table 4. This time was obtained by measuring all operations of the day and averaging them out. The simulation model produced an average job length of 76.83 sec. for this day (column ØSIM) resulting in 0.73% deviation (column %DEV). Deviations are bound to occur since always some interruptions and manual interactions occur that are not taken into account by the simulation model; see column ØDEV for average and column max DEV for maximun deviations that occured throughout a day. The time for executing the simulation model for a whole day was about 12 seconds on a PC.

|   | ØRL | ØSIM | ØDEV | max DEV | % DEV | Time |
|---|-----|------|------|---------|-------|------|
| 1 | 76.27 | 76.83 | 2.42 | 20 | 0.73 | 11.47 |
| 2 | 75.45 | 76.34 | 1.96 | 11 | 1.17 | 11.60 |
| 3 | 78.81 | 78.95 | 2.01 | 9 | 0.18 | 11.17 |
| 4 | 76.19 | 76.58 | 2.09 | 9 | 0.51 | 11.08 |
| 5 | 77.11 | 76.73 | 2.06 | 20 | 0.49 | 12.30 |

Table 4: Validation of the simulation model

The results of the validation process, which was more elaborate than indicated here, were considered very satisfactory by our industry partners and it was decided to accept proposals that are based on the use of this simulation model.

At any point in time, the stacker crane control system has to decide which of the jobs to execute next. In the old system this was done by a priority based FIFO rule. We proposed to replace this heuristic by solving the following problem. Whenever a set of jobs is given or modified, schedule the jobs in such a way that the unloaded travel time (or positioning time, i.e., the time the stacker crane moves without carrying a box) is as short as possible. This problem turns out to be a directed Hamiltonian path problem; it is trivially equivalent to an asymmetric TSP. We solve this directed Hamiltonian path problem as follows. At a general point in time, the stacker crane is executing some job or idle and there are some jobs that the stacker crane has to execute in the future. If one job is finished, the crane starts the first job of the sequence. Due to calls from the assembly line or deliveries of the driverless vehicles new jobs are created. Whenever a new job comes up, we run a very fast insertion heuristic to schedule the new job and call the resulting sequence our new solution. Then a more elaborate but still fast heuristic tries to improve the present solution. Finally, a branch-and-bound code is activated that determines an optimum solution of the present problem.

It may happen that a new job is created while we are still computing. In that case we stop the optimization process and turn to the new enlarged problem.

|   | # TT | uTr-P | uTr-0 | I % | max-TT | ∅# TT |
|---|------|-------|-------|------|--------|-------|
| 1 | 416  | 8599  | 8325  | 3.18 | 6      | 2.31  |
| 2 | 421  | 8655  | 8141  | 5.93 | 8      | 1.94  |
| 3 | 405  | 8238  | 7956  | 3.42 | 6      | 2.27  |
| 4 | 398  | 8017  | 7634  | 4.77 | 8      | 1.93  |
| 5 | 447  | 9411  | 8951  | 4.88 | 8      | 2.13  |

Table 5: Minimizing the unloaded travel time
(normal conditions)

The asymmetric TSPs that come up in this system are very small under normal conditions (see max-TT and ∅#TT in Table 5 for largest and the average instance sizes). Even under heavy load conditions (see max-TT and ∅#TT in Table 7) the sizes of the assymetric TSP instances arising have never exceeded 40 "cities" during the days for which we collected data. Thus we decided to use a branch-and-bound code instead of investing into the extensive effort of implementing a cutting plane algorithm. Currently we are using the code of Fischetti and Toth [FT92] that solves the instances arising here in reasonable time. We used the simulation model to compare our approach with the old system. Table 5 shows the results for the week that was used for the simulation model validation. The running time improvements (see column I%) of 3% to 6% were very disappointing. By analyzing the data it turned out that the week considered was a week of low production volume. On the average, there were only two jobs to schedule at a time (see column ∅#TT) and thus almost nothing to optimize.

Tables 6 and 7 report about two different periods of heavy load. These usually occur after the breakdown of some part of the production system. In such cases jobs to be executed start to pile up. The running time improvements (see columns I%) in these cases ranged from 15% to 40%. In fact, this is exactly what is needed in the system: quick recovery from "catastrophies". The results of our optimization approach were considered so satisfactory that our software for scheduling the moves of the stacker crane has been put into use at the Siemens Nixdorf Werk für Arbeitsplatzsysteme. It has considerably improved the capacity of the storage system.

|   | # TT | uTr-P | uTr-O | I %   |
|---|------|-------|-------|-------|
| 1 | 18   | 344   | 239   | 30.52 |
| 2 | 10   | 194   | 143   | 26.28 |
| 3 | 18   | 347   | 211   | 39.19 |
| 4 | 26   | 507   | 330   | 34.91 |
| 5 | 20   | 388   | 283   | 27.06 |

Table 6: Minimizing the unloaded travel time
(heavy load conditions)

|   | # TT | uTr-P | uTr-O | I %   | max-TT | ∅# TT |
|---|------|-------|-------|-------|--------|-------|
| 1 | 50   | 917   | 693   | 24.42 | 29     | 13.32 |
| 2 | 49   | 974   | 749   | 23.10 | 20     | 8.32  |
| 3 | 50   | 1007  | 783   | 22.24 | 26     | 12.23 |
| 4 | 49   | 889   | 662   | 25.53 | 31     | 15.24 |
| 5 | 50   | 985   | 839   | 14.82 | 25     | 13.08 |

Table 7: Minimizing the unloaded travel time
(heavy load conditions)

| | | |
|---|---|---|
| # TT | : | number of transportation tasks (TT) |
| uTR-P | : | unloaded travel time with priority rule (in sec.) |
| uTr-O | : | unloaded travel time with optimization (in sec.) |
| I % | : | improvement in % ((( uTR-P) − (uTr-O) ) / (uTr-P) ) |
| max-TT | : | maximal number of TT handled at the same time |
| Ø# TT | : | average number of TT handled at the same time |

We are aware of the fact that our approach may not be the best way to handle on-line decisions of the type described above. But the scientific literature on handling complex on-line situations such as this is one is almost nonexisting. We were happy that we could improve a bottleneck situation in the factory considerably, but we know that deeper theoretical investigations are necessary to get a good understanding of such cases. We are currently experimenting with other approaches that lead to new combinatorial optimization problems and are going to compare these on a practical and theoretical basis with the approach outlined above.

More details on the results of our stacker crane optimization will appear soon in joint papers with the Siemens Nixdorf partners. The mathematics involved and a thorough analysis of the practical achievements will be described in [As93].

**5. More Applications and Some Mathematics.** This paper consists of a list of a few problems arising in the world of manufacturing and my attempts to describe the use of mathematics in their solution. I have tried to indicate that the line of attack taken is a general approach, not confined to the special cases discussed here. But I also have to admit that there is no guarantee of success. In every particular case, substantial **joint work** of practitioners and theoreticians is necessary to contribute significantly to the solution of the problems coming up. Due to lack of space many important areas in manufacturing where discrete mathematics plays or could play an essential role could not be mentioned at all. I will indicate a few further fruitful topics in the subsequent subsection. I also note that the mathematical theory involved in the solution of the problems outlined above has been purposely avoided here. A glimpse at it will be provided in Section 5.2.

**5.1. A Few Further Applications.** A vast application area for discrete mathematics is the field of logistics and transportation. Factory internal questions such as where to store parts and how to move them are vital for production speed. Due to the increasing emergence of just-in-time manufacturing processes and the high traffic volume associated with these production techniques, the planning and operating of shipments of parts and components have become a major topic. This involves questions of what to order when and in which quantities. This affects the planning of the fleets of trucks that are necessary to handle the transportation volume and concerns the actual planning of tours to satisfy all requirements and minimize transportation costs. Moreover, shifts of drivers have to be determined such that all work regulations are met and the labor costs are as low as possible. Clearly, the latter question extends to plant management. Finding a cost-effective mix of the work force, assigning the right people to the right job, and planning the work so that the idle time is minimum is of substantial importance. Problems of this type lead to integer, mixed-integer or combinatorial optimization problems of very large scale.

Lot sizing problems have always been a major concern in industry. They play an increasing role since many advanced machines (for instance in the chemical industry) can be operated cheaply at high production volume but have very high set-up costs. Balancing these with the cost for inventory and the desire of keeping the stocks at a level such that all expected demands can be supplied is a challenging task.

I would also like to mention a topic that is typically neglected. There are discussions in the public press about whether or not airplanes or nuclear power plants are safe. Machine safety is rarely a public issue in general. What I have in mind here is not only the security of workers or the ecological systems, but also the safeguarding of machines against unwanted moves or self-destruction. The ordinary machines on factory floors, the automatic assembly and transfer lines are becoming more and more complicated. Tools for engineers are scarcely available to help them design machines that do not only operate correctly in standard situations, but machines that are also safe in case certain components or control sensors break. Severe accidents or the breakdown of the production process may occur if the control system of a machine is unable to handle a complicated mix of errors.

I have worked on such a topic together with Klaus Truemper (University of Texas at Dallas) jointly with the Grob Corporation (Mindelheim, Germany). Problems as indicated above can often be formulated as problems in algorithmic logic. A main task here is to design algorithms for the satisfyability problem of a logic formula (say, given in conjunctive normal form) that decide very quickly if a formula has a solution or not. It is not enough to have a fast answer in case there is a solution, but one also has to know quickly if there is none. This second requirement makes the design of a different type of algorithm for the $\mathcal{NP}$-complete satisfyability problem necessary. The standard algorithmic approaches do a reasonable job if solutions of a logic formula exist but have unpredictable running times otherwise. Predictability of the solution time here is paramount for the safety of the systems. For example, if a pilot learns after 5 minutes that an operation he performed was dangerous, it may be too late. Klaus Truemper has developed a very effective software system, called Leibniz, to handle such situations. It is based on new results on ternary matroids; see [T90], [T93].

Let me close this journey through applications with an important question that I am often asked by students and that I am unable to answer satisfactorily. How does one model practical situation mathematically? Are there guidelines for this approach? I simply don't know how to teach this methodology other than by giving examples, by describing basic situations that frequently arise, by showing "little tricks" that help, by outlining approaches that are hopeless, and by giving heuristic reasons why some models work and why others don't work. The work described in this paper reflects this fact.

**5.2. A Glimpse at the Mathematics.** Due to lack of space it is impossible to describe all the mathematical theory that has been developed to solve the problems mentioned in the previous sections. I will list here a few of the techniques used and point to the literature where in-depth treatments can be found.

It is indeed fortunate when one encounters a well-studied and polynomially solvable combinatorial optimization problem in a real-world application. Easy problems, such as shortest path, max-flow or min-cost flow problems do, in fact, frequently arise as subproblems of more general problem types. In such cases, one can look in standard textbooks like [Sch86] or [NW88] or recent survey papers such as [AMO89] or [GoTT90] for flow problems to get information about available theory and existing fast algorithms.

For example, among the cases discussed here, two of the box assignment problems described in Section 4.2 turned out to be ordinary assignment problems. We knew that Peter Kleinschmidt (Passau) had recently implemented a fast assignment code. He subsequently tuned his algorithm for the instance structure that arises in our application. It easily solves all of the instances of interest to us.

In the via minimization case reported in Section 2.4, a certain setting of the parameters (occurring in practice) results in max-cut problems for planar graphs. There exist a beautiful theory and several algorithmic approaches for solving this problem, but no implementation was available. In her master's thesis, Petra Mutzel [Mu90] closed this gap and came up with a practically and theoretically fast code for the solution of such cases. This involved the implementation of a planarity testing and the associated embedding algorithm, of a graph dualization algorithm and a certain transformation procedure, and of a shortest path and a matching algorithm.

$\mathcal{NP}$-hard problems need special treatment. What to do depends on the sizes of the problem instances that arise in the practical situation under investigation, the computers available and on the running time bounds given.

For instance, we noticed that in the stacker crane optimization problem of Section 4.2 asymmetric travelling salesman problems arise that have small to moderate size. Branch and bound algorithms based on assignment relaxations, such as the one described in, are able to solve such instances reasonably well. Standard techniques that can be found in any textbook on combinatorial optimization suffice. There is no need to use more involved machinery in such cases.

The design of heuristics for large-scale problems is of particular importance. In some cases known basic approaches work well only for small problems. Further investigation of the problem structure, of suitable data structures and their running time analysis is required to extend these heuristics to large-scale problems. For example, in TSP applications like the plotting and drilling problems discussed in Chapter 3, where points are given by their coordinates and the distances are defined by a metric, the use of techniques from computational geometry (e.g., Voronoi diagrams and convex hulls) helped to speed up existing algorithms tremendously and to achieve good quality solutions in very short time. Such heuristic techniques were thoroughly investigated in [R92]. Even the design of relatively simple heuristics may lead to mathematical problems of general interest.

Substantial mathematical work is necessary if one wants to design algorithms that produce provably optimal solutions or provide very good quality guarantees. At present, the most successful general approach in this case is the use of polyhedral combinatorics as the backbone of LP-based cutting plane or branch-and-cut algorithms. This technique associates with every instance of a combinatorial optimization problem a polyhedron such that the optimum solutions of the instance of the combinatorial optimization problem are precisely the optimum vertex solutions of the corresponding programs over the associated polyhedron. The difficulty is that the construction of the polyhedron is nothing but a theoretical device. A complete and nonredundant description of the associated polyhedron by means of linear inequalities and equations has to be found by special case investigations. Moreover, turning these theoretical investigations into efficient algorithmic tools requires further work and insight.

Many combinatorial optimization problems are currently investigated by following this approach. The survey papers [GP85], [PaG85] describe this approach for the travelling salesman problem. A more general survey of the area of polyhedral combinatorics is [Pu89]. I would like to demonstrate this technique using the max-cut problem as an example.

If $G = (V, E)$ is a graph and $W \subseteq V$ a node set, then the edge set $\delta(W) := \{ij \in E | i \in W, j \in V \setminus W\}$ is called a cut. (If $\emptyset \neq W \neq V$, the removal of this edge set will disconnect the nodes in $W$ from the nodes in $V \setminus W$.) An instance of the max-cut problem is given by a graph $G = (V, E)$ with weights $c_e \in \mathbb{R}$ for all edges $e \in E$. The task is to find a cut $\delta(W)$ of $G$ of maximum weigth $c(\delta(W)) := \sum_{e \in \delta(W)} c_e$. The max-cut problem comes up, for instance, in via minimization, see Section 2.4, and is used in statistical mechanics to calculate ground states of spinglasses, see [BaGJR88].

To associate a polyhedron with an instance of the max-cut problem we proceed as follows. Given $G = (V, E)$ we consider the vector space $\mathbb{R}^E$, where each component of a vector $x = (x_e)_{e \in E} \in \mathbb{R}^E$ is indexed by an edge of $E$. If $F \subseteq E$, we define the incidence vector $\chi^F \in \mathbb{R}^E$ of $F$ by setting $\chi^F_e = 1$ if $e \in F$ and $\chi^F_e = 0$ if $e \notin F$. The cut polytope $\mathrm{CUT}(G)$ associated with $G$ is nothing but the convex hull of all incidence vectors of cuts of $G$, i.e.,

$$(5.1) \qquad \mathrm{CUT}(G) = \mathrm{conv}\{\chi^{\delta(W)} \in \mathbb{R}^E \mid W \subseteq V\}.$$

It follows from this construction that the cuts of $G$ are in one-to-one correspondence with the vertices of the cut polytope. This implies that the optimum vertex solutions of the linear program

$$(5.2) \qquad \begin{array}{ll} \max & c^T x \\ \text{s.t.} & x \in \mathrm{CUT}(G) \end{array}$$

are precisely the optimum solutions of the max-cut problem defined by $G = (V, E)$ and the edge weights $c_e$. Hence, we could solve the max-cut problem by linear programming techniques if we knew a description of $\mathrm{CUT}(G)$ by means of linear inequalities. To find inequalities that are valid for $\mathrm{CUT}(G)$ and define facets of $\mathrm{CUT}(G)$ is an interesting research topic. The paper [DL91] of over 80 pages length surveys its state-of-the-art. The cut polytope has received so much attention because it arises in so many different ways. For instance, it is related to Eulerian subgraphs by graph duality; embedding problems of semi-metric spaces in functional analysis lead to investigations of cut cones [AsD82]; and cycles in binary matroids form a far reaching generalization [BaG86], [GT89].

A first step of investigation is usually to formulate (5.2) as an integer linear program. One of the first theorems taught in graph theory states that a cut and a circuit meet in an even number of edges. Thus, if $C \subseteq E$ is a circuit and $F \subseteq C$ of odd cardinality, then every incidence vector of a cut must satisfy the so-called "odd circuit inequality"

$$\sum_{e \in F} x_e - \sum_{e \in C \setminus F} x_e \leq |F| - 1 .$$

Every incidence vector obviously satisfies the "trivial constraints" $0 \leq x_e \leq 1$ for all $e \in E$ and thus it follows that all inequalities,

(5.3)
(i) $0 \leq x_e \leq 1$ for all $e \in E$,
(ii) $\sum_{e \in F} x_e - \sum_{e \in C \setminus F} x_e \leq |F| - 1$ for all circuits $C \subseteq E$, and all $F \subseteq C, |F|$ odd

are valid for CUT($G$). Let us set

$$P(G) := \{x \in \mathbb{R}^E | \; x \text{ satisfies (5.3) (i) and (ii)}\}.$$

Then, clearly, CUT($G$) $\subseteq$ P($G$). In fact, one can easily prove the following.

(5.4) PROPOSITION. *For every graph* $G = (V, E)$

$$\mathrm{CUT}(G) = conv\{x \in P(G) | \; x \;\; integral\}.$$

This implies that

(5.5)
max $c^T x$
(i) $0 \leq x_e \leq 1$ for all $e \in E$,
(ii) $\sum_{e \in F} x_e - \sum_{e \in C \setminus F} x_e \leq |F| - 1$ for all circuits $C \subseteq E$, and for all $F \subseteq C, |F|$ odd
(iii) $x_e \in \{0, 1\}$ for all $e \in E$

is an integer programming formulation of (5.2).

An immediate question arises. Do we really need the integrality conditions in (5.5), i. e., does CUT($G$) =P($G$) hold? A result due to Barahona and Majhoub [BaM86] answers this question.

(5.6) THEOREM. *Let* $G = (V, E)$ *be a graph. Then* P($G$) = CUT($G$) *if and only if* $G$ *is not contractible to the complete graph* $K_5$.

Theorem (5.6) together with the Wagner-Kuratowski theorem for planar graphs yields that P($G$) = CUT($G$) holds for planar graphs. A considerable generalization of (5.6) is the polyhedral characterization of those binary matroids that have the sums of circuits property; see [Sey81] and [GT89].

Let us return from this theoretical line of investigation, which is of interest in its own right, to more algorithmic issues. If we can solve the linear program that arises from (5.5) by dropping the integrality constraints, i. e., the so-called LP-relaxation of (5.5), then we will obtain an upper bound on the maximum weight of a cut. A quick count shows that the number of odd circuit inequalities grows exponentially with the graph size. Hence, there is no way to input these inequalities into a computer in polynomial time. Do we have to give up?

An obvious idea now is to check whether we really need all odd circuit inequalites and to determine which are redundant. This amounts to characterizing those inequalities of the system (5.3) (i), (ii) that define facets of the cut polytope. The following is shown in [BaM86].

(5.7) THEOREM. *Let* $G = (V, E)$ *be a graph.*
*(a) The dimension of* CUT($G$) *is equal to* $|E|$.
*(b) For every edge* $e \in E$, *the following statements are equivalent:*
$(b_1)$ $x_e \geq 0$ *defines a facet of* CUT($G$),

($b_2$) $x_e \leq 1$ defines a facet of $CUT(G)$,
($b_3$) $e$ does not belong to a triangle.
(c) Let $C \subseteq E$ be a circuit and $F \subseteq C$, $|F|$ odd, then the odd circuit inequality
$$\sum_{e \in F} x_e - \sum_{e \in C \setminus F} x_e \leq |C| - 1$$
defines a facet of $CUT(G)$ if and only if $C$ has no chord.

This result reduces the number of necessary inequalities considerably. For instance, for the complete graph $K_n$, no trivial constraint is facet-defining and only triangles are chordless circuits. Thus, only $O(n^3)$ inequalities remain, i.e., in this case the LP- relaxation has polynomial size. But in general, there are still exponentially many facet-defining odd circuit constraints.

Nevertheless, we do not have to give up. Help comes from powerful results that are based on the ellipsoid method. To formulate these we have to introduce further concepts.

Let $P$ be a polyhedron. By definition, there exist a matrix $A$ and a vector $b$ such that P= $\{x | Ax \leq b\}$, and, equivalently, there are finite sets $S$, $T$ of vectors such that P= $conv(V) + cone(T)$, where $cone(T)$ denotes the set of all points that are nonnegative linear combinations of elements of $T$. We call P rational if all entries of $A$ and $b$ (or equivalently, all entries of the vectors in $S$ and $T$) are rational. The encoding length of an inequality $a^T x \leq \alpha$, $a$ and $\alpha$ rational, is the number of binary digits needed to encode $a$ and $\alpha$, where a rational number is encoded by encoding its numerator and denominator. We say that a rational polyhedron P has facet-complexity at most $\varphi$ ($\varphi \in \mathbb{N}$) if there is an inequality system $Ax \leq b$ such that $P = \{x | Ax \leq b\}$ and each inequality of the system has encoding length at most $\varphi$.

Note that polyhedra with very many facets may have small facet-complexity. Consider, for example, the inequality system (5.3). The encoding length of the inequalities in (i) is 4, while the maximum encoding length of the inequalities in (ii) is $2|E| + \lceil log|E| \rceil + 1$. Hence, the facet-complexity of P($G$) is at most $3|E|$, say. This implies that the facet-complexity of the polytopes P($G$) is polynomial in the encoding length of $G$. One can similarly prove that the facet-complexity of cut polytopes is polynomial in the encoding length of the associated graphs.

Let $\mathcal{P}$ be a class of rational polyhedra. We say that the optimization problem for $\mathcal{P}$ can be solved in polynomial time if, for any polyhedron $P \in \mathcal{P}$ and any rational vector $c$, the linear program

$$\max \quad c^T x$$
$$\text{s.t.} \quad x \in P$$

can be solved in time polynomial in the encoding length of $c$ and the facet-complexity of $P$.

The separation problem for a rational polyhedron $P \subseteq \mathbb{R}^n$ is the following. Given a rational vector $y \in \mathbb{R}^n$, decide whether $y \in P$ and if not, determine a vector $c \in \mathbb{R}^n$ such that $c^T y > max\{c^T x | x \in P\}$. Observe that $c$ yields a hyperplane separating $y$ from P.

Let $\mathcal{P}$ be a class of rational polyhedra. We say that the separation problem for $\mathcal{P}$ can be solved in polynomial time if, for any polyhedron $P \in \mathcal{P}$ and any vector $y$, the separation problem for P and $y$ can be solved in time polynomial in the encoding length of $y$ and the facet-complexity of $P$.

The following result, using the ellipsoid method, was proved in [GLS81], see also [GLS88] for an in-depth treatment of this subject.

(5.8) THEOREM. *Let $\mathcal{P}$ be a class of rational polyhedra. Then the following two statements are equivalent:*
*(a) The optimization problem for $\mathcal{P}$ can be solved in polynomial time.*
*(b) The separation problem for $\mathcal{P}$ can be solved in polynomial time.*

What help does (5.8) provide in the study of cut polytopes? Note that the classes $\mathcal{P}_1 := \{CUT(G) | G$ a graph$\}$, $\mathcal{P}_2 := \{P(G) | G$ a graph$\}$ are classes of rational polyhedra where the polyhedra in $\mathcal{P}_1$ and $\mathcal{P}_2$ have a facet-complexity that is polynomial in the encoding length of the associated graph. We really would like to prove that the optimization problem for $\mathcal{P}_1$ can be solved in polynomial time and that means in time polynomial in $|E| + |V|$. Due to the $\mathcal{NP}$-hardness of the max-cut problem this seems to be impossible. But we can optimize over $\mathcal{P}_2$ using the equivalence of (a) and (b) in (5.8). Namely, it was shown in [BaM86] that the separation problem for $P(G)$, $G$ a graph, can be solved in polynomial time. The hard part, of course, is, given a vector $y$, to check whether $y$ satisfies all odd circuit constraints and if not to find one of these inequalities that is violated by $y$.

The result follows by a tricky transformation of this problem to a sequence of shortest path problems as follows.

Let $G = (V, E)$ be a graph and $y \in \mathbf{Q}^E$. We first check whether $0 \leq y_e \leq 1$ for all $e \in E$ by substitution. If not, a separating hyperplane is at hand. Otherwise we construct a new graph $H = (V' \cup V'', E' \cup E'' \cup E''')$ consisting of two disjoint copies $G' = (V', E')$ and $G'' = (V'', E'')$ of $G$ and an additional edge set $E'''$ that contains, for each $uv \in E$, the two edges $u'v'', u''v'$. The edges $u'v' \in E'$ and $u''v'' \in E''$ get the weight $y_{uv}$, while the edges $u'v'', u''v' \in E'''$ get the weigth $1 - y_{uv}$. For each node $u \in V$, we calculate a shortest (with respect to the weights just defined) path in $H$ from $u' \in V'$ to $u'' \in V''$. Such a path contains an odd number of edges of $E'''$ and corresponds to a closed walk in $G$ containing $u$. Clearly, if the shortest of these $[u', u'']$-paths has length at least 1 then $y$ satisfies all odd circuit constraints, otherwise there exists a circuit $C$ and a set $F \subseteq C, |F|$ odd, such that $y$ violates the corresponding inequality.

Summing all these observations up, we obtain that for any graph $G$, the LP-relaxation of (5.5) can be solved in polynomial time and thus the max-cut problem for graphs not contractible to $K_5$ (and hence for planar graphs) can be solved in polynomial time. One has to admit, though, that these algorithmic results are not practical. If one replaces the ellipsoid method (that is needed to derive the polynomial time bounds) by the simplex method and the use of cutting plane techniques one obtains a practically useful algorithm for solving the LP-relaxation of (5.5).

This is the the central idea for a relatively efficient branch-and-cut algorithm for the max-cut problem that is based on the LP-relaxation of (5.5), that uses further classes of facet-defining cutting planes for $CUT(G)$ heuristically and that employs additional heuristic techniques. This algorithm solves max-cut problems on graphs with several thousand nodes to optimality, see [BaGJR88].

Techniques like the one described above form the backbone of many recent successful attempts to solve hard large scale combinatorial optimization problems to optimality, see for example [GH91], [PaR91]. There are further approaches based, for instance, on Lagrangian relaxations and the use of nondifferentiable convex minimization or based on geometry of numbers and basis reduction for lattices, see [LS90], [CRSS92]. It is impossible to cover these approaches here.

**5.3. Final Remarks.** The purpose of this paper was to show, by means of a few examples, how and where interesting mathematical problems of a discrete nature arise in the field of manufacturing and to indicate the important role mathematics could play if all parties involved in manufacturing were aware of the contributions mathematics is able to make. The focus was on describing a few real-world applications and to indicate the difficulties that arise in mathematically modelling complex situations. There are cases where we have a solid understanding of the practical problems, a rich associated mathematical theory and sophisticated algorithmic tools. In other cases suitable mathematical models are still missing, sound theories have not yet emerged and substantial, probably quite difficult mathematical work still lies before us.

REFERENCES

[AH92] A. Abdel Hamid, *Optimization aspects of automatic storage systems*, Dissertation, Technische Universität Berlin, 1992.

[AKP89] H. Achatz, P. Kleinschmidt, K. Paparrizos, *A dual algorithm for the assignment problem*, Working Paper, Universität Passau, 1989.

[AMO89] R.K. Ahuja, T.L. Magnanti, J. B. Orlin, *Network flows*, in: G. L. Nemhauser, A. H. R. Rinnooy Kan, M. J. Todd (eds.), *Optimization*, Handbook in Operations Research and Management Science, North-Holland, Amsterdam, 1989, pp. 211-369.

[As93] N. Ascheuer, *On-line-optimization of flexible manufacturing systems*, Dissertation, Technische Universität Berlin, 1993, to appear.

[AsD82] P. Assouad. M. Deza, *Metric subspaces of $L^1$*, Publications Mathématiques d'Orsay, vol. 3, 1982.

[BaG86] F. Barahona, M. Grötschel, *On the cycle polytope of a binary matroid*, Journal of Combinatorial Theory B, 40 (1986), pp. 40-62.

[BaGJR88] F. Barahona, M. Grötschel, M. Jünger, G. Reinelt, *An application of combinatorial optimization to statistical physics and circuit layout design*, Operations Research, 36 (1988), pp. 493-513.

[BaM86] F. Barahona, A.R. Mahjoub, *On the cut polytope*, Mathematical Programming, 36 (1986), pp. 157-173.

[Bau90] P. Bauer, *Optimale Steuerung des FALKE-Automaten*, Projektdokumentation, Internal Report, Augsburg, 1990.

[Br77] M.A. Breuer, *Min Cut Placement*, J. Design Aut. &. Fault Tol. Comp., 1 (1977), pp. 343-362.

[CKC83] R.-W. Chen, Y. Katjitani, S.-P. Chan, *A graph-theoretic via minimization algorithm for two-layer printed circuit bourds*, IEEE Trans. Circuits and Syst., 30 (1983), pp. 284-299.

[CRSS92] W. Cook, T. Rutherford, H. Scarf, D. Shallcross, *Integer Programming using Lovász-Scarf basis reduction*, in preparation.

[DL91] M. Deza, M. Laurent, *A survey of the known facets of the cut cone*, Institut für Ökonomie und Operations Research, Universität Bonn, Report No. 91722-OR, 1991.

[FT92] M. Fischetti, P. Toth, *An additive bounding procedure for the asymmetric TSP*, Mathematical Programming, 53 (1992), pp. 173-197.

[GaJ77] M.R. Garey, D.S. Johnson, *The rectilinear Steiner Tree problem is NP-complete*, SIAM J. Appl. Math., 32 (1977), pp. 826-834.

[GoTT90] A. V. Goldberg, É. Tardos, R. E. Tarjan, *Network flow algorithms*, in: B. Korte, L. Lovász, H. J. Prömel, A. Schrijver (eds.), *Paths, Flows, and VLSI-Layout*, Springer, Berlin, 1990, pp. 101-164.

[GH91] M. Grötschel, O. Holland, *Solution of large-scale symmetric travelling salesman problems*, Mathematical Programming, 51 (1991), pp. 141-202.

[GJR89] M. Grötschel, M. Jünger, G. Reinelt, *Via minimization with pin preassignments and layer preference*, Zeitschrift für Angewandte Mathematik und Mechanik, 69 (1989), pp. 393-399.

[GJR91] M. Grötschel, M. Jünger, G. Reinelt, *Optimal control of plotting and drilling machines: a case study*, Zeitschrift für Operations Research, 35 (1991), pp. 61-84.

[GLS81] M. Grötschel, L. Lovász, A. Schrijver, *The ellipsiod method and its consequences in combinatorial optimization*, Combinatoria, 1 (1981), pp. 169-197.

[GLS88] M. Grötschel, L. Lovász, A. Schrijver, *Geometric Algorithms and Combinatorial Optimization*, Springer, Berlin, 1988.

[GP85] M. Grötschel, M. W. Padberg, *Polyhedral theory* in: E. L. Lawler, J. K. Lenstra, A. H. G. Rinnooy Kan, and D. B. Shmoys (eds.), *The Traveling Salesman Problem*, Wiley, Chichester, 1985, pp. 251-305.

[GT89] M. Grötschel, K. Truemper, *Decomposition and optimization over cycles in binary matroids*, Journal of Combinatorial Theory B, 46 (1989), pp. 306-337.

[JMRW92] M. Jünger, A. Martin, G. Reinelt, R. Weismantel, *Quadratic 0/1 optimization and a decomposition approach to the placement of electronic circuits*, Mathematical Programming, 1992, to appear.

[Ka72] R. M. Karp, *Reducibility among combinatorial problems*, R. E. Miller, J. W. Thatcher (eds.), *Complexity of Computer Computations*, Plenum Press, New York, 1972, pp. 85-103.

[KlSJ88] J. M. Kleinhans, G. Sigl, F. M. Johannes, *Gordian: A new global optimization/ rectangle dissection method for cell placement*, IEEE Int. Conference on CAD ICCAD-88, 1988, pp. 506-509.

[KPS90] B. Korte, H. J. Prömel, A. Steger, *Steiner trees in VLSI-Layout*, in: B. Korte, L. Lovász, H. J. Prömel, A. Schrijver (eds.), *Paths, Flows, and VLSI-Layout*, Springer, Berlin, 1990, pp. 185-214.

[KrL84] M. R. Kramer, J. van Leeuwen, *The complexitiy of wire-routing and finding minimum area layout for arbitrary VLSI circuits*, in: F.P. Preparata (ed.), *Advances in Computing Research*, 1984, pp. 129-146.

[LLRS85] E. L. Lawler, J. K. Lenstra, A. H. G. Rinnooy Kan, and D. B. Shmoys (eds.), *The Traveling Salesman Problem* Wiley, Chichester, 1985, pp. 251-305.

[L90] T. Lengauer, *Combinatorial Algorithms for Integrated Circuit Layout*, Wiley, Chichester, 1990.

[LS90] L. Lovász, H. Scarf, *The generalized basis reduction algorithm*, Cowles Foundation, Discussion Paper No. 946, Yale University, June 1990.

[M92] A. Martin, *Packen von Steinerbäumen: Polyedrische Studien und Anwendungen*, Dissertation, Technische Universität Berlin, 1992.

[Mu90] Petra Mutzel, *Implementierung und Analyse eines Max-Cut-Algorithmus*, Diplomarbeit, Universität Augsburg, 1990.

[NW88] G. Nemhauser, L, A. Wolsey, *Integer and Combinatorial Optimization*, Wiley, Chichester, 1988.

[PaG85] M. Padberg, M. Grötschel, *Polyhedral computations*, in: E. L. Lawler, J. K. Lenstra, A. H. G. Rinnooy Kan, and D. B. Shmoys, (eds.), *The Traveling Salesman Problem*, Wiley, Chichester, 1985, pp. 307-360.

[PaR91] M. Padberg, G. Rinaldi, *A branch-and-cut-algorithm for the resolution of large-scale symmetric traveling salesman problems*, SIAM Review, 33 (1991) pp. 60-100.

[Pi84] R.Y. Pinter, *Optimal layer assignment for interconnect*, J. VLSI Comput. Syst., 1 (1984), pp. 123-137.

[Pu89] W. R. Pulleyblank, *Polyhedral combinatorics*, in: G. L. Nemhauser, A. H. R. Rinnooy Kan, M. J. Todd (eds.), *Optimization*, Handbook in Operations Research and Management Science, North-Holland, Amsterdam, 1989, pp. 371-446.

[R91] G. Reinelt, *TSPLIB - A traveling salesman problem library*, ORSA Journal on Computing, 3 (1991), pp. 43-49.

[R92] G. Reinelt, *Contributions to Practical Traveling Salesman Problem Solving*, Springer, Heidelberg, 1992, pp. 43-49.

[Sch86] A. Schrijver, *Theory of Linear and Integer Programming*, Wiley, Chichester, 1986.

[Sey81] P.D. Seymour, *Matroids and multicommodity flows*, European Journal of Combinatorics, 2 (1981), pp. 257-290.

[SK72] D.G. Schweikert, B. W. Kernighan *A proper model for the partitioning of electrical circuits*, Proceedings Design Automation Conference, 1972, pp. 56-62.

[T90] K. Truemper, *Leibniz System: User's Manual*, Leibniz Company, Plano, Texas, 1990.
[T93] K. Truemper, *Logic Decomposition*, 1993, in preparation.
[W92] R. Weismantel, *Plazieren von Zellen: Analyse und Lösung eines quadratischen 0/1-Optimierungsproblems*, Dissertation, Technische Universität Berlin, 1992.

Chapter 10
# ANALYSIS OF HYPERSONIC FLOWS IN THERMOCHEMICAL EQUILIBRIUM BY APPLICATION OF THE GALERKIN/LEAST-SQUARES FORMULATION

FRÉDÉRIC L. CHALOT
Stanford University
Stanford, California

THOMAS J. R. HUGHES
Stanford University
Stanford, California
(pictured)

**1. Introduction.** In extending the Galerkin/least-squares finite element method to hypersonic flows involving chemistry and high-temperature effects, the entropy variables approach may have been expected to engender complications. In fact, not only was no fundamental impediment encountered, but also what seemed to be a consequence of the perfect gas assumption, proves to be quite general.

Although all the material presented subsequently pertains to what we call a general divariant gas, this paper is aimed primarily at the description of equilibrium flows. Before outlining the contents of the paper, we feel it useful to spend some time explaining what we mean by "equilibrium flow" and "general divariant gas." In the thermodynamic sense, equilibrium is defined as the combination of mechanical, thermal and chemical equilibria. Mechanical equilibrium is achieved when there are no unbalanced forces within the considered system or between the system and its surroundings. In the absence of body forces this leads to uniform velocity and pressure distributions. Thermal equilibrium requires all parts of the system to have the same temperature equal to the temperature of the surroundings. Of course not too many interesting flows have constant pressure, temperature, and velocity. However, in many practical instances to the engineer, the assumption of local equilibrium can be made. Under such an assumption, the system can be divided into a collection of microsystems, small on the thermodynamic scale, but large enough to allow certain equilibration processes to take place through particle collisions. Each microsystem is considered in mechanical and thermal equilibrium internally, so that local values of pressure and temperature can be defined. Note that a particular microsystem is not necessarily in equilibrium with the surrounding ones. Molecular processes associated with rotation, vibration, and electronic excitation are assumed to be in equilibrium at the translational temperature. This supposes that the corresponding time scales are small compared to that pertinent to the flow field. Obviously this condition cannot be satisfied in regions of the flow where large gradients exist, such as behind strong shock waves or in the boundary layer. In addition, small departures from translational equilibrium, such as viscous dissipation and thermal conduction, are taken into account by the Navier-Stokes terms. Hence, in the absence of chemical reactions, the system can be described as a general divariant gas: given an equation of state, its thermodynamic properties are completely defined by any pair of state variables, say pressure and temperature. The equation of state is not limited to that of a thermally perfect gas; in fact, one is not even restricted to the sole description of gases. If the system is chemically reactive, the equilibrium flow assumption requires that the chemical reactions be instantaneous. In other words, each microsystem has a uniform chemical composition which responds instantly to any change in pressure or temperature. Neglecting the chemical kinetics also precludes account

---

*This work was supported by NASA Langley Research Center under Grant NASA-NAG-1-361 and Dassault Aviation, St Cloud, France.

of mass diffusion, another translational nonequilibrium phenomenon. However, it still permits the description of the system as a general divariant gas.

An outline of the paper follows. In the next section, we derive the symmetric form of the Euler and Navier-Stokes equations for a general divariant gas. In section 3, we describe briefly the Galerkin/least-squares finite element formulation. In section 4, we propose a simple equilibrium chemistry model for air. In section 5, before giving some concluding remarks, we present a few numerical examples which confirm the practical computer implementation of the method.

**2. Symmetric Euler and Navier-Stokes equations for a general divariant gas.** As a starting point, we consider the Euler and Navier-stokes equations written in conservative form:

$$(2.1) \qquad U_{,t} + F^{\text{adv}}_{i,i} = F^{\text{diff}}_{i,i}$$

where $U$ is the vector of conservative variables; $F^{\text{adv}}_i$ and $F^{\text{diff}}_i$ are, respectively, the advective and the diffusive fluxes in the $i^{\text{th}}$-direction. Inferior commas denote partial differentiation and repeated indices indicate summation. In three dimensions, $U$, $F^{\text{adv}}$, and $F^{\text{diff}}$ read

$$(2.2) \qquad U = \rho \left\{ \begin{array}{c} 1 \\ u \\ e^{\text{tot}} \end{array} \right\}$$

$$(2.3) \qquad F^{\text{adv}}_i = u_i U + p \left\{ \begin{array}{c} 0 \\ \delta_i \\ u_i \end{array} \right\} \qquad F^{\text{diff}}_i = F^{\text{visc}}_i + F^{\text{heat}}_i$$

$$(2.4) \qquad F^{\text{visc}}_i = \left\{ \begin{array}{c} 0 \\ \tau_{ij}\delta_j \\ \tau_{ij}u_j \end{array} \right\} \qquad F^{\text{heat}}_i = \left\{ \begin{array}{c} 0 \\ 0_3 \\ -q_i \end{array} \right\}$$

where $\rho$ is the density; $u = \{u_1, u_2, u_3\}^T$ is the velocity vector; $e^{\text{tot}}$ is the total energy per unit mass, which is the sum of the internal energy per unit mass, $e$, and of the kinetic energy per unit mass, $|u|^2/2$; $p$ is the thermodynamic pressure; and $\delta_i = \{\delta_{ij}\}$ is a generalized Kronecker delta vector, where $\delta_{ij}$ is the usual Kronecker delta (viz., $\delta_{ii} = 1$, and $\delta_{ij} = 0$ for $i \neq j$). The diffusive flux, which constitutes a first order correction taking into account translational nonequilibrium effects, splits up into two parts: a viscous stress part, $F^{\text{visc}}_i$, and a heat conduction part, $F^{\text{heat}}_i$. Furthermore, $\tau = [\tau_{ij}]$ is the viscous-stress tensor; $q = \{q_1, q_2, q_3\}^T$ is the heat-flux vector; and $0_3$ is the null vector of length three.

The definition of the diffusive flux is completed by the following constitutive relations:

i) The viscous stress tensor $\tau$ is given by

$$(2.5) \qquad \tau_{ij} = \lambda^{\text{visc}} u_{k,k} \delta_{ij} + \mu^{\text{visc}}(u_{i,j} + u_{j,i})$$

where $\lambda^{\text{visc}}$ and $\mu^{\text{visc}}$ are the viscosity coefficients. $\lambda^{\text{visc}}$ may be defined in terms of $\mu^{\text{visc}}$ and the bulk viscosity coefficient $\mu^{\text{visc}}_B$ by

$$(2.6) \qquad \lambda^{\text{visc}} = \mu^{\text{visc}}_B - \frac{2}{3}\mu^{\text{visc}}.$$

For perfect monatomic gases, kinetic theory predicts that $\mu^{\text{visc}}_B = 0$. Stokes' hypothesis states that $\mu^{\text{visc}}_B$ can be taken equal to zero in the general case. However, as shown by Vincenti and Kruger [18], behaviors such as small departures from rotational equilibrium can be represented by means of bulk viscosity. In the present discussion, where thermal equilibrium is assumed, Stokes' hypothesis is valid.

ii) The heat flux is given by the usual Fourier law,

$$(2.7) \qquad q_i = -\kappa T_{,i}$$

where $\kappa$ is the coefficient of thermal conductivity.

Equation (2.1) can be rewritten in so-called quasi-linear form:

$$\tag{2.8} \boldsymbol{U}_{,t} + \boldsymbol{A}_i \boldsymbol{U}_{,i} = (\boldsymbol{K}_{ij} \boldsymbol{U}_{,j})_{,i}$$

where $\boldsymbol{A}_i = \boldsymbol{F}^{\text{adv}}_{i,\boldsymbol{U}}$ is the $i^{\text{th}}$ advective Jacobian matrix, and $\boldsymbol{K} = [\boldsymbol{K}_{ij}]$ is the diffusivity matrix, defined by $\boldsymbol{F}^{\text{diff}}_i = \boldsymbol{K}_{ij} \boldsymbol{U}_{,j}$. The $\boldsymbol{A}_i$'s and $\boldsymbol{K}$ do not possess any particular property of symmetry or positiveness.

We now introduce a new set of variables,

$$\tag{2.9} \boldsymbol{V}^T = \frac{\partial \mathcal{H}}{\partial \boldsymbol{U}}$$

where $\mathcal{H}$ is the generalized entropy function given by

$$\tag{2.10} \mathcal{H} = \mathcal{H}(\boldsymbol{U}) = -\rho s$$

and $s$ is the thermodynamic entropy per unit mass. Under the change of variables $\boldsymbol{U} \mapsto \boldsymbol{V}$, (2.8) becomes:

$$\tag{2.11} \widetilde{\boldsymbol{A}}_0 \boldsymbol{V}_{,t} + \widetilde{\boldsymbol{A}}_i \boldsymbol{V}_{,i} = (\widetilde{\boldsymbol{K}}_{ij} \boldsymbol{V}_{,j})_{,i}$$

where

$$\tag{2.12} \widetilde{\boldsymbol{A}}_0 = \boldsymbol{U}_{,\boldsymbol{V}}$$
$$\tag{2.13} \widetilde{\boldsymbol{A}}_i = \boldsymbol{A}_i \widetilde{\boldsymbol{A}}_0$$
$$\tag{2.14} \widetilde{\boldsymbol{K}}_{ij} = \boldsymbol{K}_{ij} \widetilde{\boldsymbol{A}}_0.$$

The Riemannian metric tensor $\widetilde{\boldsymbol{A}}_0$ is symmetric positive-definite; the $\widetilde{\boldsymbol{A}}_i$'s are symmetric; and $\widetilde{\boldsymbol{K}} = [\widetilde{\boldsymbol{K}}_{ij}]$ is symmetric positive-semidefinite. In view of these properties, (2.11) is referred to as a symmetric advective-diffusive system.

For a general divariant gas, the vector of so-called (physical) entropy variables, $\boldsymbol{V}$, reads

$$\tag{2.15} \boldsymbol{V} = \frac{1}{T} \left\{ \begin{array}{c} \mu - |\boldsymbol{u}|^2/2 \\ \boldsymbol{u} \\ -1 \end{array} \right\}$$

where $\mu = e + pv - Ts$ is the chemical potential per unit mass; $v = 1/\rho$ is the specific volume. In order to derive (2.15), we used Gibbs' equation written for a divariant gas ($s$ is a function of $e$ and $v$ only):

$$\tag{2.16} ds = \frac{1}{T}(de + p dv).$$

The Riemannian metric tensor $\widetilde{\boldsymbol{A}}_0$ and the advective Jacobian matrices $\widetilde{\boldsymbol{A}}_i$ require an additional equation of state to complete their definitions. For that purpose, we will assume given a relation which provides the chemical potential of the gas in terms of its thermodynamical state, e.g.,

$$\tag{2.17} \mu = \mu(p, T).$$

All thermodynamic quantities relevant to the formation of (2.11) can then be computed:

$$\tag{2.18} s = -\left(\frac{\partial \mu}{\partial T}\right)_p \qquad v = \left(\frac{\partial \mu}{\partial p}\right)_T$$
$$\tag{2.19} h = \mu + Ts \qquad e = h - pv$$

(2.20) $$\alpha_p = \frac{1}{v}\left(\frac{\partial v}{\partial T}\right)_p = \frac{1}{v}\left(\frac{\partial^2 \mu}{\partial p \partial T}\right) \qquad \beta_T = -\frac{1}{v}\left(\frac{\partial v}{\partial p}\right)_T = -\frac{1}{v}\left(\frac{\partial^2 \mu}{\partial p^2}\right)_T$$

(2.21) $$c_p = \left(\frac{\partial h}{\partial T}\right)_p = -T\left(\frac{\partial^2 \mu}{\partial T^2}\right)_p \qquad c_v = \left(\frac{\partial e}{\partial T}\right)_v = c_p - \frac{\alpha_p^2 v T}{\beta_T}$$

where $\alpha_p$ is the coefficient of volume expansion, $\beta_T$ is the isothermal compressibility, and $c_p$ and $c_v$ are the specific heats at constant pressure and volume. As an example, the Riemannian metric tensor $\widetilde{A}_0$ reads:

(2.22) $$\widetilde{A}_0 = \frac{\beta_T T}{v^2}\begin{bmatrix} 1 & u_1 & u_2 & u_3 & h + \frac{|u|^2}{2} - \frac{v\alpha_p T}{\beta_T} \\ & u_1^2 + \frac{v}{\beta_T} & u_1 u_2 & u_1 u_3 & u_1(h + \frac{|u|^2}{2} - \frac{v(\alpha_p T - 1)}{\beta_T}) \\ & & u_2^2 + \frac{v}{\beta_T} & u_2 u_3 & u_2(h + \frac{|u|^2}{2} - \frac{v(\alpha_p T - 1)}{\beta_T}) \\ & \text{symm.} & & u_3^2 + \frac{v}{\beta_T} & u_3(h + \frac{|u|^2}{2} - \frac{v(\alpha_p T - 1)}{\beta_T}) \\ & & & & a_{55} \end{bmatrix}$$

where

(2.23) $$a_{55} = \left(h + \frac{|u|^2}{2}\right)^2 + \frac{v}{\beta_T}(c_p T - 2h\alpha_p T - |u|^2(\alpha_p T - 1)).$$

All other coefficient matrices can be found in [1].

Taking the dot product of (2.11) with the vector $V$ yields the Clausius-Duhem inequality, which constitutes the basic nonlinear stability condition for the solutions of this (2.11). This fundamental property is inherited by appropriately defined finite element methods, such as the one described in the next section.

**3. The Galerkin/least-squares formulation.** The Galerkin/least-squares formulation introduced by Hughes [6, 7, 9] and Johnson [11, 12], is a full space-time finite element technique employing the discontinuous Galerkin method in time (see [13]). The least-squares operator improves the stability of the method while retaining accuracy. A nonlinear discontinuity-capturing operator is added in order to enhance the local behavior of the solution in the vicinity of sharp gradients.

We consider the time interval $I = ]0, T[$, which we subdivide into $N$ intervals $I_n = ]t_n, t_{n+1}[$, $n = 0, \ldots, N-1$. Let

(3.1) $$Q_n = \Omega \times I_n$$

and

(3.2) $$P_n = \Gamma \times I_n$$

where $\Omega$ is the spatial domain of interest, and $\Gamma$ is its boundary. In turn, the space-time "slab" $Q_n$ is tiled by $(n_{\text{el}})_n$ elements $Q_n^e$. Consequently, the Galerkin/least-squares variational problem can be stated as

Within each $Q_n$, $n = 0, \ldots, N-1$, find $V^h \in S_n^h$ (trial function space), such that for all $W^h \in \mathcal{V}_n^h$ (weighting function space), the following equation holds:

$$
\begin{aligned}
\text{(3.3)} \quad & \int_{Q_n} \left( -\boldsymbol{W}^h_{,t} \cdot \boldsymbol{U}(\boldsymbol{V}^h) - \boldsymbol{W}^h_{,i} \cdot \boldsymbol{F}_i^{\text{adv}}(\boldsymbol{V}^h) + \boldsymbol{W}^h_{,i} \cdot \widetilde{\boldsymbol{K}}_{ij} \boldsymbol{V}^h_{,j} \right) dQ \\
& + \int_{\Omega} \left( \boldsymbol{W}^h(t_{n+1}^-) \cdot \boldsymbol{U}(\boldsymbol{V}^h(t_{n+1}^-)) - \boldsymbol{W}^h(t_n^+) \cdot \boldsymbol{U}(\boldsymbol{V}^h(t_n^-)) \right) d\Omega \\
& + \sum_{e=1}^{(n_{\text{el}})_n} \int_{Q_n^e} \left( \mathcal{L} \boldsymbol{W}^h \right) \cdot \boldsymbol{\tau} \left( \mathcal{L} \boldsymbol{V}^h \right) dQ \\
& + \sum_{e=1}^{(n_{\text{el}})_n} \int_{Q_n^e} \nu^h g^{ij} \boldsymbol{W}^h_{,i} \cdot \widetilde{\boldsymbol{A}}_0 \boldsymbol{V}^h_{,j} \, dQ \\
& = \int_{P_n} \boldsymbol{W}^h \cdot \left( -\boldsymbol{F}_i^{\text{adv}}(\boldsymbol{V}^h) + \boldsymbol{F}_i^{\text{diff}}(\boldsymbol{V}^h) \right) n_i \, dP.
\end{aligned}
$$

The first and last integrals represent the Galerkin formulation written in integrated-by-parts form. The solution space consists of piecewise polynomials which are continuous in space, but are discontinuous across time slabs. Continuity in time is weakly enforced by the second integral in (3.3), which contributes to the jump condition between two contiguous slabs, with

$$
\text{(3.4)} \qquad \boldsymbol{Z}^h(t_n^{\pm}) = \lim_{\epsilon \to 0^{\pm}} \boldsymbol{Z}^h(t_n + \epsilon).
$$

The third integral constitutes the least-squares operator where $\mathcal{L}$ is defined as

$$
\text{(3.5)} \qquad \mathcal{L} = \widetilde{\boldsymbol{A}}_0 \frac{\partial}{\partial t} + \widetilde{\boldsymbol{A}}_i \frac{\partial}{\partial x_i} - \frac{\partial}{\partial x_i} \left( \widetilde{\boldsymbol{K}}_{ij} \frac{\partial}{\partial x_j} \right).
$$

$\boldsymbol{\tau}$ is a symmetric matrix for which definitions can be found in [4] and [16]. The fourth integral is the nonlinear discontinuity-capturing operator, which is designed to control oscillations about discontinuities, without upsetting higher-order accuracy in smooth regions. $g^{ij}$ is the contravariant metric tensor defined by

$$
[g^{ij}] = [\boldsymbol{\xi}_{,i} \cdot \boldsymbol{\xi}_{,j}]^{-1}
$$

where $\boldsymbol{\xi} = \boldsymbol{\xi}(\boldsymbol{x})$ is the inverse isoparametric element mapping, and $\nu^h$ is a scalar-valued homogeneous function of the residual $\mathcal{L}\boldsymbol{V}^h$. The discontinuity capturing factor $\nu^h$ used in the present work is an extension of that introduced by Hughes and Mallet [5], Mallet [14], and Shakib et al. [16].

A key ingredient to the formulation is its consistency: the exact solution of (2.1) satisfies the variational formulation (3.3). This constitutes an essential property in order to attain higher-order spatial convergence. In addition, it must be noted that the numerical method presented here does not rely on the advective fluxes being homogeneous in the conservative variables, which is true only for a thermally perfect gas (see [1]). More complex equations of state such as those needed for describing equilibrium chemistry, do not require any particular approximation to be introduced, which is often the case with other techniques. One can say that the formulation is consistent with the equation of state. For further details about the method, the reader is referred to the works mentioned in this section.

**4. A simple chemistry model for equilibrium air.** In this section, we describe a chemistry model for equilibrium air. Although it is simple, it encompasses all the ingredients necessary to compute the composition of the gas mixture and the quantities (2.18)–(2.21). The state of the system is given by the vector of entropy variables, from which the chemical potential of the mixture $\mu$ and its temperature $T$ can be extracted trivially. On the other hand, a solver based on conservative variables would typically have the density $\rho$ and the internal energy $e$ at its disposal to define the thermodynamic state of the system. We find the entropy variables advantageous here, since temperature is a more convenient variable than density, especially to express quantities such as energies (internal energy, Gibbs' free energy, etc.).

We consider air as a mixture of five thermally perfect gases: $N_2$, $O_2$, $NO$, $N$, and $O$. Given the thermodynamic state of the system $(\mu, T)$, we propose to compute the equilibrium partial pressure of each component, and the quantities (2.18)–(2.21). In order to solve for the five $p_s$'s, we need five independent equations.

First, we can write the chemical potential as a function of the $p_s$'s and $T$:

$$\sum_s y_s(\mathbf{p})\mu_s(p_s, T) = \mu \quad (4.1)$$

where $y_s$ and $\mu_s$ are respectively the mass fraction and the chemical potential of species $s$, and $\mathbf{p} = \{p_s\}$ is the vector of partial pressures. The mass fraction $y_s$ is related to the mole fraction $x_s$ and the molar mass $\hat{M}_s$ of species $s$, and to the molar mass $\hat{M}$ of the mixture by

$$y_s = \frac{\hat{M}_s}{\hat{M}} x_s. \quad (4.2)$$

In turn, $x_s$ and $\hat{M}$ are given in terms of the partial pressures by

$$x_s = \frac{p_s}{p} \quad (4.3)$$

and

$$\hat{M} = \sum_s x_s \hat{M}_s. \quad (4.4)$$

The pressure is provided by Dalton's law of partial pressures:

$$p = \sum_s p_s. \quad (4.5)$$

Each species being considered as a thermally perfect gas, we have

$$p_s = \rho_s R_s T \quad (4.6)$$

where $\rho_s$ and $R_s$ are respectively the density and the specific gas constant of species $s$. $R_s$ is linked to the universal gas constant $\hat{R} = 8.31441$ J/mol·K through

$$R_s = \frac{\hat{R}}{\hat{M}_s}. \quad (4.7)$$

The chemical potential of species $s$ is

$$\mu_s = h_s - T s_s \quad (4.8)$$

where, in the thermally perfect gas case,

$$h_s = e_s(T) + R_s T \quad (4.9)$$

$$s_s = \int \frac{de_s}{T} + R_s \ln T - R_s \ln p_s + s_{0s}. \quad (4.10)$$

The internal energy $e_s$ is a function of temperature only, and $s_{0s}$ is the specific reference entropy upon which we will elaborate later. We adopt the rigid-rotator and harmonic-oscillator model. Under this assumption, a simple closed form expression exists for the internal energy. It splits up into a translational, a rotational, and a vibrational contribution, to which the heat of formation $h_s^0$ must be added:

$$e_s(T) = e_s^{\text{trans}} + e_s^{\text{rot}} + e_s^{\text{vib}} + h_s^0 \quad (4.11)$$

$$e_s^{\text{trans}}(T) = 3 \times \frac{1}{2} R_s T \quad (4.12)$$

$$e_s^{\text{rot}}(T) = \begin{cases} 0, & \text{for atoms} \\ 2 \times \frac{1}{2} R_s T, & \text{for diatomic molecules} \end{cases} \quad (4.13)$$

$$e_s^{\text{vib}}(T) = \begin{cases} 0, & \text{for atoms} \\ \dfrac{R_s \Theta_s^{\text{vib}}}{\exp(\Theta_s^{\text{vib}}/T) - 1}, & \text{for diatomic molecules.} \end{cases} \quad (4.14)$$

We ignore any electronic contribution to the internal energy. This makes the model simpler, but does not limit the generality of the present development. Equation (4.10) can now be integrated exactly, yielding

$$s_s = \frac{e_s^{\text{vib}}}{T} - R_s \ln\left[1 - \exp(-\Theta_s^{\text{vib}}/T)\right] + c_{ps} \ln T - R_s \ln p_s + s_{0s} \tag{4.15}$$

where

$$c_{ps} = c_{vs} + R_s = \begin{cases} \dfrac{5}{2} R_s, & \text{for atoms} \\ \dfrac{7}{2} R_s, & \text{for diatomic molecules.} \end{cases} \tag{4.16}$$

Finally, $\mu_s$ is given by

$$\mu_s = c_{ps} T(1 - \ln T) + R_s T \ln p_s + h_s^0 + R_s T \ln\left[1 - \exp(-\Theta_s^{\text{vib}}/T)\right] - T s_{0s}. \tag{4.17}$$

We introduce the molar chemical potentials

$$\hat{\mu}_s = \hat{M}_s \mu_s = \hat{c}_{ps} T(1 - \ln T) + \hat{R} T \ln p_s + \hat{h}_s^0 + \hat{R} T \ln\left[1 - \exp(-\Theta_s^{\text{vib}}/T)\right] - T \hat{s}_{0s} \tag{4.18}$$
$$= \hat{\mu}_s^0(T) + \hat{R} T \ln p_s$$

where $\hat{\mu}_s^0$ is the molar chemical potential of species $s$ in the pure state and at unit pressure. Equation (4.1) can be restated as

$$\sum_s \rho_s \left(\hat{\mu}_s - \hat{M}_s \mu\right) = 0. \tag{4.19}$$

This constitutes the first equation of our system.

The second equation is obtained by stating that the local proportion of nitrogen atoms relative to oxygen atoms is constant, viz.,

$$\frac{2x_{\text{N}_2} + x_{\text{NO}} + x_{\text{N}}}{2x_{\text{O}_2} + x_{\text{NO}} + x_{\text{O}}} = \frac{79}{21} \tag{4.20}$$

where we have assumed that air is a mixture of 79% of nitrogen and 21% of oxygen by volume. In terms of partial pressures, (4.20) can be rewritten as

$$\frac{2p_{\text{N}_2} + p_{\text{NO}} + p_{\text{N}}}{2p_{\text{O}_2} + p_{\text{NO}} + p_{\text{O}}} = \frac{79}{21}. \tag{4.21}$$

In addition to (4.19) and (4.21), we need three more equations. These are provided by three independent chemical reactions,

$$\text{N}_2 \rightleftharpoons 2\text{N} \tag{4.22}$$
$$\text{O}_2 \rightleftharpoons 2\text{O} \tag{4.23}$$
$$\text{NO} \rightleftharpoons \text{N} + \text{O}. \tag{4.24}$$

We can write the law of mass action for each of these. For consistency, we do not state the equilibrium condition for reaction $R$ in the usual form, i.e.,

$$\prod_s p_s^{\nu_{sR}} = K_{pR}(T) \tag{4.25}$$

where $\nu_{sR}$ is the stoichiometric coefficient of species $s$ in reaction $R$, and $K_{pR}(T)$ the equilibrium constant of reaction $R$. The latter is a function of temperature which is often given in the form of a curve fit of experimental results. In place of (4.25), we write for each reaction the following statement which is equivalent, but does not require any extraneous data:

$$\sum_s \nu_{sR}\hat{\mu}_s = 0. \tag{4.26}$$

Once a model has been chosen for the internal energies of the different species, the system is closed, and the addition of any superfluous piece of information, such as equilibrium constants, can only introduce inconsistencies. However, in order to use the chemical potentials given by (4.18) in the equation for reaction equilibrium (4.26), the absolute entropy must be computed carefully, and in particular the integration constant $\hat{s}_{0s}$. It is provided by statistical mechanics, and is the sum of four terms:

$$\hat{s}_{0s} = \hat{s}_{0s}^{\text{trans}} + \hat{s}_{0s}^{\text{rot}} + \hat{s}_{0s}^{\text{vib}} + \hat{s}_{0s}^{\text{el}} \tag{4.27}$$

where

$$\hat{s}_{0s}^{\text{trans}} = \hat{R}\left\{\ln\left[\left(\frac{2\pi m_s}{h^2}\right)^{3/2} k^{5/2}\right] + \frac{5}{2}\right\} \tag{4.28}$$

$$\hat{s}_{0s}^{\text{rot}} = \hat{R}(1 - \ln \sigma_s \Theta_s^{\text{rot}}) \tag{4.29}$$

$$\hat{s}_{0s}^{\text{vib}} = 0 \tag{4.30}$$

$$\hat{s}_{0s}^{\text{el}} = \hat{R}\ln g_{0s}. \tag{4.31}$$

Equation (4.28) is known as the Sackur-Tetrode formula, in which $m_s$ is the mass of one particle of species $s$:

$$m_s = \frac{\hat{M}_s}{\hat{N}}; \tag{4.32}$$

$\hat{N} = 6.022045 \times 10^{23}$ is Avogadro's number; $h = 6.626176 \times 10^{-34}$ J·s is Planck's constant; and Boltzmann's constant $k$ is given by

$$k = \frac{\hat{R}}{\hat{N}}. \tag{4.33}$$

In the rotational part $\hat{s}_{0s}^{\text{rot}}$, $\sigma_s$ is the symmetry factor of the molecule, and $\Theta_s^{\text{rot}}$ is its characteristic temperature for rotation. Although we have neglected any electronic excitation, we must take into account the degeneracy of the ground level, which yields zero energy, but is crucial for a correct evaluation of the reference entropy. The constants used in the present model are gathered in the table below. For the most part, they were taken from [2].

|  | $N_2$ | $O_2$ | NO | N | O |
|---|---|---|---|---|---|
| $\hat{M}_s$ (kg/mol) | $28 \times 10^{-3}$ | $32 \times 10^{-3}$ | $30 \times 10^{-3}$ | $14 \times 10^{-3}$ | $16 \times 10^{-3}$ |
| $\hat{h}_s^0$ (J/mol) | 0 | 0 | 89,775 | 470,820 | 246,790 |
| $\Theta_s^{\text{vib}}$ (K) | 3,393.50 | 2,273.56 | 2,738.87 | 0.00 | 0.00 |
| $\Theta_s^{\text{rot}}$ (K) | 2.87 | 2.08 | 2.45 | 0.00 | 0.00 |
| $\sigma_s$ | 2 | 2 | 1 | 0 | 0 |
| $g_{0s}$ | 1 | 3 | 4 | 4 | 9 |

Finally, for the three independent chemical reactions (4.22)–(4.24), equation (4.33) reads

$$\hat{\mu}_{N_2} = 2\hat{\mu}_N \tag{4.34}$$
$$\hat{\mu}_{O_2} = 2\hat{\mu}_O \tag{4.35}$$
$$\hat{\mu}_{NO} = \hat{\mu}_N + \hat{\mu}_O. \tag{4.36}$$

The resulting system of five nonlinear equations for the $p_s$'s in terms of $\mu$ and $T$ ((4.19), (4.21), (4.34)–(4.36)) can formally be expressed as

$$\tag{4.37} \boldsymbol{f}(\boldsymbol{p}, \mu, T) = \boldsymbol{0}$$

where

$$\tag{4.38} \boldsymbol{f} = \left\{ \begin{array}{c} \sum_s p_s(\hat{\mu}_s - \hat{M}_s \mu) \\ 42 p_{N_2} - 158 p_{O_2} - 58 p_{NO} + 21 p_N - 79 p_O \\ 2\hat{\mu}_N - \hat{\mu}_{N_2} \\ 2\hat{\mu}_O - \hat{\mu}_{O_2} \\ \hat{\mu}_N + \hat{\mu}_O - \hat{\mu}_{NO} \end{array} \right\}.$$

The system (4.37) is solved using the Newton-Raphson method: given an initial guess $\boldsymbol{p}^{(0)}$ for $\boldsymbol{p}$, the $(n+1)^{\text{st}}$ iterate is defined by

$$\tag{4.39} \boldsymbol{p}^{(n+1)} = \boldsymbol{p}^{(n)} + \Delta \boldsymbol{p}^{(n)}$$

where

$$\tag{4.40} \Delta \boldsymbol{p}^{(n)} = -\boldsymbol{J}^{-1}\left(\boldsymbol{p}^{(n)}, \mu, T\right) \boldsymbol{f}\left(\boldsymbol{p}^{(n)}, \mu, T\right)$$

and

$$\tag{4.41} \boldsymbol{J} = \left(\frac{\partial \boldsymbol{f}}{\partial \boldsymbol{p}}\right)_{\mu, T}.$$

A good initial guess assures quadratic convergence of the process. Typically, the $p_s$'s are computed up to ten significant digits in two iterations. Initial values for the partial pressures are obtained from a table look-up.

Once convergence of the Newton scheme has been achieved, the Jacobian $\boldsymbol{J}$ satisfies

$$\tag{4.42} \boldsymbol{J} d\boldsymbol{p} + \left(\frac{\partial \boldsymbol{f}}{\partial \mu}\right)_{\boldsymbol{p}, T} d\mu + \left(\frac{\partial \boldsymbol{f}}{\partial T}\right)_{\boldsymbol{p}, \mu} dT = 0$$

which is the differential of (4.37). Thus,

$$\tag{4.43} d\boldsymbol{p} = -\boldsymbol{J}^{-1} \left(\frac{\partial \boldsymbol{f}}{\partial \mu}\right)_{\boldsymbol{p}, T} d\mu - \boldsymbol{J}^{-1} \left(\frac{\partial \boldsymbol{f}}{\partial T}\right)_{\boldsymbol{p}, \mu} dT$$

and

$$\tag{4.44} \left(\frac{\partial \boldsymbol{p}}{\partial \mu}\right)_T = -\boldsymbol{J}^{-1} \left(\frac{\partial \boldsymbol{f}}{\partial \mu}\right)_{\boldsymbol{p}, T}$$

$$\tag{4.45} \left(\frac{\partial \boldsymbol{p}}{\partial T}\right)_\mu = -\boldsymbol{J}^{-1} \left(\frac{\partial \boldsymbol{f}}{\partial T}\right)_{\boldsymbol{p}, \mu}.$$

These derivatives are obtained at essentially no extra cost, since in practice we use the $\boldsymbol{LU}$-decomposition of the last iteration Jacobian. From (4.44) and (4.45), we can calculate any thermodynamic derivative. For instance, the partial derivatives of the mass fractions with respect to pressure and temperature are given by

$$\tag{4.46} \left(\frac{\partial y_s}{\partial p}\right)_T = y_s \left[ \frac{1}{p_s} \left(\frac{\partial p_s}{\partial p}\right)_T - \sum_r \frac{y_r}{p_r} \left(\frac{\partial p_r}{\partial p}\right)_T \right]$$

$$\tag{4.47} \left(\frac{\partial y_s}{\partial T}\right)_p = y_s \left[ \frac{1}{p_s} \left(\frac{\partial p_s}{\partial T}\right)_p - \sum_r \frac{y_r}{p_r} \left(\frac{\partial p_r}{\partial T}\right)_p \right]$$

where

$$(4.48) \quad \left(\frac{\partial p_s}{\partial p}\right)_T = \frac{\left(\frac{\partial p_s}{\partial \mu}\right)_T}{\sum_r \left(\frac{\partial p_r}{\partial \mu}\right)_T}$$

$$(4.49) \quad \left(\frac{\partial p_s}{\partial T}\right)_p = \left(\frac{\partial p_s}{\partial T}\right)_\mu - \frac{\sum_r \left(\frac{\partial p_r}{\partial T}\right)_\mu}{\sum_r \left(\frac{\partial p_r}{\partial \mu}\right)_T} \left(\frac{\partial p_s}{\partial \mu}\right)_T .$$

We now have everything at our disposal to compute the quantities required to form (2.11). For instance, we have

$$(4.50) \quad c_p = \left(\frac{\partial h}{\partial T}\right)_p = \sum_s y_s \left(\frac{\partial h_s}{\partial T}\right)_p + \sum_s \left(\frac{\partial y_s}{\partial T}\right)_p h_s$$

$$(4.51) \quad \alpha_p = \frac{1}{v}\left(\frac{\partial v}{\partial T}\right)_p = \frac{1}{T} + \frac{1}{R}\sum_s \left(\frac{\partial y_s}{\partial T}\right)_p R_s$$

$$(4.52) \quad \beta_T = -\frac{1}{v}\left(\frac{\partial v}{\partial p}\right)_T = \frac{1}{p} - \frac{1}{R}\sum_s \left(\frac{\partial y_s}{\partial p}\right)_T R_s$$

$$(4.53) \quad c_v = c_p - \frac{\alpha_p^2 v T}{\beta_T}$$

$$(4.54) \quad a^2 = \frac{v c_p}{c_v \beta_T}$$

where $a$ is the speed of sound.

The techniques portrayed in this section may seem elaborate for a simple chemistry model. In fact, all the ingredients necessary for dealing with the most complex situation are contained within the previous description.

**5. Numerical examples.** We now present two sets of two-dimensional computations which illustrate the procedures described in the preceding sections. The first one contrasts the perfect gas and the equilibrium chemistry solutions for the same inviscid case; the second one compares viscous and inviscid treatments of the flow past a space-shuttle like configuration.

The spatial domains are meshed using unstructured combinations of bilinear quadrilaterals and linear triangles. Adaptive refinement is introduced in both the shock and the boundary layer regions; we will elaborate on this strategy when we describe the individual test cases and especially the Navier-Stokes one.

Convergence to steady state is achieved with the help of a first-order-in-time low-storage fully implicit iterative solver based on the preconditioned GMRES algorithm (see [10, 15, 16]).

*5.1. Flow over a blunt body.* The geometry is a simple circular cylinder of unit radius extended with two planes at 15° angle. The body faces an inviscid Mach 17.9 flow at zero angle of attack. The free stream density and temperature are respectively $10^{-4}$ kg/m$^3$ and 231 K. This test case is described by Desideri et al. in [3]. Figure 1 compares the equilibrium chemistry solution (top) with the perfect gas one (bottom) for the same inflow conditions. In view of the symmetry of the problem, the computation is performed on half the domain only. The "chemistry" mesh contains 4,378 nodes and 8,573 elements; the "perfect gas" one 4,856 nodes and 9,527 elements. Both meshes consist of triangular elements, and are adaptively enriched in order to better capture the detached bow shock. Figure 1 presents both meshes (a), pressure (b) and temperature (c) contours. The difference between the two equations of state appears quite clearly: the stand-off distance of the shock is much reduced in the more realistic equilibrium chemistry case. In addition, the temperature rise through the shock goes down by a factor of almost 3. The numerical results are found in remarkable agreement with the theoretical solutions. Typically, the relative error on all stagnation values is under 0.5%, with minima still an order of magnitude below this value (e.g., the stagnation temperature in the equilibrium case is overestimated by a mere 0.0582%). In an industrial setting, the additional cost due to the chemistry routine might be a real concern. In fact, it turns out that the cost of an equilibrium chemistry computation is only about 20% higher than that of a perfect gas computation. Specially designed curve fits may further reduce this figure.

*5.2. Flow over a double ellipse.* With this example, we compare the Euler and the Navier-Stokes solutions of the same flow over a generic space-shuttle geometry given by a double ellipse. The inflow Mach number is 25; the angle of attack is 30°. The free stream conditions simulate a 75 km altitude in the U.S. Standard Atmosphere [17]: $T_\infty = 205.3$ K and $p_\infty = 2.52$ Pa. For the viscous case, the Reynolds number is 22,000 per meter; the geometry measures 0.76 m in length, with the major half axis of the larger ellipse being 0.6 m long; the wall temperature is fixed at 1500 K. These test cases are two-dimensional variants of cases proposed at the Workshop on Hypersonic Flows for Reentry Problems – Part II, held in Antibes, France, April 15–19, 1991.

The meshes employed are shown in Figures 2 and 3. The Euler mesh consists of 8,307 nodes and 16,231 triangular elements. The Navier-Stokes mesh which contains 10,613 nodes, is an unstructured combination of 13,605 triangles in the main flow and of 3,620 quadrilaterals along the body. The structured strip is made of 20 layers of quadrilateral elements; its thickness is adapted to match that of the boundary layer. This strategy, while maintaining the advantages of unstructuredness, facilitates capturing the fine features of the boundary layer. In addition, both meshes are enriched in the shock region. Figures 4–11 present the pressure, temperature, N and O mass fraction contours for the inviscid (left) and the viscous (right) cases. As one would expect, the pressure contours are quite similar for the two solutions. The Navier-Stokes solution shows however a clear recombination of nitrogen and oxygen at the wall. The canopy shock present in the Euler solution has nearly completely vanished in the viscous calculation. Finally, one must note the extreme thinness of the boundary layer on the windward side of the body.

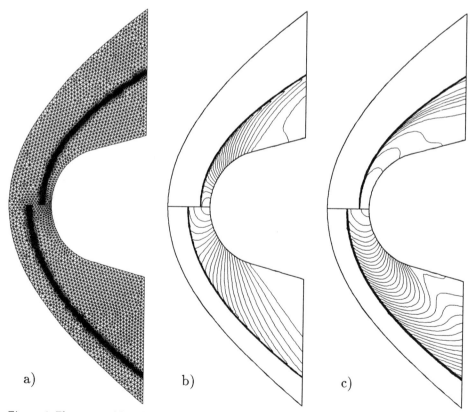

*Figure 1.* Flow past a blunt body. Top: equilibrium chemistry; bottom: perfect gas. a) Mesh; b) Pressure contours; c) Temperature contours.

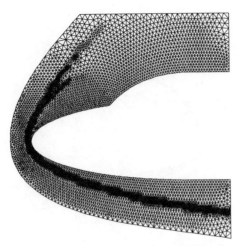

*Figure 2.* Mesh: 8,307 nodes, 16,231 elements (inviscid case).

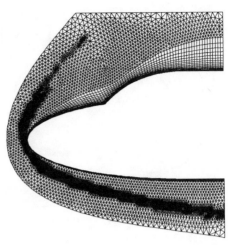

*Figure 3.* Mesh: 10,613 nodes, 17,225 elements (viscous case).

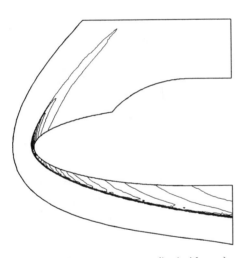

*Figure 4.* Pressure contours (inviscid case).

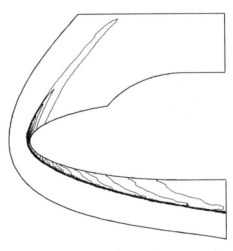

*Figure 5.* Pressure contours (viscous case).

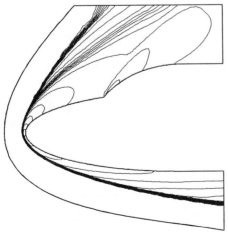

*Figure 6.* Temperature contours (inviscid case).

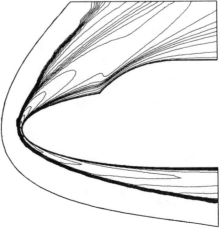

*Figure 7.* Temperature contours (viscous case).

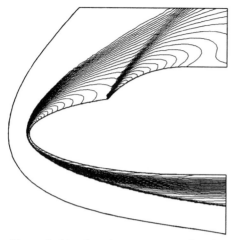

*Figure 8.* Atomic nitrogen mass fraction contours (inviscid case).

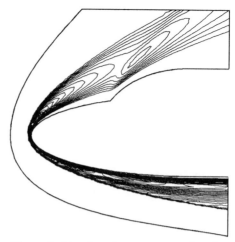

*Figure 9.* Atomic nitrogen mass fraction contours (viscous case).

*Figure 10.* Atomic oxygen mass fraction contours (inviscid case).

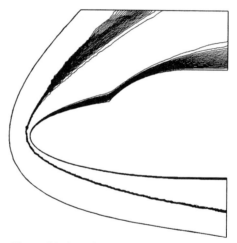

*Figure 11.* Atomic oxygen mass fraction contours (viscous case).

**6. Conclusion.** Consistency has been the leitmotiv of this work, and was indeed one of the main concerns in designing the method. First, the Galerkin/least-squares finite element method is consistent in that it is a residual method: the solution of the initial problem is a solution of the actual numerical problem. Then, the discretization of the problem is performed consistently, in the sense that no alteration to the physical model nor any additional approximation are required by the numerical method. Finally, the equilibrium chemistry model is consistent, since it uses only the minimum number of theoretical and experimental constants, and thus eliminates any dangerous redundancy. These ingredients result in a mathematically sound flow solver. Clearly, Numerical results, such as those presented herein, strongly benefit from this firm mathematical basis without compromising with performance.

### REFERENCES

[1] F. Chalot, T.J.R. Hughes, and F. Shakib, "Symmetrization of conservation laws with entropy for high-temperature hypersonic computations," *Computing Systems in Engineering*, Vol. 1, pp. 465–521, 1990.

[2] M.W. Chase, Jr., C.A. Davies, J.R. Downey, Jr., D.J. Frurip, R.A. McDonald, and A.N. Syverud, "JANAF thermochemical tables, third edition," *Journal of Physical and Chemical Reference Data*, Vol. 14, Suppl. 1, 1985.

[3] J.-A. Désidéri, N. Glinsky, and E. Hettena, "Hypersonic reactive flow computations," *Computers and Fluids*, Vol. 18, pp. 151–182, 1990.

[4] T.J.R. Hughes and M. Mallet, "A new finite element method for computational fluid dynamics: III. The generalized streamline operator for multidimensional advective-diffusive systems," *Computer Methods in Applied Mechanics and Engineering*, Vol. 58, pp. 305–328, 1986.

[5] T.J.R. Hughes and M. Mallet, "A new finite element method for computational fluid dynamics: IV. A discontinuity-capturing operator for multidimensional advective-diffusive systems," *Computer Methods in Applied Mechanics and Engineering*, Vol. 58, pp. 329–336, 1986.

[6] T.J.R. Hughes, "Recent progress in the development and understanding of SUPG methods with special reference to the compressible Euler and Navier-Stokes equations," *International Journal for Numerical Methods in Fluids*, Vol. 7, pp. 1261–1275, 1987.

[7] T.J.R. Hughes and G.M. Hulbert, "Space-time finite element methods for elastodynamics: formulations and error estimates," *Computer Methods in Applied Mechanics and Engineering*, Vol. 66, pp. 339–363, 1988.

[9] T.J.R. Hughes, L.P. Franca, and G.M. Hulbert, "A new finite element method for computational fluid dynamics: VIII. The Galerkin/least-squares method for advective-diffusive equations," *Computer Methods in Applied Mechanics and Engineering*, Vol. 73, pp. 173–189, 1989.

[10] Z. Johan, T.J.R. Hughes, and F. Shakib, " A globally convergent matrix-free algorithm for implicit time-marching schemes arising in finite element analysis in fluids," *Computer Methods in Applied Mechanics and Engineering*, Vol. 87, pp. 281–304, 1991.

[11] C. Johnson and J. Pitkäranta, "An analysis of the discontinuous Galerkin method for a scalar hyperbolic equation," *Technical Report MAT-A215*, Institute of Mathematics, Helsinki University of Technology, Helsinki, Finland, 1984.

[12] C. Johnson, U. Nävert, and J. Pitkäranta, "Finite element methods for linear hyperbolic problems," *Computer Methods in Applied Mechanics and Engineering*, Vol. 45, pp. 285–312, 1984.

[13] P. Lesaint and P.-A. Raviart, " On a finite element method for solving the neutron transport equation," in: C. de Boor (ed.), *Mathematical Aspects of Finite Elements in Partial Differential Equations*, Academic Press, New York, pp. 89–123, 1974.

[14] M. Mallet, *A Finite Element Method for Computational Fluid Dynamics*, Ph.D. Thesis, Division of Applied Mechanics, Stanford University, 1985.

[15] F. Shakib, T.J.R. Hughes, and Z. Johan, "A multi-element group preconditioned GMRES algorithm for nonsymmetric systems arising in finite element analysis," *Computer Methods in Applied Mechanics and Engineering*, Vol. 75, pp. 415–456, 1989.

[16] F. Shakib, T.J.R. Hughes, and Z. Johan, *A new finite element method for computational fluid dynamics: X. The compressible Euler and Navier-Stokes equations*, Computer Methods in Applied Mechanics and Engineering, Vol. 89, pp. 141–219, 1991.

[17] United States Committee on Extension to the Standard Atmosphere, *U.S. Standard Atmosphere Supplements 1966*, U.S. Govt. Print. Off., Wsahington, 1967.

[18] W.G. Vincenti and C.H. Kruger, Jr., *Introduction to Physical Gas Dynamics*, Krieger, 1965.

Chapter 11
# INTERIOR-POINT METHODS IN OPTIMIZATION

**NARENDRA KARMARKAR**
AT&T Bell Laboratories
Murray Hill, New Jersey

**1. Combinatorial optimization—the origin of interior-point methods.** As a typical real-life example of combinatorial optimization, consider the problem of operating a flight-schedule of an airline at minimum cost. A flight schedule consists of many flights with specified arrival and departure times connecting many cities. There are several operating constraints: Each plane must fly a round-trip route. Each pilot must also fly a round-trip route, but not necessarily the same route taken by the plane, since at each airport the pilot can change planes. There are obvious timing constraints interlocking the schedules of pilots, schedules of planes and schedules of flights. There must be adequate rest built into pilot-schedule and periodic maintenance built into the plane-schedule. Only certain crews are qualified to operate certain types of planes. The operating cost consists of many components, some of them more subtle than others. For example, in an imperfect schedule, a pilot may have to fly as a passenger on some flight. This results in lost revenue not only because a passenger seat is taken up but also because the pilot has to be paid even for the time he is riding as a passenger. How does one make an operating plan for an airline that minimizes the total cost, meeting all the constraints? A problem of this type is called a combinatorial optimization problem, since there are only a finite number of combinations possible, and in principle, one can enumerate all of them, eliminate the ones that do not meet the conditions and among those that do, select the one that has the least operating cost. Needless to say, one needs to be more clever than simple enumeration due to the vast number of combinations involved. In this respect, the interior-point methods have made major advances over earlier, more straightforward combinatorial approaches. On one hand, the computational time of interior-point methods is coming down, but on the other hand, the mathematical sophistication of what goes behind the creation of these algorithms is going up, requiring concepts from deeper areas of mathematics such as Riemannian geometry.

As another example, consider a communication network consisting of switches interconnected by trunks (e.g. terrestrial, oceanic, satellite) in a particular topology. A telephone call originating in one switch can take many different paths of switches to terminate in another switch, using up trunks along the path. The problem is to design a minimum cost network that can carry the expected traffic. After a network is designed and implemented, operating the network involves various other combinatorial optimisation problems such as dynamic routing of telephone calls.

As a third example, consider inductive inference, a central problem in artificial intelligence and machine learning. Inductive inference is the process of hypothesizing a general rule from examples. Inductive inference involves the following steps: *(i)* Inferring rules from examples, finding compact abstract models of data or hidden patterns in the data; *(ii)* Making predictions based on abstractions; *(iii)* By comparing predictions with actual results, one can modify the abstraction, which is the process of *learning*; *(iv)* Designing questions to generate new examples. Consider the first step of the above process, i.e. discovering patterns in data. For example, given the sequence $2, 4, 6, 8, \ldots$, we may ask "What comes next?" One could pick any number and justify it by fitting a fourth degree polynomial through the 5 points. However, the answer "10" is considered the most "intelligent". That is so because it is based on the first-order polynomial $2n$, which is linear and hence *simpler* than a fourth degree polynomial. The answer to an inductive inference problem is not unique. In inductive inference, one wants a *simple* explanation that fits a given set of observations. Simpler answers are considered better answers. One therefore needs a way to measure simplicity. For example, in finite automaton inference, the number of states could be a measure of simplicity. In Boolean circuit inference, the measure could be the number of gates and wires. Inductive inference, in fact leads to a discrete optimization problem, where one wants to maximize simplicity, or find a model, or set of rules, no more complex than some specified measure, consistent with already known data.

Even though the three examples given above come from three different facets of life and look superficially to be quite different, they all have a common mathematical structure and can be described in a common mathematical notation called integer programming. In integer programming the unknowns are represented by variables that take on a finite or discrete set of values. The various constraints or conditions on the problem are captured by algebraic expressions of these variables. For example, in the airline crew assignment problem discussed above, let us denote by variable $x_{ij}$ the decision quantity that assigns crew $i$ to flight $j$. Let there be $m$ crew and $n$ flights. If the variable $x_{ij}$ takes on a value 1, then we say that crew $i$ is assigned to flight $j$, and the cost of that assignment is $c_{ij}$. If the value is 0, then crew $i$ is *not* assigned to flight $j$. Thus, the remuneration component of the total operating cost for the airline is expressed by the expression

$$\text{(1.1)} \qquad \min \sum_{i=1}^{m} \sum_{j=1}^{n} c_{ij} x_{ij} \, .$$

The condition that every flight should have one crew is expressed by the equation

$$\text{(1.2)} \qquad \sum_{i=1}^{m} x_{ij} = 1, \text{ for every flight } j = 1, \ldots, n \, .$$

We should also stipulate that the variables should take on only values 0 or 1. This condition is denoted by the notation

$$\text{(1.3)} \qquad x_{ij} \in \{0,1\}, \ 1 \leq i \leq m; \ 1 \leq j \leq n \, .$$

Other conditions on the crew can be expressed in a similar fashion. Thus, an integer programming formulation of the airline crew assignment problem is to minimize the operating cost given by (1.1) subject to various conditions given by other algebraic equations and inequalities. The formulations of the network design problem and the inductive inference problem look mathematically similar to the above problem.

Linear programming is a special and simpler type of combinatorial optimisation problem in which the integrality constraints of the type (1.3) are absent and we are given a linear objective function to be minimised subject to linear inequalities and equalities. A standard form of linear program is stated as follows:

$$\text{(1.4)} \qquad \min_{x} \{ c^T x \mid Ax = b; \ l \leq x \leq u \} \, ,$$

where $c, u, l, x \in \Re^n$, $b \in \Re^m$ and $A \in \Re^{m \times n}$. In (1.4) $x$ is the vector of decision variables, $Ax = b$ and $l \leq x \leq u$ represent constraints on the decision variables, and $c^T x$ is the linear objective function to be minimized.

Linear programming has a wide range of applications, including personnel assignment, production planning and distribution, refinery planning, target assignment, medical imaging, control systems, circuit simulation, weather forecasting, signal processing and financial engineering. Many polynomial-time solvable combinatorial problems are special cases of linear programming (e.g. matching and maximum flow). Linear programming has also been the source of many theoretical developments, in fields as diverse such as economics and queueing theory.

Combinatorial problems occur in diverse areas. These include graph theory (e.g. graph partitioning, network flows, graph coloring), linear inequalities (e.g. linear and integer programming), number theory (e.g. factoring, primality testing, discrete logarithm), group theory (e.g. graph isomorphism, group intersection), lattice theory (e.g. basis reduction), and logic and artificial intelligence (e.g. satisfiability, inductive and deductive inference boolean function minimisation). All these problems, when abstracted mathematically have a commonality of discreteness. The solution approaches for solving these problems also have a great deal in common. In fact, attempts to come up with solution techniques revealed more commonality of the problems than was revealed from just the problem formulation. The solution of combinatorial problems has been the subject of much research. There is a continuously evolving body of knowledge, both theoretical and practical, for solving these problems.

**2. Main concepts underlying interior-point methods.** Solution techniques for combinatorial problems can be classified into two groups: combinatorial approaches and continuous approaches.

**Combinatorial approach.**

The combinatorial approach creates a sequence of states drawn from a discrete and finite set. Each state represents a suboptimal solution or a partial solution to the original problem. It may be a graph, a vertex of a polytope, collection of subsets of a finite set or some other combinatorial object. At each major step of the algorithm, the next state is chosen in an attempt to improve the current state. The improvement may be in the quality of the solution measured in terms of the objective function, or it may be in making the partial solution more feasible. In any case, the improvement is guided by *local* search. By local search we mean that the solution procedure only examines a neighboring set of configurations and selects one greedily that improves the current solution. Thus the local search is quite myopic; no consideration is given to evaluate whether this move may make any sense globally. Indeed, a combinatorial approach often lacks the information needed for making such an evaluation. In many cases, the greedy local improvement may trap the solution in a local minimum that is qualitatively much worse than a true global minimum. In order to escape from a local minimum, the combinatorial approach needs to resort to techniques such as backtracking or abandoning the sequences of states created so far altogether and restarting with a different initial state. Most combinatorial problems suffer from the property of having a large number of local minima when the search space is confined to a discrete set. However, as we shall see later, the situation changes dramatically when a property designed continuous approach is followed. Thus, for a majority of combinatorial optimisation problems, the phenomenon of multiple local minima makes the combinatorial approach less effective.

On the other hand, for a limited class of problems, one can rule out the possibility of local minima and show that local improvement also leads to global improvement. For many problems in this class, polynomial-time algorithms (i.e. algorithms whose running time can be proven to be bounded from above by polynomial functions of the lengths of the problems) were known for a long time. Examples of problems in this class are bipartite matching and network flows. It turns out these problems are special cases of linear programming, which is also a polynomial-time problem. However, the simplex method, which employs a combinatorial approach to solving linear programs, has been shown to be an exponential-time algorithm. In contrast, all polynomial-time algorithms for solving the general linear programming problem employ a continuous approach. These algorithms use either the ellipsoid method or one of the variants of the Karmarkar method.

**Continuous approach.**

In the continuous approach to solving discrete problems, the set of candidate solutions to a given combinatorial problem is embedded in a larger continuous space. The topological and geometric properties of the continuous space play an essential role in the construction of the algorithm as well as in the analysis of its efficiency. The algorithm involves creation of a sequence of points in the enlarged space that converges to the solution of the original combinatorial problem. At each major step of the algorithm, the next point in the sequence is obtained from the current point by making a good *global* approximation to the entire set of relevant solutions and solving it. Usually it is also possible to associate a continuous trajectory or a set of trajectories with the limiting case of the discrete algorithm obtained by taking infinitesimal steps. Topological properties of the underlying continuous space such as connectivity of the level sets of the function being optimised are used for bounding the number of local minima and choosing an effective formulation of the continuous optimisation problem. The geometrical properties such as distance, volume or curvature of trajectories are used for analysing the *rate* of convergence of an algorithm, whereas the topological properties help determine if a proposed algorithm would converge at all. We now elaborate further on each of the main concepts involved in the continuous approach.

**Examples of embedding.**

Suppose the candidate solutions to a discrete problem are represented as points in the $n$-dimensional real space, $\mathbb{R}^n$. This solution set can be embedded into a larger continuous space by forming the convex hull of these points. This is the most common form of continuous embedding and is used for solving linear and integer programming problems. As another example, consider a discrete problem whose candidate solution set is a finite cyclic group.

This can be embedded in a continuous lie group $\{e^{i\theta}|0 \leq \theta < 2\pi\}$. A lie group embedding is useful for the problem of graph isomorphism or automorphism. In this problem, let $A$ denote the adjacency matrix of the graph. Then the discrete solution set of the automorphism problem is the permutation group given by $\{P|AP = PA; P$ is a permutation matrix $\}$. This can be embedded in a larger continuous group given by $\{U|AU = UA; U$ is a complex unitary matrix$\}$.

**Global approximation.**

At each major step of the algorithm, a subproblem is solved to obtain the next point in the sequence. The subproblem should satisfy two properties: (a) the subproblem should be a *global* approximation to the original problem; and (b) the subproblem should be efficiently solvable. In the context of linear programming the Karmarkar method contains a way of making global approximation having both of the above desirable properties and is based on the following theorem: Given any polytope $P$ and an interior point $\mathbf{x} \in P$, there exists a projective transformation $T$, that transforms $P$ to $P'$ and $\mathbf{x}$ to $\mathbf{x}' \in P'$ so that it is possible to find in the transformed space a circumscribing ball $B(\mathbf{x}', R) \supseteq P'$, of radius $R$, center $\mathbf{x}'$, containing $P'$ and a inscribing ball $B(\mathbf{x}', r) \subseteq P'$ of radius $r$, center $\mathbf{x}'$ contained in $P'$ such that the ratio $R/r$ is at most $n$. The inverse image (under $T$) of the inscribed ball is used as the optimisation space for the subproblem and satisfies the two properties stated above, leading to a polynomial-time algorithm for linear programming. The effectiveness of this global approximation is also borne out in practice since Karmarkar's method and its variants take very few approximation steps to find the global optimum of the original problem. Extension of this global approximation step to integer programming have fed to powerful new algorithms for solving NP-complete problems, as we shall see later.

Another example of continuous embedding and global approximation applicable to the *dual* of NP-complete problem or to the problem of showing *non-existence* of combinatorial structure with specified properties is given in [KK92] and [KT92]. Here the global approximation involves a certain tensor optimisation problem.

**Continuous trajectories.**

Suppose an interior-point method produces an iteration of the following type, where $\alpha$ is the step-length parameter.

(2.1) $$\mathbf{x}^{(k+1)} \leftarrow \mathbf{x}^{(k)} + \alpha \mathbf{f}^{(k)} + 0(\alpha^2) \ .$$

Then by taking limit as $\alpha \to 0$, we get the "infinitesimal version" of the algorithm whose continuous trajectories are given by the differential equation

(2.2) $$\frac{d\mathbf{x}}{d\alpha} = \mathbf{f}(\mathbf{x})$$

where $\mathbf{f}(\mathbf{x})$ defines a vector field. Thus, the infinitesimal version of the algorithm can be thought of as a non-linear dynamical system. For the projective method for linear programming, the differential equation is given by

$$\frac{d\mathbf{x}}{dt} = -[D - \mathbf{x}\mathbf{x}^T]P_{AD} \cdot D\mathbf{c} \quad \text{where}$$
$$P_{AD} = I - DA^T(AD^2A^T)^{-1}AD$$
(2.3) $$D = \text{diag}\{x_1, x_2, \cdots, x_n\}$$

Similarly, continuous trajectories and the corresponding differential equations can be derived for other interior-point methods. These trajectories have a rich mathematical structure in them. Many times they also have algebraic descriptions and alternative interpretations. The continuous trajectory given above for the linear programming problem converges to an optimal solution of the problem corresponding to the objective function vector $\mathbf{c}$. Note that the vector field depends on $\mathbf{c}$ in a smooth way and as the vector $\mathbf{c}$ is varied one can get to each vertex of the polytope as limit of some continuous trajectory. If one were to attempt a direct combinatorial description of the discrete solution set of a linear programming problem, it would become enormously complex since the number of solutions can be exponential with respect to the size of the problem. In contrast, the simple differential equation given above implicitly encodes the complex structure of the solution set. Another important fact to be noticed is that the differential equation is written in terms of the *original input matrix A* defining the problem. Viewing combinatorial objects as limiting cases of continuous objects often makes them more accessible to mathematical reasoning and also permits construction of more efficient algorithms.

The power of the language of differential equations in describing complex phenomena is rather well-known in natural sciences. For example, if one were to attempt a direct description of the trajectories involved in planetary motion, it would be enormously complex. However, a small set of differential equations written in terms of the *original parameters* of the problem are able to describe the same motion. One of the most important accomplishments of Newtonian mechanics was *finding a simple description* of the apparently complex phenomena of planetary motion, in the form of differential equations.

In the context of combinatorial optimisation, the structure of the solution set of a discrete problem is often rather complex. As a result, a straightforward combinatorial approach to solving these problems has not succeeded in many cases and has led to a belief that these problems are intractable. Even for linear programming, which is one of the simplest combinatorial optimisation problem, the best known method, in both theory and practice, is based on the continuous approach rather than the combinatorial approach. Underlying this continuous approach is a small set of differential equations that are capable of encoding the complicated combinatorial structure of the solution set. As this approach is extended and generalized, we expect to find new and efficient algorithms for many other combinatorial problems.

**Topological properties.**

There are many ways of formulating a given discrete problem as a continuous optimisation problem, and it is rather easy to make a formulation that would be difficult to solve even by means of continuous trajectories. How does one make a formulation that is solvable? The most well-known class of continuous solvable problems is the class of convex minimisation problems. This leads to a natural question: Is convexity the characteristic property that separates the class of efficiently solvable minimisation problems from the rest? To explore this question we need to look at topological properties.

Topology is a study of properties invariant under any continuous, one-to-one transformation of space having a continuous inverse. Sometimes it is desirable to impose additional conditions of differentiability on the transformation, or to even require the transformation to be a diffeomorphism, because interior-point methods often use higher-order derivatives of the continuous trajectory for accelerating the algorithm.

Suppose we have a continuous optimisation problem that is solvable by means of continuous trajectories. It may be a convex problem, for example. Suppose we apply a nonlinear transformation to the space that is a diffeomorphism. The transformed problem need not be convex, but it will continue to be solvable by means of continuous trajectories. In fact the image of the continuous trajectories in the original space, obtained by applying the *same* diffeomorphism gives us a way of solving the transformed problem. Conversely, if the original problem was unsolvable, it could not be converted into a solvable problem by any such transformation. Hence any diffeomorphism maps solvable problems onto solvable problems and unsolvable problems onto unsolvable problems. This argument shows that the property characterising the solvable class must be a purely topological property and could not possibly be a geometric property such as convexity.

The simplest topological property relevant to the performance of interior-point methods is connectivity of the level sets of the function being optimised. Intuitively, a subset of continuous space is connected if any two points of the subset can be joined by a continuous path lying entirely in the subset. In the context of function minimisation, significance of connectivity lies in the fact that functions having connected level sets do not have spurious local minima. In other words every local minimum is necessarily a global minimum. A continuous formulation of NP-complete problems having such desirable topological properties is given in [KAR90]. The approach described there provides a theoretical foundation for constructing efficient algorithms for discrete problems on the basis of a common principle. Algorithms for many practical problems can now be developed which differ mainly in the way combinatorial structure of the problem is exploited to gain additional computational efficiency.

In sections 3, a few examples of the application of this approach are described. It is also necessary to re-examine problems which were not pursued in the past due to the belief that they are extremely difficult. e.g. the Inductive Inference problem in Artificial Intelligence. [KKRR91]

**Geometric properties.**

Analysis of algorithms based on continuous approach makes use of geometric concepts like distance, volume, curvature of trajectories, etc. For example, proof of polynomial-time convergence of the ellipsoid method is based on ratio of volumes of ellipsoids at two successive steps. Most commonly used proof-techniques in interior-point methods view an algorithm as an iterative process in Euclidean geometry. However, such an approach does not give the best possible result, since the most natural geometry associated with a given interior point method is generally a Riemannian geometry.

In a Riemannian geometry, the length $ds$ of an infinitesimal line segment $d\mathbf{x}$ situated at a point $\mathbf{x}$ is given by

$$(2.4) \qquad ds^2 = \sum_{i,j=1}^{n} g_{ij}(\mathbf{x}) d\mathbf{x}_i d\mathbf{x}_j \ .$$

The quadratic form $g$ depends on the point $\mathbf{x}$. Angles between infinitesimal line segments are defined using inner-product based on the same quadratic form $g$. Length and curvature of a finite curve are obtained by integrating the corresponding infinitesimal quantity along the curve. In a Riemannian geometry, geodesics play a role analogous to the role played by straight-lines in Euclidean geometry. e.g. they give shortest-path between two nearby points. In case the space used by the interior-point method is a lie group, geodesics also have additional property of being one-parameter subgroups of the lie group. Curvature of the manifold used by an interior point method is represented by means of a fourth order tensor called the Riemann curvature tensor. Previous major applications of Riemannian geometry (outside of pure mathematics) have been in natural sciences. e.g. in Einstein's theory of general relativity and Yang-Mills theory. Now similar concepts seem applicable in the entirely new domain of interior-point methods.

To understand the reasons behind this, consider the projective algorithm for linear programming. The behavior of this algorithm is invariant with respect to any projective transformation. If the method of measuring the progress of the algorithm is itself not invariant with respect to such transformations, then one can not get the best possible performance bound. Euclidean metric is not invariant with respect to projective transformations, and there is a unique Riemannian metric having the required invariance. The curvature of trajectories of the algorithm measured with respect to this Riemannian metric determines the number of steps required by the algorithm. Also, the vector field of the algorithm has a variational interpretation in terms of the geodesics of the underlying geometry. Riemannian geometry is important not only for analysing of interior-point methods but also for designing better interior-point methods.

**3. The scope and computational efficiency of interior-point methods.** We illustrate with some examples the broad scope of applications of the interior-point techniques and their computational effectiveness. Since the most widely used combinatorial optimisation problem is linear programming, we begin with this application. Each step of the interior-point method as applied to linear programming involves solution of a linear system of equations. If a higher-order acceleration technique is employed, the same linear system with multiple right-hand sides is solved at each iteration. While a straightforward implementation of solving these linear systems can still outperform the simplex method, more sophisticated implementations have achieved orders of magnitude improvement over the simplex method. These advanced implementations make use of techniques from many disciplines such as linear algebra, numerical analysis, computer architecture, advanced data structures, differential geometry, etc. The following table shows the performance comparison between simplex and interior-point method on a variety of linear programming problems drawn from many disciplines such as operations research, electrical engineering, computer science, statistics, etc. As the table (3.1) shows, the relative superiority of interior-point method over the combinatorial simplex method grows as the problem size grows and the speed-up factor exceeds 200. Larger problems in the table could be solved only by the interior-point method because of impracticality of running simplex method. Furthermore, the interior-point method has been used for solving many large-scale real-life problems in diverse fields such as telecommunication, transportation, defense, which could not be solved earlier by the simplex method.

From the point of view of efficient implementation, interior-point methods have another important property: they can exploit parallelism rather well. [HHL89] Parallel architecture based on multi-dimensional finite projective geometries, particularly well suited for interior-point methods have been proposed. [KAR91]

We now illustrate computational experience with integer programming. Here again, the main computational task at each iteration, is the solution of one or more systems of linear equations. These systems have a structure similar to the system solved in each iteration of interior point algorithms for linear programming and therefore code written for linear programming can be used in integer programming implementations.

Consider, as a first example, the Satisfiability Problem (SAT) in propositional calculus, a central problem in mathematical logic. A Boolean variable $x$ can assume only values 0 or 1. Boolean variables can be combined by the logical connectives OR ($\vee$), AND ($\wedge$) and NOT ($\bar{x}$) to form Boolean formulae (e.g. $x_1 \wedge \bar{x}_2 \vee x_3$). A variable or a single negation of the variable is called a *literal*. A Boolean formula consisting of only literals combined by the $\vee$ operator is called a *clause*. SAT can be stated as follows: Given $m$ clauses $C_1, \ldots, C_m$ involving $n$ variables $x_1, \ldots, x_n$, does the formula $C_1 \wedge \cdots \wedge C_m$ evaluate to 1 for some Boolean input vector $[x_1, \ldots, x_n]$? If so, the formula is said to be satisfiable. Otherwise it is unsatisfiable.

Comparison of Simplex and Projective Algorithms

| Problem Name | Problem Size | | | CPU TIMES (Mins.) | | **Speedup** |
|---|---|---|---|---|---|---|
| | Rows | Columns | Nonzeros | Simplex | Projective | |
| 300rden | 300 | 900 | 180300 | 256 | 14 | 18 |
| 50lnord | 1225 | 41650 | 120050 | 292 | 13 | 22 |
| cntrl1 | 1045 | 3889 | 139800 | 610 | 28 | 22 |
| 30hcov | 2700 | 8928 | 26784 | 2673 | 63 | 43 |
| 75lnord | 2775 | 140600 | 410700 | 4615 | 77 | 60 |
| 600rden | 600 | 1800 | 720600 | 5625 | 91 | 60 |
| 100pde | 10000 | 20000 | 60000 | 10652 | 123 | 87 |
| cntrl2 | 2205 | 8209 | 580128 | 39265** | 360 | > 109 |
| 100minc | 10001 | 20000 | 39900 | 2700 | 22 | 123 |
| 200minc | 40001 | 80000 | 159800 | 50820 | 194 | 262 |
| 40hcov | 4800 | 15917 | 47751 | 26100 | 91 | 287 |
| 50hcov | 7500 | 25097 | 75294 | * | 319 | * |
| 200pde | 40000 | 80000 | 240000 | * | 2619 | * |
| 100lnord | 4950 | 333300 | 980100 | * | 86 | * |
| 1000rden | 1000 | 3000 | 2001000 | * | 364 | * |
| cntrl3 | 4105 | 15809 | 1159628 | * | 1452 | * |

\* Simplex was not run due to projected cpu time requirements.
\*\* Simplex run was terminated before optimal solution was reached.

TABLE 3.1
*Comparison of computational efficiency of the simplex method and the interior-point method for linear programming [KR91].*

SAT can be formulated as an integer programming feasibility problem. Denote

$$\mathcal{I}_\mathcal{C} = \{j \mid \text{literal } x_j \text{ appears in clause } \mathcal{C}\}$$

$$\mathcal{J}_\mathcal{C} = \{j \mid \text{literal } \bar{x}_j \text{ appears in clause } \mathcal{C}\} \ .$$

Associate with each Boolean literal $x_j$, the integer variable $w_j$, such that

$$w_j = \begin{cases} 1 & \text{if } x_j = 1 \\ 0 & \text{if } x_j = 0 \ , \end{cases}$$

and with each Boolean literal $\bar{x}_j$ the integer variable $\bar{w}_j = 1 - w_j$. To satisfy each clause $\mathcal{C}$, it is necessary that at least one literal with index in $\mathcal{I}_\mathcal{C}$ or in $\mathcal{J}_\mathcal{C}$ be set to 1, i.e.

$$\sum_{j \in \mathcal{I}_c} w_j + \sum_{j \in \mathcal{J}_c} \bar{w}_j \geq 1, \ \mathcal{C} = \mathcal{C}_1, \ldots, \mathcal{C}_m \ .$$

Substituting for $\bar{w}_j$, results in

(3.1) $$\sum_{j \in \mathcal{I}_c} w_j - \sum_{j \in \mathcal{J}_c} w_j \geq 1 - |\mathcal{J}_c|, \ \mathcal{C} = \mathcal{C}_1, \ldots, \mathcal{C}_m \ .$$

If an integer vector $w \in \{0,1\}^n$ is produced satisfying (3.1), the corresponding SAT problem

| SAT Problem Size | | Speed |
|---|---|---|
| Variables | Clauses | Up |
| 50 | 100 | 5 |
| 100 | 200 | 22 |
| 200 | 400 | 66 |
| 400 | 800 | 319 |

TABLE 3.2
*SAT: Comparison of Simplex and interior point algorithm [KKRR90].*

is said to be satisfiable.

In [KKRR90], an interior point implementation was compared with an approach based on the Simplex Method to prove satisfiability of randomly generated instances of SAT. It was shown that the interior point approach can handle instances of SAT that are much larger than those previously solved by other algorithms. Instances with up to 1000 variables and 32,000 clauses were solved. Compared with the Simplex Method approach on small problems (Table 3.2), speedups of over two orders of magnitude were observed. Furthermore, the interior point approach was successful in proving satisfiability in over 250 instances that the Simplex Method approach failed.

As a second example, consider inductive inference, a central problem in artificial intelligence and machine learning. Inductive inference is the process of hypothesizing a general rule from examples. In [KKRR91], the interior point approach was applied to a basic model of inductive inference. In this model there is a black box (Fig. 3.1) with $n$ Boolean input variables $x_1, \ldots, x_n$ and a single Boolean output variable $y$. The black box contains a hidden Boolean function $\mathcal{F} : \{0,1\}^n \to \{0,1\}$ that maps inputs to outputs. Given a limited number of inputs and corresponding outputs, we ask: Does there exist an algebraic sum-of-products expression with no more than $K$ product terms that matches this behavior? If so, what is it? It turns out that this problem can be formulated a SAT problem (see [KKRR91] for details).

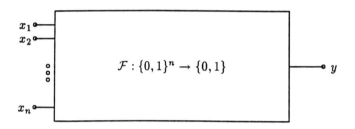

Figure 3.1: Black box with hidden logic.

Consider the hidden logic described by the 32-input, 1-output Boolean expression $y = x_4 x_{11} x_{15} \bar{x}_{22} + x_2 x_{12} \bar{x}_{29} + \bar{x}_3 x_9 x_{20} + \bar{x}_{10} x_{11} \bar{x}_{29} x_{32}$. This function has $2^{32} \simeq 4.3 \times 10^9$ distinct

| I/O Samples | SAT Prob Size | | Iter | CPU Time* | Inferred Logic | Prediction Accuracy |
|---|---|---|---|---|---|---|
| | Vars | Clauses | | | | |
| 50 | 332 | 2703 | 49 | 66s | $y = \bar{x}_{22} x_{28} \bar{x}_{29} + x_{12} \bar{x}_{17} \bar{x}_{25} x_{27} + \bar{x}_3 x_9 x_{20} + x_{11} x_{12} \bar{x}_{16} \bar{x}_{32}$ | .74 |
| 100 | 404 | 5153 | 78 | 2m58s | $y = x_9 x_{11} \bar{x}_{22} \bar{x}_{29} + x_4 x_{11} \bar{x}_{22} + \bar{x}_3 x_9 x_{20} + x_{12} \bar{x}_{15} \bar{x}_{16} \bar{x}_{29}$ | .91 |
| 400 | 824 | 19478 | 147 | 20m27s | $y = x_4 x_{11} \bar{x}_{22} + \bar{x}_{10} x_{11} \bar{x}_{29} x_{32} + \bar{x}_3 x_9 x_{20} + x_2 x_{12} \bar{x}_{15} \bar{x}_{29}$ | exact |

TABLE 3.3
Inductive inference SAT problems.

| Variables Hidden Logic | SAT Problem | | Interior Method | | Davis-Putnam time |
|---|---|---|---|---|---|
| | vars | clauses | itr | time | |
| 8 | 396 | 2798 | 1 | 9.33s | 43.05s |
| 8 | 930 | 6547 | 13 | 45.72s | 11.78s |
| 8 | 1068 | 8214 | 33 | 122.62s | 9.48s |
| 16 | 532 | 7825 | 89 | 375.83s | 20449.20s |
| 16 | 924 | 13803 | 98 | 520.60s | * |
| 16 | 1602 | 23281 | 78 | 607.80s | * |
| 32 | 228 | 1374 | 1 | 5.02s | 159.68s |
| 32 | 249 | 2182 | 1 | 9.38s | 176.32s |
| 32 | 267 | 2746 | 1 | 9.76s | 144.40s |
| 32 | 450 | 9380 | 71 | 390.22s | * |
| 32 | 759 | 20862 | 1 | 154.62s | * |

* Did not find satisfiable assignment in 43200s.

TABLE 3.4
Computational Efficiency of an Interior Point approach and a Combinatorial Approach on Inductive Inference Problems

input-output combinations. Table 3.3 summarizes the computational results. Subsets of input-output examples of size 50, 100 and 400 were considered. The number of terms in the expression to be synthesized was fixed at $K = 4$. In all instances the interior point algorithm synthesized a function that described completely the behavior of the sample. With a sample of only 400 input-output patterns the approach succeeded in exactly describing the hidden logic. The prediction accuracy given in the table was computed with Monte Carlo simulation, where 10,000 random vectors were input to the black box and to the inferred logic and their outputs compared.

As another example of application of the continuous approach to combinatorial problems consider the wire routing problem for gate arrays, an important subproblem arising in VLSI design.

As shown in Figure 3.2, a gate array can be abstracted mathematically as a grid graph.

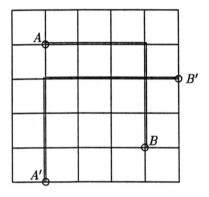

Figure 3.2.

Input to the wire routing problem consists of a list of wires specified by end points on a rectangular grid. Each edge of the graph, also known as a channel, has a prespecified capacity representing the maximum number of wires it can carry. The combinatorial problem is to find a wiring pattern without exceeding capacity of horizontal and vertical channels. This problem can be formulated as 0-1 integer programming problem. Interior-point approach has successfully obtained **provably optimal global** solutions to **large-scale** problems of this type having more than 20,000 wires [PKR90]. On the other hand, combinatorial heuristics, such as simulated annealing do not come anywhere close either in the quality of the solution they can find or in terms of the computational cost.

In case of many other combinatorial problems, numerous heuristic approaches have been developed. Many times, the heuristic merely encodes the prior knowledge or anticipation of the structure of solution to a specific class of practical applications into the working of the algorithm. This may make a limited improvement in efficiency without really coming to grips with the problem of exponential growth that plagues the combinatorial approach. Besides, one needs to develop a lot of different heuristics to deal with different situations. Interior-point methods have given us a unified approach to create efficient algorithm to many different combinatorial problems.

**4. Application to Continuous Computational Problems..** Although interior-point methods originated in the context of discrete optimisation problems, they are applicable to a vast number of continuous computational problems in science and engineering, such as finding minimum energy state in quantum mechanics, finding equilibrium points of chemical systems, $L_1$-estimation in statistics, fourier transform problems with inequality constraints, electrical circuit simulation, optimal control problems, partial differential equations with inequality constraints, etc.

Many of these problems result in mathematical programming problems in infinite dimensional space, e.g. Hilbert space. The theoretical framework of Riemannian geometry and lie groups can be extended to such spaces. Actual numerical computation is performed on the discretized version of the problem. Application to control systems problem can be found in [KR91]. We now describe some other examples in more details.

**Partial Differential Equations with Inequality Constraints.**

As an example of this type consider the following problem. Suppose we want to measure a function

$$f(x,y), \qquad 0 \leq x \leq 1, \quad 0 \leq y \leq 1$$

that is known to satisfy Laplace's equation

$$\frac{\partial^2 f}{\partial x^2} + \frac{\partial^2 f}{\partial y^2} = 0 \ .$$

But the process of measurement introduces *random noise*, $r(x,y)$

$$\tilde{f}(x,y) = f(x,y) + r(x,y)$$

From the measured function $\tilde{f}(x,y)$ we want to recover $f(x,y)$ exploiting the fact that it satisfies Laplace's equation

$$\min_{g:\nabla^2 g=0} \max_{(x,y)} | \tilde{f}(x,y) - g(x,y) |$$

This is an infinite dimensional linear programming problem. After discretization, we get a finite dimensional linear program.

There are other variations of this problem:

a. Instead of $L_\infty$ norm, if we use $L_1$ norm $\| \tilde{f} - g \|_1$, it also leads to a linear programming problem.

b. We can also use a weight function $w(x, y)$ to account for the fact that measurements in certain regions may be more reliable.

$$\min_{g:\nabla^2 g=0} \max_{x,y} w(x,y) \mid \tilde{f}(x,y) - g(x,y) \mid .$$

The resulting problem is still linear. Large-scale problems of this type involving more than 80,000 variables have been solved by the interior-point method [KR91]. Application of other formulations of this type to numerical weather modelling can be found in [PAR85].

**Finding Minimum Energy State of a Physical System.**

As an example of this type, consider a quantum mechanical system having a Hamiltonian $H$ and wave function denoted by $\psi(\mathbf{x})$. We are interested in the solution of the following eigenfunction problem with minimum eigenvalue $E$.

$$H\psi = E\psi .$$

In many cases, the minimum eigenfunction $\psi(\mathbf{x})$ has the same sign everywhere, $\psi_1(\mathbf{x}) \leq 0$. Imposing the inequalities $\psi_1(\mathbf{x}) \leq 0$ in the computational method has several advantages:

1. **Faster convergence**

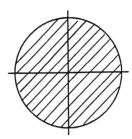 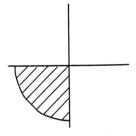

Original search space　　　　　　　Reduced search space

Although it may appear that we are making the original problem more complicated by adding inequality constraints, we are actually helping the search process by reducing the volume of search space by factor of $2^n$.

2. **Better separation between $\psi_1(\mathbf{x})$ and $\psi_2(\mathbf{x})$**

In case of nearly degenerate problems, we can obtain better separation between $\psi_1(\mathbf{x})$ and $\psi_2(\mathbf{x})$. Numerically,

$$\frac{\psi_1^\star H \psi_1}{\psi_1^\star \psi_1} \text{ is close to } \frac{\psi_2^\star H \psi_2}{\psi_2^\star \psi_2} .$$

But qualitatively, there is a major difference between $\psi_1(\mathbf{x})$ and $\psi_2(\mathbf{x})$:

$\psi_1(\mathbf{x})$ : never changes sign

$\psi_2(\mathbf{x})$ : changes sign exactly once

Interior-point approach can exploit this qualitative difference to gain robustness and speed of convergence.

**Computing Equilibrium Point of Chemical System.**

This problem can be formulated as solution of a system of non-linear algebraic equations. Standard methods for solving such equations (e.g. Newton's method) can converge to **non-physical** solutions with **negative** values of concentration. As the problem size grows, the total number of isolated solutions of the algebraic equations grow and the proportion of solutions with all non-negative components shrinks. Interior-point approach is ideally suited for solving such equations with non-negativity constraints. Due to the non-linear geometry underlying the method, non-physical solutions are at infinite distance from the starting condition, in the solution space.

**Fourier Transform with Inequality Constraints.**

There are several computational problems in which we wish to impose inequality constraints on the Fourier Transform of a function. Since the Fourier Transform is obtained essentially by applying a linear operator to the original function, we get an implicitly defined linear programming problem. Combining the mathematical properties of the interior point technique with special properties of Fourier Transform, leads to a very efficient approach to solving these problems [SK90]. Such problems arise in digital signal processing, design of digital filters, designing input waveforms to an array antenna when the desired radiation pattern is specified, band limited extrapolation of signals, numerical weather modelling, etc.

**Robust Simulation of Electronic Circuits.**

Mathematical model of an electronic circuit has two sets of coupled equations:

- Circuit Topology Equation: These are Kirchoff Voltage Law (KVL), and Kirchoff Current Law (KCL).

- Device Models: These are non-linear algebraic equations, and non-linear differential equations. Most commonly used solution approach (based on Newton's method) is known to fail on difficult circuits when starting far away from actual solution.

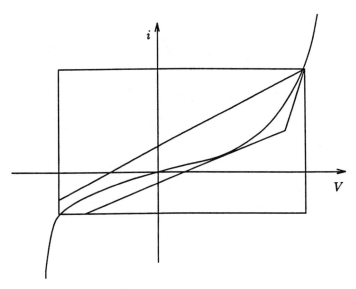

As shown in the figure, application of interior-point approach is based on enclosing the operating region of the device equations in a convex region and using the underlying non-linear geometry of this region to create a vector field. Integral curves of this vector field are then followed to arrive at the solution. This approach has significantly improved the robustness of simulation methodology developed by researchers in Bell Laboratories [SS90].

**5. Concluding remarks - the place of optimisation in natural and man-made sciences.** Optimisation is of central importance in both natural sciences such as physics, chemistry and biology as well as artificial or "man-made" sciences such as computer science and operations research. Nature inherently seeks optimal solutions. For instance, crystalline structure is the minimum energy state for a set of atoms, light travels through the shortest path etc. Behavior of nature can often be explained on the basis of variational principles. Laws of nature then simply become optimality conditions. Concepts from continuous mathematics have always played a central role in the description of these optimality conditions and in analysis of the structure of their solutions. On the other hand, in artificial sciences, the problems are stated using the language of discrete mathematics or logic, and a simple-minded search for their solution confines one to a discrete solution set. Such a restricted approach has had the effect of distorting the development of artificial sciences. The most efficient method for solving the most widely used optimisation problem in industry remained undiscovered for nearly forty years. A mistaken belief was created regarding the computational difficulty of numerous other discrete problems arising in engineering design, business decision making and artificial intelligence.

With the advent of interior point methods, the picture is changing dramatically, because these methods do not confine their working to a discrete solution set, but instead view combinatorial objects as limiting cases of continuous objects and exploit the topological and geometric properties of the continuous space. As a result, the number of efficiently solvable combinatorial problems is expanding rapidly. Also, the interior-point methods have revealed that the mathematical structure relevant to optimisation in natural and artificial sciences have a great deal in common. Differential equations and Riemannian geometry play a major role in our current understanding of both fields. This suggests that whatever we do not yet know about the structure of solutions to combinatorial problems should also have a lot to do with as yet unexplained (or partially explained) phenomena in natural sciences. This makes the subject of interior-point methods an even more exciting field of inquiry. Nowadays conferences on global optimisation [FP92] are attracting researchers from diverse fields ranging from computer science to molecular biology, thus merging the development paths of natural and artificial sciences. The phenomena of multiple solutions to combinatorial problems is intimately related to multiple configurations a complex molecule can assume. While the structure of solutions to *linear* partial differential equations occurring in nature, such as Maxwell's equations is simple and well understood, introduction of the simplest type of quadratic non-linearity as in Navier-Stokes equations, already gives the solution space a rather complex structure. In the context of combinatorial problems, analysis of topological properties such as connectivity of level sets shows [KAR90-1] that the number of connected components can increase exponentially as one goes from linear to quadratic optimisation problems. Thus, understanding the structure of solution sets of non-linear problems is a common challenge faced by both natural and artificial sciences, in order to explain natural phenomena in the former case and to create more efficient interior-point algorithms in the latter case.

**6. Suggestions for further reading.** There is a vast amount of literature on interior point methods for linear and convex programming. I direct the reader to the following two references for a survey of theoretical results and for further information on implementation:

TOD89 Todd M. Recent Developments and New Directions in Linear Programming, in Mathematical Programming, Recent Developments and Applications, Kluwer Academic Press, 1989, 108-157.

KR91 Karmarkar, N. K., Ramakrishnan, K. G., Computational Results of an Interior Point Algorithm for Large Scale Linear Programming, Mathematical Programming, 52 (1991) 555-586.

For these readers who are not already familiar with interior point methods, but would like to gain further understanding of the concepts described in this paper, I have given below a selected list of references. I have also made an attempt to classify them according to the major concept involved, although some of the papers could fit multiple classification. The purpose of this list is to provide references that can be used as convenient starting points for various concepts, rather than to provide an exhaustive bibliography.

**Global Approximation.**

YE90 Yinyu Ye, Interior-point Algorithms for Global Optimisation, Annals of Operations Research, 25, (1990) 59-74.

KAR90 Karmarkar N., An Interior-Point Approach to NP-complete Problems, Proceedings of the conference on Integer Programming and Combinatorial Optimisation, Waterloo, Canada, May 1990, 351-366.

KK92 Kamath, A., Karmarkar, N., A continuous method for computing bounds in integer quadratic optimisation problems, Journal of Global Optimisation, 1992.

KT92 Karmarkar, N., Thakur, S., An interior point approach to a tensor optimisation problem with application to upper bounds in integer quadratic optimisation, Proceedings of second conference on Integer Programming and Combinatorial Optimisation, May 1992.

**Continuous Trajectories.**

BL89-1 D. A. Bayer and J. C. Lagarias, The nonlinear geometry of linear programming I, Affine and projective scaling trajectories, Trans. Amer. Math. Soc. 314 (1989), 499-526.

BL89-2 D. A. Bayer and J. C. Lagarias, The nonlinear geometry of linear programming II, Lengendre transform coordinates and central trajectories, Trans. Amer. Math. Soc. 314 (1989) 527-581.

KSLW89 Karmarkar, N., Lagarias, J., Slutsman, L., Wang, R., Power series variants of Karmarkar type algorithms, AT&T Technical Journal, vol. 68, no. 3, May/June 1989.

MEG89 N. Megiddo, Pathways to the optimal set in linear programming, Progress in Mathematical Programming, Interior-point and Related Methods (N. Megiddo, ed.), Springer-Verlag, New York, 1989, 131-158.

SON86 G. YU. Sonnevand, An "analytic centre" for polyhedrons and new classes of global algorithms for linear (smooth, convex) programming, Proc. 12th IFIP Conf. System Modelling, Budapest, Lectures Notes in Computer Science 1986.

**Topological Properties.**

KAR90-1 Karmarkar, N., An interior-point approach to NP-complete problems – Part I, Mathematical developments arising from linear programming, Contemporary Mathematics, Vol. 114, 1990, 297-308.

**Differential Geometry.**

KAR90-2 Karmarkar, N., Riemannian Geometry Interior-Point Methods for Linear Programming, in Mathematical Developments Arising from Linear Programming, Contemporary Mathematics, Vol. 114, 1990, 51-75.

LAG90 Lagarias, J. C., The nonlinear geometry of linear programming III, Projective Legendre transform coordinates and Hilbert geometry, Trans. Amer. Math. Soc. 320 (1990), 193-225.

**Parallel Implementation.**

HHL89 Housos, E., Huang, C. and Liu, L., Parallel algorithms for the AT&T korbx system, AT&T Technical Journal, vol. 68, no. 3, May/June 1989.

KAR91 Karmarkar, N., A new parallel architecture for sparse matrix computation based on finite projective geometries, Proceedings of Supercomputing '91, IEEE computer society, Nov. 1991, 358-369.

**Integer Programming and Mathematical Logic.**

KKRR90 Kamath, A. P., Karmarkar, N. K., Ramakrishnan, K. G., Resende, M. G. C., Computational experience with an interior point algorithm on the satisfiability problem. Annals of O. R. 25: 43-58, 1990.

PKR90 Pai, R., Karmarkar, N. K., Rao, S. S. S. P., A Global Router for Gate-Arrays Based on Karmarkar's Interior Point Methods, Proceedings of the Third International Workshop on VLSI System Design, Bangalore, India, Jan. 6-9, 1990, 73-82.

KKRR91 Kamath, A. P., Karmarkar, N. K., Ramakrishnan, K. G., Resende, M. G. C., A Continuous Approach to Inductive Inference, Proceeding of the Conference on Complexity Issues in Numerical Optimisation, Cornell, March 1991.

**Continuous Computational Problems.**

TOD91 Todd, M., Interior-point algorithms for semi-infinite programming, Technical Report No. 978, School of operations research and industrial engineering, Cornell University, August 1991.

SK90 Soman, A., Karmarkar, N., Design of optimal FIR filters, Technical Memorandum no. 11211-901203-22TM, December 1990.

SS90 Sani, N., Singhal, K., Private communication.

PAR85 Parrish, D., Application of Karmarkar's algorithm to meteorological analysis, proceedings of ninth conference on Probability and Statistics in Atmospheric Sciences, October 1985.

FP92 Floudas, C., Pardalos, P., Recent Advances in Global Optimisation, Princeton series in computer science, Princeton University Press, 1992.

# Chapter 12
# VISCOSITY SOLUTIONS AND OPTIMAL CONTROL

**P. L. LIONS**
Université Paris-Dauphine
France

## I. Introduction.

We wish to illustrate here the intrinsic links that exist between Optimal Control theory and Viscosity Solutions theory. At the time of ICIAM 91, the notion of viscosity solutions was almost precisely ten years old and in this decade the field of viscosity solutions has been extremely active (probably more than 500 papers...). This is why it is almost impossible to survey the theory even if we restrict our attention to control problems and viscosity solutions.

Our goal here will be thus somewhat more limited. In section II below we shall introduce the notion of viscosity solutions (as introduced by M.G. Crandall and the author [18],[19]) starting from a model optimal stochastic control problem and we shall show how the classical Dynamic Programming argument with a slight modification yields easily the formulation of viscosity solutions. Then, in section III we will list the major achievements of the theory of viscosity solutions, recalling in particular some of its applications to optimal control or differential games problems. The list will not be exhaustive, not will the references cover all aspects of the interplay between viscosity solutions and control problems.

Before we leave this general outline of our report, a few observations might be in order. First of all, the notion of viscosity solutions is a notion of solutions for general fully nonlinear scalar second-order elliptic possibly degenerate equations in finite or infinite dimensions. It is very much related to the maximum principle, a property shared by all these equations. This notion has allowed a remarkably general understanding of these equations (see section III below). The relationship with control theory is primarily due to the fact that the so-called value functions of control or games problems are characterized as the viscosity solutions of the corresponding Dynamic Programming partial differential equations. This characterization not only allows a full justification of the somewhat formal derivation of those equations but also enables to establish many properties of control (or games) problems and their value functions. In summary, viscosity solutions can be seen from the control viewpoint as a powerful tool to analyse control problems via the Dynamic Programming equations. And it is worth emphasizing that, even if Viscosity Solutions theory is a PDE theory, it is both simple and elementary.

Our goal here is to introduce viscosity solutions and to show how "natural" they are. We shall do so on a model example : this example is a typical infinite horizon completely observable optimal stochastic control problem without state constraints involving diffusion-like processes. Therefore, we first explain the setting of this model control problem: we consider in a probability space $(\Omega, F_t, P, W_t)$ equipped with a filtration and a Brownian motion $W_t$ taking values in $\mathbb{R}^m$ (satisfying the usual conditions) a controlled system whose state obeys the following stochastic differential equation

$$(1) \quad dX_t = \sigma(X_t, \alpha_t) dW_t + b(X_t, \alpha_t) dt \quad, \quad X_0 = x \in \mathbb{R}^N$$

where $x$ (the initial position of the system) is arbitrary in $\mathbb{R}^N$. The reader feeling uncomfortable with SDE should just cancel the term $\sigma$, in which case (1) is just a (controlled) ODE and all processes $(X_t, \alpha_t)$ are mere measurable functions of $t$. The problem becomes them a deterministic control problem. The process $\alpha_t$ is the control process and is just an arbitrary progressively measurable process taking values in a given set $\mathcal{A}$ (usually taken to be a separable metric space but this is not necessary for the discussion here). Finally, $\sigma$ et $b$ are given functions over $\mathbb{R}^N \times \mathcal{A}$ taking values respectively in the space of $N \times m$ matrices, $\mathbb{R}^N$. In order to avoid technicalities, we shall assume that $\sigma$ and $b$ are bounded, Lipschitz in $x$ uniformly in $\alpha$ and continuous (say) in $\alpha$. Therefore, for a given $x$ and a given control $\alpha_t, X_t$ (the state of the system) is uniquely determined by (1).

We then define the so-called cost function

(2) $$J(x, \alpha_t) = E \int_0^\infty f(X_t, \alpha_t) e^{-\lambda t} \, dt$$

where $\lambda > 0$ is given (the so-called discount factor), $f$ is a scalar function on $\mathbb{R}^N \times \mathcal{A}$ that we assume to be (for instance) bounded uniformly continuous in $x$, uniformly in $\alpha$ and continuous in $\alpha$ (say). Finally, the value function of the problem is defined by

(3) $$u(x) = \inf_{\alpha_t} J(x, \alpha_t) \quad , \quad \forall \, x \in \mathbb{R}^N \, .$$

It is often better to consider the whole system $(\Omega, F_t, P, W_t, \alpha_t)$ to be the "parameter" with respect to which the minimization is taking place but such technicalities are irrelevant here. Of course, the determination of this function $u$ is one of the two major goals of such control problems (the other being to determine possible optimal controls or even optimal feedbacks...).

In order to achieve this goal, the so-called Dynamic Programming principle, as introduced by R. Bellman [7],[8] - see also W.H. Fleming and R. Rishel [27], D.P. Bertsekas [10]..., gives a general strategy of attack that can be decomposed into two main steps : i) obtain a general identity that reflects the dynamic character of the problem, ii) derive a nonlinear partial differential equation that $u$ should satisfy and that should characterize $u$. We shall recall the main statements corresponding to these two steps but we shall not give any details about their justification which, in full rigor, involves serious technical difficulties. These classical arguments are exposed in the references aforementioned and in A. Bensoussan and J.L. Lions [9], N.V. Krylov [35], P.L. Lions [38], W.H. Fleming and H.M. Soner [28]...

Step i) (often called the optimality principle for the identity below yields additionnal informations on optimal controls that we shall not discuss) consists in establishing the following identity for all $x \in \mathbb{R}^N$

(4) $$u(x) = \inf_{\alpha_t} E \left[ \int_0^h f(X_t, \alpha_t) e^{-\lambda t} + u(X_h) e^{-\lambda h} \right]$$

where $h \geq 0$ is arbitrary ($h$ may even be a stopping time depending on the control $\alpha_t$). This property of dynamic optimization is transparent in the very particular case when there is no control (so the infimum disappears). In that case, it is an easy consequence of the fact that $X_t$ is a Markov process (and if we are in the deterministic case i.e. $\sigma \equiv 0$, this is only the usual flow or group property of autonomous ODE's).

Step ii) is formally a simple differentiation with respect to $h = 0$ of the identity (4). More precisely, one writes for $h > 0$ and for all $x \in \mathbb{R}^N$

$$(5) \quad \sup_{\alpha_t} \left\{ \frac{u(x) - E[u(X_h)e^{-\lambda h}]}{h} - \frac{1}{h} E\left[ \int_0^h f(X_t, \alpha_t) e^{-\lambda t}\, dt \right] \right\} = 0$$

and letting $h$ go to $0_+$, one deduces if $u$ is twice differentiable at a point $x$ (or differentiable in the deterministic case $\sigma \equiv 0$):

$$(6) \quad \sup_{\alpha \in \mathcal{A}} \left[ -a_{ij}(x, \alpha) \partial_{ij} u(x) - b_i(x, \alpha) \partial_i u(x) - f(x, \alpha) \right] + \lambda u(x) = 0$$

where we use the convention of implicit summation on repeated indices, $\partial_i = \frac{\partial}{\partial x_i}$ and the $N \times N$ matrix $a$ is defined by $a = \frac{1}{2}\sigma\sigma^T$. This second-order equation (first-order in the deterministic case) is called the Hamilton-Jacobi-Bellman equation (HJB in short). It is also called sometimes the Bellman equation. In fact, it is not difficult to check that the minimal smoothness requirement on $u$ in order to deduce (6) from (5) is the existence at $x$ of a second-order expression of $u$ in which case the matrix of second-order terms replaces $\partial_{ij} u(x)$ in (6).

However, it is easily seen that this requirement is not met in general even in very simple situations. For instance, we choose $N = 1$, $\mathcal{A} = [-1, +1]$, $\sigma \equiv 0$, $b = \alpha$ and $f(x, \alpha) = \varphi(x)$ where $\varphi \in C_b^\infty$ ($\varphi$ and all its derivatives are bounded), $\varphi$ is even and $\varphi'(x) < 0$ for $x > 0$. Then, one checks easily that $u(x) = \int_0^\infty \varphi(|x| + t) e^{-t}\, dt$ so that $u$ is smooth for $x > 0$ and for $x < 0$ but $u$ is not differentiable at 0 (indeed, $u$ admits at 0 a right-derivative given by $-\varphi(0) + \int_0^\infty \varphi(t) e^{-t}\, dt$ ($> 0$) and a left-derivative given by the opposite of this quantity).

This lack of regularity which is quite general (in fact, even worse situations are possible in more complicated problems) prevents to use (6) in a naive way. However, it is tempting to believe that (6) should hold in some generalized sense and this is precisely what we shall achieve with the notion of viscosity solutions. But before we do as, we want to mention here that the Dynamic Programming method yields even more information. Indeed, at least formally, any solution (say bounded) of (6) is indeed the value function $u$ and one gets a simple rule to build optimal controls in feedback form. These formal considerations can be made rigorous if we assume (and this is unrealistic) the existence of a bounded (say) $C^2$ solution of (6). And, with some serious mathematical effort, "$\varepsilon$-optimal feedbacks" can be built.

We now go back to the derivation of (6) which, as we recalled, above holds at a point $x_0 \in \mathbb{R}^N$ if $u$ admits at this point a second-order expansion. Or in other words, (6) holds at a point if we can "accurately replace $u$ by a smooth function". Of course, this is bound to fail if $u$ is not at least differentiable. But, then we might try to make a one sided substitution of $u$ that is, whenever this is possible, consider a function $\varphi \in C^2$ such that $u(x_0) = \varphi(x_0)$, $u(x) \leq \varphi(x)$ on $\mathbb{R}^N$ (in fact we need this comparison only in a neighborhood of $x_0$ by some elementary modifications of the argument below). This is possible when $u - \varphi$ admits a global minimum at $x_0$ (or local...) and $\varphi \in C^2$ - and replacing if necessary $\varphi$ by $\varphi(\cdot) + [u(x_0) - \varphi(x_0)]$ we may always assume that $u(x_0) = \varphi(x_0)$. In that case (5) becomes

$$(7) \quad \sup_{\alpha_t} \left\{ \frac{\varphi(x_0) - E\varphi(X_h)e^{-\lambda h}}{h} - \frac{1}{h} E \int_0^h f(X_s, \alpha_s) e^{-\lambda s} ds \right\} \geq 0$$

and since $\varphi$ is smooth, we can now perform the classical derivation of the HJB equation which now yields

$$(8) \quad \sup_{\alpha \in \mathcal{A}} \left[ -a_{ij}(x_0, \alpha) \partial_{ij} \varphi(x_0) - b_i(x_0, \alpha) \partial_i \varphi(x_0) - f(x_0, \alpha) \right] + \lambda u(x_0) \geq 0.$$

Of course, we might try a "reverse" one-sided substitution and thus in the event that $u - \varphi$ admits a local maximum at $x_0$ where $\varphi \in C^2$, we deduce

$$(9) \quad \sup_{\alpha \in \mathcal{A}} \left[ -a_{ij}(x_0, \alpha) \partial_{ij} \varphi(x_0) - b_i(x_0, \alpha) \partial_i \varphi(x_0) - f(x_0, \alpha) \right] + \lambda u(x_0) \leq 0.$$

These inequalities are in fact the definition of viscosity solutions. To be more precise, we need a few notations. First of all, we are going to recall the definition of viscosity solutions for general equations of the form

$$(*) \quad F(x, u(x), Du(x), D^2 u(x)) = 0 \quad \text{in } \Omega$$

where $\Omega$ is an open set of $\mathbb{R}^N$ and $F = F(x, t, p, A)$ is a given, continuous function (say) acting on $x \in \Omega$, $t \in \mathbb{R}$, $p \in \mathbb{R}^N$, $A$ $N \times N$ symmetric matrix. The definition makes sense when $F$ satisfies the following weak form of degenerate ellipticity

$$(10) \quad F(x, t, p, A) \leq F(x, t, p, B)$$
$$\text{for all} \quad x \in \Omega, \, t \in \mathbb{R}, \, p \in \mathbb{R}^N, \, A \geq B$$

(in the sense of symmetric matrices). Another monotonicity condition which is required (and which can be sometimes relaxed) is

(11) $F(x,t,p,A)$ is nondecreasing with respect to $t$ for all $x, p, A$.

These conditions are clearly met in the case of HJB equations where $F$ is given by

(12) $$F = \sup_{\alpha \in \mathcal{A}} \left[ -Tr(a(x,\alpha)A) - b_i(x,\alpha)p_i - f(x,\alpha) \right] + \lambda t.$$

It is easily seen that these conditions allow to accomodate many classes of elliptic or degenerate elliptic equations (see M.G. Crandall, H. Ishii and P.L. Lions [17] for a list of examples). In particular, included are the Isaacs equations obtained in differential games problems by an analogous argument to the one sketched above.

We may now turn to the

DEFINITION. 1) *A scalar upper semi-continuous (resp. lower semi-continuous) function $u$ on $\Omega$ is a viscosity subsolution (resp. supersolution) of $(*)$ if we have for all $\varphi \in C^2(\Omega)$ and at each local maximum (resp. minimum) point $x_0$ of $u - \varphi$ the following inequality*

(13) $$F\bigl(x_0, u(x_0), D\varphi(x_0), D^2\varphi(x_0)\bigr) \leq 0$$

*(resp.*

(14) $$F\bigl(x_0, u(x_0), D\varphi(x_0), D^2\varphi(x_0)\bigr) \geq 0 \text{ ).}$$

2) *A scalar continuous function $u$ on $\Omega$ is a viscosity solution of $(*)$ if it is both a viscosity subsolution and a viscosity supersolution of $(*)$.*

The notion was introduced by M.G. Crandall and the author [19] - see also M.G. Crandall, L.C. Evans and P.L. Lions [15], P.L. Lions [37]... Since, in the early stages of the theory, the emphasis was set on first-order equations the terminology of viscosity solutions arose from the fact that the definition can be recovered from approximating the equation by a regularized problem where one adds a viscosity term $(-\varepsilon\Delta)$ and letting $\varepsilon$ go to 0.

The links between the notion of viscosity solutions and the classical rules of maximum principle can be illustrated by the following argument which proves a fact of interest by itself. If $u$ is twice differentiable on $\Omega$ and satisfies $(*)$ then $u$ is a viscosity solution of $(*)$. Let us check indeed that $u$ is a viscosity subsolution, the supersolution part being shown exactly in the same way. To this end, let $\varphi \in C^2(\Omega)$ and let $x_0$ be a local maximum point of $u - \varphi$. Then, one knows that

$$(15) \qquad Du(x_0) = D\varphi(x_0) \quad , \quad D^2 u(x_0) \leq D^2 \varphi(x_0)$$

and these relations are at the basis of the classical maximum principle.

Next, since $(*)$ holds at $x_0$ by assumption and since $F$ satisfies (10), we immediately deduce (13) from (15) proving thus our claim.

Of course, the way we introduced the notion has shown that $u$ is indeed a viscosity solution of the HJB equation (6) provided we check that $u$ is continuous on $\mathbb{R}^N$. This can be done directly from the definition of $u$. And the dynamic programming argument is then fully validated by the following result (first proven in P.L. Lions [39]) that shows in particular that (6) (interpreted in viscosity sense) completely characterizes $u$.

THEOREM. *The value function $u$ is the unique bounded viscosity solution of the HJB equation (6).*

The reader might worry about the boundedness restriction. It suffices to recall that some growth conditions at infinity have to be imposed in order to obtain the uniqueness of solutions of general equations (take $\frac{du}{dx} + u = 0$ in $\mathbb{R}$ for example). And boundedness is the easiest to state even if the above result remains valid under much more general growth conditions at infinity.

## III. A guided tour of viscosity solutions theory.

We embark now for a list of topics that have been successfully studied via viscosity solutions. Before we start, we would like to mention the survey article by M.G. Crandall, H. Ishii and P.L. Lions [17] which offers a self-contained presentation of most of the existing PDE part of the theory and a balanced sample of references, and the book by W.H. Fleming and H.M. Soner [28] on optimal stochastic control which discusses in particular the impact of viscosity solutions on various issues and classes of control problems. The list we present is organized by topic and a limited (somewhat biased) sample of references is provided on each topic.

1) UNIQUENESS: Uniqueness and comparison results are central to the viscosity solutions theory. The first results of that sort were obtained by M.G. Crandall and P.L. Lions [19] (see also M.G. Crandall, L.C. Evans and P.L. Lions [15]) for first-order problems (Hamilton-Jacobi equations). Analogous results were shown in a rather undirect way for the Hamilton-Jacobi-Bellman equations corresponding to optimal stochastic control in P.L. Lions [39]. A major breakthrough in the case of second-order equations was achieved by R. Jensen [34] with some further simplifications and improvements by H. Ishii [32], H. Ishii and P.L. Lions [33], M.G. Crandall [14], M.G. Crandall and H. Ishii [16]. A self-contained presentation of the general uniqueness and comparison results can be found in [17].

2) EXISTENCE: Many strategies for existence are known leading to quite general existence results. However, the simplest one is the adaptation of Perron's method introduced by H. Ishii [31] which can be basically summarized by "comparison-uniqueness implies existence".

3) BOUNDARY CONDITIONS: Viscosity solutions not only account for general equations but also for general boundary conditions which can be of Dirichlet type, Neumann or oblique derivative type even fully non-linear first-order boundary conditions. Motivated by state constraints optimal control problems, H.M. Soner [49] has introduced a new type of boundary condition which is naturally expressed in terms of viscosity solutions.

4) REGULARITY AND QUALITATIVE PROPERTIES OF SOLUTIONS: Many properties of viscosity solutions are in fact related to uniqueness issues and even striking regularity results are direct consequence of the method of proof used for uniqueness. Let us mention some surprising regularity results obtained by L. Caffarelli [11], N.S. Trudinger [52] for uniformly elliptic equations using viscosity solutions.

5) INFINITE-DIMENSIONAL EQUATIONS: Such equations are in particular consequences of the dynamic programming method for optimal control problems involving PDE's or stochastic evolutions or stachastic PDE's (such as the Zakaï equation). And a large part of the theory has been "raised" to infinite dimensions with a series of investigations on the unbounded terms that appear naturally (say in the optimal control of PDE's). And we refer to M.G. Crandall and P.L. Lions [21], P.L. Lions [40],[41],[42], D. Tataru [51]...

6) CONTROL THEORY: Not only viscosity solutions yield a complete justification of the Dynamic Programming principle but they have allowed to study various issues in Control Theory (state constraints problems [49],[12], infinite-dimensional problems, singular controls [28], the Pontrjagin principle [6]...).

7) DIFFERENTIAL GAMES: The uniqueness of viscosity solutions immediately yields the equivalence of various notions of value functions in deterministic differential games [24] and allow to study these value functions. In addition, they even provide the framework for stochastic differential games [30].

8) WEAK PASSAGE TO THE LIMIT: A general argument to pass to the limit using only bounds on viscosity solutions was discovered by G. Barles and B. Perthame [4]. It has become one of the few crucial tools in the theory.

9) APPROXIMATION, NUMERICAL SCHEMES AND CONVERGENCE ANALYSIS: A large body of works exist in those directions with applications to Control Theory [26], the convergence of monotone consistent schemes [20],[5] or even the convergence of MUSCL schemes for one-dimensional scalar conservation laws [14]...

10) ASYMPTOTIC PROBLEMS: Viscosity solutions have provided very simple and efficient approaches to many asymptotic problems such as singular perturbations, large deviations [29],[23], homogeneization [22]...

11) FRONTS AND INTERFACES: Viscosity formulations of moving fronts and interfaces have provided general long-time evolutions for such geometrical or physical processes (see [2],[46],[25],[13],[50]...). They also account for the derivation of these geometrical models from various PDE's.

12) APPLICATIONS TO ECONOMIC AND FINANCE MODELS: Many models have now been studied by viscosity solutions methods (see [53], [54] for example...).

13) INTEGRO-DIFFERENTIAL OPERATORS: It is well-known that jump diffusion processes lead to integro-differential operators that obey the maximum principle. The viscosity solutions theory has been adpated and extended to treat general fully nonlinear problems of that kind (see [36],[48],[43]...).

14) APPLICATIONS TO IMAGE PROCESSING AND VISION: It has been observed in [47],[45] that the "shapes from shading" problem in vision is naturally formulated with viscosity solutions, leading to completely

new numerical (reconstruction) schemes. Also, general image processing models (related to the so-called "Mathematical Morphology") have been treated via viscosity solutions [1].

15) APPLICATIONS TO PROBABILITY THEORY: Like the construction of general reflected diffusion processes ([3]) or large deviations of controlled processes...

Let us conclude, finally, by emphasizing the fact that this huge (and too vague) list is by no means exhaustive !

## REFERENCES

[1] L. Alvarez, P.L. Lions and J.M. Morel, *Image selective smoothing and edge detection by nonlinear diffusion*, To appear in SIAM J. Numer. Anal..

[2] G. Barles, *Remark on a flame propagation model*, Rapport INRIA, n. 464, 1985.

[3] G. Barles and P.L. Lions, Work in preparation.

[4] G. Barles and B. Perthame, *Exit time problems in optimal control and the vanishing viscosity method*, SIAM J. Control Optim., 26 (1988), pp. 1133-1148.

[5] G. Barles and P.E. Souganidis, *Convergence of approximation schemes for fully nonlinear second-order equations*, To appear in Asymp. Anal..

[6] E.N. Barron and R. Jensen, *The Pontryagin maximum principle from dynamic programming and viscosity solutions to first-order partial differential equations*, Trans. Amer. Math. Soc., 298 (1986), pp. 635-641.

[7] R. Bellman, *Dynamic Programming*, Princeton Univ. Press, Princeton, NJ, 1957.

[8] R. Bellman, *Adaptive processes. A guided tour*, Princeton Univ. Press, Princeton, NJ, 1961.

[9] A. Bensoussan and J.L. Lions, *Applications des inéquations variationnelles en contrôle stochastique*, Dunod, Paris, 1978.

[10] D.P. Bertsekas, *Dynamic programming and stochastic control*, Acad. Press, New-York, NY 1978.

[11] L. Caffarelli, *Interior a priori estimates for solutions of fully nonlinear equations*, Annals Math., 130 (1989), pp. 180-213.

[12] I. Cappuzzo-Dolcetta and P.L. Lions, *Viscosity solutions of Hamilton-Jacobi-Equations and state constraints*, Trans. Amer. Math. Soc., 318 (1990), pp. 643-683.

[13] Y.G. Chen, Y. Giga and S. Goto, *Uniqueness and existence of viscosity solutions of generalized mean curvature flow equations*, Preprint, 1989.

[14] M.G. Crandall, *Quadratic forms, semidifferentials and viscosity solutions of fully nonlinear elliptic equations*, Ann. I.H.P. Anal. Non. Lin. 6 (1989), pp. 419-435.

[15] M.G. Crandall, L.C. Evans and P.L. Lions, *Some properties of viscosity solutions of Hamilton-Jacobi equations*, Trans. Amer. Math. Soc., 282 (1984), pp. 487-502.

[16] M.G. Crandall and H. Ishii, *The maximum principle for semicontinuous functions*, To appear in Diff. and Int. Equations.

[17] M.G. Crandall, H. Ishii and P.L. Lions, *User's guide to viscosity solutions of second order partial differential equations*. To appear in Bull. A.M.S..

[18] M.G. Crandall and P.L. Lions, *Condition d'unicité pour les solutions généralisées des équations de Hamilton-Jacobi du premier ordre*, C.R. Acad. Sci. Paris, 292 (1981), pp. 183-186.

[19] M.G. Crandall and P.L. Lions, *Viscosity solutions of Hamilton-Jacobi equations*, Trans. Amer. Math. Soc., 277 (1983), pp. 1-42.

[20] M.G. Crandall and P.L. Lions, *Two approximations of solutions of Hamilton-Jacobi equations*, Math. Comp., 43 (1984), pp. 1-19.

[21] M.G. Crandall and P.L. Lions, *Hamilton-Jacobi equations in infinite dimensions*, J. Func. Anal. Part I. Uniqueness of viscosity solutions, 62 (1985), pp. 379-396. Part II. Existence of viscosity solutions,

65 (1986), pp. 368-405. Part III, 68 (1986), pp. 214-247. Part IV. Unbounded linear terms, 90 (1990), pp. 237-283. Part V. *B*-continuous solutions, to appear.

[22] L.C. Evans, *The perturbed test function technique for viscosity solutions of partial differential equations*, Proc. Royal Soc. Edinburgh, 111 A (1989), pp. 359-375.

[23] L.C. Evans and H. Ishii, *A pde approach to some asymptotic problems concerning random differential equations with small noise intensities*, Ann. Inst. H. Poincaré Analyse Non Linéaire, 2 (1985), pp. 1-20.

[24] L.C. Evans and P.E. Souganidis, *Differential games and representation formulas for solutions of Hamilton-Jacobi-Isaacs equations*, Indiana U. Math. J., 33 (1984), pp. 773-797.

[25] L.C. Evans and J. Spruck, *Motion of level sets by mean curvature, I.*, preprint.

[26] M. Falcone, *A numerical approach to the infinite horizon problem of deterministic control theory*, Appl. Math. Opt., 15 (1987), pp. 1-13.

[27] W.H. Fleming and R. Rishel, *Deterministic and Stochastic Optimal Control*, Springer-Verlag, New-York, 1975.

[28] W.H. Fleming and H.M. Soner, Book in preparation on Optimal deterministic and stochastic control.

[29] W.H. Fleming and P.E. Souganidis, *PDE-viscosity solution approach to some problems of large deviations*, Ann. Scuola Norm. Sup., 13 (1986), pp. 171-192.

[30] W.H. Fleming and P.E. Souganidis, *On the existence of value functions of two player, zero-sum stochastic differential games*, Ind. U. Math. J., 38 (1989), pp. 293-314.

[31] H. Ishii, *Perron's method for Hamilton-Jacobi equations*, Duke Math. J., 55 (1987), pp. 369-384.

[32] H. Ishii, *On uniqueness and existence of viscosity solutions of*

*fully nonlinear second-order elliptic PDE's*, Comm. Pure Appl. Math., 42 (1989), pp. 14-45.

[33] H. Ishii and P.L. Lions, *Viscosity solutions of fully nonlinear second-order elliptic partial differential equations*, J. Diff. Equa., 83 (1990), pp. 26-78.

[34] R. Jensen, *The maximum principle for viscosity solutions of fully nonlinear second-order partial differential equations*, Arch. Rat. Mech. Anal., 101 (1988), pp. 1-27.

[35] N.V. Krylov, *Controlled Diffusion Processes*, Springer-Verlag, Berlin, 1980.

[36] S. Lenhart and N. Yamada, *Perron's method for viscosity solutions associated with piecewise-deterministic processes*, To appear in Funkcial. Ekvac..

[37] P.L. Lions, *Generalized Solutions of Hamilton-Jacobi Equations*, Research Notes in Mathematics 69, Pitman, Boston, 1982.

[38] P.L. Lions, *On the Hamilton-Jacobi-Bellman equations*, Acta Applicandae, 1 (1983), pp. 17-41.

[39] P.L. Lions, *Optimal control of diffusion processes and Hamilton-Jacobi-Bellman equations. Part I : The dynamic programming principle and applications and Part 2 : Vicosity solutions and uniqueness*, Comm. P.D.E., 8 (1983), pp. 1101-1174 and 1229-1276.

[40] P.L. Lions, *Viscosity solutions of fully nonlinear second-order equations and optimal stochastic control in infinite dimensions. Part I : The case of bounded stochastic evolutions*, Acta Mathematica, 161 (1988), pp. 243-278.

[41] P.L. Lions, *Viscosity solutions of fully nonlinear second-order equations and optimal stochastic control in infinite dimensions. Part II*, in Stochastic Partial Differential Equations and Applications II, Springer Lecture Notes in Math. ♯ 1350, Berlin 1989. Proceedings of the International Conference on Infinite Dimensional Stochastic Differential Equations, held in Trento, Springer, Berlin, (1988).

[42] P.L. Lions, *Viscosity solutions of fully nonlinear second-order equations and optimal stochastic control in infinite dimensions. Part III: Uniqueness of viscosity solutions of general second-order equations,* J. Funct. Anal., 86 (1989), pp. 1-18.

[43] P.L. Lions and A. Sayah, Work in preparation.

[44] P.L. Lions and P.E. Souganidis, *Convergence of MUSCL type methods for scalar conservation laws,* Cr. Acad. Sci. Paris, 311 (1990), pp. 259-264.

[45] P.L. Lions, E. Rouy and A. Tourin, Work in preparation.

[46] S. Osher and J. Sethian, *Fronts propagating with curvature dependent speed : algorithms based on Hamilton-Jacobi formulations,* J. Comp. Physics, 79 (1988), pp. 12-49.

[47] E. Rouy and A. Tourin, *A viscosity solution approach to shape from shading,* Preprint.

[48] A. Sayah, *Equations d'Hamilton-Jacobi du premier ordre avec termes intégro-différentiels,* Parts I and II, Preprint.

[49] M. Soner, *Optimal control with state-space constraint I,* SIAM J. Cont. Opt., 24 (1986), pp. 552-562.

[50] M. Soner, *Motion of a set by the curvature of its mean boundary,* Preprint.

[51] D. Tataru, *Viscosity solutions of Hamilton-Jacobi equations with unbounded nonlinear terms,* Preprint.

[52] N.S. Trudinger, *Hölder gradient estimates for fully nonlinear elliptic equations,* Proc. Roy. Soc. Edinburgh, Sect. A 108 (1988), pp. 57-65.

[53] T. Zariphopolou and J.L. Vila, *Optimal portfolio rebalancing with credit constraints,* Preprint.

[54] T. Zariphopolou, *Investment consumption modes with borrowing,* To appear in Mathematics of Operation Research.

# Chapter 13
# DYNAMICS OF PATTERNS, WAVES, AND INTERFACES FROM THE REACTION-DIFFUSION ASPECT

**MASAYASU MIMURA**
Hiroshima University
Japan

**1. Introduction.** During the past decade, various interesting phenomena of patterns, waves and interfaces have been observed in natural sciences such as population dynamics [17], morphogenesis [13], chemical and biochemical reaction processes [23], combustion [4], solidification [22]. Interfaces or layers are present as sharp gradient with narrow width between two qualitatively different states such as liquid and solid or two quantitatively different chemical or physical states. Typical examples are spiral waves in the Belousov-Zhabotinsky reactions, flame front in combustion and dendritic crystal growth in supercooling liquid. The dynamics of interfaces and layers can be often described by reaction-diffusion model equations with layer parameter, say $\varepsilon$, which is sufficiently small. The simplest but suggestive equation describing the evolution of internal layers is

$$(1.1) \qquad u_t = \varepsilon \Delta u + \varepsilon^{-1} f(u),$$

where $\Delta$ is the Laplace operator in $\mathbb{R}^N$ and $f(u)$ has a cubic-like nonlinearity as in Fig. 1. Here $u(t,x)$ means the phase variable at time $t$ and position $x$. It is obvious that the equilibria $u \equiv \underline{u}$ and $u \equiv \overline{u}$ are both stable, while $u \equiv u_0$ between them is unstable. For this reason, (1.1) is called a bistable equation. If $\varepsilon$ is sufficiently small, one could expect that the dynamics of the solution $u(t,x)$ of (1.1) consists of two stages : In the first stage, $u(t,x)$ tends to either $\overline{u}$ or $\underline{u}$, by which the region is separated into two different subregions and there occur internal layers with width $O(\varepsilon)$ between these subregions. In the second stage, these internal layers propagate. Recently, the study of the dynamics of layers generated in (1.1) has been well developed. One approach is to regard layers as (diffusive) interfaces in the limit as $\varepsilon \searrow 0$ and to consider the dynamics of these interfaces. Let $\Gamma(t)$ be an interface in $\mathbb{R}^N$, which is the boundary between two regions where $u$ takes either $\overline{u}$ or $\underline{u}$. Then the equation for the interface $\Gamma(t)$ is approximately described by

$$(1.2) \qquad \frac{d}{dt}\Gamma(t) = \left[c - \varepsilon(N-1)\kappa\right]\nu,$$

where $\nu$ and $\kappa$ are respectively the unit normal vector and the mean curvature at $\Gamma$. Here $c$ is the velocity of the traveling front solution of the one dimensional version of (1.1)

$$u_t = u_{xx} + f(u) \quad x \in \mathbb{R}$$

with the boundary conditions

$$u(t, -\infty) = \overline{u},$$
$$u(t, \infty) = \underline{u}.$$

It is known that the velocity is uniquely determined and the corresponding traveling front solution is stable. (1.2) clearly shows that the dynamics of the interface is obeyed by the bistable kinetic effect and the geometrical one. For the validity of (1.2) to (1.1), we refer to de Mottoni and Schatzman [14]. On the other hand, for an application of (1.2), we also refer to Foerster et al. [5]. They consider the collision of expanding circular chemical waves arising in the Belousov-Zhabotinsky (B-Z) reaction. They measured the temporal evolution of cusp like structures which appear after the collision of two circular waves by using a computerized spectrophotometric video technique (Fig. 2). From a modelling view point, the spatio-temporal dynamics of B-Z reaction can be described by the simplified Oregonator including diffusion

(1.3)
$$\begin{cases} \dfrac{\partial u}{\partial t} - D\Delta u = u^2 - hv\dfrac{u-q}{u+q} \equiv f(u,v) \\ \dfrac{\partial v}{\partial t} - D\Delta v = \varepsilon(u-v) \equiv \varepsilon g(u,v) \end{cases}$$

with some constants $h$, $q$ and $\varepsilon$ which are experimentally given by $h = 3$, $q = 2 \times 10^{-4}$ and $\varepsilon = 10^{-2}$. The nulclines of $f$ and $g$ are drawn in Fig. 3 where there is only one constant equilibrium $(\overline{u}, \overline{v})$ which is asymptotically stable. Singular perturbation arguments show that $v$ approximately takes the value of $\overline{v}$ in the vicinity of the front layer of waves so that $u$ satisfies approximately the scalar equation

(1.4)
$$\varepsilon\frac{\partial u}{\partial t} = \varepsilon D\Delta u + f(u, \overline{v})$$

where $f(u, \overline{v})$ is exactly a cubic-like function. Thus, the interface equation (1.2) can be derived from (1.4) in the limit $\varepsilon \searrow 0$. Fig. 4 shows the numerical calculation of (1.2) which is in good agreement with the experimental observations. This result clearly indicates that the interface equation can be well used to understand the dynamics of chemical waves of the B-Z reaction.

The approach of interfacial dynamics can be applied to more general systems of equations of the form

(1.5)
$$\frac{\partial u}{\partial t} = \varepsilon D\Delta u + \frac{1}{\varepsilon}F(u)$$

where $u = (u_1, u_2, \cdots, u_n)$, $D$ is the diagonal matrix with elements $d_i > 0$. Here we assume that $F(u)$ is of bistable nonlinearity, that is, $F(u) = 0$ takes exactly three solutions $\underline{u}$, $u_0$ and $\overline{u}$ where $\underline{u}$ and $\overline{u}$ are both stable equilibria of (1.5).

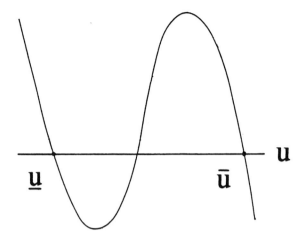

Fig. 1　　Cubic nonlinearity of $f(u)$

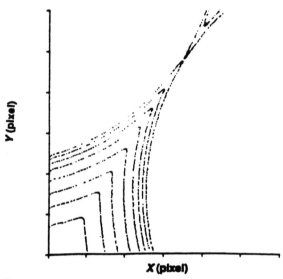

Fig. 2　　A montage of several successive contour maps ([5])

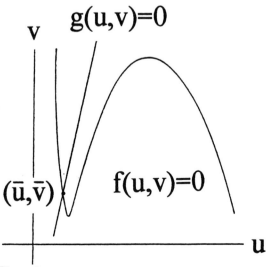

Fig. 3　　Nulclines of $f$ and $g$ in Oregonator

Let us show one example which arises in population ecology.

(1.6)
$$\begin{cases} \dfrac{\partial u_1}{\partial t} = d_1 \Delta u_1 + a_1(1 - \alpha u_1 - u_2)u_1 \\ \dfrac{\partial u_2}{\partial t} = d_2 \Delta u_2 + a_2(1 - u_1 - \beta u_2)u_2 \end{cases}$$

where $u_1$ and $u_2$ are the population densities of two competing species, $d_i$ is the diffusion rate and $a_1$, $a_2$ are the intrinsic growth rate, and $\alpha$, $\beta$ are intraspecific competition rates of $u_i$ ($i = 1, 2$). Suppose that $\alpha$ and $\beta$ are chosen to satisfy

$$\alpha < 1 < \frac{1}{\beta},$$

which means ecologically that the intraspecific competition is weaker than the interspecific one. Simple calculations show that $(a_1/b_1, 0)$ and $(0, a_2/b_2)$ are both stable, while $\left( \dfrac{1-\beta}{1-\alpha\beta}, \dfrac{1-\alpha}{1-\alpha\beta} \right)$ is unstable. We thus find that the kinetics in (1.6) is of bistable nonlinearity.

For (1.5), we assume that there is a stable 1-dimensional traveling front solution $u(z)$ ($z = x - ct$) with the velocity $c$, connecting with $\overline{u}$ and $\underline{u}$ at the infinities $x = \pm\infty$. That is, $u(z)$ satisfies

(1.7)
$$\begin{cases} Du'' + cIu' + F(u) = 0, \quad -\infty < z < \infty \\ u(-\infty) = \underline{u}, \quad u(\infty) = \overline{u}, \end{cases}$$

where $' = \dfrac{d}{dz}$. Unfortunately, the existence and uniqueness problems of traveling front solutions to the system (1.5) have not yet been solved except for special cases. For the specific system (1.6), the existence of traveling front solutions is shown in [3], but, as far as we know, the uniqueness of solutions have been still unsolved. However, numerical simulations confirm that (1.6) has only one traveling front solution which is unique and stable. Here we should note that this result is not general. In fact, for suitable nonlinear functions $F(u)$, there possibly coexist stable and unstable traveling front solutions of (1.5) even if it is a two coupled system([9]).

Let $\mathcal{L}$ be the linearized operator of (1.7) around the traveling front solution $u(z)$ with the velocity $c$,

$$\mathcal{L} = D\frac{d^2}{dz^2} + cI\frac{d}{dz} + F'(u(z))$$

where $F'$ is the Jacobian at $u = u(z)$, and let $\mathcal{L}^*$ be the adjoint operator of $\mathcal{L}$. We note that both $\mathcal{L}$ and $\mathcal{L}^*$ have the zero eigenvalue $\lambda_0 = 0$ since any spatial translation of $u(z)$ is also a solution of (1.6). Let $\xi_0$ and $\xi_0^*$ be the eigenfunction of $\mathcal{L}$ and $\mathcal{L}^*$ associated with $\lambda_0$, respectively, such that $<\xi_0^*, \xi_0> = 1$ holds. Here we define $\alpha$ by $\alpha = <\xi_0^*, D\xi_0>$. The interface equation of (1.5) in the limit $\varepsilon \searrow 0$ is

(1.8)
$$\frac{d}{dt}\Gamma(t) = \left[ c - \varepsilon(N-1)\alpha\kappa \right]\nu.$$

When $D = dI$, it is easy to see $\alpha = d$, so that (1.8) is essentially equivalent to (1.2). For the derivation, we refer to Ohta [18].

Recently, the qualitative study of (1.2) or (1.8) has been investigated by numerous authors, from theoretical and numerical view points. One global property of interfaces in $\mathbb{R}^2$ is that if the initial interface $\Gamma(0)$ is given by a smooth, embedded closed curve, $\Gamma(t)$ expands ($c > 0$), or shrinks to a point ($c \leq 0$), becoming round (Grayson [7], Osher and Sethian [21], for instance).

## 2. Activator-inhibitor systems.

For one extension of (1.5) to system versions, we meet the following reaction-diffusion equations coupled with another single species $v$:

$$(2.1) \quad \begin{cases} \tau \dfrac{\partial u}{\partial t} = \varepsilon D \Delta u + \dfrac{1}{\varepsilon} F(u, v), & t > 0, \ x \in \Omega, \\ \dfrac{\partial v}{\partial t} = d \Delta v + g(u, v), & t > 0, \ x \in \Omega, \end{cases}$$

where $u = (u_1, u_2, \cdots, u_n)$ and $v$ is a scalar function with the diffusion rate $d$. (2.1) is often used to understand spatio-temporal patterns occuring in the fields of chemistry and biology ([10], [15]).

One of the features of (2.1) is that the diffusing species $u$ reacts so fast compared with the other diffusing species $v$, since $\varepsilon$ is sufficiently small. We assume that $F$ and $g$ have the following characteristics:

(A.1) $F(u, v) = 0$ has only one solution $u = \underline{u}(v)$ for large $v$ and $u = \overline{u}(v)$ for small $v$ and both of them are stable in the sense of the first system of (2.1) for fixed $v$

$$(2.2) \quad \frac{\partial u}{\partial t} = \varepsilon D \Delta u + \frac{1}{\varepsilon} F(u, v).$$

On the other hand $F(u, v) = 0$ has three solutions $u = \underline{u}(v), u_0(v), \overline{u}(v)$ for intermediate $v$ and two of them, $\underline{u}(v), \overline{u}(v)$ are stable while $u_0(v)$ is unstable in (2.2). In biological terms, we call the $\underline{u}(v)$-branch the rest state and the $\overline{u}(v)$-branch the active state in $(u, v)$-space. Then (A.1) indicates that the activation of $u$-species is inhibited by the presence of $v$-species. For this reason, $u$ and $v$ are respectively named an activator and its inhibitor.

(A.2) $g(u, v)$ is increasing with $u$ and decreasing with $v$, that is

$$\frac{\partial g}{\partial u_i} > 0 \quad \text{(production)} \quad (i = 1, 2, \cdots, n),$$

$$\frac{\partial g}{\partial v} > 0 \quad \text{(degradation)}.$$

Let us show some examples of $F$ and $g$.

**Example 1.** ($n = 1$, Bonhoeffer–van der Pol model)

$$(2.3) \quad \begin{cases} F(u, v) = u(1 - u)(u - a) - v \\ g(u, v) = u - \gamma v + I \end{cases}$$

with positive constants $a$, $\gamma$ and $I$. It is obvious to see that (2.3) satisfies (A.1) and (A.2) when $0 < a < 1$. The nonlinearity in (2.3) is essentially the same one as in (1.5).

**Example 2.** ($n = 2$, 2 prey–1 predator population model)

$$F(u_1, u_2, v) = \begin{pmatrix} a_1(1 - \alpha u_1 - u_2 - kv)u_1 \\ a_2(1 - u_1 - \beta u_2 - v)u_2 \end{pmatrix}$$

$$g(u_1, u_2, v) = (-r + k\gamma u_1 + u_2)v,$$

where $u_1$, $u_2$ and $v$ are respectively the population densities of two competing species and their predator. $a_1, a_2, \alpha$ and $\beta$ are positive constants used in (1.6). $k$ is the ratio of predation rates, $\gamma$ is the transformation rate of predation and $r$ is the death rate of $v$. For the ecological explanation, we refer to [8]. We note that $F(u_1, u_2, 0)$ is exactly the same form as the one in (1.6). We now assume that $\alpha$, $\beta$ and $k$ satisfy $\alpha < 1 < \beta$ and $\alpha\beta < 1$ and $k > 1$. Under these assumptions, the lines of $(1 - kv) - \alpha u_1 - u_2 = 0$ and $(1 - v) - u_1 - \beta u_2 = 0$ are classified into three cases as in Fig. 5. For fixed $v$, the analysis of

$$\begin{cases} \dfrac{\partial u_1}{\partial t} = d_1 \Delta u_1 + a_1(1 - \alpha u_1 - u_2 - kv)u_1 \\ \dfrac{\partial u_2}{\partial t} = d_2 \Delta u_2 + a_2(1 - u_1 - \beta u_2 - v)u_2 \end{cases}$$

shows that for $\alpha < \dfrac{1}{\beta} < \dfrac{1-kv}{1-v}$ (small $v$), $(u_1, u_2) = \left(\dfrac{1}{\alpha}, 0\right)$ stable, for $\alpha < \dfrac{1-kv}{1-v} < \dfrac{1}{\beta}$ (intermediate $v$), $\left(\dfrac{1}{\alpha}, 0\right)$ and $\left(0, \dfrac{1}{\beta}\right)$ are both locally stable and $\dfrac{1-kv}{1-v} < \alpha < \dfrac{1}{\beta}$ (large $v$), $\left(0, \dfrac{1}{\beta}\right)$ is stable.

Recently it has been reported that systems (2.1) in the framework of activator-inhibitor interaction satisfying (A.1) and (A.2) exhibit a surprising variety of spatio-temporal patterns. So far almost all studies of (2.1) are concentrated on the case when $d$ is very small or at most of the order $\varepsilon$. In this case, (2.1) exhibits a traveling pulse solutions in the situation where the kinetics of $(F, g)$ is of monostable. The typical examples are the Fitzhugh–Nagumo equations describing an exciting pulse propagating along the nerve axon ($d = 0$) and the Oregonator (1.3) describing the B-Z reaction. Fig. 6 shows numerical spiral waves. On the other hand, the opposite case where $\varepsilon \ll d$, that is, the inhibitor $v$ diffuses so fast compared with the activator $u$. Because of big difference between the diffusion rates, the behavior of the solution drastically changed. Suppose that the activator is initially distributed in some finite area. The inhibitor is, then, produced by the reaction but diffuses so fast that it flows out of the area. Therefore, if the ratio of diffusion rates and the interaction between the activator and its inhibitor are suitably balanced, the localized pattern could neither shrink or expand. This suggests us the appearance of a stationary localized pattern. Fig. 7a shows two pulses propagation to the opposite directions when $d$ is small, while Fig. 7b shows one localized pulse when $d$ is large compared

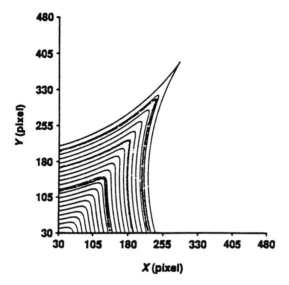

Fig. 4   Superposition of the numerical contour (continuous lines) and experimental determined contour (dots) maps ([5])

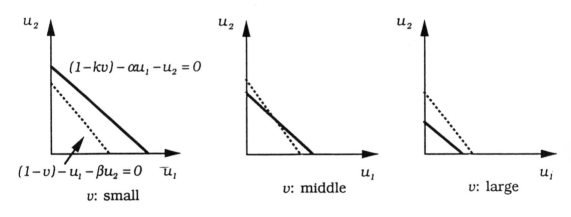

Fig. 5   Lines of $(1-kv) - \alpha u_1 - u_2 = 0$ and $(1-v)u_1 - \beta u_2 = 0$

Fig. 6   Numerical spiral waves in 2 dimensional space

with $\varepsilon$. By the above intuitive explanation and numerical simulations, one could think that the localized pattern due to lateral inhibition would be stable. However, this is not necessarily true. In fact, for small $\tau$, the localized pulse undergoes sustained oscillation like a breathing motion as in Fig. 8. The analysis of these phenomena has been investigated in one dimensional space, by using singular perturbation methods ([12], [16]).

We now derive the interface equation from (2.1). Using the discussion on the scalar equation (1.1), we know that the region is separated into two subregions, as $\varepsilon$ tends to zero : one is $\overline{\Omega}(t)$ where $(u,v)$ satisfies $u = \overline{H}(v)$ and the other is $\underline{\Omega}(t)$ where $(u,v)$ satisfies $u = \underline{H}(v)$. Here $u = \overline{H}(v)$ and $\underline{H}(v)$ define respectively two branches of $u$ as functions of $v$ as in Fig. 9. Let $\Gamma(t)$ be the interface which divides these two subregions. Then $v$ satisfies

$$(2.4) \qquad v_t = \begin{cases} d\Delta v + g(\overline{H}(v), v) & \text{in } \overline{\Omega}(t) \\ d\Delta v + g(\underline{H}(v), v) & \text{in } \underline{\Omega}(t), \end{cases}$$

with $v \in C^1(\mathbb{R}^N)$ as the continuity condition ([2]). Here we note that the value of $v$ on the interface $\Gamma(t)$, say $v_I$, is not known *a priori*. By (1.8), the interface equation is given by

$$(2.5) \qquad \tau \frac{d\Gamma}{dt} = \left[ c(v_I) - \varepsilon(N-1)\alpha(v_I)\kappa \right] \nu,$$

where $c(v_I)$ is the velocity of the 1-dimensional traveling front solution of the system for $u$

$$\frac{\partial u}{\partial t} = D u_{xx} + F(u, v_I) \quad t > 0,\ x \in \mathbb{R},$$

with the boundary conditions

$$u(t, -\infty) = \overline{H}(v_I) \quad t > 0$$

$$u(t, +\infty) = \underline{H}(v_I) \quad t > 0,$$

and $\alpha(v_I)$ is defined in a similar way to $\alpha$ in (1.8). For the 2 prey–1 predator population model in Example 2, the dependency of $c(v_I)$ on $v_I$ is shown in Fig. 10. We thus found that the dynamics of interfaces is described by (2.4), (2.5). The essentially different point of (2.5) from (1.8) is that $c$ as well as $\nu$ are not constants but functions of $v_I$, that is, the interface $\Gamma(t)$ is truely coupled with the inhibitor $v$ so that even if $\Gamma(0)$ is smooth in $\mathbb{R}^2$, complicated singularities may develop in $\Gamma(t)$ ([27]). This phenomenon is totally different from the scalar version (1.1) which does not interact with the inhibitor.

Recently, the existence of global solutions to (2.4), (2.5) has been proved in weak formulation, although the uniqueness problem is still unsolved ([6]). On the other hand, the study of dynamics of interfaces is only just beginning for higher dimensions. As the first step, we consider the existence and stability of radially symmetric stationary solution in $\mathbb{R}^N$. To do it, we assume that $F = 0$ and $g = 0$

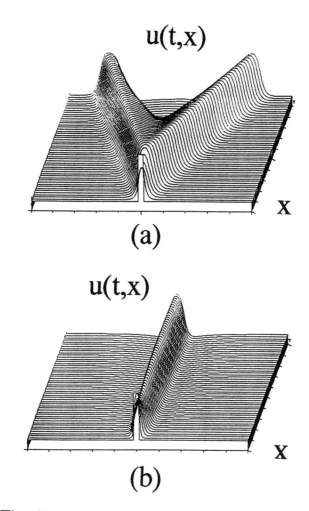

Fig. 7  (a) propagating pulses  (b) localized pulse

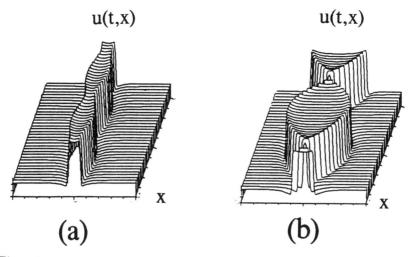

Fig. 8  breathing motions (a) small amplitude  (b) large amplitude

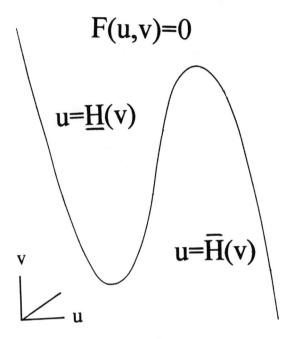

Fig. 9   Stable branches $u = \overline{H}(v)$, $u = \underline{H}(v)$ of $F(u,v)$

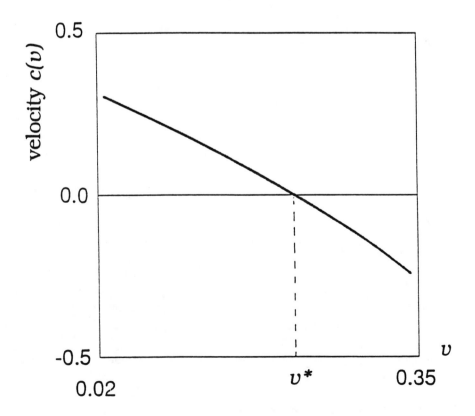

Fig. 10   Dependency of $c(v)$ on $v$

intersects at one point $(\underline{u}, \underline{v})$ as in Fig. 11. With the variable $r$, two unknowns $(v(r), R)$ satisfy

(2.6)
$$\begin{cases} 0 = \left[\dfrac{d^2}{dr^2} + \dfrac{N-1}{r}\dfrac{d}{dr}\right]v + g(\overline{H}(v), v) & \text{in } 0 < r < R \\ 0 = \left[\dfrac{d^2}{dr^2} + \dfrac{N-1}{r}\dfrac{d}{dr}\right]v + g(\underline{H}(v), v) & \text{in } R < r < +\infty \\ 0 = -\dfrac{\varepsilon\alpha(v(R))(N-1)}{R} + c(v(R)), \end{cases}$$

and the boundary conditions

(2.7)
$$\begin{cases} \dfrac{du}{dr} = 0 = \dfrac{dv}{dr} & \text{at } r = 0 \\ v \in C^1 & \text{at } r = R \\ \lim_{r \to \infty}(u, v) = (\underline{u}, \underline{v}). \end{cases}$$

If $N = 2$, the interface takes a circle with the radius $R$. For a simple example, we consider two-components system $(u, v)$ where $F$ and $g$ are specified as

(2.8)
$$\begin{cases} F(u, v) = H(u-a) - u - v \\ g(u, v) = u - \gamma v \end{cases}$$

where $H(\xi)$ is the Heaviside step function, $a$ and $\gamma$ are constants satisfying $0 < a < 1$ and $\gamma > 0$. For this specified $F$, $c(v)$ is given by

$$c(v) = 2\left(\dfrac{1}{2} - a - v\right)\left\{(a+v)(1-a-v)\right\}^{-\frac{1}{2}}.$$

Then, the solution of (2.7) can be explicitly solved. Fig. 12 demonstrates the solution branch ($N = 2$) with $\gamma = 3$ when $a$ is varied, which shows that the radius $R$ is uniquely determined, depending on the value of $a$. For the stability of this radially symmetric solution, we can show that (i) for large values of $\tau$, it is stable when $R$ is small, while it is unstable and radially asymmetric stationary solutions appear when $R$ is large (ii) even if $R$ is small, the solution becomes unstable and it undergoes a sustained (breathing) oscillation. Fig. 13 is the stability diagram when $\tau$ and $R$ are used as the bifurcation parameters. A remarkable fact is the occurrence of codimension 2 bifurcation for suitable values of $\tau$ and $R$. This suggests us the possibility that interfaces become complicated curves. Although the conclusions above are proved for specific nonlinearities (2.8), numerical simulations confirm that these phenomena possibly occur for more general nonlinearities ([20], [27]).

Our discussion above is restricted to the linear stability of radially symmetric stationary solutions. Quite recently, we have observed complicated dynamics of interfaces in 1 dimensional finite interval with the Neumann boundary conditions when $\tau$ is sufficiently small. The nonlinearities used there are the ones in Example 2. Fig. 14

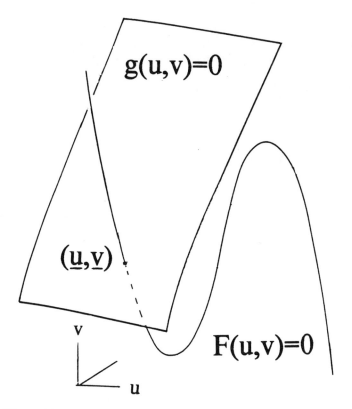

Fig. 11   Mono stable state $(\underline{u}, \underline{v})$ of $F = 0$ and $g = 0$

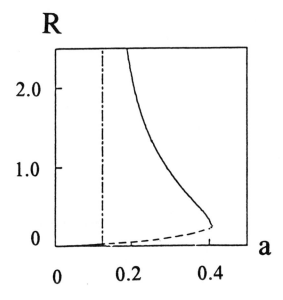

Fig. 12   solution branch of (2.6) with ($N = 0$). The dashed line indicates the unstable branch which is given by solving the stationary problem of (2.1) with (2.8).

clearly shows the appearance of complex oscillations interfaces through period doubling cascades ([8]). Unfortunately these are unsolved even in 1 dimensional space. Such phenomena is beyond the scope of the linear stability analysis.

**3. Discussion.** The characteristic property of the interface equation is that it truely coupled with the fast diffusing but slowly varying inhibitor $v$. This results in the nonlocal and time-delayed interaction in the dynamics of interfaces. In this sense, the present problem has a sort of similarity, to some extent, to the phase-field approach to crystal growth ([1], [11]), where the local temperature plays the role of the inhibitor ([19]). As a real experiment closely related to the present work, we should mention the formation of chemical patterns in the chlorite-iodide-malonic acid reaction ([10]).

**Acknowledgement.** The author thanks R. Kobayashi and T. Ikeda and many colleagues in Applied Analysis group in the Department of Mathematics, Hiroshima University for nice collaboration with him on numerical simulations and visualizations into video tapes related to the present work.

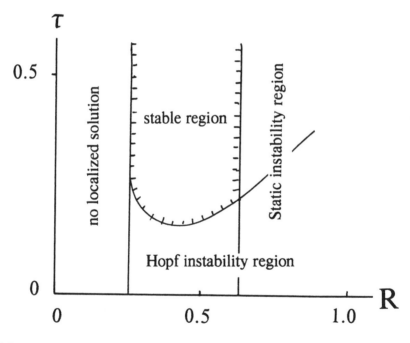

Fig. 13   Stability diagram of a radially symmetric stationary solution ($N = 2$).

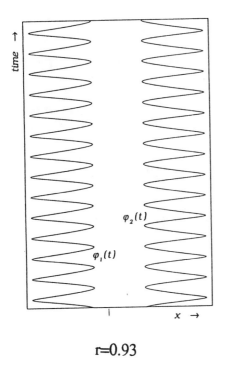

r=0.93

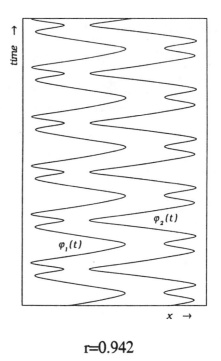

r=0.942

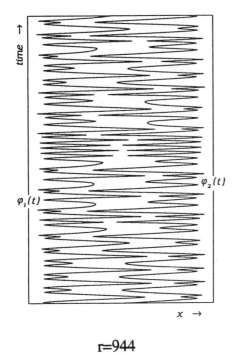

r=944

$\beta = 1.01$, d=1.0, D=0.2,
$\tau = 0.36$ k=1.35, $\alpha = 0.818$,
$\gamma = 1.0$, $a_1 = a_2 = 1.0$

Fig. 14  Dynamics of interfaces $x = \varphi_i(t)$ ($i = 1, 2$) for different values of $r$.

# References

[1] G. Caginalp, *An analysis of phase field model of free boundary*, Arch. Rat. Mech. Anal., 92 (1986), pp. 205–245.

[2] X. Y. Chen, *Dynamics of interfaces in reaction diffusion systems*, Hiroshima Math. J., 21 (1991), pp. 47–84.

[3] C. Conley and R. Gardner, *An application of the generalized Morse index to travelling wave solutions of a competitive reaction- diffusion model*, Indiana Univ. Math. J., 23 (1984), pp. 321–343.

[4] P. C. Fife, *Dynamics of internal layers and diffusive interfaces*, CBMS-NSF. Regional Conf. Series in Appl. Math. 53, 1988.

[5] P. Foerster, S. C. Müller and B. Hess, *Curvature and Propagation Velocity of Chemical Waves*, Science, 241 (1988), pp. 685–687.

[6] Y. Giga and S. Goto, *Motion of hypersurfaces and geometric equations*, preprint.

[7] M. A. Grayson, *The heat equation shrinks embedded plane curves to round points*, J. Differential Geometry, 26 (1987), pp. 285–314.

[8] T. Ikeda and M. Mimura, *An interface approach to regional segregation of two competing species mediated by a predator*, to appear in J. Math. Biol.

[9] H. Ikeda, M. Mimura and Y. Nishiura, *Global bifurcation phenomena of traveling wave solutions for some bistable reaction-diffusion systems*, Nonlinear Analysis, 13 (1989), pp. 507–526.

[10] P. De Kepper, V. Castets, E. Dulos and J. Boissonade, *Turing-type chemical patterns in the chlorite-iodide-malonic acid reaction*, Physica D, 49 (1991), pp. 161–169

[11] R. Kobayashi, *Modeling and numerical simulations of dendritic crystal growth*, to appear in Physica D.

[12] S. Koga and Y. Kuramoto, *Localized patterns in reaction-diffusion systems*, Prog. Theor. Phys., 63 (1980), pp. 106–121.

[13] H. Meinhardt, *Models of Biological Pattern formation*, Academic Press, New York, 1982.

[14] P. de Mottoni and M. Schatzman, *Evolution geometrique d'interfaces*, C.R. Acad. Paris, 309 (1989), pp. 453–458.

[15] J. D. Murray, *Mathematical Biology*, Springer, Berlin, 1989.

[16] Y. Nishiura and M. Mimura, *Layer oscillations in reaction- diffusion systems*, SIAM J. Appl. Math., 49 (1989), pp. 481–514.

[17] A. Okubo, *Diffusion and Ecological Problems : Mathematical Models*, Springer, Berlin, 1980.

[18] T. Ohta, *Euclidean invariant formulation of phase dynamics. I-Nonpropagating periodic pattern-*, Prog. Theor. Phys., 73 (1985), pp. 1377–1389.

[19] T. Ohta and M. Mimura, *Pattern dynamics in excitasble reaction-diffusion media*, Formation Dynamics and Statistics of Patterns, K. Kawasaki et al. ed., World Scientific Publishing Co., 1980, pp. 55–112.

[20] T. Ohta, M. Mimura and R. Kobayashi, *Higher-dimensional localized patterns in excitable media*, Physica D, 34 (1989), pp. 115–144.

[21] S. Osher and J. A. Sethian, *Front propagation with curvature-dependent speed : algorithms based on Hamilton-Jacobi formulations*, J. Comput. Phys., 79 (1988), pp. 12–49.

[22] P. Pelcé (ed.) *Dynamics of Curved Fronts, Perspectives in Physics*, Academic Press, New York, 1988.

[23] A. T. Winfree, *The Geometry of Biological Times*, Springer, Berlin, 1980.

**Video tapes**

[24] M. Mimura, R. Kobayashi, H. Okazaki and H. Ikeda, *Pattern formation in excitable media*, (1986) (Instabilities of traveling pulses, targets, spirals, breathing motions).

[25] M. Mimura, R. Kobayashi, H. Okazaki and T. Yamanoue, *Dynamics of interfaces in systems of reaction-diffusion equations*, (1988) (Solidifications, targets, spirals).

[26] R. Kobayashi and M. Mimura, *Pattern dynamics in reaction-diffusion systems far from equilibrium* (1989) (Phase separation and coarsening).

[27] T. Ikeda and R. Kobayashi, *Numerical simulations to interfacial dynamics*, (1989).

## Chapter 14
## COMPLEX PATTERN FORMATION IN EMBRYOLOGY: MODELS, MATHEMATICS, AND BIOLOGICAL IMPLICATIONS

**J. D. MURRAY**
University of Washington
Seattle, Washington

**Biological background.** The development of spatial pattern during the various stages of embryogenesis is a central issue. Model mechanisms - morphogenetic models - for biological pattern generation can suggest to the embryologist possible scenarios as to how, and sometimes when, pattern is laid down and how the embryonic form might be created. Although genes control pattern formation, genetics says nothing about the actual mechanisms involved nor how the vast range of pattern and form that we see evolves from a homogeneous mass of dividing cells. The area of pattern formation mechanisms in biology has been extensively discussed in the book by Murray (1989).

Because of the remarkable similarity between the embryos of a wide spectrum of species, as shown in Fig. 1, it is clear that an understanding of the patterning process of any aspect of almost any species would be a major step forward in our understanding of morphogenesis.

Broadly speaking, there are two prevailing views of pattern generation that embryologists have recognized in the past few years. One is the long standing and well known Turing (1952) chemical pre-pattern approach and the more recent continuum mechanochemical approach developed by G. F. Oster and J. D. Murray and their colleagues (see, for example, Odell *et al.* 1981, Murray *et al.* 1983, Oster *et al.* 1983, Murray and Oster 1984a,b. General descriptions have been given by Murray and Maini (1986) and Oster and Murray (1989), with a general review by Murray *et al.* (1988).

Turing's (1952) theory of morphogenesis involves hypothetical chemicals - morphogens - which react and diffuse in such a way that if the chemical kinetics and the diffusion coefficients have certain properties, steady state heterogeneous spatially patterned solutions in chemical concentrations can evolve. These models give rise to parabolic systems of nonlinear partial differential equations. Morphogenesis then proceeds by the cells reacting to the chemical prepattern and differentiating according to some bauplan, such as Wolpert's (1981) positional information concept. This considers the cells to have been pre-programmed to differentiate according to the underlying morphogenetic pre-pattern. Turing's theory has stimulated a vast amount of research, both mathematical (for example, the books by Fife 1979 and Britton 1986) and experimental (for example, Wolpert 1981). Such reaction diffusion models have been widely studied and applied to a variety of biological problems: see, for example, Murray (1977, 1981a,b, 1989), Meinhardt (1982).

---

* This work was in part supported by Grant DMS 9003339 from the U.S. National Science Foundation.

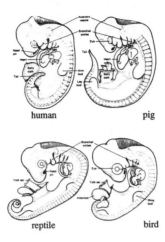

**Fig.1** Embryos of different species at similar developmental stages. The similarity justifies studying patterning problems and pattern formation mechanisms associated with a wide variety of species. An understanding of any one of these would be of fundamental importance to human morphogenesis. (Drawings from Haeckel, 1866)

The Belousov-Zhabotinskii reaction is a well known and widely studied reaction which can generate steady state concentration patterns, among other complex spatial behaviors: the book of articles edited by Field and Burger (1985) gives a good overview of this reaction. Recent experimental work by de Kepper *et al.* (1991) has dramatically demonstrated these spatial structures.

The mechanochemical approach directly brings mechanical forces and known properties of cells and biological tissue into the process of morphogenetic pattern formation. Here pattern formation and morphogenesis go on simultaneously as a single process. The form-shaping movements of the cells and the embryological tissue interact continuously to produce the observed pattern. An important aspect of this approach is that the models are formulated in terms of measurable quantities such as cell densities, elastic forces, cell traction forces, tissue deformation, known chemicals and so on. This focuses attention on the morphogenetic process itself and is more amenable to experimental investigation.

The two approaches are basically quite different. In the chemical prepattern approach, pattern formation and morphogenesis take place sequentially. First the chemical pattern is formed, then the cells 'read out' the chemical concentration pattern, and the various cell differentiations, cell movement, and so on, are assumed to follow from this chemical blueprint. So, in this reaction diffusion approach morphogenesis is essentially a slave process which is determined once the chemical pattern has been established. Mechanical shaping of form which occurs during embryogenesis is not addressed in any chemical theory of morphogenesis. The elusiveness of the chemical morphogens is also proving a considerable drawback in the biologist's acceptance of such a theory of morphogenesis. There is, however, no question that chemicals play important roles in embryogenesis and experimental work by Wolpert (personal communication 1988) on limb cartilage formation suggests that there may indeed be a chemical prepattern.

The principal use of any theory is in its predictions and, even though each theory might be able to create similar patterns, they are mainly distinguished by the different experiments they suggest. A major point in favour of simultaneous development is that such mechanisms have the potential for self correction. Embryonic development, which proceeds sequentially, is usually a stable process with the embryo capable of adjusting to many outside disturbances. The process whereby a prepattern exists and then morphogenesis takes place is effectively an open loop system. These are potentially unstable processes and make it difficult for the embryo to make the necessary corrective adjustments as development proceeds.

This article is mainly concerned with a mechanical models which involve coordinated movement or patterning of populations of cells. The type of early embryonic cells we shall be concerned with are fibroblast (or dermal or mesenchymal) cells. Fibroblast cells are capable of independent movement, due to long finger-like protrusions - lamellipodia - which grab on to adhesive sites in the tissue matrix and pull themselves along (Oster 1984). These cellular forces deform the tissue within which the cells move and this in turn influences their movement. Depending on the relative size of the various parameters, spatial patterns appear, such as in cell number density and variation in tissue matrix density (Oster et al. 1983). Epidermal cells, on the other hand, in general do not move but are packed together in sheets and spatial patterns in their population are manifested by cell or sheet deformations.

In animal development the basic body plan is more or less laid down in the first few weeks. It is during this crucial early period that we expect these pattern and form - generating mechanisms to be operative.

**Pattern formation mechanisms.** A large number of pattern formation models involve the concept of a local activation and lateral or long-range inhibition. If we denote morphogenetic variables (cells, tissue and so on) by **m**(x,t) then most (but not in fact the one we discuss in more detail below) continuous pattern generator mechanisms fall into the general evolution equation category

$$\frac{\partial \mathbf{m}}{\partial t} = \mathbf{L}(\mathbf{m}) + \mathbf{S}(\mathbf{m};x,t)$$

rate of     local     spatial effects (convection,
change of **m**   dynamics   diffusion, chemotaxis, ....)

where **L**(**m**) represents the local dynamics, such as the reaction kinetics in a reaction diffusion system, while the **S**-term includes the spatial effects. The combination of the local dynamics and spatial effects can generate a spatial pattern in **m**, typically as in Fig. 2.

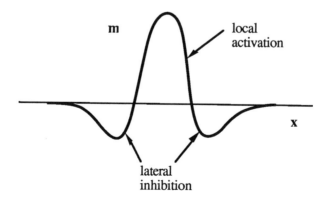

**Fig. 2.** A typical spatial pattern which evolves from a local activation and lateral inhibition system.

Typical examples of this class are the usual Turing (1952) reaction diffusion systems and chemotaxis models (for pedagogical discussions see, for example, Segel 1984 and Murray 1989). Chemotaxis is when there is movement up a concentration gradient rather than down it, like diffusion processes.

The model we now describe here does not fall into the above category. It is the continuum mechanical theory developed by Oster and Murray and their colleagues (Oster *et al.* 1983, 1985; Murray and Oster 1984a,b). The morphogenetic variables are the cell density, n, the extracellular tissue matrix, $\rho$, and the displacement of the tissue, **u**, which is caused by the forces deriving from the cell traction. A full discussion and derivation together with the background biology behind these mechanochemical models is given in Murray (1989). An example of the typical field equations are:

*Cell conservation:*

$$\frac{\partial n}{\partial t} + \nabla.(n\frac{\partial \mathbf{u}}{\partial t}) = \nabla.(D_n \nabla n) - \nabla.(H \nabla c) + f(n;\tau) \qquad (1)$$

$\qquad\quad$ convection $\quad$ diffusion $\quad$ haptotaxis $\quad$ cell division

*Matrix conservation:*

$$\frac{\partial \rho}{\partial t} + \nabla.(\rho\frac{\partial \mathbf{u}}{\partial t}) = g(\rho,n) \qquad (2)$$

*Force balance:*

$$\nabla.[\sigma_{viscous} + \sigma_{elastic} + \sigma_{cells}] + \rho s \mathbf{u} = 0 \;\Rightarrow\; \psi(n,\rho,\mathbf{u}) = 0. \qquad (3)$$

$\qquad\qquad$ stress tensors $\qquad\quad$ external forces

In the cell conservation equation there is convective transport of cells because of matrix displacement **u**. Cell division is affected by cell deformation and this is reflected in the traction parameter $\tau$. Haptotaxis, with parameter H, is a kind of chemotaxis: it represents the facility for the cells to move up a concentration gradient in tissue. The matrix secretion term g represents the secretion by the cells; that is the mechanism has a built-in facility for enlarging its pattern forming domain. Since movement during development is slow the forces are in equilibrium with contributions to the stress tensors from the viscous shear, the elastic resistance of the tissue and, crucially, the forces generated by the cells themselves. The external forces come from tethers which pass through the dermis giving it tensile support with s a parameter. Substituting for the various tensor forms used in these models results in a third order vector equation $\psi = 0$ involving n, $\rho$ and **u**.

The pattern formation scenario is the following. The cells exert traction on the tissue matrix. It is known experimentally that the cell traction increases with time during morphogenesis (up to a factor of 4 in some experiments). These forces are resisted by the elastic properties of the matrix and the external tethers. If the cell traction force is less than the resistive forces no pattern is formed. As the cell traction increases the a bifurcation point is reached when the cell forces are equal and then greater than the tissue forces and cell aggregation takes place. A small aggregation of cells forms with a corresponding increase locally in the forces generated by the cells. This deforms the matrix and creates a gradient in tissue which, because of the haptotaxis term, further encourages cell aggregation. The specific patterns created depend, of course, on the geometry and scale of the domain. The cell traction is the key element in pattern formation. Fig.3 below illustrates qualitatively the patterning process.

**Linear analysis for pattern formation.** A standard linear analysis about the uniform steady state highlights the pattern formation potential of these pattern generators. If we denote the general system by $L(m) = 0$ this gives a uniform steady state solution $m = m_0$. Linearizing the system about the steady state we look for solutions in the form $m - m_0 \propto \exp[\lambda t + i k \cdot x]$, where $k$ is the wave vector of the linear disturbances. This gives a dispersion relation $\lambda = \lambda(k, p)$ where $p$ represents the parameter set for the system. It can be shown that for a range of wavenumbers k and appropriate parameters, $Re\lambda > 0$: that is, wave vectors with wavenumbers in the range where $Re\lambda > 0$ grow linearly with time. These linearly unstable modes evolve into steady state spatially heterogeneous solutions, which are patterns in cell and matrix densities and tissue deformation. These cell aggregations in turn are precursors of the embryonic unit such as cartilage, stripes on alligators, feather or scale primordia and so on. This type of analysis gives the usual parabolic dispersion relation for $Re\lambda$ as a function of the wavenumber for a range of parameters - the first of

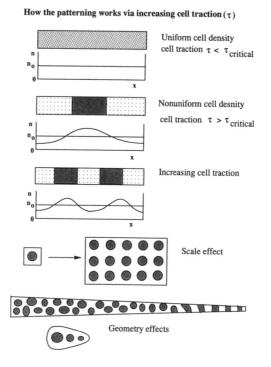

**Fig.3** As the cell traction increases it eventually overcomes the elastic forces in the tissue and cell aggregation can take place. The form of the space operators in the equations result in more complex patterns as the scale and geometry of the domain in creases and changes.

those in Fig.4 below. In a one-dimensional situation such a dispersion relation is a reasonable predictor of the type of spatially patterned solution which ultimately obtains from simulations of the full nonlinear system. Unlike other pattern generators, however, there is a range of other much less usual dispersion relations some of which are shown in Fig.4. With several of these we still do not know how to use them to predict final nonlinear solutions: some depend crucially on the initial conditions. The system of equations for these highly nonlinear mechanical mechanisms give rise to a remarkable wide spectrum of solution behaviour (see, for example, Murray 1989 for a pedagogical survey). Their structure and pattern formation potential is only just beginning to be explored. In the following two sections we discuss two specific real biological applications of this theory.

**Formation of fingerprint patterns.** The study and classification of fingerprint patterns (dermatoglyphics) has a long history and their widespread use has produced an extensive descriptive literature. The descriptive methods used may loosely be described as topological (Penrose 1979) and statistical (Sparrow 1985). Topological methods have been particularly useful for genetic and diagnostic purposes since ridges appear in their definitive forms during embryogenesis and the patterns do not change under continuous deformation during growth. Topological classification adopts rules to quantify features of fingerprints (see, for example, Loesch, 1983). There are certain generalities underlying fingerprint pattern development (Cherrill 1954, Elsdale and Wasoff 1976).

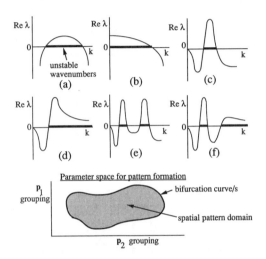

**Fig.4** Linear analysis gives dispersion relations which determine the range of wavenumbers which are linear unstable. In one-dimension this often predicts the final steady state solution; it is the mode with the fastest linear growth as in the case of the first of these dispersion relations. With others such as (d)-(f) it is difficult to make predictions. The final solution often depends crucially on the initial conditions. In all these cases however, linear analysis gives the parameter space: that is, with parameters **p**, grouped appropriately and in the appropriate space, spatially inhomogeneous solutions will be obtained.

During embryogenesis the palms and soles of primates develop ridges which are preserved throughout adulthood and these patterns are unique to an individual. In humans, the development of epidermal ridges seems to start around 4 months of gestation when epidermal folds are formed. These epidermal ridges form a variety of patterns including loops, whorls, triradii and so on (Fig.5), and it is these which form the basis for classification. The mechanism which produces these ridges is not known. Okajima (1982) concluded from his experimental study of the patterns on exposed dermis that ridged structures there reflect those on the epidermis. Since the formation of these ridges involves a mechanical deformation we suggest they could be formed by a mechanical pattern formation process and thus consider the possibility of forming them with the above model mechanism for fibroblast dermal cells.

In their experimental work on fibroblast cultures, Elsdale and Wasoff (1976) obtained typical dermatoglyphic patterns. Dramatic spiral patterns and other typical dermatoglyph forms were obtained by Green and Thomas (1978) using cultured human epidermal cells. A particularly challenging aspect of dermatoglyphics from a modelling viewpoint is the appearance of unusual ridge bifurcation patterns such as sketched in Fig.5 above. These singularities have no directionality and have half-power indices (Penrose, 1979).

During pattern formation the deformation of the tissue is small and the variation in cell density between the aggregations and the surroundings is also small. Further, during the patterning process there seems to be little cell division and tissue creation, diffusion and haptotaxis play negligible roles in cell transport. We thus consider the simpler model which

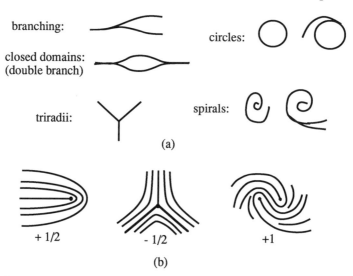

**Fig.5** (a) Typical ridge curves in fingerprints. (b) These singularities have half-power indices.

has only cell and tissue convection. The cell and matrix conservation equations (1) and (2) thus become

$$\frac{\partial n}{\partial t} + \nabla \cdot (n \frac{\partial \mathbf{u}}{\partial t}) = 0, \quad \frac{\partial \rho}{\partial t} + \nabla \cdot (\rho \frac{\partial \mathbf{u}}{\partial t}) = 0$$

Since variations in $\mathbf{u}$ and $\rho$ are small we can linearise about the steady state $n_0$ and $\rho_0$ and integrate the linear equations to obtain $n = n_0(1-\theta)$, $\rho = \rho_0(1-\theta)$ where $\theta = \nabla \cdot \mathbf{u}$ is the dilation. Substituting these forms into the above force balance equation (3) with appropriate forms for the various stress tensors $\sigma$ (see, Murray, 1989) we obtain a single scalar equation for the dilation, namely

$$\mu \nabla^2 \theta_t + \nabla^2 \theta - \beta \nabla^4 \theta + \nabla^2 [\tau(\theta)\{(1-\theta) - \beta \nabla^2 \theta\}] - s\theta = 0 \qquad (4)$$

where $\beta$ and $s$ are positive constants and $\tau(\theta)$ is a monotonic decreasing function of $\theta$. The $\beta$ and $s$ reflect respectively the strength of the long range elastic forces (because the tissue is strand-like) and the tethers that pass through the tissue. Typical numerical simulations of this equation give the strain field from which we can derive the displacement field $\mathbf{u}$ of the matrix and the cell density. An interesting general problem associated with pattern formation models in general is how to choose the parameter range to generate specific patterns. Bentil and Murray (1991) present a very simple and logical way to do this. Some examples of the typical results obtained are shown in Fig.6. The solutions exhibit patterns which are remarkably like those in human dermatoglyphs. They are also similar to those found by Green and Thomas (1978) for *in vitro* experiments with human fibroblasts (Bentil and Murray, 1991). We feel that the results obtained with this simplified mechanical model support the claim of a mechanical theory for producing fingerprint patterns.

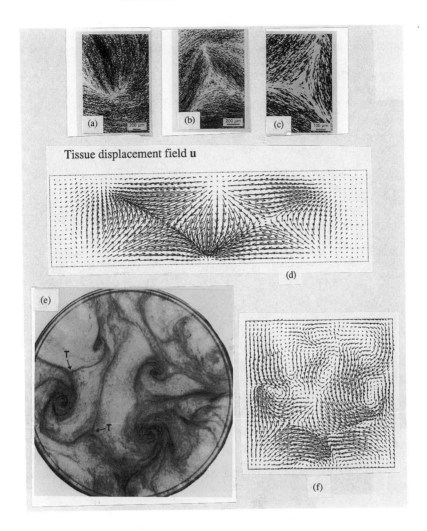

**Fig.6** (a)-(c) Patterns of fibroblast cells in culture (from Elsdale and Wasoff, 1976; photograph courtesy of Dr. J.B.L. Bard). (d) Typical displacement field solutions of the small strain mechanical model for pattern formation. (e) Patterns in cultured human epidermal cells : note the spirals (from Green and Thomas, 1978; photograph courtesy of Professor H. Green). (f) Numerical simulations of the small strain model (4) which exhibit spiral solutions. Note the singularities, which directly relate to cell density, and compare them with those in Fig.5 and the experimental forms in (a). (Numerical simulations by Dr. D. E. Bentil)

**Cartilage formation and the formulation of morphogenetic laws.** The cartilage patterns in the vertebrate limb has been widely studied because of its seminal role in embryology and evolutionary biology: the basic limb is illustrated in Fig.7. The evolutionary interest is how natural selection can act on the developmental programmes which create the limb patterns. What is lacking is a view of morphological evolution which takes into account the mechanism on which evolution can act. We have to move beyond the level of observation to a mechanistic explanation of morphological diversification. With this in mind we consider a model for generating the patterns of cells which become limb cartilage. We again suggest that a mechanical mechanism could be the pattern generator for such cartilage patterns. There are however, other candidates, such as cell-chemoattractant and reaction diffusion models. Our purpose in this section, however, is not simply to show that a specific mechanism can produce the sequential pattern observed in the developing limb but rather to show how mathematical modelling can result in what we might call "morphogenetic laws". That is, does the study of pattern generators suggest that certain patterns or morphologies are not possible. One example of such a developmental constraint arose in the study of animal coat patterns (Murray, 1981a,b). There it was

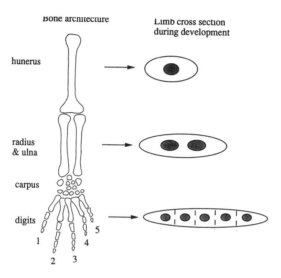

**Fig.7** Cartilage condensations in the general vertebrate limb. The cross sections at each stage approximate the geometric domain for the pattern generator with dark areas representing cell aggregations of presumptive cartilage.

shown that a spotted animal could have a striped tail but not the other way round: refer also to Fig.3 where there is an example of patterns on such a tail-like tapering domain. This conclusion was a direct consequence of the domain and geometry constraints on the solutions of the model equations - reaction diffusion equations in that case. Although the numerical simulations were for a reaction diffusion system the conclusion is effectively model independent. and as such can be formulated as a kind of morphological "law" or rule. In this section we shall use the mechanical model to derive other developmental "laws" which were verified experimentally.(Oster et al. , 1988) as a direct consequence of the mathematical predictions.

Although morphogenesis seems to be deterministic on a macroscopic scale there is, at the cellular level, considerable randomness. The highly structured and ordered patterns which emerge during development comes about by an average outcome of various possibilities. We should really only talk of the probability of of specific outcomes. Certain morphologies are simply highly unlikely. In this section we shall discuss an example of this. Although we shall call it a developmental constraint it is rather a developmental bias. This, together with extensive experimental evidence, is discussed in some detail by Oster et al. (1988) and more specifically by Oster and Murray (1989).

Let us start again with the full model system involving cells, tissue and tissue deformation. The cells in question are considered to be dermal cells. The differentiation in the biological sense of dermal cells into chondroblasts - cells which eventually become cartilage - seems to be intimately tied in with the process of condensation. Experimental evidence (Solursh, 1984) with cell-cultures shows that the cells in high-density cultures form chondroblasts whereas they do not in low-density cultures. The crucial aspect of patterning here is therefore the distribution of patterns of cell density: these areas of cell aggregation become cartilage. Although we have the mechanical model of morphogenesis in mind the following discussion is in effect mechanism independent.

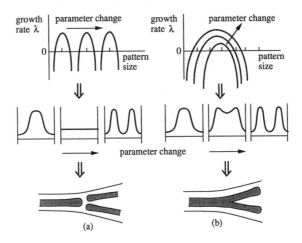

**Fig.8** The possible effects of a parameter variation on the dispersion relation giving the growth rate of specific modes with wavenumber k together with the biological predictions as it relates to limb cartilage patterning.

A linearisation about the uniform steady state gives a dispersion relation. Let us consider the simplest form which gives rise to spatially heterogeneous solutions, namely that in Fig.4(a). As a parameter varies the dispersion relation varies, with a corresponding change in the patterned solution which will be obtained. In one-dimension for example, the fastest growing linear mode changes. Such a parameter could be the cell traction or the domain geometry and scale. Certainly in the developing limb bud at the very least the latter occurs. There are several possible variations but let us concentrate on two and investigate the biological implications. Fig.8 shows two possible dispersion curve variation scenarios.

In Fig.8(a) as the parameter changes the dispersion relation shows that between each patterned solution there is a region of spatial homogeneity. If the first patterned solution has one hump with the next bifurcation being to two humps there is a parameter range where no spatially heterogeneous solution is possible. This is shown qualitatively with a fixed domain. Alternatively we can think of scale as the parameter in which case the three domains represent progressively larger domains with the third being effectively twice the size of the first. This interpretation can also be used with the increase in pattern complexity shown in the one-dimensional sketches in Fig.3: there we used cell traction as the bifurcation parameter. Another possible solution variation is shown in Fig.8(b). In this case although the solution evolves from one hump to two there is no parameter range in between when there is no pattern.

If we now consider the developing limb bud and consider a parameter variation during limb growth which effects a change from one aggregation to two, that is from the humerus to the radius and ulna; refer to Fig.7. The implications for limb development of the two scenarios in Fig.8 are quite different and are shown in the lower drawings. In (a) there is a distinct break with no presumptive chondroblast cells between the bifurcation from one to two bones while in (b) there is no break. At this stage the mathematics can say nothing further as to which is biologically correct. The problem must be resolved experimentally. It is interesting that two experimentalists had quite different views. The controversy was sufficiently interesting that one of these (Professor P. Alberch, Harvard University) initiated an extensive experimental programme to investigate sequential cartilage condensations in the vertebrate limb. The result is unequivocably that in Fig.8(b) (Shubin and Alberch, 1986). Such a result in turn puts constraints on the type of pattern formation mechanism that is possible for cartilage formation. The point about this example is that the mathematics suggested two possible developmental pathways which were resolved only by experiment. It further emphasizes the necessity of an interdisciplinary approach.

If we now consider the more distal patterning in the limb, the cross section in Fig. 7 shows a bifurcation from two aggregations to several. How is this achieved? Consider first the transition from two aggregations to three. If we consider the parameter spaces in which one, two and three humps are predicted from a linear analysis then, with parameters in the appropriate range random initial conditions will, in one dimension, usually evolve to the spatially structured solution indicated. A systematic method for generating such parameter spaces is given by Murray (1982) and this has been carried out for a reaction diffusion system by Arcuri and Murray (1986). Theoretically it is possible to go from two humps to three but the question arises as to what happens in the transition. We can also ask if it is possible to go from one hump to three. Again in principle this is often possible. Returning to the discussion above regarding which scenario is more likely we found from extensive simulations that it was very difficult to have a reliable transition from one hump to three. In fact, as the appropriate parameter was varied the transition was from one to two and then one of the legs in the latter bifurcated again to produce the requisite three humps in cross section as shown in Fig.9(d). Finally we know from the results shown in Fig.3 that simply lengthening the domain results in further aggregations: these are longitudinal bifurcations.

The conclusion from this analysis and the small probability that seems to exist for bifurcation from one to three condensations is that during development of the limb the more complex sequence of aggregations are likely to be constructed from a combination of a limited number of very simple bifurcations. That is, when there are three or more condensations, they could be made up from a combination of a single aggregation, its bifurcation to two followed by the longitudinal bifurcations shown in Fig.3. We hypothesize therefore a set of "morphogenetic laws" for constructing all vertebrate limb cartilage architecture. These are encapsulated in Fig.9(a)-(c) below. It is easy to demonstrate that complex structures can be developed from the use of such bifurcations as shown in Fig.9(f).

As before such morphogenetic laws can only be a hypothesis when based solely on a mathematical analysis of a possible pattern generator. In the final analysis a theory can only be confirmed by experiment. This was subsequently done in a thorough study of the sequential cartilage patterning in the limb of a wide variety of animals, such as the salamander, the crocodile and the turtle. One specific example, namely for a salamander, of the sequence of events in cartilage formation is shown in Fig.9(e). In each case the sequence of patterning followed the above morphogenetic laws (Oster *et al.* (1988). Once again it was a genuinely interdisciplinary approach which made the most significant scientific progress.

The mathematical problems associated with these morphogenetic laws are numerous and highly nonlinear. The conclusions were reached on the basis of a simple linear analysis, some nonlinear analysis near bifurcation but, in the nonlinear regime, mainly through numerical simulation. These morphogenetic rules point to a new approach to the study of limb morphology and show how pattern formation can constrain morphological evolution. There is widespread current interest in the role of developmental constraints in

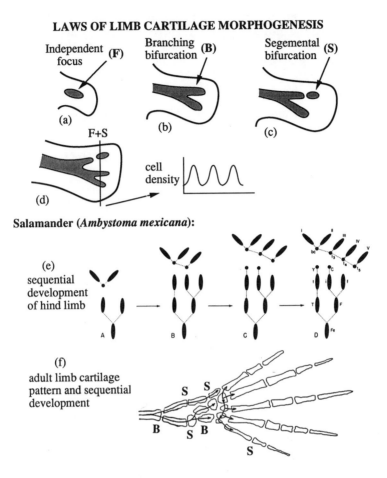

**Fig.9** There are three possible types of cartilage bifurcations during limb development. (a) a focal condensation, F; (b) a branching bifurcation, B; and (c) a segmental bifurcation S. Use of only these can generate the complex cartilage architecture shown in Fig.7. The sequential evolution of the bone structure in the salamander, found experimentally, is shown in (e). The sequential evolution of the pattern is summarized in (f) and limits itself only to the three bifurcation possibilities in (a)-(c).

evolution. These can only be understood in the context of the patterning process, that is in terms of the mechanisms which produce spatial pattern.

There is an interesting hypothesis which we can propose from this analysis if we consider the developmental constraints and morphogenetic laws to be applicable to a wider spectrum of patterning problems. It arises in teratology (the study of monsters). There is little evidence of the existence of three-headed monsters! Two headed snakes are not uncommon; there is no recorded evidence of three-headed snakes. There is, however, an example from the 19th century of a child with three heads. It is clear though from the evidence presented that first there was a Y-bifurcation in the back to two backbones after which one of them again Y-bifurcated.

**Conclusions.** What is presented here is only a fraction of what was contained in the ICIAM lecture. It suffices to show how the study of model mechanisms for pattern formation in biology can suggest possible and real developmental scenarios for specific developmental processes and hence, in conjunction with experimentalists, further the progress of biology. The examples show how mathematics can pose highly relevant biological questions and how, vice versa, biological problems can pose fascinating mathematical questions. It seems that the most interesting mathematical questions and problems arise from a close connection with real biology.

## References

P. Arcuri, J. D. Murray, *Pattern sensitivity to boundary conditions in reaction-diffusion models* . J. Math. Biol. 24, (1986), pp. 141-165.

D. E. Bentil, Murray, J. D., *Pattern selection in biological pattern formation*. Appl. Math. Lett. 4, (1991), pp. 1-5.

D. E. Bentil, Murray, J. D., *On the generation of dermatoglyphic patterns in primates*. (To appear)

N.F. Britton, *Reaction-Diffusion Equations and their applications in Biology*. Academic: New York 1986.

F. R. Cherrill, *The Fingerprint Systems at Scotland Yard*. London: H.M. Stationary Office 1954.

T. Elsdale, F. Wasoff, *Fibroblast cultures and dermatoglyphics: the topology of two planar patterns*. Wilhelm Roux's Arch. 180, (1976), pp. 121-147.

R. J. Field, M. Burger (eds.): *Oscillations and Travelling Waves in Chemical Systems. Wiley*, Wiley: New York 1985.

P. C. Fife, *Mathematical Aspects of Reacting and Diffusing Systems*. Lect. Notes in Biomath. 28. Springer: Heidelberg 1977.

H. Green, J. Thomas, *Pattern formation by cultured human epidermal cells: development of curved ridges resembling dermatoglyphics*. Science 200, (1978), pp. 1385-1388.

P. de Kepper, V. Castets, E. Duclos, J. Boissonade, *Turing-type chemical patterns in the chloride-iodine-malonic acid reaction*. Physica D 49 (1991), pp. 161-169.

E. Haeckel, *Genrerelle Morphologie der Organismen: Allgemeine Grundzüge der organischen Formen-Wissenschaft mechanisch begründet durch die von Charles Darwin reformitrte Descendenz-Theorie. (2 vols.)* Georg Reiner: Berlin 1866.

D. Z. Loesch, *Quantitative Dermatoglyphics*. Oxford University Press: Oxford 1983.

H. Meinhardt, *Models of Biological Pattern Formation*. Academic Press: London 1982.

J. D. Murray, *A pre-pattern formation mechanism for animal coat patterns*. J. theor. Biol. 88, (1981a), pp.161-199.

J. D. Murray, *On pattern formation mechanisms for lepidopteran wing patterns and mammalian coat markings*. Phil. Trans. Roy. Soc. (Lond.) B295, (1981b), pp. 473-496 .

J. D. Murray, *Parameter space for Turing instability in reaction diffusion mechanisms: a comparison of models*. J. theor. Biol. 98, (1982), pp. 143-163.

J. D. Murray, G. F. Oster, *Generation of biological pattern and form*. IMA J. Maths. Appl. in Medic. and Biol. 1, (1984a), pp. 51-75.

J. D. Murray, G. F. Oster, *Cell traction models for generating pattern and form in morphogenesis*. J. Math. Biol. 19, (1984b), pp. 265-279

J. D. Murray, G. F. Oster, A. K. Harris, *A mechanical model for mesenchymal morphogenesis*. J. Math. Biol. 17, (1983), pp. 125-129.

J. D. Murray, P. K. Maini, *A new approach to the generation of pattern and form in embryology*. Science Progress 70, (19??), pp. 539-553.

J. D. Murray, P. K. Maini, R. T. Tranquillo, *Mechanical models for generating biological pattern and form in embryology*. Physics Reports ??(1986).(in press).

J. D. Murray, *Mathematical Biology*, Springer-Verlag, Heidelberg, 1989.

G. Odell, G. F. Oster, B. Burnside, P. Alberch, *The mechanical basis for morphogenesis*. Dev. Biol. 85,( 1981), pp. 446-462.

M. Okajima, *A methodological approach to the development of epidermal ridges viewed on the dermal surface of fetuses*. Prog. in Dermat. Res. 20, (1982), pp. 175-188.

G. F. Oster, *On the crawling of cells*. J. Embryol. exp. Morph. 83, (1984) Supplement, pp. 329-364.

G, F, Oster, J. D. Murray, *Pattern formation models and developmental constraints*. [J.P. Trinkaus Anniversary Volume] J. exp. Zool. 252, (1989), pp. 186-202.

G, F, Oster, J. D. Murray, A. K. Harris, *Mechanical aspects of mesenchymal morphogenesis*. J. Embryol. exp. Morphol. 78, (1983), pp. 83-125.

G. F. Oster, J. D. Murray, P. K. Maini, *A model for chondrogenic condensations in the developing limb: the role of extracellular matrix and cell tractions*. J. Embryol. exp. Morphol. 89, (1985), pp. 93-112.

G. F. Oster, N. Shubin, J. D. Murray, P. Alberch, *Evolution and morphogenetic rules: the shape of the vertebrate limb in ontogeny and philogeny*. Evolution 42(5), (1988), pp. 862-884.

R. Penrose, *The topology of ridge systems*. Ann. Human Genet. (Lond.) 42, (1979), pp.435-444.

L. A. Segel, *Modelling Dynamic Phenomena in Molecular and Cellular Biology*. Cambridge University Press: Cambridge 1984.

N. Shubin, P. Alberch, *A morphogenetic approach to the origin and basic organisation of the tetrapod limb*. Evol. Biol. 20, (1986), pp. 319-387.

M. Solursh, *Ectoderm as a determinant of early tissue pattern in the limb bud*. Cell Differentiation, 15 (1984), pp. 17-24.

M. K. Sparrow, *Fingerprints - Mathematical models (A topological approach)*. U.S. Govt. Printing
 Off. Wash. D.C. 1985.

A. M. Turing, The chemical basis of morphogenesis. Phil. Trans. Roy. Soc. (Lond.) B237, (1952), pp. 37-72.

L. Wolpert, *Positional information and pattern formation*. Phil. Trans. Roy. Soc. (Lond.) B295, (1981), pp. 441-450.

# Chapter 15
# TRENDS IN RADAR ARCHITECTURES

**GABRIEL RUGET**
Thomson-CSF
France

**Introduction**

Global parameters for dimensioning a radar architecture appear in the so called radar equation, which is a simple energetic balance :

$$S/N = 1/4\pi \cdot R_1^2 \cdot R_2^2 \cdot T/\Omega \cdot PA/kTFL \cdot \sigma_\lambda$$

where:  - S/N is the signal to noise ratio at the input of processing,
- $R_i$ are the distance from target to transmitter and receiver antennas,
- T/$\Omega$ is the time per steradian dedicated to observation,
- P is the mean transmitted power, A the receiving antenna area,
- T the receiver temperature, F its noise factor, and L global losses due to microwave guides;
- $\sigma_\lambda$ is the target radar cross-section, depending on the carrier wavelenght $\lambda$, but also on the transmitter/ target/ receiver geometry.

In this paper, we will concentrate on the **PA T/$\Omega$** terms, since the trend is to diminish the expensive **P**, the technology of which is largely radar specific, and to improve the information involved in **S**, in its very content, if S/N is good enough.

In real life, it appears that each $\Omega \ni \omega$ space cell deserves a specific treatement (choice of pulse and burst durations [*] , pulse repetition frequencies, carrier frequencies, scanning frequencies... for search, track, reconnaissance, illumination), whence, if technology permits it (with phased array radars for instance), there is a huge problem in sharing time : space versus time balance is the problem, opening the way to real-time constraint logic programming.

Evolution of needs include increase of air traffic in terminal area as well as saturating threats in military applications, stealth targets (microbursts or windshear in civilian applications !) and/or reconnaissance demands, all of them forcing to take more and more into account a background in itself more complex. The answers to that challenge are <u>increased resolution</u>, especially in angular dimensions where to-day resolution is dramatically poor compared to range resolution (100 is a typical ratio) , <u>continous observation</u> of space permitting to detect stealth targets as soon as possible, (time overheads of several seconds are unacceptable at short ranges), and to avoid combinatorial explosion when associating new plots to old tracks. One further possibility for saving time is ultra agile antijamming, or, more generally speaking, <u>analysis</u> of the background <u>and adaptation</u> of processing <u>within the burst</u>.

---

[*] the radar burst, fractioned into pulses in order to allow echoes detection, is the total coherent wave form which is delivered by the transmitter to explore one elementary sector of space.

Coming back to the **PA T/Ω** parameters, the technical mean to implement is increase of **A** while fractioning and distpersng the reception antenna ; one can then reduce the mean power **P** but also the instantaneous power w.r.t. time and space: this is obtained through continous waves (favoured by multistatic configurations) and orthogonalization of the waves issued from several simple omni-directional antennas. The spatial extension of the dispersed transmission and reception antennas enables the use of waveforms of the same level of sophistication in the space (angular) and time (ramge) axes.

To sum up, trends will be :

. to consider transmission and reception geometries as separate variables in the optimization of radar architectures, with many degrees of freedom for both, most of them accessible in real or near real time. This is the best response to the stealth challenge and also to the electromagnetic warfare challenge.

. to add a huge memory capacity at the receiving side, as close as possible to the original signal, so as to implement the strongest possible forms of closed loop processing.

**Part one : From the pencil beam radar to the RIAS concept.**

Beamforming in the pencil beam radar is achieved with largely different objectives and means in the angular and range variables. Considering angular variables, the wavefront issued from a compact transmitter is simply phase shifted at different aperture locations by passive means (mirrors, waveguides) or active ones (electronically switched microwave circuits, analog magnetic devices). The primary purpose is to approximate some kind of narrow Gaussian pulse at infinity, and to steer it electronically. As we cannot vary the transmitted frequency at different aperture locations, we cannot get omnidirectionality without losing directivity; moreover, this directivity is modest as aperture size is low.

In the range dimension, we create the desired (wide band) pulse directly in the temporal domain, by varying the transmitted frequency with time (that is "pulse compression" - the spacial equivalent of which we are looking for)."Optimum" detection is obtained through the use of a linear matched filter: optimality refers to Gaussian assumptions which are certainly not valid for modern jammers; hence the opportunity we will not discuss here to use higher order methods, for instance complex Volterra filters optimized through linear calculus [CHEVALIER]. In fact, bursts are coherent bursts long enough to reach sensitivity to radial speed as well as to target position; if the transmitted pulse **u** lives in $L^2(E)$, **E** being the 1D range space, the density **d** describing the radar cross-section landscape lives in $L^2$ ($E \oplus E^*$), the Doppler space being conveniently identified to the dual $E^*$. Optimum detection result in

$$A_d(u)(x,\xi) = \int \int d(y,\eta) \, A(u)(y-x,\eta-\xi) \, \exp\pi i <y+x,\xi-\eta> \, dy d\eta$$

where A(u) is the ambiguity function of **u**

$$A(u)(z,\zeta) = \int u(z'-z/2) \, u^*(z'+z/2) \, \exp-2\pi i z'\zeta \, dz'$$

which is the Fourier transform of the Wigner-Ville distribution

$$H(u)(x,\xi) = \int u(x+z/2) \, u^*(x-z/2) \, \exp-2\pi i z\xi \, dz$$

For **u** gaussian and **d** a Dirac mass at $x,\xi$ (resp. a linear combination of sparse Dirac masses at $x_i, \xi_i$ ), $A_d(u)$ reaches its unique maximum at $x,\xi$ (resp. local maxima the $x_i, \xi_i$) .

EXACT INVERSION FORMULAS

The previous result is not well suited for high resolution applications, as in Synthetic Aperture Radar imagery (where indeed Doppler represents azimuth ). Relying on an interpretation of $A_d$ in terms of pseudo differential operators:

$$A_d(u)(0,0) = (\text{Op}(d^\sim) u, u)$$

where $\text{Op}(d^\sim)$ has for symbol the total Fourier transform of **d**, LERNER proves an exact inverse formula involving a two dimensional family of pulses :

$$d(y,\eta) = \int \int A_d(\tau_Y u)(y,\eta) \, dY$$

with

$$\tau_{(x,\xi)} u(t) = u(t-x) \exp 2\pi i < t - x/2, \xi >$$

A more classical setting for exact reconstruction was given by FEIG-GRUNBAUM : the pulses they use are chirps, that is, for u ultra-short gaussian, $\tau_{S^{-1}}(u) = u * \exp \pi i < t S t >$. Then, we have

$$A_d(\tau_{S^{-1}}(u))(y,\eta) = \int d(y,\eta) \, 1_{y-x-S(\eta-\xi)=0} \exp\pi i < y+x, \xi-\eta > dy d\eta$$
$$= \exp-\pi i < xS^{-1}x > \int d(S\eta, \eta+\xi - S^{-1}x) \exp-\pi i < \eta S \eta > d\eta$$

Playing with the S and $\xi$ parameters of the chirp, one reconstructs **d** from $A_d(\tau_{S^{-1}} u)(0,\xi)$ through a variant of the Radon inverse problem (Feig-Greenleaf). These exact formulas suffer from the same limitation w.r.t. high resolution : increasing resolution requires the use of a wideband family of pulses, which means :
. modifying the model and the ambiguity formula, because Doppler effect is indeed time stretch, not frequency shift,
. facing dispersive propagation conditions which destroy the wideband coherency of received signals.

Moreover, the model of radar cross-section depends largely on the spectral content of the pulse : one promising way to take it into account [ALTES] is to modelize the target impulse response as a distribution in the so-called Schwartz meaning, for instance in the finite support case

$$d(x) = \Sigma \, d_{k,m} \delta^{(m)} (x-x_k)$$

Transmitting **u**, one then detects with the matched filters $u^{*(m)}(-t)$, the problem being to reassemble the different responses. As far as computation is concerned, it is attractive [FLANDRIN - MAGAND - ZAKHARIA] to select **u** so thats the receivers are of equal complexity (in terms of bandwidth times duration product) : this is obtained if **u** has a linear period modulation and a lognormal envelope. It is worth noting that the result is then the same as analysing **d** with the wavelet $\Gamma_u$, which is of the Grossmann-Morlet type :

$$\Gamma_u(\omega) = 1_{\omega > 0} \exp\text{-a} \ln^2 \omega$$

## UNCERTAINTY PRINCIPLES

For a pulse **u** of unit energy, the ambiguity function A(**u**) is bounded by 1 in module, with equality at zero where its second derivatives are related to the bandwidth and time extension of **u**. Moreover, A is of unit $L^2$ norm [WILCOX], what forbids recovering **d** from $A_d(u)$ with a single **u**.

Having explained why A(**u**) is away from $\delta$, it is of interest, in view of the super resolution results quoted in the second part, to show that the Wigner Ville transform itself is away from $\delta$: FLANDRIN optimizes the "energy" concentration in a time-frequency domain ; for a wheighting bearing elliptical symmetry, he gives explicit results in terms of Laguerre polynomials ; for a gaussian wheighting

$$\exp-\pi( (x/T)^2 + (\xi/B)^2 ), \qquad BT = \sigma,$$

optimal pulses are the gaussian pulses, with score $2\sigma/(2\sigma+1)$;
($\sigma = 1/2$ corresponds to the Gabor-Heisenberg cell).

Deeper results are given by DONOHO-STARCK, where the bandwidth and duration constraints are no longer of the interval type :
**Theorem** : let T and B be measurable sets and suppose f of unit $L^2$ norm, f$^\sim$ its Fourier transform, be such that f is concentrated on T mod $\varepsilon_T$ (in $L^2$ norm) and f$^\sim$ concentrated in B mod $\varepsilon_B$ (idem). Then :

$$|B| \, |T| > (1 - \varepsilon_T - \varepsilon_B)^2$$

Corollary : if $|B||T| < 1/2$, less than half the energy of a B band limited f lies within T.

The proof goes through the inequality $|B||T| > |P_B P_T|^2$, where $P_B$, $P_T$ are the projectors on $\{f, B \supset \text{supp } \hat{f}\}$, resp. $\{f, T \supset \text{supp } f\}$. Practical bounds are significantly better for dispersed T. Super resolution results rely on a $L^1$ analog of the theorem, with minorant $(1 - \varepsilon_T - \varepsilon_B)(1 + \varepsilon_B)$

## AMBIGUITY FUNCTIONS AND GROUP REPRESENTATION THEORY

We briefly indicate some simple transformations of the pulse space $L^2(E)$, which are described for instance in [PAPOULIS] chapter eight, in the framework of group theory. We begin with the 1D situation [MILLER], and then move to the multidimensional case, following [LERNER].

The $\exp(2\pi i \lambda w)$ representation of the $\{z=0\}$ subgroup of the Heisenberg group $H = \begin{pmatrix} 1 & z & w \\ & 1 & \zeta \\ & & 1 \end{pmatrix}$

induces the following representation of H in the pulse space:

$$T^\lambda(z,\zeta,w)u(t) = \exp(2\pi i \lambda(w+t\zeta))u(t+z)$$

The ambiguity function appears then as $A(u)(z,\zeta) = \exp(\pi i z \zeta) < T^1(z,\zeta,0)u, u >$

In fact, the multiplicative factor disappears if we embed $E \oplus E^*$ into H through

$$(z,\zeta) \rightarrow (z,\zeta, w=-z\zeta/2)$$

which amounts equipping $E \oplus E^* \oplus R$ with the natural Lie group structure inherited from the Lie bracket.

The Schrödinger group is the semi direct product of H and a two fold covering $SL^\sim(2,R)$ of the special linear group, which acts on H as a group of outer automorphisms; it inherits a representation in the pulse space from the $T^1$ representation of H and from the following representation of $SL^\sim(2,R)$:

$$\tau_\Sigma u(t) = \begin{pmatrix} 1 & 0 \\ \Sigma & 1 \end{pmatrix} u(t) = \exp(\pi i <t\Sigma t>)u(t) \qquad \tau_S u(t) = \begin{pmatrix} 1 & S \\ 0 & 1 \end{pmatrix} u(t) = \exp(\pi i <tS^{-1}t>)*u(t)$$

$$\begin{pmatrix} A & 0 \\ 0 & A^{-1} \end{pmatrix} u(t) = A^{1/2}u(At) \qquad \begin{pmatrix} \cos\alpha & \sin\alpha \\ -\sin\alpha & \cos\alpha \end{pmatrix} u(t) = \exp(-i\alpha/2)U(\alpha)u(t)$$

where $U(\alpha)$ is the one parameter group with infinitesimal generator $i/2 \, (d^2/dt^2 - t^2 + 1)$, $U(\pi/2)$ being simply the Fourier transform. Interesting pulse tranforms are $\tau_\Sigma$ and $\tau_S$, which are linear frequency modulation and quadratic phase filtering respectively, pulse compression being obtained by applying $\tau_S * \tau_\Sigma$ with $S \cong \Sigma^{-1}$ [PAPOULIS].

For the multidimensional case, the metaplectic group $Mp(E)$ operating on $L^2(E)$ generalizes the Schrödinger group; it is fibered onto the symplectic group $Sp(E)$ which is generated by translations, dilatations, already mentioned transforms associated with symmetrical $\Sigma$ or $S$ in $E^* \oplus E^*$ or $E \oplus E$, and finally exchanges $(x_i, \xi_i) \rightarrow (\xi_i, -x_i)$. We already gave formulas for the relevant liftings in $Mp(E)$, which are $\tau_Y$, $\tau_A$, $\tau_S$, $\tau_\Sigma$ and partial Fourier transform. The cross-Wigner transform commutes with $Mp(E)$ and $Sp(E)$ actions.

In that setting, it is interesting to interpret some 2D pulses arising as $\tau_S$ or $\tau_\Sigma$ transforms involving "space/time" coupling in the S or $\Sigma$ matrix. As an example, NICOLAS-PIROLLI described a radar with a transmitting equispaced array and a linear progression in the aperture of transmitted frequencies : this is the antidiagonal-S pulse in the aperture plane, producing at infinity the spatial sweep of a short pulse, within the duration of the long pulse radiated by each array's element.

RIAS DESCRIPTION :

Like unidimensional pulses of practical use are not limited to chirps , the Nicolas-Pirolli pulse only represents a limit situation. The RIAS (Radar à Impulsion et Antenne Synthétiques) implements the "omnidirectivity in the pulse" concept with a highly lacunar antenna : transmitters are regularly installed on a circle and transmit orthogonal pulses of equal contiguous bandwidths; receivers are located on another circle and each receives the whole bandwidth. This amounts to space coding at transmission, allowing for instance a cooperative target to locate itself within the transmitted electromagnetic field. Detection is through direction dependent correlation or equivalently through double phase shifting by subband (w.r.t. target /$T_x$ /$R_x$ antennas geometry).

Physical dimensions of the RIAS result from a mix of operational requirements, which lead to the choice of <u>low carrier frequency</u> (VHF, permitting to work at long ranges without practical Doppler ambiguity), <u>moderate range resolution</u> associated with the <u>very high Doppler resolution</u> the excellent stability of targets in the VHF band makes possible to obtain.

We show in Fig.1 the way the distribution of frequencies amongst the $T_x$ antennas modifies the antenna diagram. One has in fact to separate transmit and receive contributions in the process of optimizing the total diagram through geometries. Due to lacunarity, the level of spurious lobes is moderately good, so that it is mandatory to rely on Doppler filters and furthermore on adaptive beamforming : Fig. 2 shows the result of standard versus adaptive processing applied to a target plus jammer situation.

Even with fixed or slowly adaptive algorithms, the computational burden for RIAS processing is huge : number of complex operations per second is

$$N_{scan/s} \cdot N_{Doppler} \cdot N_{range} \cdot N_{beam} \cdot (N_{R_x} + N_{T_x}),$$

typical values of which are 10, 100, 400, 10000, 25, 40 respectively ; this amounts to one teraoperation ($10^{12}$) per second.

There are several ways to reduce the computing power by more than one order of magnitude. One way is to use sequential sweep of Doppler bins, which amounts to reduce $N_{scan/s}$. It is more clever to detect in only one out of $T_x$ range bins, leading to no practical information loss, but computing reduction is small (1/2). Space under observation is in fact almost empty except for some specific Doppler bins occupied by clutter. A better solution would be to replace systematic beam forming by equation solving ; this is possible with parametric methods like MUSIC or other super resolution methods. However we will describe here a new promising method [DESODT-MULLER] which is independent components analysis (ICA) : we see in Fig. 3 the output of five RIAS receivers for one Doppler bin (that is after recombination of responses within a burst) . ICA takes advantage of the statistical independence, as time series, of the RIAS waveforms reflected from different directions in space (this is practically achieved although the spatial arrangement of transmitted frequencies has not been optimized in that respect). The result of ICA is shown in Fig 4, where two long pulses emerge from noise and jammers. To know what linear combinations of received signals do convey independent information is to know direction, which in turn allows to recover range resolution through pulse compression.

INDEPENDENT COMPONENT ANALYSIS

We start with the linear statistical model y = Dx + n , x, y, n being random vectors and D a regular matrix. Realizations of y being given, the problem is to estimate both D and the corresponding realizations of x, assuming that the components of x are statistically independent. This is a "cocktail-party" problem, a neuromimetic approach of which was given by JUTTEN-HERAULT.

Let $V\Delta^{1/2}\Lambda$ be is a singular value decomposition of D, and suppose the correlation $\rho_x$ is 1, so that

$\rho_y = V\Delta V^* + \rho_n$

LACOUME-RUIZ remarked that knowledge of $\rho_y$ and $\rho_n$ gives access to V but not to $\Lambda$; if one has extra information at hand (e.g. independence of x-components), they proposed to play with $\Lambda$ to take profit of it.

The CARDOSO approach is of the <u>identification</u> type : if y is a random vector with values in Y, $\rho_y$ makes Y* (anti)isomorphic to Y, and then the $\Omega_y$ tensor of 4-cumulants of y appears as an hermitian operator on the space of hermitian operators on Y.

**Lemma**: if $y = \Sigma\ x_i d_i + n$, $x_i$ being independent random variables of kurtosis $k_i$, and n a gaussian noise independent of x, then $\Omega_y = \Sigma\ k_i\ (d_i \otimes d_i^*) \otimes (d_i \otimes d_i^*)^*$. If moreover the $d_i$ are linearly independent, then ipso facto they are orthonormal ( w.r.t. the $\rho_y$ metric, which is the practical problem as one has only an estimator of it ).

The first way to use this lemma is a variant of MUSIC : let us call "source manifold" the collection of $d_i$ which can occur for a given antenna geometry; MUSIC recovers the individual $d_i$ by trying to intersect this source manifold with Im V, while 4-MUSIC looks at the vectors in the source manifold which are the closest possible to Im $\Omega_y$. This allows identification of more sources than sensors.

In the less sources than sensors situation, $\Sigma\ k_i\ (d_i \otimes d_i^*) \otimes (d_i \otimes d_i^*)^*$ appears as the eigen vector decomposition of $\Omega_y$ : this gives the $d_i$ without any use of the source manifold: selfcalibration is then for free.

This property is also shared by the COMON method which was the one used in the RIAS example. Comon starts with the concept of <u>contrast</u>, that is a real function $\Psi$ on (square integrable) random vectors such that :

$\Psi(z)$ depends only on the probability law $p_z$
$\Psi$ is invariant by scale change: $\qquad\qquad \Psi(Lx) = \Psi(x) \quad \forall\ L$ diagonal
If x has independent components, then $\qquad \Psi(Ax) < \Psi(x) \quad \forall\ A$ regular

A natural candidate is minus the Kullback-Leibler distance between $p_z$ and the product of marginals $\Pi p_{z_i}$, where $\underline{z}$ is any orthonormalized version of z. A more tractable one is $\Psi_4(z) = \Sigma\ |\Omega_z(\text{iiii})|^2$ which appears in the expansion of the K-L distance resulting from an Edgeworth expansion of the ratio $p_{\underline{z}}$/Normal  The Comon algorithm is initialized with $\underline{y}$ and then proceeds by successive rotations in 2-variable planes , maximizing the $\Psi_4$ contrast at each step. For natural cyclic sweeps of pairs of variables, one proves that $\Psi_4$ is monotonically increasing under the previous procedure ; if n = 0, the limit is $\Psi_4(x)$.

Apart from the already quoted difficulty in estimating $\rho_y$ (i.e. computing $\underline{y}$ ), which is a common prerequisite to all ICA methods and constitutes the practical limitation to the accuracy of direction recovery, more investigation is to be devoted to the wideband situation, as well as to a mean for introducing some a priori knowledge of the source manifold.

## RIAS, EAR AND POLYPHONY.

Let us try a somewhat unusual comparison: RIAS transmission uses colouring of space through space dependent time filtering ; so does ear reception, with the head-pinna arrangement. In ICA as well as in human audition, primary interest is in distinguishing messages rather than locating sources, those messages being issued for instance from leading "instruments", above some orchestral and noise background. At the turn of the XVI[th] century (as we learned from Barry Vercoe), the viol family acquired this leading capability: this happened by removing the frets from the fingerboard and by complexifying the resonant body, giving it steep angular formants; both facts enabled the introduction of instrumental vibrato (human voices madee use of vibrato for a long tim). What I suggest is the equivalence of a radar's target with an instrument body, and of transmitter + antenna with hand + bow + finger + string. We have to get more vibrato through extension of antennas in frequencies and space, and to exploit more dimensions of targets variability, for instance differential Doppler and structure resonances, by using Gjessing type arrangements [GJESSING -HJELMSTAD]. Studies must be devoted to determine whatever kind of lacunarity is good for this purpose.

TRENDS IN RADAR ARCHITECTURES   **233**

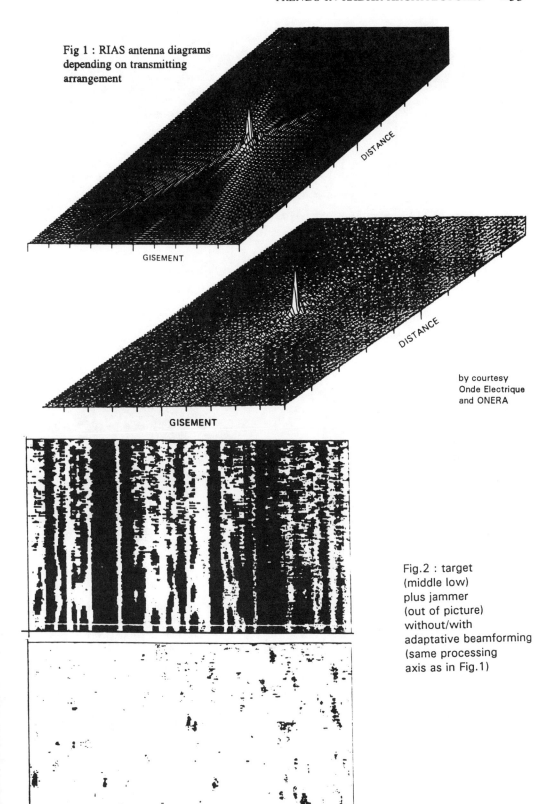

Fig 1 : RIAS antenna diagrams depending on transmitting arrangement

by courtesy Onde Electrique and ONERA

Fig.2 : target (middle low) plus jammer (out of picture) without/with adaptative beamforming (same processing axis as in Fig.1)

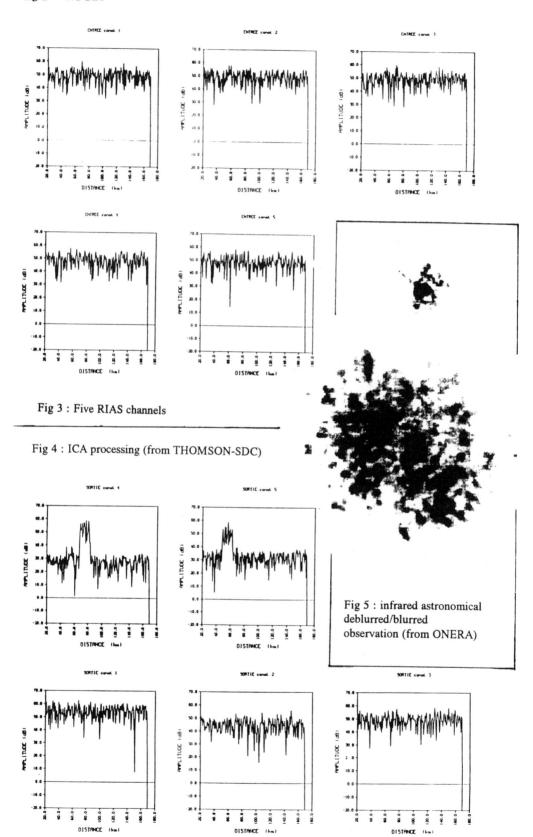

Fig 3 : Five RIAS channels

Fig 4 : ICA processing (from THOMSON-SDC)

Fig 5 : infrared astronomical deblurred/blurred observation (from ONERA)

**Part two : Selfcalibration and related matters.**

The need for real time calibration of an antenna (selfcalibration if the user wants the process to be transparent) may have two unrelated causes: a changing propagation medium (turbulent atmosphere in optics, varying multipaths) or a non rigid antenna with poorly known geometry. Both of them affect the beamforming in the same way, producing speckle or even blur (fig 5). Extended coherent sources give also speckle which, instead of being an artifact, bears information about the source.

For correcting speckle of blur by adapting the beamforming, without any reference source, one needs instantaneous images, and not a succession of pencil beam echoes. There appears to be a difference between radar and sonar on one hand, and optics on the other : whatever the way beams are formed in radar, measurements for each beam are amplitude and phase measurements, so that they provide detailed information on what happens in the aperture. This permits to compute many modified beams. Detection in optics is usually amplitude detection and gives partial information only about what happens within the aperture. Then one may have to work in the aperture and focal planes simultaneously.

In sonar or radar, if we think of fully developed digital beamforming (omnidirectionnal radiating elements, each one with an encoded output), a mathematically sound approach is to rely on parametric methods by increasing the number of degrees of freedom in the antenna description, and then by proceeding to some identification or likelihood maximization. Examples are VEZZOSI for MUSIC type algorithms, FUCHS for state variable methods and the "time differences of arrival problem" (wide bandwidth), MUNIER for the propagator method, NICOLAS-VEZZOSI for the broadband situation again, in a multifrequency and phase unwrapping setting. This is the clean way to cope with many unknown parameters, and without extra information.

In optics or radar, one may have a limited number of parameters to control: relative positions of otherwise rigid sub-arrays, main deformation modes of a compact antenna. Moreover, relying on subaperture measurements of a lower resolution, for which selfcalibration is not a necessity, one gets indications for identifying the aperture.geometry This is what STERNBERG describes in the radar (radio camera) context : finding a linear phase adjustment for each sub-array is simply adjusting their pointing. Nevertheless, synchronization of the dispersed sub-arrays requires one non extended target with a good S/N ratio. In optics, for compensating phase distorsion generated by atmospheric turbulence, PRIMOT-ROUSSET-FONTANELLA evaluate wavefront reconstruction from Hartmann-Shack sensing and through the hierarchy of Zernike polynomials [NOLL] [FRIED]. They point to the quality of reconstruction being limited by the accuracy of the model rather than by measurement noise. For recovering an unblurred high resolution image, one image/wavefront pair may not be enough (fading effects may annihilate the transmission gain in some directions). P-R-F. propose to use joint deconvolution of several short exposure image/wawefront pairs instead of long exposures with expensive deformable mirrors.

When playing with very few unknown parameter, adaptation may be a blind procedure relying on the sharpness of the image $I$ of an object $r$. This was introduced by MULLER-BUFFINGTON with various sharpness functions $S(I)$ like kurtosis $I^2$ or neguentropy $I.\log I$ : if the adjusted image $I_a$ is $r*P_a$, with an adjustable point spread function $P_a = |H_a^-|^2$ ( $H_a$ is the product of a positive wheighting function H standing for the ultimate limitations of the aperture device by an adjustable phase distorsion term $A_a$ ), the sharpness criterion $S(I)$ can sometimes be proven to be maximum when the phase of $A_a$ is linear within the aperture. This is referred to by STEINBERG for the radio camera, as a fine grain adjustment after wavefront sensing.

Simulation results are given by KOPP-AIACH in the sonar context (circular or sinuous towed array); it is interesting to compare the sharpness functions of Kurtosis type related to adaptive beamforming resp. MUSIC : if $d(x)$ is the true (as opposed to $d_0(x)$ the believed) source manifold, if $\rho$ is the autocorrelation matrix of observations and $\Pi$ the projector onto the "noise space" ($\rho$ and $\Pi$ depends on **d**), one has, for adaptive beam forming

$$S(d_0, d) = \int [ d_0^*(x) \Pi d_0(x) ]^{-2} dx$$

and for MUSIC

$$S(d_0, d) = \int [ d_0^*(x) \rho^{-1} d_0(x) ]^{-2} dx$$

Experimental results are given by PRATI for a SAR application where the unknown 1D parameter is the platform speed, with kurtosis as criterion of image sharpness.

## SELFCALIBRATION WITH A PRIORI INFORMATION

We will just point to the LOHMANN-WEIGELT-WIRNITZER phase-closure imaging method, which was amongst the first ones making use of higher order statistics. The idea is that it is possible to recover the bispectrum of the object, not only in amplitude as speckle interferometry gives it, but also in phase, relying on multiple image-bispectra measurements and on a simple statistical model of the variability of propagation conditions : this is speckle masking. How to get phase restoration from bispectral data is examined in depth in LANNES, especially in the case of partially redundant arrays.

Blind deconvolution uses a statistical model of the object: for instance, **r** is supposed to be i.i.d. (discrete space model) with a priori p.d.f. $\pi$, and **r** is $I_{*AR}$ ( AR for:"autoregressive model"); a quite indirect application is described by PRATI, once again in a SAR context; AR is computed relying on a crude approximation of $\pi$; one then establishes a relationship between AR and the unknown velocity error, for improving the calibration. Direct use of blind deconvolution for 2D images with a "sparse bright spots in a dark background" model $\pi$ is described in JACOVETTI-NERI, but we don't know of a radar/sonar application of it.

A last class of algorithms is constrained deconvolution, or more generally speaking, use of a priori constraints (positivity, support... ) on **r**. For active sonar towed arrays, FLORIN (private communication) proposes to calibrate the array with clutter echoes, which are known to be bound within the transmitted beam; with a very few shots, this method gets rid of the spurious optima of the Muller-Buffington approach. In optics, constrained deconvolution is another candidate for fine tuning after wavefront sensing. In sonar as well as in Over The Horizon radar applications, it may improve the result of beamforming through forward scattering computations.

The setting of selfcalibration in inverse methods is the following one: a linear transform # has for coordinates some values of the Fourier transform:

$$\rho = \#\mathbf{r} = \int \varphi(x) r(x) dx, \qquad x \text{ in } X \text{ compact metrizable separable;}$$

\# conveys a priori knowledge about array geometry, and it operates on a convex subset C of the r-space which expresses support or energy constraints. The image #C in the Fourier space is a convex set, which may not contain the measured $\rho_m$. If we are looking for high resolution in a definite direction, the unknown discrepancy between the real and believed geometries can be taken into account by enlarging $\rho_m$ into a "selfcalibration" manifold of low dimension, within which we select the autofocused $\rho_{foc}$ the closest possible to #C . Let

$$\rho = \text{proj}_{\#C} \rho_{foc}.$$

We will finally choose **r** in $\#^{-1} \rho$ according to one of the following rules :

1) $\max_{\rho_{foc}} \min_r S(r)$
2) $\min_r S(r)$ with $\rho_{foc} = \text{argmin } d(\rho, \rho_{foc})$

The "entropy" function S may be designed so as to include the constraints: for instance, if $a(x) < r(x) < b(x)$, take $S(r) = h_x(r(x))$,

with $h_x$ the Cramer transform of a measure $\mu_x$ the support of which is $[a(x),b(x)]$.

The criterion (1) is in the Muller-Buffington spirit. The criterion (2) interprets $d(\rho,\rho_{foc})$ as a residual measurement error; it is implemented by using the algebraic structures of phase closing [LANNES].
A priori knowledge of #C could help real time implementation: DACUNHA-CASTELLE GAMBOA give indeed a characterization of #C which is an extension of a Krein-Nudelman's result. With large deviation methods, they prove the

**Theorem** : define $L_\mu(\Lambda) = \int \text{Log Laplace } \mu_x (<\Lambda,\varphi(x)>) \, dx$
$H_\mu(\rho) = -\inf[-<\theta,\rho> + L_\mu(\Lambda)]$
if $\{ x \mid <v,\varphi(x)> = 0 \}$ has an empy interior for any v non zero
then $H_\mu(\rho) = -\inf_{\#r = \rho} S(r)$,
Moreover, for any $\mu$ adapted to the a,b constraints, $\#C \ni \rho <=> H_\mu(\rho) < +\infty$

The proof gives $r_{mem} = (\arg \min S(r) \mid \#r = \rho)$ as a limit of solutions of discretized problems ( X finite ), in which case $r_{mem}$ appears to be $E_Q(r_{discr})$, with Q the probability law on $\{r_{discr}\}$ the closest possible to $\Pi\mu_x$ which satisfies the constraint in the mean $E_Q (\Sigma \varphi(x) r_{discr}(x)) = \rho$; whence the "maximum entropy on the mean" name of the method [NAVAZA]. Complete proofs, in a more general setting, can be found in GAMBOA-GASSIAT.

SUPER-RESOLUTION

The GAMBOA-GASSIAT method leads to the following

**Theorem** : for the [0,b] constraint, the $r^*$ in #C such that $\#^{-1} r^*$ is a singleton are the boundary points of #C, and their counterimages are saturated objects $b1_A$. One then has furthermore a stability property, that is $\#^{-1}$ is continuous at $r^*$

This has to be compared with results by DONOHO and al. which give also unicity and stability conditions for recovery of 1D r ( X interval of **R** ) with bandlimited observations, and with a variety of support constraints: beginning with unicity (the no noise situation) in the discrete Z/nZ case, one has :

**Theorem** : Let B be the set of observed frequencies, that is $\rho = P_B r$, and suppose that supp r has cardinality less than $N_T$. Let $r^\sim = \text{argmin} |r'|_{L^1}$ for $P_B r' = \rho$
If $(N-N_B) \cdot N_T < N/2$, then $r^\sim = r$

The first instance of such a result was given by SANTOSA-SYMES, using a $L^1$-penalized algorithm; it is a consequence of the $L^1$ uncertainty principle (Logan phenomenon).

As for the GAMBOA-GASSIAT super-resolution theorem, coping with noise requires a stability property which results from a compacity argument, but evaluating the module of continuity is the problem ; DONOHO gives a deep theoretical result when a full interval $|\omega| < \Omega$ of low frequencies is observed, in the following setting :
Let $\Delta$ be a 1D lattice with mesh $\Delta$, and let's suppose that the support of r has a "density" less than 1. If we were glad enough with a resolution of 1 , $\Omega = \pi$ would be enough (Rayleigh limit); on the other hand, a $\Delta$-resolution is surely obtained with $\Omega > \pi/\Delta >> \pi$. What DONOHO proves is :

**Theorem:** For $\Omega > 2\pi$, knowledge of the Fourier transform $\#r(\omega)$ for $\omega < \Omega$ determines a unique r.
Moreover, if $\Omega > 4\pi$ and if both of $r_i$ belong to
sparse $(R,\Delta) = \{ 1^1(\Delta) \ni r \mid \forall t, \text{card}(\text{supp } r \cap [t,t+R[)/R < 1 \}$,
then
$\sup \{ |r_1 - r_2|_{l^2} \mid |\#r_1 - \#r_2|_{l^2(-\Omega, \Omega)} < \varepsilon \} < \varepsilon \Delta^{-4R-1} b(R,\Omega)$

Practically speaking, the constraints on **r** involved in the previous theorem bear some resemblance with the "less sources than sensors" MUSIC setting; but here the support of **r** is bounded in density rather than in cardinality. Moreover, knowledge of the wavefronts, which is hidden indeed behind the use of the Fourier transform, is taken into account from the beginning.
DONOHO-JOHNSTONE-HOCH-STERN is an excellent survey of maximum entropy or alternative methods, for the sake of signal to noise ratio improvement and of super-resolution, in the recovery of "nearly black" objects. One finds there references to practical methods which will probably spread from the fields of optics and seismic exploration to radar.

REFERENCES

G. S.AGARWAL, A. T. FRIBERG, E. WOLF, *Elimination of distorsions by phase conjugation without losses or gains*, Opt. Com 43 (6), 446-450 (1981).

J. F. CARDOSO, *Localization and identification with the quadricovariance*, Traitement du Signal 7 (5), 397-406 (1990).

P. CHEVALIER, *Antenne adaptative: d'une structure linéaire à une structure non linéaire de Volterra*, Thèse, Université de Paris-sud Orsay (juin 1991)

P. COMON, *Independent component analysis*, Int. Signal Processing Workshop on High Order Statistics Chamrousse (1991) 111-120

D. DACUNHA-CASTELLE, F. GAMBOA, *Maximum d'entropie et problèmes des moments*, Ann. Inst. Henri Poincaré 26 (4) 567-596 (1990)

G. DESODT, D. MULLER, *Complex independent components analysis applied to the separation of radar signals*, Int. Conf. EUSIPCO-90, Barcelona.

D. L. DONOHO, *Super-resolution via sparsity constraints*, Technical report, Dpt of Statistics, University of California, Berkeley (1990)

D. L. DONOHO, I. M. JOHNSTONE, J. C. HOCH, A. S. STERN, *Maximum entropy and the nearly black object*, J. R. Statist. Soc. B 53 (3) 1-26 (1991)

D. L. DONOHO, P. B. STARK, *Uncertainty principles and signal recovery*, SIAM J. Appli. Math., 49 (3) 906-931 (1989)

J. DOREY, G. GARNIER, G. AUVRAY, *RIAS, Synthetic Impulse and Antenna Radar*, Int. Conference on Radar, Paris (1989) vol. 2, 556-562

E. FEIG, F. A. GRUNBAUM, *Tomographic methods in range- Doppler radar*, Inverse Problems 2, 185-195, (1986)

P. FLANDRIN, *Maximum signal energy concentration in a Time Frequency Domain*, IEEE Int. Conf. on Acoust, Speech and Signal Proc. ICASSP-88 New-York 2176-2179

P. FLANDRIN, F. MAGAND, M. ZAKHARIA, *Generalized target description and wavelet decomposition*, IEEE Trans on Acoust, Speech and Signal Proc. ASSP-38 (2), 350-352 (1990)

D. L. FRIED, *Least square fitting a wavefront distorsion estimate to an array of phase difference measurements*, J. Opt. Soc Am 67, 370, (1977)

J. J. FUCHS, *State-space modeling and estimation of time differences of arrival*, IEEE Trans. on Acoust. Speech and Signal Proc. ASSP 34 (2), 232-244, (1986)

F. GAMBOA, E. GASSIAT, *M.E.M. techniques for solving moment problems*, prépublications Université de Paris Sud, Mathématiques, 91-23

F. GAMBOA, E. GASSIAT, *Maximum d'entropie et problème des moments, cas multidimensionnel*, Probability and Mathematical Statistic (1990)

D.T. G. JESSING, J. HJELMSTAD, *Artificial perception. Characterization and recognition of objects and scattering surfaces based on electromagnetic waves. Target Adaptive Matched Illumination Radar*. Int. Conference on Radar, Paris (1989), vol 1, 115-126

G. JACOVETTI, A. NERI, *A Bayesian approach to 2D non minimum phase AR identification*, Proc V ASSP workshop on spectrum estimation and modeling, Rochester, New-York (1990) 79-83

C. JUTTEN, J. HERAULT, *Blind separation of sources*, Signal Processing 24 (1) (1991)

L. KOPP, M. AIACH, *Spatial processing with distorted arrays*, GRETSI - Huitième colloque sur le traitement du signal et ses applications 365-370 (1981)

J. L. LACOUME, P. RUIZ, *Extraction of independent components from correlated inputs, a solution based on cumulants*, Proc. Workshop Higher Order Spectral Analysis, Vail, Colorado (1989) 146-151

A. LANNES, *Backprojection mechanisms in phase closure imaging. Bispectral analysis of the phase restoration process*, Experimental Astronomy 1 (1989) 47-76

N. LERNER, *Wick-Wigner functions and tomographic methods*, SIAM J. Math Anal. 21 (4) 1083-1092 (1990)

A. W. LOHMANN, G. WEIGELT, B. WIRNITZER, *Speckle masking in astronomy : triple correlation theory and applications*, Applied Optics 22 (24) 4028-4037 (1983)

W. MILLER Jr., *Topics in harmonic analysis with applications to radar and sonar*, Radar and sonar part I, Springer-Verlag (1991) 66-168

R. A. MULLER, A BUFFINGTON, *Real-time correction of atmospherically degraded telescope images through image sharpening*, J. Opt. Soc. America 64 (9) 1200-1210 (1974)

J. MUNIER, *Identification of correlated and distorted wavefronts*, Traitement du Signal 4 (4), 281-296, (1987)

J. NAVAZA, *On the maximum-entropy estimate of the electron density function*, Acta cryst A 41, 232-244 (1985)

M. NICOLAS, C. PIROLLI, *Balayage rapide par traitement de signal*, L'onde électrique 49 (2) (1969)

M. NICOLAS, G. VEZZOSI, *Localisation of broasband sources with an array of unknown geometry*, GRETSI (1989)

R. J. NOLL, *Zernike polynomicals and atmospheric turbulence*, J. Opt. Soc. Am. 66, 207, (1976)

A. PAPOULIS, *Signal Analysis*, Mc Graw Hill (1984)

C. PRATI, *Autofocusing synthetic aperture radar images*, Int. conference on radar Paris (1989) Vol 2, 314-319

J. PRIMOT, G. ROUSSET, J. C. FONTANELLA, *Deconvolution from wavefront sensing : a new technique for compensating turbulence-degraded images*, J. Opt. Soc. America (1990)

W. H. SOUTHWELL, *Wavefront estimation from wavefront slope measurements*, J. Opt. Soc. Am. 70, 998, (1990)

B.D. STERNBERG, *On the design of a radio-camera for high resolution in microwave imaging*, J. Franklin Inst. (1973)

G. VEZZOSI, *Estimation of phase angles from the cross spectral matrix*, IEEE Trans on Acoust, Speech and Signal Proc. ASSP - 34 (3) 405-422 (1986)

C. H. WILCOX, *The synthesis problem for radar ambiguity functions*, Radar and sonar part I, Springen-Verlag (1991) 229-259

# Chapter 16
# MASSIVELY PARALLEL COMPUTING: STATUS AND PROSPECTS

**D. J. WALLACE**
University of Edinburgh
Scotland

**1. Introduction.** At the simplest level, the conventional serial approach to computing involves breaking the problem of interest down into a sequence of steps which are coded in a standard language such as C or Fortran; the resulting statements are then performed in sequence by a single processor. By contrast, in parallel computing, one first identifies those parts of the computation which can be performed simultaneously, and organises the computation over a collection of communicating processors.

Parallel hardware architectures can be characterised under a number of different headings. Flynn's classification [8] is based on the instruction and data streams in the machine: are identical instructions broadcast to each processor (Single Instruction-stream Multiple Data-stream SIMD) or is each processor independently programmable (Multiple Instruction-stream Multiple Data-stream MIMD)? Do the processors access a global shared memory (through multiple ports, or bus or switch), or is there distributed memory, with each processor having its own local memory? In the distributed memory case, the local data of any processor must be accessible by any other: what is the topology of processor interconnection? Is there a hidden level of parallelism in each processor through vector or long instruction word pipelines? While these, and other, features have been critical issues for the applications programmer, the more relevant issue of software models and environments is beginning to crystallise now, and it is this aspect which we would like to emphasise in this paper; it is discussed in section 2. Those interested in an overview of parallel systems are recommended to consult Hockney and Jesshope [16] and Trew and Wilson [31]. Among the many books on parallel computing, Fox et al [10] is a particularly useful reference, as is the collection of reviews, bibliography and indexes, primarily to Computer Science work, in the ACM Resources in Computing series [9].

The potential advantages of parallel computing are many. In principle, the power of the system scales as the number of processors exploited, and (at least in distributed memory) the von Neumann bottleneck of data access in serial computing is removed, as each processor can access local data, so that the bandwidth can also scale with the power of the system. The technology can thus enable some problems to be tackled which would otherwise be impossible, either because of their real-time requirements, or because of the scale of the computation. Parallel machines also generally have better cost performance and much lower memory costs than traditional supercomputers, because they can be built from mass-produced commodity silicon chips.

Many problems in industrial and applied mathematics have enormous potential parallelism. They may contain many independent tasks, which can be farmed out to the processors; or have a spatial structure which is naturally decomposed, in the framework of eg finite difference or finite elements, into domains which are allocated to different processors; or the algorithm may decompose functionally into different components. These ideas are not new. In 1842, Menabrea [19] discussed how the computation of mathematical tables could be done on an array of mechanical computers, such as Babbage's Analytical Engine [1,2]; see also Morrison and Morrison [21]. In his remarkable text [27] on Weather Prediction by Numerical Process, published in 1922, Lewis Fry Richardson envisages a machine of 64,000 computers exploiting spatial and fucnctional decomposition to "race the weather for the whole globe"; his computers were people of course in those days.

So why have computational science and engineering remained dominated by serial computing for so long? There are many factors. The first is the existence of a standard programming paradigm, based on the von Neumann model, which has permitted the emergence of stable serial languages in which a huge body of tools and applications software has been developed. By contrast, no single programming model has yet emerged for parallel computing, so the software environment has been to date relatively poor, without the standards of a mature technology. The second is the astonishing increase in conventional computer performance, sustained by a combination of increases in intrinsic speed of components and improvements in VLSI design, so that single silicon chips (the Intel i860 is an example) now exist with a peak performance of 80 million floating point operations per second (Mflops). Coupled with improvements in (serial and vector) algorithms, this increase has met both the aspirations of many application scientists and matched the funds available to them.

The goal of this talk is to provide some insight into the accepted wisdom that increasing levels of parallelism will become the norm in high performance computing systems during the 90s, in systems from workstations to the highest performance supercomputers. I will begin by further exploring some of the software and environment issues, give a brief description of Edinburgh Parallel Computing Centre as an illustration of current parallel computing resources and activities, give a glimpse into future technology, and close with a few examples of real applications in which massively parallel systems are now being sucessfully exploited.

**2. Algorithms, software and programming models.** As discussed in the introduction, the take-up of massively parallel computing is now gated by these issues. In this section we illustrate the challenges from two points of view. The first revolves around domain (or spatial) decomposition. As in Richardson's 'race the weather' example, in principle this provides embarrassingly massive parallelism in all problems involving flow, diffusion, properties of mechanical structures etc. In practice, the existence of sophisticated serial algorithms make competitive parallelism more difficult to achieve. The second is the more general question of practical programming models, which are the primary requirement for a stable software industry. We include also a parable to point out that progress in general algorithm development can also be stimulated by the availability of parallel systems.

**2.1 Beyond naïve domain decomposition.** Explicit time-stepping, updating, or relaxation approaches exploit the parallelism of nature and are natural to implement on a parallel machine. The main message of this section is that there are now many examples of algorithms which are intrinsically faster than these. In particular, these faster algorithms have the attraction of much better scaling behaviour than the naïve algorithms as the size of the system increases.

The classic example is provided by systems with significant phenomena occurring on many length-scales. In such problems, the time-step must be adjusted to be small enough to ensure stability and accuracy in the integration of the smallest length-scales in the problem, which typically prescribe the scale of the discretisation. For large features of scale-size $L$ greater than this, significant evolution is achieved only after $L$ time-steps, or (sometimes approximately) $L^2$, in the case of diffusive processes. The same correlation-time effects are evident in stochastic (eg Monte Carlo) updating. The prime example is at continuous phase transitions, or critical points, where the spatial correlation length diverges, and the correlation time with it (for a general introduction see Ma [18] and Bruce and Wallace [5]); this critical slowing down is a real physical phenomenon, requiring equilibration times of order hours or days in laboratory experiments.

The relevant point for this discussion is that there are now cluster updating algorithms which largely eliminate the critical slowing down characteristic of the standard Monte Carlo update procedure [28,29]. One may refer to these as 'supernatural acceleration' algorithms, in the sense that they compress the range of physical time-scales in the simulation. It is imperative therefore that future parallel systems are able to support these algorithms efficiently, or else the performance advantage of the parallel system will be seriously eroded. It is not untypical of these algorithms that, by updating clusters or coarse-grained variables, they reduce the level of naïvely available parallelism.

Some other examples of algorithms of this kind (in the widest sense) will underline the importance of this point (Many of them are discussed in detail in *Numerical Recipes* by Press et al [25]):

- Conjugate gradient can have many advantages over, for example, Gauss-Seidel or Jacobi relaxation. The problem for parallelisation is not so much the global sums required in the conjugate gradient algorithm; rather it is the implementation of preconditioning, which can accelerate convergence dramatically and which tends to be of a serial nature.

- It is well known in engineering simulations of stiff systems (i.e. systems with a large range of time constants) that implicit integration rather than explicit time-stepping can enable much larger time-steps to be used while still maintaining stability.

- Multi-grid [26] and cluster algorithms (as well as fast Fourier transform (FFT) in some cases) can all be used to address the problem of critical slowing down and of stiff systems. In the case of long-range interactions (such as Coulomb or gravitational forces), the same techniques can also be used to reduce the computational complexity for $N$ degrees of freedom from $O(N^2)$ to $O(N\ln(N))$ or even $O(N)$.

- Finally, we note that there is a wide range of problems in which naïve domain decomposition appears natural, but would be inefficient, because the computations are localised or inhomogeneous across the data. Examples include:
  - the need for irregular grids in e.g. computational fluid dynamics;
  - the existence of locally different time-scales in the simulation of stellar clusters;
  - the tracking and display of surfaces in three-dimensional data-sets [23].

The last is an extreme example. One wants to 'ping' a point on the surface, from which the surface feature detector can create a wave-front spreading through the data-set over the surface. At any instant therefore, the computation takes place only on a one-dimensional subset of the data (the wave-front at that instant), which cannot be known in advance. A naïve domain decomposition of the data-set, with $M$ grains being mapped onto $M$ processors would inevitably be inefficient, with absolute failure to achieve any speed up over serial code in many instances. One trick here is to use scattered domain decomposition, in which the data-set is divided into a much larger number of much smaller domains, which are then scattered into the processors, so that there is an increasing probability of achieving statistical load-balancing. Of course, if too high a granularity is chosen, the computation time will increase because of increased overheads of communication. The need in this application for a general process-to-process message passing system is self-evident.

Figure 1 shows the results of such computations on the Meiko Computing Surface, a transputer array, for NMR tomographic data with several hundred thousand voxels, running on a 64-node machine. Naïve domain decomposition would imply a grain size of around 10,000 voxels. In fact the optimum grain size is a few hundred. In addition to NMR data, the code has been successfully applied in parallel with a distributed graphics pipeline [24,35] to display for example Riedel shears in experiments [22] to model earthquake faults (the 3-d visualisation is reproduced in [34]), and to display 3-d structures in simulations of percolation in soils [14].

With grain sizes this small, the need for a message passing system with low latency is manifest. The particular system used here is Tiny [6], written by Lyndon Clarke, and available on a range of transputer systems. This utility explores the multi-processor configuration at run-time and sets up process-to-process communications and broadcasts. Code does not have to be recompiled to run on different configurations. The utility also has fault tolerant capabilities; provided a booting path is available, efficient routing between pairs will be set up even if some of the links are defective - they will simply not be utilised in setting up the routing tables. It can be called from Fortran or C as well as used in an occam program. An option to provide deadlock-free communication is also provided.

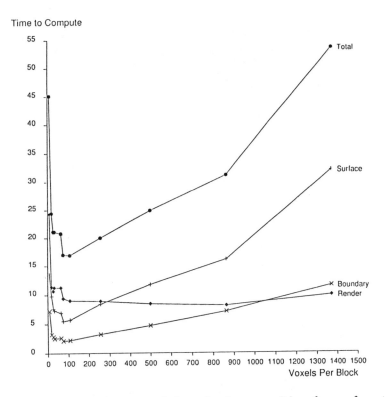

Figure 1: Performance of scattered domain decomposition for surface tracking, as a function of the grain size, illustrating the competition between communication overheads for small grain size and load imbalance for large grain size.

From the user point of view, the emergence of a standard message passing interface building on the experience of systems like Express, Meiko's CS-Tools and Tiny is important to ensure portability.

To summarise, parallelisation must be achieved which is competitive with the best serial algorithms; while this is not necessarily ruled out in any of the cases above, it may be non-trivial, and will almost inevitably place additional demands on the communication capability of future multi-processor systems.

**2.2 Algorithms: a parable.** Although the efficient parallel implementation of existing algorithms can present a major challenge, it is worth pointing out that the stimulation of thinking parallel can also lead to intrinsically faster algorithms. We illustrate this in the context of the Hybrid Monte Carlo Algorithm (HMC).

In many simulations we need to explore a phase space of a large number of variables, and to work out averages of quantities of interest over this space. Behind the movement of the system through this space there may be detailed dynamics in which we are explicitly interested, or the problem may simply be one of calculating the phase-space average as quickly as possible. The HMC algorithm [7] is of the latter kind; it is helpful to think of it in the context of parallelisation, which was a major motivation in its discovery.

We have already mentioned the importance of the Monte Carlo algorithm. From the viewpoint of parallel computing, it suffers from the defect of implementations being dependent on the range of the interactions, since variables cannot be simultaneously updated if they interact directly with one another. In the case of real-valued variables, the Langevin equation (first order in time, including a stochastic force) can be thought of as a Monte Carlo update which allows infinitesimal changes in the variables at each update. It has the advantage that the coupled differential equations for each of the variables are intrinsically parallel. Molecular dynamics is also intrinsically parallel, and uses the 'real' forces and Newton's equations to drive the system round its phase space, and hence to obtain the averages of interest; for weakly coupled or approximately harmonic systems, there is the disadvantage that the real motion then tends to follow closed orbits in the phase space (ellipses in the case of a harmonic oscillator system) and unless the system is kicked into many different orbits, the phase space is explored very slowly.

A problem common to both Langevin and molecular dynamics approaches is that there are systematic errors introduced by the discretisation of time and hence it is in principle necessary to perform several simulations for different time-steps, and to extrapolate the results to zero step size. There is also the difficulty that as the step size is reduced, the speed at which phase space is traversed (in computer time) is correspondingly reduced.

The HMC algorithm combines these ideas: it is intrinsically parallel, has no systematic error associated with the step-size and allows large time-steps which permit a more rapid exploration of phase space than the above methods.

The idea is a hybrid of the Monte Carlo and molecular dynamics approaches, as follows. Consider a molecular dynamics simulation. For any finite time step, at each update there will be a small change in the total energy of the system. These violations of energy conservation accumulate, and eventually invalidate a naïve molecular dynamics approach. In the HMC approach one utilises this drift off the 'energy shell' by performing a **global** Monte Carlo accept or reject of the whole system after $n$ time-steps. The global accept/reject ensures that there is no systematic dependence on the step-size, which can thus be tuned to give an optimal global acceptance rate. The molecular dynamics calculation is then restarted, with **new** velocities drawn from the Gaussian distribution appropriate to the conventional kinetic term in e.g. conventional kinetic theory. Thus as well as large time-steps, the simulation gets kicked in a big way at regular intervals. The method can produce an order of magnitude speed up, with improved volume dependence, so that the benefits become even larger the larger the system under study.

The method was first used in lattice field theory [7]. Figure 2 shows results of calculations in Quantum Electrodynamics (the theory of the electromagnetic force), illustrating the time-step ($\delta\tau$) independence of the HMC algorithm compared to the dependence in an earlier Hybrid algorithm. It has recently been shown to give an order of magnitude speed-up in molecular modelling, in a simulation of pancreatic trypsin inhibitor [4].

The morals to be drawn from this example include: we can expect rare but dramatic algorithmic gains; such developments may be stimulated by thinking in parallel; and they may be of wider applicability than the problem for which they are initially designed.

**2.3 Programming models.** In many respects this is the most fundamental prob-

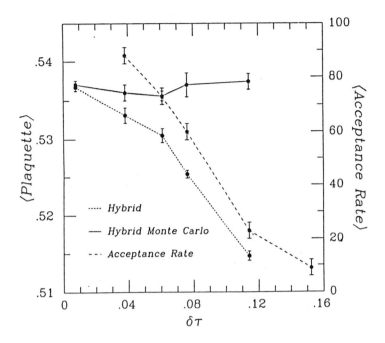

Figure 2: A correlation function, the average plaquette, as a function of the integration step-size $\delta\tau$ for Quantum Electrodynamics. The HMC results are statistically independent of the step-size, in contrast to the strong dependence in the earlier Hybrid method. The figure is from Duane et al [7]

lem facing parallel computing at present, because the emergence of stable accepted programming models is a necessary condition for software standards and portability. The issue is the conceptual framework within which the application programmer thinks and works, and the associated software interface which shields the details of the hardware architecture which are outlined in the Introduction. It is appropriate that this issue will replace the focus to date on the such details. Many aspects of the subject remain research activities, but the following general comments are probably not controversial.

First, it is clearly desirable that the model or interface should, from the programming point of view, be at as high a level as possible, with maximal abstraction from hardware. Even in serial programming, where one might think, for example, in terms of a functional specification of the equations, as opposed to a Fortran implementation, there are issues of efficiency. In parallel programming, the compromise between use of high level constructs and efficiency is much more acute. One ideal towards which one might strive is a model in which to the programmer the data-space looks like a homogeneous shared memory, both in the access mechanism and in the latency of access. The former is now possible. Achieving the latter efficiently is a research issue; even in shared memory machines it is non-trivial, increasing in difficulty as one moves from the less scalable CRAY architectures through more scalable bus-based systems (eg Sequent, Encore and Alliant) to swith-based systems, such as the BBN Butterfly. So in practice what can be expected in truly scalable distributed memory architectures?

The bottom line is a process-to-process message passing system, which provides transparent data access independently of the underlying hardware configuration or details of any through-routing etc. Specification of specific hardware channels should be unnecessary: by analogy with a telephone call, we don't want to know all the exchanges through which the call is connected. Such systems are available now on distributed memory MIMD machines such as the Intel iPSC and Meiko Computing Surface, but there is as yet no standard, and communications latency and bandwidth are typically not well balanced against the computational power of each node, so that the efficiency is low unless there is high achieved locality of data-processing. New products currently being or about to be announced should improve this aspect.

There are two rather obvious avenues above the message passing level. One revolves around Fortran90 [20] and extensions, which is appropriate for the array-based problems common in numerical computation. The potential of this approach was recognised by ICL in the Distributed Array Processor (DAP) now manufactured by Active Memory Technology, and further developed by Thinking Machines Corporation for the Connection Machine and by Maspar. All these systems are SIMD machines with regular processor topology, and acceptable efficiencies (say of the order of a $few \times 10\%$ of peak) are achieved on a range of "data-parallel" problems. In principle such an environment can be implemented on distributed memory MIMD systems, but it requires more system software investment; examples may emerge over the next two years or so.

The recent announcement by Thinking Machines of the CM5, with its dual MIMD and SIMD characteristics which permit it to support both message passing and array-based programming models, is an interesting marker of likely future trends.

The second area in which one may expect to see developments is in the provision of virtual shared memory environments on distributed memory machines. There are many potential candidates: the Bulk Synchronous Parallel model of Valiant and coworkers, based on the P-RAM and related ideas [32,33]; the Linda model of Gelernter (see [13] and references therein), and so on. None of these is in any way yet a de-facto standard, and this is still an active area for fundamental research. The issue of efficiency is particularly relevant. Scalable commercial systems providing this functionality have yet to prove themselves. The BBN effort on the Butterfly and the TC2000 has been discontinued, and the Kendall Square system is only now appearing. The goal is an important one for the maturing of parallel computing technology.

**3. Edinburgh Parallel Computing Centre.** Work on applications of parallel computing at Edinburgh dates back to 1980, when members of the Physics Department began exploiting the ICL DAP at Queen Mary College, London. Now the activites represent a significant focus in Europe, with some 35 staff on research and development contracts. A brief overview of these may therefore give some impression of the current status of parallel computing.

The procurement of large state-of-the-art parallel systems has been an integral part of the strategy from the earliest work. Access to such systems is essential to gain experience and it attracts strong groups. The procurements were driven primarily by Physicists, who recognised that parallel systems were the only way forward on realisable budgets. Capital funding came primarily from Government Agencies such as the Department of Trade and Industry (DTI), the Science and Engineering Research Council (SERC) and the Computer Board (now part of the Universities Funding Council), with essential contributions from the vendors concerned. Edinburgh University Computing Service (EUCS) plays a vital role in hosting the systems and providing networked service on them. The net effect has been a positive feedback loop of broadening interest and utilisation. The major systems presently include:

- A Meiko Computing Surface with more than 400 nodes (Inmos T800 transputers) and 1.8 Gbytes of memory; this MIMD system provides a national multi-user resource with an active user community of around 250, claiming around 1.5M processor hours per year.

- A 4K-processor AMT DAP 608; a 1K processor DAP510 is hosted in the Biocomputing Research Unit, and provides a national service for molecular sequencing.

- A 64-node i860 Computing Surface with 5 Gflops peak performance and 1 Gbyte of memory; this system is hosted at Edinburgh primarily on behalf of two UK "Grand Challenge" collaborations, in Quantum chromodynamics (QCD) and material structure calculations using the Car-Parrinello method. The first sustains up to 1.5 Gflops, the second around 5 times (single processor) Cray XMP.

- A 16K processor CM-200 was installed by Thinking Machines in September 1991; it has a peak of 8 Gflops, 0.5 Gbytes of memory and a 10 Gbyte data vault.

A significant feature of these systems is their relatively large memory. It is also remarkable that high levels of utilisation of all machines have been obtained with a maximum of 4 people in service provision and user support (excluding operator cover and infrastructure provide by EUCS as part of their University service, and the substantial informal user support which has been provided by other experienced personnel in the Centre). The burden of service provision is being significantly eased as SunOS becomes the standard operating system in the Centre. The annual Project Directory summarises around 100 projects; there is no doubt that this would be even higher if more technical user support were available.

The Centre has a number of major programmes covering research, development, technology transfer and contract work for industrial applications. To date most of the recurrent support has come from industry, through Affiliation and Subscription to the Key Technology Programmes (currently in message passing, networked visualisation and parallel utility libraries), as well as through contracts. Recently the Centre has been awarded $3.5M under the DTI/SERC Parallel Applications Programme, to stimulate a $9M industrial collaboration programme over 4 years. The development with industrial funding of a Summer Scholarship Programme for undergraduates has been particularly rewarding, with 25 awards in 1991; applications are considered from around the world. Further details of these and other activities are contained in the quarterly Newsletter and Annual Report available from EPCC at the University of Edinburgh.

**4. A glimpse into the future.** The only certainty about future predictions in high performance computing is that they will surely be wrong. Even the extrapolation of trends is fraught with uncertainty, because of the many different factors which have influenced the subject in the past thirty years.

Nevertheless, as a guide it is worth noting the consistent exponential increase in performance by close to two orders of magnitude per decade in the systems utilised by the UK Meteorological Office [36] since 1953. Broadly half of this performance increase has come from intrinsic speed of devices (including valves to solid state) and half from improved circuit design culminating in the still-advancing silicon VLSI (including very significant hidden parallelism in eg individual integer and floating point operations).

This improvement has also resulted from modest levels of explicit parallelism, exemplified for example by the current 8-processor YMP of CRAY. Limited parallelism has been an essential feature in a number of domains other than numerical computation, for example for multi-user Unix throughput (eg Sequent), for large data-base systems (eg Terradata), and in graphics workstations (eg Silicon Graphics).

So what can one envisage for this decade?

First, while it seems increasingly difficult to push down the cycle time of conventional supercomputer systems, there appears to be no foreseeable limit (on this timescale) to the remarakable march of VLSI on silicon. Feature sizes are set to decrease from around 1 micron to 0.2 micron, or perhaps even 0.1 micron, implying an increase in component density by a factor of more than 25. At the same time, clock speeds can be expected to improve by a factor of three or more, and chip area by perhaps four. It is also suggested that interconnect bandwidth will increase to over 1 Gbit/sec/wire. The implications are therefore that the trend in component performance will continue, and even accelerate, with single chips containing multi-processors and other concurrent functionality. This is reflected in the current improvements in workstation performance. It takes no account of new technologies such as optoelectronics.

Second, there can be no doubt that the scale of explicit parallelism in high performance systems will increase dramatically, and it seems likely that Teraflops systems will be available from a number of vendors by around 1995, with commensurate memory increases (100 Gbytes or more). There seems no reason why another factor of at least ten should not be available by the year 2000. Closing, or at least controlling, the potentially increasing gulf between peak and realised performance will be an important factor for commercial success.

Finally, one should not forget the impact of (unpredictable) algorithm development; as noted in the US federal report [Grand Challenges: High Performance Computing and Comunications, The FY 1992 U.S. Research and Development Programme], software and algorithm improvements actually made a greater contribution to wall-clock speed-up than hardware improvements in the period 1970 – 1990. The importance of this aspect will increase, since improved scaling behaviour of algorithms brings greater rewards the larger the simulation.

All the pointers are therefore to a decade of even more remarkable progress [12], and this is widely recognised in Government reports and funding proposals [see for example the US HPCC report above, and the Report of the Working Group on High-Performance Computing, Commission of the European Communities, 1991]. But even with advanced algorithms such as Hybrid Monte Carlo, where the computation in the theory of the strong nuclear force (QCD) scales as roughly $N^{10}$ (where $N^{-1}$ measures the linear resolution – previous algorithms were even worse!) one may safely predict that demand will always outstrip supply.

**5. Some applications.** A comprehensive discussion of applications is clearly beyond the scope of this article. Some impression of their breadth and scale can be obtained for example from the review by Fox of activities at Caltech [11], from the EPCC Annual Report and Project Directory, and from the "Navigator" data-base of Thinking Machines.

Instead we limit ourselves to seven examples chosen from a range of companies and industries. Many provide excellent visual displays.

**Radiosity** [15,17] can produce 3-d visualisation of very high quality. The method is based on (i) dividing the surfaces in the scene into N patches and determining line of sight visibility between pairs of patches, (ii) solving the N simultaneous equations for the brightness of each patch, and (iii) rendering the scene using z-buffering or ray-casting. Rendering the scene from different view-points requires only stage three, and changing the lighting only stages two and three. The natural parallelism in the calculations of the form-factors (the coefficients relating the contribution of the illumination of one patch due to the light from another) and in the iteration of the matrix of these form factors to determine the solution enables the method to run effectively on hundreds of processors. It has been developed as a demonstrator and package on the Parsys transputer system.

**Car-body design:** Designers evaluate colours and shapes very strictly, and their requirements are very severe, including the checking of shapes and colours in various simulated environments and weather conditions. A 256-node Meiko Computing Surface was procured by Toyota to demonstrate successfully that parallel computing offered sufficient power to meet these demanding criteria, in a cost-effective way. Their work involved development of ray tracing to incorporate diffuse and specular reflection, direct sunlight and various conditions of sky lighting, as well as regular transmission through glass [30]. The vehicle configuration was prepared by the Toyota Styling CAD system. The values generated by the model are in excellent agreement with measurements on actual cars.

**Cellular automata** use a regular lattice of sites each with a current state and a rule determining how the state changes with time and with the states of the other local sites (for an introduction and further references, see Boghosian[3]). A cellular automaton model for fluid flow models the movement of simple particles through the lattice with collisions occuring at each site so that the macroscopic behaviour required in the fluid is correctly captured by the mean momenta averaged over suitably large cells of the lattice. Cellular automata are intrinsically parallel, are natural to implement on an array of processors, and have been exploited on a wide range of machines. Perhaps the most significant commercial sale in this connection has been the installation of a 400 processor Parsytec system in Shell Research in Amsterdam

(KSLA) whose dedicated application areas include cellular automata studies of flow and diffusion through porous media.

**EGS4** is a Monte Carlo simulation package for electron gamma showers in a material geometry specified by the user's application, which may cover fields from high energy physics to radiation dosimetry. It provides straightforward task-farming to improve the statistics, and does not gain significantly from vectorisation. On a 128-node BBN TC2000, an HEP calorimeter application generates more than 35000 showers per minute, which is claimed to be some 35 times faster than a Cray YMP (single processor).

**Seismic processing:** The oil and gas industry is one of the heaviest industrial users of supercomputing cycles, with applications which include reservoir modelling and analysis of seismic data. Both areas are potential candidates for massively parallel machines, and significant progress has already been made in the latter. In addition to the Mobil seismic modelling code which won the Gordon Bell prize for absolute performance on a CM-2 in 1990 (with a sustained performance of more than 5 Gflops), Schlumberger Austin are now providing a commercial bureau service for 3-d seismic migration on a Connection Machine, because the system is cheaper and faster, and has larger memory and higher availability than a conventional supercomputer; the time-scale from prototype, through productising of code and system procurement, to beginning of the commercial service was around one year.

**High temperature superconductivity:** A 128-node Intel iPSC860 won the 1990 Gordon Bell Award for best price/performance. The code calculates the electronic structure of alloys and materials using the KKR-CPA method. In the particular application in superconductivity, the system at Oak Ridge achieved 2.6 Gflops, in comparison with around 1.6 Gflops on code optimised for an 8-processor Cray YMP.

**Text retrieval:** Several manufacturers of distributed memory systems (Meiko, Ncube and Parsys among them) have announced an Oracle product, and the data-vault of the Connection Machine gives impressive data-base demonstrations. The particular application on which we focus here is provided by the AMT DAP; the customer is a News Agency. The problem is a keyword search of a 50 Gbyte data-base, with up to 3000 active users seeking a sub-second response-time. The solution leaves the user interface and full text on the existing VAX host, with a compressed data-base on a large memory DAP. The DAP does the keyword search and gives pointers to articles back to the VAX which then provides the full text. The system responds to 50 queries per second, and provides a speed-up over a VAX 6000 by a factor of about 400.

These examples fail to do justice to the wealth of work currently being undertaken. They are not even illustrative, in the sense that this paper cannot convey the visualisation quality which they deliver. I hope nevertheless that they confirm the impression of an emerging technology which will grow in importance in the coming years.

**Acknowledgements.** I am very grateful for helpful and stimulating discussions with colleagues in the Physics Department, in Edinburgh Parallel Computing Centre and elsewhere, without whose work this talk could not have been given; I thank M. G. Norman particularly for comments on the manuscript. I thank the vendors for their prompt and helpful responses to my request for information about applications, and apologise to them that I have been able to include so little. Edinburgh Parallel Computing Centre is supported by major grants and contracts from the Department of Trade and Industry, the Information Systems Committee of the Universities Funding Council, the Science and Engineering Research Council, and industry.

## REFERENCES

1. C. BABBAGE, *A note respecting the application of machinery to the calculation of astronomical tables*, Mem. Astron. Soc. 1 (1922) p. 309.

2. H. P. BABBAGE, *Babbage's analytical engine*, Mon. Not. Roy. Astron. Soc. 70 (1910) pp. 517-526, 645.

3. B. M. BOGHOSIAN, *Lattice gases illustrate the power of cellular automata in physics*, Computers in Physics, Nov 1991, pp. 585-590.

4. A. BRASS, B. J. PENDLETON, Y. CHEN and B. ROBSON, *Hybrid Monte Carlo Simulations of Pancreatic Trypsin Inhibitor*, Manchester preprint (Biochemistry and Molecular Biology) 1990.

5. A. D. BRUCE and D. J. WALLACE, *Critical point phenomena: universal physics on large length scales*, in The New Physics, P.Davies ed., Cambridge University Press, Cambridge (1989), pp. 236-267.

6. L. J. CLARKE and G. V. WILSON, *Tiny: an efficient routing harness for the Inmos transputer*, Concurrency: Practice and Experience 3 (1991) pp. 221-245.

7. S. DUANE, A. D. KENNEDY, B. J. PENDLETON and D. ROWETH, *Hybrid Monte Carlo*, Phys. Lett. B195 (1987) pp. 216-222.

8. M. J. FLYNN, *Some computer organisations and their effectiveness*, IEEE Trans. Comput. C-21 (1972) pp. 948-960.

9. E. A. FOX, editor in chief, ACM Resources in Computing, *Resources in parallel and concurrent systems*, ACM Press, New York, 1991.

10. G. C. FOX, M. A. JOHNSON, G. A. LYZENGA, S. W. OTTO, J. SALMON and D. WALKER, *Solving Scientific Problems on Concurrent Processors*, Prentice Hall, New Jersey, 1988.

11. G. C. FOX, *Parallel computing comes of age: supercomputer level parallel computations at Caltech*, Concurrency: Practiceand Experience, 1 (1989) 63-103.

12. GARTNER GROUP Inc, *High Performance Computing and Communications: Investment in American competitiveness*, prepared for U.S. Department of Energy and Los Alamos National Laboratory, 1991.

13. D. GELERNTER, *Information management in Linda* in Parallel processing and artificial intelligence, M. Reeve and S. E. Zenith eds., Wiley, Chichester,1989, pp. 23-35.

14. C. A. GLASBEY and G. W. Horgan, *Computer visions of the future*, in Scottish Agricultural Statistics Service Annual Report, 1991, pp. 103-106, available from SASS, University of Edinburgh.

15. C. M. GORAL, K. E. TORRANCE, D. P. GREENBERG and B. BATTAILE, *Modelling the interaction of light between diffuse surfaces*, Computer Graphics 18 (1984) no.3.

16. R. W. HOCKNEY and C. R. JESSHOPE, *Parallel computers 2*, Adam Hilger, Bristol, 1988.

17. D. S. IMMEL, M. F. COHEN and D. P. GREENBERG, *A radiosity method for non-diffuse environments*, Computer Graphics 20 (1986) no.6.

18. S. K. MA, *Modern theory of critical phenomena*, W. A. Benjamin, Massachusetts, 1976.

19. L. F. MENABREA, *Sketch of the analytical engine invented by Charles Babbage*, Bibliotheque Universelle de Geneve, October, 1842.

20. M. METCALF and J. REID, *Fortran 90 Explained*, Clarendon Press, Oxford, 1989.

21. P. MORRISON and E. MORRISON, *Charles Babbage and his calculating engines*, Dover, New York, 1961, p. 244.

22. M. A. NAYLOR, G. MANDL and C. H. K. SIJBESTEIJN, *Fault geometries in basement-induced wrench faulting under different initial stress states*, J. Struct. Geol. 8 (1986) pp. 737-52.

23. M. G. NORMAN and R. B. FISHER, *Surface tracking in three-dimensional data sets using a generalised message-passing sub-system*, in Developments using occam: Proc. of the 8th Technical Meeting of the Occam User Group, IOS, Amsterdam 1988, pp. 77-82.

24. M. G. NORMAN, *A Parallel 3D Graphics Utility for Parallel Programs*, in Proc., Conf. on Applications of Transputers, (Liverpool), eds T. L. Freeman and C. Phillips, IOS, Amsterdam (1990).

25. W. H. PRESS, B. P. FLANNERY, S. A. TEUKOLSKY and W. T. VETTERLING, *Numerical Recipes: the art of scientific computing*, Cambridge University Press, Cambridge, 1988, and related volumes.

26. W. H. PRESS and S. A. TEUKOLSKY, *Multigrid methods for boundary value problems*, Computers in Physics (1991) pp. 514-519.

27. L. F. RICHARDSON, *Weather Prediction by Numerical Process*, Cambridge University Press, 1922, (republished by Dover Publications, New York, 1965) Chapter 11.2.

28. R. H. SWENDSEN and J.-S. WANG, *Non-universal critical dynamics in Monte-Carlo simulations*, Phys. Rev. Lett. 58 (1987) pp. 86-88.

29. R. H. SWENDSEN, J.-S. WANG and A. M. Ferrenberg, *New Monte-Carlo methods for improved efficiency of computer simulations in statistical mechanics*, Physics Department preprint, Carnegie-Mellon University (1991).

30. A. TAKAGI, H. TAKAOKA, T. OSHIMA and Y. OGATA, *Accurate rendering technique based on colorimetric conception*, Computer Graphics 24 (1990) no.4.

31. A. S. TREW and G. V. WILSON, *Past, Present, Parallel: A survey of available parallel computing systems*, Springer-Verlag, London, 1991.

32. L. G. VALIANT, *Optimally Universal Parallel Computers*, in Scientific Applications of Multiprocessors eds. R. J. Elliott and C. A. R. Hoare, Prentice Hall International Series in Computer Science 1989, pp. 17-20.

33. L. G. VALIANT, *Bulk-synchronous parallel computers*, in Parallel processing and artificial intelligence, M. Reeve and S. E. Zenith eds., Wiley, Chichester,1989, pp. 15-22.

34. D. J. WALLACE, *Supercomputing with transputers*, Computing Systems in Engineering, 1 (1990) pp. 131-141.

35. A. J. S. WILSON, J. G. MILLS and M. G. NORMAN, *Bodyscan: a transputer based 3D image analysis package*, in Transputer Applications 91, IOS Press (1991), pp. 130-135.

36. R. J. WILEY, *Parallel processing and numerical weather prediction*, in Proc. Second Intern. Specialist Seminar. on Design and Application of Parallel Digital Processors, IEE publications, London, 1992.

# Chapter 17
# HIERARCHICAL BASES

**HARRY YSERENTANT**
Universität Tübingen
Germany

**1. Introduction.** Today the idea of the *self-similarity of structures* plays a dominant role in the natural sciences and in mathematics. The physicist may think of renormalization techniques, the biologist of the beautiful pictures of plants and the models for their growth, that can be generated using Lindenmayer systems [31], the computer scientist of the principle of recursion, the reader of "Scientific American" perhaps of the beauty of fractals, and a numerical analyst like me of multigrid methods [20].

But probably most people are not aware the fact that many of these techniques have roots going back to the beginning of this century. Here, I shall describe an idea which can be traced back to a paper [19] of Georg Faber published 1909 in the "Mathematische Annalen".

Faber's idea, *the hierarchical representation of functions* can best be illustrated by a picture taken from his own paper; see. Fig. 1 for a facsimile.

Meanwhile Faber's idea spread out through many parts of applied mathematics. In computer graphics, it is used for the construction of "rough" curves and surfaces. The wavelet transformation [24] is a consequent extension of the idea which is widely used in signal analysis and image processing.

After a short introduction in Section 2, I shall restrict my attention to the use of Faber's technique for the construction of fast adaptive solvers for partial differential equations. This idea goes back to a paper [40] of Zienkiewicz, Kelly, Gago and Babuška who analyzed the one-dimensional case. I myself gave a complete theory for two space dimensions and have been able to construct extremely simple fast solvers for second order elliptic boundary value problems [34]. A related technique, which is very successful in practice, has been developed in a joint paper [7] with R. Bank and T. Dupont.

Today, subspace decompositions of finite element spaces like Faber's decomposition became an essential, if not the essential tool in the construction and analysis of fast solvers.

$$\delta_0 = f(1),$$
$$\delta_{\frac{1}{2}} = f\left(\frac{1}{2}\right) - \frac{f(0)+f(1)}{2},$$
$$\delta_{\frac{1}{4}} = f\left(\frac{1}{4}\right) - \frac{f(0)+f\left(\frac{1}{2}\right)}{2},$$
$$\delta_{\frac{3}{4}} = f\left(\frac{3}{4}\right) - \frac{f\left(\frac{1}{2}\right)+f(1)}{2},$$
(2) $$\delta_{\frac{1}{8}} = f\left(\frac{1}{8}\right) - \frac{f(0)+f\left(\frac{1}{4}\right)}{2},$$
$$\delta_{\frac{3}{8}} = f\left(\frac{3}{8}\right) - \frac{f\left(\frac{1}{4}\right)+f\left(\frac{2}{4}\right)}{2},$$
$$\delta_{\frac{5}{8}} = f\left(\frac{5}{8}\right) - \frac{f\left(\frac{2}{4}\right)+f\left(\frac{3}{4}\right)}{2},$$
$$\delta_{\frac{7}{8}} = f\left(\frac{7}{8}\right) - \frac{f\left(\frac{3}{4}\right)+f(1)}{2},$$
. . . . . . . . . . .

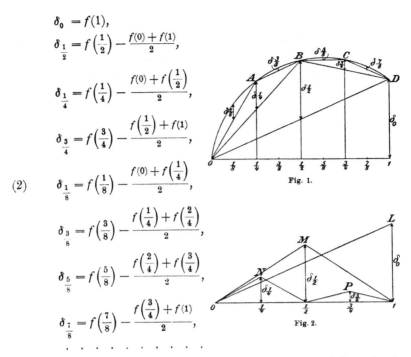

Fig. 1.

Fig. 2.

Allgemein ist $\delta_{\frac{2m+1}{2^n}}$ (wobei $m$ und $n$ ganze Zahlen sind und $2m+1<2^n$ ist) der Unterschied zwischen dem Funktionswert $f\left(\frac{2m+1}{2^n}\right)$ und dem arithmetischen Mittel $\frac{f\left(\frac{m}{2^{n-1}}\right)+f\left(\frac{m+1}{2^{n-1}}\right)}{2}$:

(3) $$\delta_{\frac{2m+1}{2^n}} = f\left(\frac{2m+1}{2^n}\right) - \frac{f\left(\frac{m}{2^{n-1}}\right)+f\left(\frac{m+1}{2^{n-1}}\right)}{2}.$$

Durch Auflösung der Gleichungen (2) kann man die Funktionswerte $f(x_i)$ (unter $x_i$ immer eine dyadisch rationale Zahl verstanden) durch die $\delta$ ausdrücken; man kann dies zunächst in *rekurrenter* Weise tun, indem man aus der ersten dieser Gleichungen den Funktionswert $f(1) = \delta_0$ entnimmt und dann mit Berücksichtigung von $f(0) = 0$ aus der zweiten und dritten der Gleichungen (2) die Funktionswerte $f\left(\frac{1}{4}\right)$ und $f\left(\frac{3}{4}\right)$ durch die $\delta$ ausdrückt, sodann mittels der nächsten vier Gleichungen $f\left(\frac{1}{8}\right), f\left(\frac{3}{8}\right), f\left(\frac{5}{8}\right), f\left(\frac{7}{8}\right)$ usw.

6*

FIG. 1:. *Faber's idea to represent functions*

## 2. Continuous 1D–functions.

Let $I$ be a fixed interval; without restriction we assume that $I = [0,1]$. As usual, let $C(I)$ be the space of the continuous functions $u : I \to \mathbb{R}$, and $C^q(I)$ be the space of $q$-times continuously differentiable functions.

We define a family of partitions $\mathcal{P}_0, \mathcal{P}_1, \mathcal{P}_2, \ldots$ of $I$. $\mathcal{P}_0$ consists of only one interval, $I$ itself. If one subdivides every interval in $\mathcal{P}_k$ into two subintervals of equal length, one obtains the intervals in $\mathcal{P}_{k+1}$. $\mathcal{N}_k$ is the set of the endpoints of the intervals in $\mathcal{P}_k$.

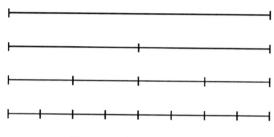

FIG. 2:. *The partitions $\mathcal{P}_k$*

$\mathcal{S}_k$ is the subspace of $C(I)$ consisting of the functions the restriction of which to the intervals in $\mathcal{P}_k$ is linear. $\mathcal{S}_k$ is a subspace of $\mathcal{S}_{k+1}$, and a function in $\mathcal{S}_k$ is uniquely determined by its values at the nodes in $\mathcal{N}_k$.

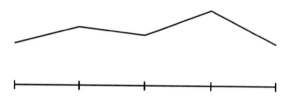

FIG. 3:. *A function in $\mathcal{S}_k$*

The interpolation operators $I_k : C(I) \to \mathcal{S}_k$ are given by

$$(2.2) \qquad (I_k u)(x) = u(x) , \quad x \in \mathcal{N}_k .$$

Faber's idea is to represent a function $u$ in terms of $I_0 u$ and the *differences*

$$(2.3) \qquad I_k u - I_{k-1} u , \quad k = 1, 2, \ldots .$$

These differences are functions in $\mathcal{S}_k$ and vanish at all nodes of $\mathcal{N}_{k-1}$. Therefore they can be represented by their values at the nodes $x \in \mathcal{N}_k \backslash \mathcal{N}_{k-1}$.

The interpolation error in $C(I)$ can be estimated as

$$(2.4) \qquad \|u - I_k u\|_{C(I)} \leq \omega(u, 2^{-k}) .$$

As the modulus of continuity $\omega(u, \delta)$ of a function $u \in C(I)$ tends to zero for $\delta \to 0+$, we have

$$(2.5) \qquad u = \lim_{j \to \infty} I_j u = \lim_{j \to \infty} \{ I_0 u + \sum_{k=1}^{j} (I_k u - I_{k-1} u) \}$$

in the maximum norm $\|\cdot\|_{C(I)}$.

(2.6) $$u = I_0 u + \sum_{k=1}^{\infty}(I_k u - I_{k-1}u)$$

is Faber's decomposition of the continuous functions $u$.

The crucial point is that the differences (2.3) become rapidly very small. For $u \in C^1(I)$

(2.7) $$\|I_k u - I_{k-1} u\|_{C(I)} \leq 2^{-k}\|u'\|_{C(I)}$$

holds, and for $u \in C^2(I)$ even one has

(2.8) $$\|I_k u - I_{k-1} u\|_{C(I)} \leq \tfrac{1}{2} 4^{-k}\|u''\|_{C(I)}.$$

Therefore, for smooth functions $u$, the series (2.6) converges very fast, and $u$ is well represented by a finite part of this series.

On the other hand, if the sequence

(2.9) $$2^k \|I_k u - I_{k-1} u\|_{C(I)}, \quad k = 0, 1, 2, \ldots$$

is unbounded, $u \in C(I)$ cannot be a $C^1$-function. In this way one can construct continuous, but in no subinterval of $I$ continuously differentiable functions. This was Faber's original aim.

Consider, as an example, the function

(2.10) $$u = \sum_{k=1}^{\infty} v_k$$

in $C(I)$ where the $v_k \in \mathcal{S}_k$ are given by

(2.11) $$v_k(x) = \begin{cases} (\tfrac{2}{3})^k, & x \in \mathcal{N}_k \setminus \mathcal{N}_{k-1} \\ 0, & x \in \mathcal{N}_{k-1}. \end{cases}$$

Then $I_k u - I_{k-1} u = v_k$ so that

$$u|I' \notin C^1(I')$$

for any subinterval $I'$ of $I$.

The properties of the decomposition (2.6) of $u$ are reflected in the *hierarchical basis expansion* of $u$.

To introduce the hierarchical basis functions, we number the nodes

(2.12) $$x \in \mathcal{N}_0 \cup \bigcup_{k=1}^{\infty} (\mathcal{N}_k \setminus \mathcal{N}_{k-1})$$

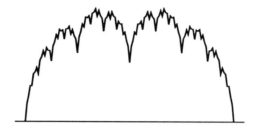

FIG. 4:. $u_7 = \sum_{k=1}^{7} v_k$ as an approximation of the function (2.10)

beginning with the nodes $x \in \mathcal{N}_0$ and continuing with the nodes in $\mathcal{N}_1 \backslash \mathcal{N}_0, \mathcal{N}_2 \backslash \mathcal{N}_1, \ldots$
With $x_i \in \mathcal{N}_k$ we associate the *nodal basis function* $\psi_i^{(k)} \in \mathcal{S}_k$ given by

(2.13) $$\psi_i^{(k)}(x_l) = \delta_{il}, \quad x_l \in \mathcal{N}_k.$$

The *hierarchical basis functions* are

(2.14) $$\widehat{\psi}_i = \psi_i^{(0)}, \quad x_i \in \mathcal{N}_0,$$

and

(2.15) $$\widehat{\psi}_i = \psi_i^{(k)}, \quad x_i \in \mathcal{N}_k \backslash \mathcal{N}_{k-1}.$$

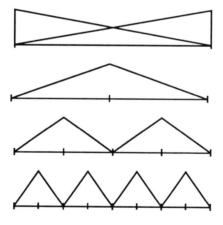

FIG. 5:. *Hierarchical basis functions*

The *hierarchical basis expansion* of $I \in C(I)$ is

(2.16) $$u = \sum_{x_i \in \mathcal{N}_0} a_i \widehat{\psi}_i + \sum_{x_i \in \mathcal{N}_k \backslash \mathcal{N}_{k-1}} a_i \widehat{\psi}_i$$

where, for $x_i \in \mathcal{N}_0$,

(2.17) $$a_i = (I_0 u)(x_i) = u(x_i)$$

and, for $x_i \in \mathcal{N}_k \backslash \mathcal{N}_{k-1}$,

(2.18) $$a_i = (I_k u - I_{k-1} u)(x_i) = (u - I_{k-1} u)(x_i) \ .$$

Depending on the *local smoothness* of $u$, the contribution $a_i$ of a basis function $\hat{\psi}_i$ can be very small and can be neglected without an essential loss of accuracy. Therefore this expansion is a perfect tool for the *adaptive representation* of functions.

**3. 2D–finite element functions.** Of course, the approach is not restricted to one space dimension. In this section we extend our construction to the two-dimensional case, but, contrary to the last section, we restrict our attention to finite element functions.

Let $\bar{\Omega}$ be a bounded polygonal domain. The partitions of the interval $I$ in Section 2 are replaced by triangulations of $\bar{\Omega}$. By a *triangulation* $\mathcal{T}$ of $\bar{\Omega}$ we mean a set of triangles such that the union of these triangles is $\bar{\Omega}$ and such that the intersection of two such triangles is either empty or consists of a common side or a common vertex of both triangles.

We start with an intentionally coarse triangulation $\mathcal{T}_0$ of $\bar{\Omega}$. This triangulation is refined several times, giving a family of nested triangulations $\mathcal{T}_0, \mathcal{T}_1, \mathcal{T}_2, \ldots$ A triangle of $\mathcal{T}_{k+1}$ is generated by subdividing every triangle of $\mathcal{T}_k$ into four congruent subtriangles.

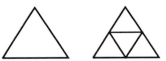

FIG. 6:. *The subdivision of a triangle into four congruent subtriangles*

Corresponding to the triangulations $\mathcal{T}_k$ we have finite element spaces $\mathcal{S}_k$. $\mathcal{S}_k$ consists of all functions which are piecewise continuous on $\bar{\Omega}$ and linear on the triangles $T \in \mathcal{T}_k$. Clearly, $\mathcal{S}_k$ is a subspace of $\mathcal{S}_l$ for $l > k$.

Let $\mathcal{N}_k = \{x_1, \ldots, x_{n_k}\}$ be the set of vertices of the triangles in $\mathcal{T}_k$. Then $\mathcal{S}_k$ is spanned by the *nodal basis functions* $\psi_i^{(k)} \in \mathcal{S}_k$, $i = 1, \ldots, n_k$, which are defined by

(3.2) $$\psi_i^{(k)}(x_l) = \delta_{il} \ , \quad x_l \in \mathcal{N}_k \ .$$

The *hierarchical basis functions* are

(3.3) $$\hat{\psi}_i = \psi_i^{(0)} \ , \quad x_i \in \mathcal{N}_0 \ ,$$

and

(3.4) $$\hat{\psi}_i = \psi_i^{(k)} \ , \quad x_i \in \mathcal{N}_k \backslash \mathcal{N}_{k-1} \ .$$

$\hat{\psi}_i$, $i = 1, \ldots, n_k$, is the *hierarchical basis* of $\mathcal{S}_k$.

We fix a final level $j$ and set $\mathcal{S} = \mathcal{S}_j$. The interpolation operators $I_k : \mathcal{S} \to \mathcal{S}_k$ are defined by

$$(3.5) \qquad (I_k u)(x) = u(x), \quad x \in \mathcal{N}_k .$$

Because of $u = I_j u$ one has the splitting

$$(3.6) \qquad u = I_0 u + \sum_{k=1}^{j} (I_k u - I_{k-1} u)$$

of the functions $u \in \mathcal{S}$, corresponding to Faber's decomposition (2.6).

The subspace $\mathcal{V}_k$ of $\mathcal{S}_k$ is defined as the span of the hierarchical basis functions $\widehat{\psi}_i$, $x_i \in \mathcal{N}_k \setminus \mathcal{N}_{k-1}$, or alternatively as the range of $I_k - I_{k-1}$. The space $\mathcal{S}_k$ is the direct sum

$$(3.7) \qquad \mathcal{S}_k = \mathcal{S}_0 \oplus \mathcal{V}_1 \oplus \ldots \oplus \mathcal{V}_k .$$

If $u \in \mathcal{S}$ has the representation

$$(3.8) \qquad u = u_0 + \sum_{k=1}^{j} v_k, \quad u_0 \in \mathcal{S}_0, \quad v_k \in \mathcal{V}_k ,$$

then

$$(3.9) \qquad u_0 = I_0 u, \quad v_k = I_k u - I_{k-1} u .$$

As in the one-dimensional case, a function can be represented *adaptively* by a selected subset of the hierarchical basis.

Alternatively to this approach, and essentially equivalent to it, one can consider *nonuniformly refined triangulations* and the associated hierarchical bases of the corresponding finite element spaces; see [36].

The most popular of these refinement schemes for triangulations has been developed by Bank, Sherman and Weiser [8]. It is also described in [7] and used in Bank's finite element package PLTMG [4].

In three space dimensions, the construction of appropriate refinement schemes is a much more complicated task. Possible solutions are presented in [3] and [38].

Similar constructions as for piecewise linear finite element functions are possible for higher order functions. Consider, as an example, the space $\mathcal{S}_Q$ of piecewise quadratic functions with respect to the given final triangulation. This space can be represented as the direct sum

$$(3.10) \qquad \mathcal{S}_Q = \mathcal{S}_L \oplus \mathcal{V}_Q$$

of the corresponding space $\mathcal{S}_L = \mathcal{S}$ of piecewise linear functions and the subspace $\mathcal{V}_Q$ of piecewise quadratic functions which vanish at the vertices of the triangles. With $\mathcal{S}_L = \mathcal{S}_j$ as in (3.7), $\mathcal{V}_Q$ behaves like $\mathcal{V}_{j+1}$. The functions in $\mathcal{V}_Q$ can be represented by their values at the midpoints of the edges. Thus the splitting (3.10) induces a hierarchical basis of $\mathcal{S}_Q$.

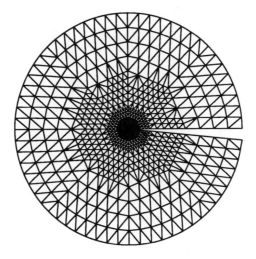

FIG. 7:. *A nonuniform triangulation*

We remark that splittings like (3.10) can be used very successfully for the construction of error estimators; we refer to [15].

Also for nonconforming and $C^1$-elements similar constructions are known; we refer to [16], [17] and [30] and to [29] and [14], respectively.

One of the most fascinating developments in the field are Zenger's *sparse grid subspaces* [37] of tensor-product finite element spaces.

Consider, as an example, a square $\bar{\Omega}$ of sidelength 1, which is subdivided into $n^2 = 4^j$ small squares of sidelength $1/n = 1/2^j$, and the corresponding space of continuous and piecewise bilinear functions which vanish on the boundary of $\bar{\Omega}$. A hierarchical basis of this space can be constructed as a tensor-product of two 1D–hierarchical bases for spaces of piecewise linear functions.

A careful examination shows that most of these basis functions give only a very little contribution in the corresponding expansion of a smooth function. Only $n \log_2 n$ of the $n^2$ basis functions are necessary to maintain nearly the full accuracy. In the $d$–dimensional case, the possible reduction of the dimension is even more impressive, instead of

(3.11) $$n^d$$

basis functions one needs only

(3.12) $$n(\log_2 n)^{d-1}$$

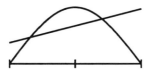

FIG. 8:. *The splitting $S_Q = S_L \oplus V_Q$*

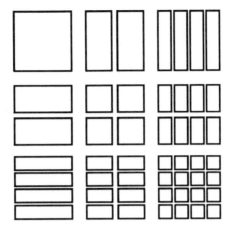

FIG. 9:. *The supports of Zenger's basis functions*

basis functions.

**4. Algorithms.** The hierarchical basis can be handled very easily. Consider, for example, the conversion

$$(4.2) \qquad \alpha \to S\alpha$$

of the hierarchical basis coefficient vector $\alpha$ of a function $u \in \mathcal{S}$, $\mathcal{S} = \mathcal{S}_j$ as in the last section, into its nodal basis coefficient vector $S\alpha$. The algorithm for evaluating $S\alpha$ has already been described in Faber's paper as can be seen in the facsimile above. In the presentation here, we follow the lines given in [34].

The transformation (4.2) means that one has to evaluate the function $u$ at the nodes $x_i \in \mathcal{N}_j$. This can be done recursively, beginning with the nodes $x_i \in \mathcal{N}_0$, where the values of $u$ are given by the corresponding hierarchical basis coefficients. Then one proceeds from one level to the next, utilizing that, for $x_i \in \mathcal{N}_k \setminus \mathcal{N}_{k-1}$,

$$(4.3) \qquad u(x_i) = (u - I_{k-1}u)(x_i) + (I_{k-1}u)(x_i) \ .$$

The value $(u - I_{k-1}u)(x_i)$ is the known hierarchical basis coefficient associated with $x_i$. $I_{k-1}u$ is completely determined by the already known values of $u$ at the nodes in $\mathcal{N}_{k-1}$. If $x_i$ is the midpoint of the edge connecting the nodes $x_{I1(i)}, x_{I2(i)} \in \mathcal{N}_{k-1}$, then

$$(4.4) \qquad (I_{k-1}u)(x_i) = [u(x_{I1(i)}) + u(x_{I2(i)})]/2 \ ,$$

where the values of $u$ at the right-hand side have already been computed.

To be more specific, assume that the indices $I1(i)$ and $I2(i)$ are stored in the arrays $I1$ and $I2$ and that, at the beginning of the computation, the entry $X(i)$ of the array $X$ contains the hierarchical basis coefficient of the given function $u$ corresponding to the node with index $i$. Then the following algorithm returns the value of the function

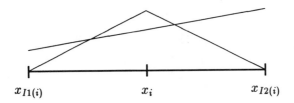

FIG. 10:. *The operation (4.3), (4.4)*

$u$ in the gridpoint with index $i$ in the entry $X(i)$ of the array $X$.

(4.5)
```
for k = 1 to j do
    for x_i ∈ N_k\N_{k-1} do
        X(i) ← X(i) + [X(I1(i)) + X(I2(i))]/2
    end i
end k
```

The outer loop is a loop on the refinement levels whereas the inner loop performs the operations which are necessary to get the values at the nodes in $\mathcal{N}_k \backslash \mathcal{N}_{k-1}$ when the values at the nodes in $\mathcal{N}_{k-1}$ are already known. Note that no additional storage for the new coefficient vector is needed; the hierarchical basis coefficients are overwritten by the nodal values. The whole algorithm requires at most $2n$ additions and $n$ divisions by 2 where $n = n_j$ is the dimension of the finite element space.

The operations necessary to compute $S\alpha$ can be performed in many different orders, giving much freedom to the programmer. One has only to guarantee that the midpoint of an edge is not treated before the values at the endpoints are known. Especially one can use very elegant tree-like data structures [21] reflecting the recursive structure.

Mathematically, the given algorithm corresponds to a factorization

$$(4.6) \qquad S = S_j S_{j-1} \ldots S_1$$

of the matrix $S$. The matrix $S_k$ represents the inner loop in the algorithm (4.5). It describes the evaluation of the function at the new nodes in $\mathcal{N}_k \backslash \mathcal{N}_{k-1}$, when the values at the nodes in $\mathcal{N}_{k-1}$ are already known. $S_k$ is the identity matrix except for the entries

$$(4.7) \qquad S_k|_{i,I1(i)} = \tfrac{1}{2}, \qquad S_k|_{i,I2(i)} = \tfrac{1}{2}$$

in the $i$-th row and columns $I1(i)$ and $I2(i)$ for $x_i \in \mathcal{N}_k \backslash \mathcal{N}_{k-1}$.

The representation (4.6) of $S$ leads to the factorization

$$(4.8) \qquad S^T = S_1^T \ldots S_{j-1}^T S_j^T$$

of $S^T$. This means that the computation of $S^T \alpha$ is as cheap as the computation of $S\alpha$.

The following algorithm overwrites the vector $\alpha$ stored in the array $X$ by $S^T\alpha$:

(4.9)
```
for k = j down to 1 do
    for x_i ∈ N_k\N_{k-1} do
        X(I1(i)) ← X(I1(i)) + X(i)/2
        X(I2(i)) ← X(I2(i)) + X(i)/2
    end i
end k
```

For different purposes also the computation of $S^{-1}\alpha$ and $S^{-T}\alpha$ may be necessary. Using the factorization

$$(4.10) \qquad S^{-1} = S_1^{-1} \ldots S_{j-1}^{-1} S_j^{-1}$$

one obtains the algorithm

(4.11)
```
for k = j down to 1 do
    for x_i ∈ N_k\N_{k-1} do
        X(i) ← X(i) - [X(I1(i)) + X(I2(i))]/2
    end i
end k
```

which overwrites $\alpha$ by $S^{-1}\alpha$, and utilizing

$$(4.12) \qquad S^{-T} = S_j^{-T} S_{j-1}^{-T} \ldots S_1^{-T}$$

the algorithm

(4.13)
```
for k = 1 to j do
    for x_i ∈ N_k\N_{k-1} do
        X(I1(i)) ← X(I1(i)) - X(i)/2
        X(I2(i)) ← X(I2(i)) - X(i)/2
    end i
end k
```

overwriting $\alpha$ with $S^{-T}\alpha$.

As an application we consider the finite element discretization of a second order elliptic boundary value problem with respect to the space $\mathcal{S}$.

Let $A$ be the corresponding nodal basis– and $\hat{A}$ be the hierarchical basis discretization matrix. $A$ is sparse and can easily be obtained as the sum of the local element matrices. Contrary, $\hat{A}$ has a much more complicated structure and contains more nonzero elements. But, with the transformation matrix $S$ above, $\hat{A}$ has the factorization

$$(4.14) \qquad \hat{A} = S^T A S$$

so that $\hat{A}x$ can be evaluated without much more effort than $Ax$ utilizing

(4.15) $$\hat{A}x = S^T(A(Sx)).$$

Algorithms for the direct storage and manipulation of $\hat{A}$ have been discussed in [23].

**5. More mathematical properties.** The weak formulation of the simple 1D–boundary value problem

(5.2) $$-u''(x) = f(x), \quad 0 \le x \le 1,$$

(5.3) $$u(0) = u(1) = 0,$$

is to find a function $u \in H_0^1(I)$ with

(5.4) $$a(u,v) = \int_0^1 fv \, dx.$$

The bilinear form on the left hand side is

(5.5) $$a(u,v) = \int_0^1 u'v' \, dx \ ;$$

it induces the seminorm

(5.6) $$|u|_1 = (\int_0^1 |u'(x)|^2 \, dx)^{1/2}$$

on $H^1(I) \subseteq C(I)$, which is a norm on $H_0^1(I)$.

The interesting point is that

(5.7) $$u = I_0 u + \sum_{k=1}^{\infty}(I_k u - I_{k-1}u)$$

is an orthogonal decomposition of $u \in H^1(I)$ with respect to the bilinear form (5.5); for all $u \in H^1(I)$ one has

(5.8) $$|u|_1^2 = |I_0 u|_1^2 + \sum_{k=1}^{\infty} |I_k u - I_{k-1}u|_1^2,$$

a fact that has already been observed by Zienkiewicz, Kelly, Gago and Babuška [40]. The reason for that property is simple: As a direct computation shows, a hat-function is orthogonal to a linear function with respect to the inner product (5.5).

As a consequence, the discretization matrix of the two point boundary value problem above with respect to the hierarchical basis of one of the finite element spaces $S_k$ of Section 2 is diagonal.

For two space dimensions, very similar properties hold as I could show in [34].

The reason is that for any triangle $T$ of a diameter $H$, which is subdivided into nondegenerate triangles of a diameter $\geq h$, and for any function $u$, which is continuous on $T$ and linear on each of these triangles,

$$(5.9) \qquad \max_{x,y \in T} |u(x) - u(y)| \leq c_0 (\log \frac{H}{h} + 1)^{1/2} |u|^2_{1;T}$$

holds. Here $|u|_{1;T}$ denotes the seminorm which is induced by the inner product

$$(5.10) \qquad D(u,v)|_T = \sum_{i=1}^{2} \int_T D_i u D_i v \, dx \, .$$

We don't claim to prove (5.9) but refer to [34], for example.

A simple consequence of (5.9) and of the scaling properties of 2D–functions is the estimate

$$(5.11) \qquad |I_k u|_{1;T} \leq c_1 (j - k + 1)^{1/2} |u|_{1;T}$$

for the functions in the space $\mathcal{S} = \mathcal{S}_j$ of Section 3 and $T \in \mathcal{T}_k$. The estimate

$$(5.12) \qquad |I_0 u|^2_{1;T} + \sum_{k=1}^{j} |I_k u - I_{k-1} u|^2_{1;T} \leq c_2 (j+1)^2 |u|^2_{1;T}$$

for the functions $u \in \mathcal{S}$ and $T \in \mathcal{T}_0$ follows.

Utilizing the representation (3.6) of $u \in \mathcal{S}$ and the triangle– and the Cauchy-Schwarz inequality, on the other hand one obtains

$$(5.13) \qquad |u|^2_{1;T} \leq (j+1) \{ |I_0 u|^2_{1;T} + \sum_{k=1}^{j} |I_k u - I_{k-1} u|^2_{1;T} \}$$

where $T \in \mathcal{T}_0$. The estimate (5.13) can be improved. It has been shown in [34] that for $v_k \in \mathcal{V}_k$, $v_l \in \mathcal{V}_l$ and all triangles $T \in \mathcal{T}_k$,

$$(5.14) \qquad D(v_k, v_l) \leq c_3 (\frac{1}{\sqrt{2}})^{|k-l|} |v_k|_{1;T} |v_l|_{1;T} \, .$$

If one applies this strengthened Cauchy-Schwarz inequality to the functions $v_0 = I_0 u$ and $v_k = I_k u - I_{k-1} u$ for $k \geq 1$, one can replace (5.13) by

$$(5.15) \qquad |u|^2_{1;T} \leq c_4 \{ |I_0 u|^2_{1;T} + \sum_{k=1}^{j} |I_k u - I_{k-1} u|^2_{1;T} \} \, .$$

Finally we observe that, for $v \in \mathcal{V}_k$ and $T \in \mathcal{T}_{k-1}$,

$$(5.16) \qquad |v|^2_{1;T} \sim \sum_{x_i \in T \cap \mathcal{N}_k \setminus \mathcal{N}_{k-1}} |\hat{\psi}_i|^2_{1;T} |v(x_i)|^2 \, .$$

The reason is quite simple: $v$ vanishes at the nodes $x_i \in T \cap \mathcal{N}_{k-1}$.

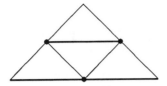

FIG. 11:. *The nodes determining a function in $\mathcal{V}_k$ on a triangle in $\mathcal{T}_{k-1}$*

Now, consider a second order elliptic boundary value problem on $\Omega$ which induces an energy norm

(5.17) $$\|u\|^2 \sim |u|^2_{1;\Omega} \; .$$

On $\mathcal{S}$, we define the discrete norm

(5.18) $$\|u\|^2 = \|I_0 u\|^2 + \sum_{k=1}^{j} \sum_{x_i \in \mathcal{N}_k \setminus \mathcal{N}_{k-1}} \|\widehat{\psi}_i\|^2 \, |(I_k u - I_{k-1} u)(x_i)|^2 \; .$$

Up to the part coming from $I_0 u$, that means from the initial level, this norm is a weighted Euclidean norm of the hierarchical basis coefficient vector of $u$.

The estimates above say that there are positive constants $K_1$ and $K_2$ with

(5.19) $$\frac{K_1}{(j+1)^2} \|u\|^2 \leq \|u\|^2 \leq K_2 \|u\|^2$$

for all functions $u \in \mathcal{S}$. As the proof shows, (5.19) is essentially a local estimate. The constants $K_1$ and $K_2$ neither depend on global regularity properties of the boundary value problem nor on jumps of the coefficient functions across the boundaries of the triangles in the initial triangulation.

(5.19) is the key property of the hierarchical basis representation of 2D-functions.

## 6. The hierarchical basis solver.

Let $D$ be the matrix inducing the norm (5.18) on $\mathcal{S}$ with respect to the hierarchical basis representation of this space. As it has been stated in Section 5, $D$ is diagonal up to a small block corresponding to the initial level.

Let $\widehat{A}$ be the hierarchical basis discretization matrix of the boundary value problem inducing the energy norm (5.17). Then the estimate (5.19) states that the spectral condition number of $D^{-1/2} \widehat{A} D^{-1/2}$ can be estimated by

(6.2) $$\kappa(D^{-1/2} \widehat{A} D^{-1/2}) \leq \frac{K_2}{K_1} (j+1)^2 \; .$$

Using the representation (4.14)

(6.3) $$\widehat{A} = S^T A S$$

of $\widehat{A}$ in terms of the nodal basis discretization matrix $A$, this means that

(6.4) $$B^{-1} = S D^{-1} S^T$$

is a nearly perfect approximate inverse for $A$; the ratio of the maximum and the minimum eigenvalue of $B^{-1}A$ is bounded by the right hand side of (6.2).

For the case of a uniform refinement, the condition number of $A$ is reduced from
$$4^j \sim h^{-2}$$
to
$$j^2 \sim |\log h|^2,$$
a astonishing fact taking into account the simplicity of the preconditioner. Apart from the solution of a level zero problem, an application of $B^{-1}$ requires less than $4n$ additions and $3n$ divisions, where $n$ is the dimension of $A$.

In a conjugate gradient framework, only
$$\sim j\,|\log \varepsilon|\quad \text{iteration steps}$$
and
$$\sim nj\,|\log \varepsilon|\quad \text{operations}$$
are necessary to reduce the energy norm of an initial error by the factor $1/\varepsilon$.

**7. A multiplicative version.** The method described in the last section can be seen as a *cg*–accelerated Jacobi–type iteration with respect to the hierarchical basis functions of the discrete boundary value problem. The corresponding Gauß–Seidel iteration is the *hierarchical basis multigrid method* which has been introduced and analyzed in a joint paper [7] with R. Bank and T. Dupont.

From a computational point of view, the hierarchical basis multigrid method is a $V$–cycle multigrid method with a Gauß–Seidel smoother, with the difference that only the points in $\mathcal{N}_k \backslash \mathcal{N}_{k-1}$ and not all points in $\mathcal{N}_k$ are smoothed on level $k$. Due to this modification, the operation count remains $O(n)$ independent of the distribution of the unknowns among the level.

The hierarchical basis multigrid method exhibits qualitatively the same convergence behavior as the Jacobi–version. The experience says that the additional effort pays of and that, in terms of the number of iteration steps, the hierarchical basis multigrid method is faster. On the other hand, it requires an explicit knowledge of parts of the discretization matrices on the different levels and not only a routine for evaluating $Ax$.

From a more general point of view, the hierarchical basis multigrid method can be seen as an approximate multiplicative subspace correction method in the sense of Xu [32], with the hierarchical basis solver as the corresponding additive version. Using the estimates of Section 5, Xu's theory gives the same estimates for the rate of convergence as we could obtain in [7]. In fact, Xu's theory generalizes the algebraic theory developed in [7], he does not need the property that the subspaces $\mathcal{V}_k$ form a direct sum (3.7), as we did.

The hierarchical basis multigrid method is the iterative solver behind Bank's finite element package PLTMG [4]. PLTMG works well for slightly indefinite problems, and especially also for problems with a convection term. Theoretically this is backed by the work of Bank and Benbourenane [5].

## 8. W–cycles.

The main drawback of the hierarchical basis solvers presented in Section 6 and Section 7 is that they are dimension-dependent and that they deteriorate for the three-dimensional case, because the estimate (5.9) does not hold any longer. For a detailed analysis of 3D–hierarchical basis solvers we refer to the work of Ong [25], [26].

One can overcome these difficulties as follows. First, we consider a two level version of the hierarchical basis multigrid methods of the last section with a coarse $C$-level and a fine $F$-level. For this case, the discrete problem reads

$$(8.2) \quad \left( \begin{array}{c|c} A_{CC} & A_{CF} \\ \hline A_{FC} & A_{FF} \end{array} \right) \left( \begin{array}{c} u_C \\ u_F \end{array} \right) = \left( \begin{array}{c} f_C \\ f_F \end{array} \right).$$

The two-level hierarchical basis method is a block Gauß–Seidel iteration for this system where the well–conditioned $A_{FF}$-system itself is solved only approximately.

In the multi–level version of the hierarchical basis multigrid method, the $A_{CC}$-equation is approximately solved by *one* recursive call of the method, resulting in a $V$-cycle iteration. In the $W$-cycle version, *two or more* steps are applied.

It is easy to show that the $W$-cycle version converges always as fast as the $V$-cycle version.

A convergence analysis for the two–level case requires only a so–called *strengthened Cauchy-Schwarz inequality* [6], [9],[18] for the $C$- and the $F$-subspace. Such a strengthened Cauchy–Schwarz inequality can be proven very easily by simple local arguments, for three as well as for two space dimensions.

If the constant in the strengthened Cauchy-Schwarz inequality is small enough (which is the case for many applications), the convergence of the $W$-cycle with a rate of convergence independent of the number of levels can be proven by a simple induction.

The analysis of such schemes began with a paper of Bank and Dupont [6] and continued with the work of Braess [9], Maitre and Musy [22], Axelsson and Gustafsson [1] and many others. A definitive treatment including arbitrary cycling strategies has been given by Axelsson and Vassilevski [2].

The main drawback of all these schemes is that the operation count for a single iteration step can explode for nonuniformly refined grids, if the dimensions of the subspaces don't increase with a sufficient speed.

## 9. An alternative decomposition.

A solution of these problems has been found by Xu [33] and Bramble, Pasciak and Xu [11]. They replace our decomposition (3.6)

$$(9.2) \quad u = I_0 u + \sum_{k=1}^{j} (I_k u - I_{k-1} u)$$

of $\mathcal{S} = \mathcal{S}_j$ by the related decomposition

$$(9.3) \quad u = Q_0 u + \sum_{k=1}^{j} (Q_k u - Q_{k-1} u)$$

where the $Q_k : L_2(\Omega) \to S_k$ are orthogonal projections with respect to a weighted $L_2$-inner product inducing a norm

$$(9.4) \qquad \|v\|_0 \sim \|v\|_{L_2(\Omega)}.$$

The discrete norm

$$(9.5) \qquad \|u\|^2 = \|Q_0 u\|^2 + \sum_{k=1}^{j} 4^k \|Q_k u - Q_{k-1} u\|_0^2$$

on $S$ behaves like the energy norm induced by a second order boundary value problem. As one can show there are positive constants $K_1$ and $K_2$ with

$$(9.6) \qquad \frac{K_1}{j+1} \|u\|^2 \leq \|u\|^2 \leq K_2 (j+1) \|u\|^2.$$

We refer to [33], [11] and, for a careful examination of nonuniform meshes, to [35].

Utilizing the same arguments as they have been used in [34] for the decomposition (9.2), Zhang [39] could improve the upper bound in (9.6) to

$$(9.7) \qquad \|u\|^2 \leq K_2^* \|u\|^2.$$

For uniformly refined families of triangulations, but without assuming elliptic regularity, Oswald [27], [28] and Dahmen and Kunoth [13] improve the lower bound to

$$(9.8) \qquad K_1^* \|u\|^2 \leq \|u\|^2.$$

In [10] it is shown that the estimate (9.8) for the case of a uniform refinement implies the corresponding estimate for the nonuniform case; note that the norm (9.5) (contrary to the norm (5.18)) changes, if the space $S$ is replaced by a subspace spanned by a subset of its hierarchical basis. Thus, the discrete norm (9.5) and the energy norm become equivalent. Assuming full $H^2$-regularity, a proof of (9.8) for the uniform case has already been given in [11].

Contrary to the estimate (5.19), the estimates (9.6), (9.7) and (9.8) are dimension independent; they hold also for the analogous 3D-case.

Norms like (9.5), respectively their continuous counterparts for $j \to \infty$, are well-known and widely used in approximation theory; Oswald's papers are based on these connections. We refer to the book [12] of Butzer and Scherer.

The norm equivalence (9.6), (9.7), (9.8) can be utilized to construct simple, but very efficient preconditioners (or iterative methods, as one likes it) for second order elliptic boundary value problems. Here we describe only the result; for a derivation we refer to [33], [11], [35].

If $S^*$ denotes the space of linear functionals on $S$, the discrete boundary value problem can be written as

$$(9.9) \qquad Au = f$$

with a linear operator

(9.10) $$A : \mathcal{S} \to \mathcal{S}^* .$$

The nodal basis preconditioner developed in [33] and [11] is an approximate inverse

(9.11) $$C : \mathcal{S}^* \to \mathcal{S}$$

of this operator $A$. It is given by

(9.12) $$C_X r = u_0 + \sum_{k=1}^{j} \sum_{i=1}^{n_k} \frac{1}{d_i^{(k)}} < r, \psi_i^{(k)} > \psi_i^{(k)}$$

where $u_0 \in \mathcal{S}_0$ is the solution of the level zero problem

(9.13) $$< A u_0, v_0 > = < r, v_0 >, \quad v_0 \in \mathcal{S}_0 ,$$

corresponding to (9.9) and where the $d_i^{(k)}$ are scaling factors. The hierarchical basis preconditioner of Section 6 can be given a very similar form. It is

(9.14) $$C_H r = u_0 + \sum_{k=1}^{j} \sum_{x_i \in \mathcal{N}_k \setminus \mathcal{N}_{k-1}} \frac{1}{d_i} < r, \hat{\psi}_i > \hat{\psi}_i ,$$

or, reformulated in terms of the nodal basis functions

(9.15) $$C_H r = u_0 + \sum_{k=1}^{j} \sum_{i=n_{k-1}+1}^{n_k} \frac{1}{d_i} < r, \psi_i^{(k)} > \psi_i^{(k)}$$

with the scaling factors $d_i = d_i^{(k)}$.

The computation of $C_H r$, as defined by (9.14) and (9.15), is cheap. Beginning with the values

(9.16) $$< r, \psi_i^{(j)} >, \quad i = 1, \ldots, n_j,$$

which are the components of the right hand side in a matrix formulation with respect to the nodal basis fo $\mathcal{S}_j$, first one computes

(9.17) $$< r, \hat{\psi}_i >, \quad i = 1, \ldots, n_j,$$

recursively advancing from one level to the next coarser level. As $\hat{\psi}_i$, $i = 1, \ldots, n_0$, is the nodal basis of the initial space $\mathcal{S}_0$, the computation of $u_0 \in \mathcal{S}_0$ additionally requires only the solution of a low-dimensional level zero matrix equation. Finally, all terms must be summed up. This is done recursively beginning with the values of $C_H r$ at the nodes $x_i$, $i = 1, \ldots, n_0$, of the initial level and ending with

(9.18) $$(C_H r)(x_i), \quad i = 1, \ldots, n_j .$$

As exactly one hierarchical basis function $\hat{\psi}_i$ is associated with every node $x_i$, the overall work and storage for evaluating $C_H r$ remains strictly proportional to the number of unknowns, independent of the distribution of the unknowns among the levels. In fact, this algorithm is equivalent to the algorithm for computing the product of the matrix $SD^{-1}S^T$ with a vector along the lines given in Section 4.

The nodal basis preconditioner (9.12) can be realized along the same lines. Note, however, that the double sum in (9.14) and (9.15), respectively, consists of

$$(9.19) \qquad \sum_{k=1}^{j}(n_k - n_{k-1}) = n_j - n_0$$

terms, whereas in (9.12)

$$(9.20) \qquad n_j \leq \sum_{k=1}^{j} n_k \leq j n_j$$

terms occur. The reason is that a changing number of basis functions $\psi_i^{(k)}$ is associated with every node $x_i$. In the general framework of adaptively refined grids, this leads to considerable complications. For a naive realization, both the work and storage can grow like $O(j n_j)$.

We remark that the norm equivalence (9.6), (9.7), (9.8) is a also very useful tool for the analysis of classical multigrid methods; as one can show it implies a nearly optimal or even optimal computational complexity without assuming elliptic regularity; we refer to Xu's paper [32].

**10. Conclusions.** Subspace decompositions of finite element spaces became an essential, if not *the* essential tool for the construction and analysis of fast solvers for elliptic partial differential equations.

Two of these decompositions are of special importance, the $L_2$-like decomposition of Bramble, Pasciak and Xu described in Section 9, and the hierarchical decomposition by interpolation operators already invented at the beginning of this century.

For three-dimensional applications, the $L_2$-like Fourier-type decompositions have the superior mathematical properties. The hierarchical decomposition has the striking advantage of an extreme simplicity allowing the construction of very elegant, compact algorithms.

Beside of its application for the construction of fast solvers for partial differential equations, it can also be used for the efficient and flexible representation, storage and manipulation of smooth and nonsmooth functions.

## REFERENCES

[1] Axelsson, O., Gustafsson, I.: *Preconditioning and two-level multigrid methods of arbitrary degree of approximation*. Math. Comp. 40, 219-242 (1983).

[2] Axelsson, O., Vassilevski; P.: *Algebraic multilevel preconditioning methods I*. Numer. Math. 56, 157-177 (1989).

[3] Bänsch, E.: *Local mesh refinement in 2 and 3 dimensions*. Report Nr. 6, SFB 256, Bonn 1989
[4] Bank, R.E.: *PLTMG: A Software Package for Solving Elliptic Partial Differential Equations.* Philadelphia, SIAM (1990).
[5] Bank, R.E., Benbourenane, M.: *The hierarchical basis multigrid method for convection-diffusion equations.* Preprint, Department of Mathematics, University of California at San Diego (1991).
[6] Bank, R.E., Dupont, T.: *Analysis of a two-level scheme for solving finite element equations.* Report CNA-159, Center for Numerical Analysis, University of Texas at Austin (1980).
[7] Bank, R.E., Dupont, T., Yserentant, H.: *The Hierarchical Basis Multigrid Method.* Numer. Math. **52**, 427–458 (1988).
[8] Bank, R.E., Sherman, A.H., Weiser, A.: *Refinement algorithms and data structures for regular local mesh refinement.* In: Scientific computing, 3-17 (R. Stepleman, ed.) Amsterdam, IMACS/North Holland 1983.
[9] Braess, D.: *The contraction number of a multigrid method for solving the Poisson equation.* Numer. Math. **37**, 387-404 (1981).
[10] Bramble, J.H., Pasciak, J.E.: *New estimates for multilevel algorithms including the V-cycle.* Report, Mathematical Sciences Institute, Cornell University (1991)
[11] Bramble, J.H., Pasciak, J.E., Xu, J.: *Parallel multilevel preconditioners.* Math. Comput. **55**, 1-22 (1990).
[12] Butzer, P.L., Scherer, K.: *Approximationsprozesse und Interpolationsmethoden.* Bibliographisches Institut, Mannheim 1968.
[13] Dahmen, W., Kunoth, A.: *Multilevel precondtioning.* Preprint, Fachbereich Mathematik, Freie Universität Berlin (1991).
[14] Dahmen, W., Oswald, P., Shi, X.Q.: $C^1$-*hierarchical bases.* (Manuscript).
[15] Deuflhard, P., Leinen, P., Yserentant, H.: *Concepts of an adaptive hierarchical finite element code.* IMPACT of Computing in Science and Engineering **1**, 3-35 (1989).
[16] Dörfler, W.: *Hierarchical bases for elliptic problems.* Report 123, SFB 256, Universität Bonn (1990).
[17] Dörfler, W.: *The conditioning of the stiffness matrix for certain finite elements approximating the incompressibility condition in fluid dynamics.* Numer. Math. **58**, 203-214 (1990)
[18] Eijkhout, V., Vassilevski, P.: *The role of the strengthened Cauchy-Buniakowski-Schwarz inequality in multilevel methods.* SIAM Review **33**, 405-419 (1991)
[19] Faber, G.: *Über stetige Funktionen.* Mathematische Annalen **66**, 81-94 (1909).
[20] Hackbusch, W.: *Multigrid Methods and Applications.* Springer–Verlag Berlin, Heidelberg, New York 1985.
[21] Leinen, P.: *Ein schneller adaptiver Löser für elliptische Randwertaufgaben auf Seriell- und Parallelrechnern.* Dissertation, Universität Dortmund 1990.
[22] Maitre, F., Musy, F.: *The contraction number of a class of two-level methods; an exact evaluation for some finite element subspaces and nodal problems.* In: Multigrid Methods, (W. Hackbusch and U. Trottenberg, eds.) Lecture Notes in Mathematics 960, Springer–Verlag, Berlin, Heidelberg, New York 1982
[23] Maubach, J.: *On the sparsity patterns of hierarchical finite element matrices.* In: Proc. of the Conf. on Preconditioned Conjugate Gradient Methods, Nijmegen 1989.
[24] Meyer, Y.: *Ondelettes et Opérateurs I, II.* Hermann, Paris 1990.
[25] Ong, M.E.G.: *Hierarchical basis preconditioners for second order elliptic problems in three dimensions.* Technical Report 89-3, Department of Applied Mathematics, University of Washington, Seattle (1989).
[26] Ong, M.E.G.: *Linear hierarchical basis preconditioners in three dimensions.* Technical Report CAM 91-15, Department of Mathematics, University of California, Los Angeles (1991).
[27] Oswald, P.: *On function spaces related to finite element approximation theory.* Zeitschrift für Analysis und ihre Anwendungen **9**, 43-64 (1990).
[28] Oswald, P.: *On discrete norm estimates related to multilevel preconditioners in the finite element method.* In: Proc. Int. Conf. Constr. Theory of Functions, Varna 1991 (to appear).
[29] Oswald, P.: *Hierarchical conforming finite element methods for the biharmonic equation.* SIAM J. Numer. Anal. (to appear).

[30] Oswald, P.: *On a hierarchical basis multilevel method with nonconforming $P_1$ elements.* (submitted).
[31] Prusinkiewicz, P., Lindenmayer, A.: *The Algorithmic Beauty of Plants.* Springer–Verlag, Berlin, Heidelberg, New York 1990.
[32] Xu, J.: *Iterative methods by space decomposition and subspace correction: a unifying approach.* Report AM67, Department of Mathematics, Pennsylvania State University (1990).
[33] Xu, J.: *Theory of Multilevel Methods.* Report AM48, Department of Mathematics, Pennsylvania State University (1989).
[34] Yserentant, H.: *On the multi-level splitting of finite element spaces.* Numer. Math. **49**, 379-412 (1986).
[35] Yserentant, H.: *Two preconditioners based on the multi-level splitting of finite element spaces.* Numer. Math. **58**, 163-184 (1990).
[36] Yserentant, H.: *Hierarchical bases give conjugate gradient type methods a multigrid speed of convergence.* Applied Mathematics and Computation **19**, 347-358 (1986).
[37] Zenger, Chr.: *Sparse grids.* In: Parallel Algorithms for Partial Differential Equations, Proc. of the Sixth GAMM-Seminar Kiel 1990 (W. Hackbusch, ed.). Vieweg, Braunschweig, Wiesbaden 1991.
[38] Zhang, S.: *Multilevel Iterative Techniques.* Report, Department of Mathematics, Pennsylvania State University (1988)
[39] Zhang, S.: *Multi-level additive Schwarz methods.* Report, Courant Institute, New York (1991)
[40] Zienkiewicz, O.C., Kelly D.W., Gago, J., Babuška, I.: *Hierarchical finite element approaches, error estimates and adaptive refinement.* In: The Mathematics of Finite Elements and Applications IV. (J.R. Whiteman ed.) Academic Press, London 1982

Part II

**MINISYMPOSIA**

# Chapter 18
# APPLIED PROBABILITY AND STATISTICS

## Approximation and Adaptation for Stochastic Systems

In recent years, there has been rapid progress in approximation and adaptation for stochastic systems. The theoretical findings, in turn, have significant impact on various applications, such as estimation, system identification, adaptive control, and many other related fields. Many numerical algorithms have been successfully implemented. The applications, on the other hand, bring about new and exciting theoretical results. Approximation and adaptation have historically played and will increasingly continue to play an important role in the development of stochastic systems theory.

In this minisymposium, the speakers addressed several current research topics. They discussed perturbation analysis techniques for discrete event systems, strong consistency and asymptotic efficiency issues of self-tuning regulators, converse theorems to stochastic approximation, asymptotic properties of stochastic averaging principles, and adaptive control of diffusion processes with small parameters. In addition, they described the impact of the underlying study on many applications in science and engineering.

**Organizer:** George Yin, Wayne State University

**A Taxonomy of Perturbation Analysis Techniques**
Yu-Chi Ho, Harvard University; S. Strickland, University of Virginia

**Nonatomic Martingale Transform with Applications to Stochastic Approximation**   Ching Zong Wei, University of Maryland, College Park and Institute of Statistical Science, Academia Sinica, Taiwan, R.O.C.

**On Convergence Properties of the Stochastic Averaging Principle**   Andrew Heunis, University of Waterloo, Canada

**Stochastic Adaptive Control with Small Observation Noise**   Qing Zhang, University of Toronto, Canada

## Asymptotic Analysis of Stochastic Processes in Physics and Chemistry

This minisymposium addressed three applications of stochastic processes in physics and chemistry to activation rate processes, disordered materials, and mesophysics. The following models and applications were discussed.

Rate processes such as dissociation of molecules, isomerization, surface desorption, and other chemical reactions are often modeled as multistable dynamical systems driven by small noise. Such models lead to singular perturbation problems for the Fokker-Planck equation.

The physical properties of disordered materials often differ drastically from those of ordered materials. Models for disordered materials involve Schrodingers equation with a random potential, the wave equation with a random index of refraction, and so on. Models with small disorder lead to singular perturbation problems for these equations.

In extremely small capacitance devices (e.g., the point Josephson junction), the random tunneling of an electron causes a measurable change in voltage, giving rise to new physical phenomena such as the charging effect. The dynamics of the charge on the capacitor is modeled as a Markovian jump process, giving rise to a master equation, which depends in a singular way on several parameters.

**Organizer:** Malgorzata M. Klosek, University of Wisconsin, Milwaukee

**A Singular Perturbation Approach to Products of Random Matrices and Applications to Disordered Media**   Rachel Kuske, Northwestern University; Z. Schuss, Tel-Aviv University, Israel; I. Goldhirsch, Tel-Aviv University, Israel; S.H. Noskowicz, Tel-Aviv University, Israel

**Transport and Fluctuations in Small Open Conductors**   Markus Buttiker, IBM T.J. Watson Research Center

**Two Parameter Singular Perturbation in the Fokker-Planck Equation**   Malgorzata M. Klosek, Organizer; Bernard J. Matkowsky, Northwestern University; Zeev Schuss, Tel-Aviv University, Israel

Washington's National Cathedral.

## Computationally Efficient Methods in Statistics

The enhanced raw computing power and parallel architectures of modern computers now allow the application of computationally intensive statistical methodologies that were previously not feasible. This is not to say that the need for efficient computing has passed. Some important large-scale (multi-variate) problems are beyond even modern mainframes. In addition, the power of personal computers is rapidly expanding to accommodate larger-scale problems, creating a new arena for the application of efficient methodologies.

The speakers in this minisymposium discussed an unusual way to exploit the workings of a parallel processor, a method for significantly increasing the efficiency of a large-scale stochastic approximation algorithm, and a view of the relationship of non-iterative algorithms to modern computing systems.

**Organizer:** John L. Maryak, Johns Hopkins University

**Asynchronous Iterative Computation of Fixed Points** William F. Eddy, Carnegie-Mellon University

**Multivariate Stochastic Approximation Using a Simultaneous Perturbation Gradient Approximation** James C. Spall, Johns Hopkins University

**Twilight of the Kalman Paradigm** Fred E. Daum, Raytheon Company, Wayland

**Discussion** William M. Sallas, IMSL, Inc., Houston

## Nonuniform Sampling Theory and Practical Reconstruction Methods

This minisymposium focused on recent developments in sampling theory (nonuniform and random). Special attention was given to signal recovery from a set of irregular samples. Practical reconstruction methods were discussed, including the relationship between sampling theory and coding theory, and how signal recovery from samples can be an alternative to coding theory. The speakers also showed equivalence of the sampling theory and the fundamental theorem of information theory.

**Organizer:** Farokh A. Marvasti, Illinois Institute of Technology

**Equivalence of the Sampling Theorem and the Fundamental Theorem of Information Theory** Farokh A. Marvasti, Organizer; Chuande Lia, Illinois Institute of Technology

**A Proof of New Summation Formulae by Using Sampling Theorems** Ahmed I. Zayed, University of Central Florida

**Iterative Reconstruction of Sigma Delta Encoded Signals** Soren Hein, University of California, Berkeley; Avideh Zakhuor, University of California, Berkeley

**General Sampling Theorems for Functions in Reproducing Kernel Hilbert Spaces** M. Zuhair Nashed, University of Delaware; Gilbert G. Walter, University of Wisconsin, Milwaukee

## Numerical Approximation of Stochastic Differential Equations with Applications

Stochastic differential equations (SDEs) model a variety of phenomena in science and engineering, thus obtaining effective (asymptotic or numerical) methods to approximate such solutions, which is of primary importance. Further, it is possible to solve certain related partial differential equations (PDEs), whenever the solutions to the associated SDEs are obtained. The latter approach is related to a Monte Carlo-type procedure (based on simulating independent realizations of the white-noise processes) that can naturally exploit massively parallel computers, and might perform better than any other algorithm that directly solves the corresponding PDEs in higher dimensions.

The speakers presented an overview of current research and applications in these areas.

**Organizer:** Renato G.C. Spigler, University of Padova, Italy

**New Developments for Stochastic Particle Methods** Denis Talay, INRIA, France; L. Tubaro, University of Trento, Italy

**Numerical Stability for Stochastic Implicit Runge-Kutta Methods** Diego Bricio Hernandez, CIMAT, Mexico; R. Spigler, Organizer

**High Order Stochastic Taylor Approximations** Peter E. Kloeden, Murdoch University, Australia; E. Platen, Akademie der Wissenschuften, Germany

**Implications of Supercomputers for Simulations of Stochastic Differential Equations** Wesley P. Petersen, ETH, Switzerland

## Statistical Methods for Inverse Problems and Maximum Entropy

Statistical methods for inverse problems are essential tools for applications such as signal theory, image processing, geophysics, crystallography, and communication theory. These applications are connected with mathematical problems such as deconvolution, phase problems, and generalized moment problems. Nonlinear statistical methods such as maximum of entropy, L1-minimum, are very powerful for special situations, typically "concentrated" signals. Generalizing max-entropy methods allows one to include in the same optimization problem methods such as least squares, Burg prediction, max-entropy, and the more interesting classical methods of regularization of inverse problems.

The speakers presented an overview of research in these areas and discussed ideas concerning how to manage difficult numerical situations.

**Organizer:** Didier Dacunha-Castelle, Université Paris-Sud, France

**Nonlinear Statistical Methods for Inverse Problems: Concentration of Signals** David L. Donoho, University of California, Berkeley

**Deconvolution Methods in Blind or Non-Gaussian Situations** Elizabeth Gassiat, Université Paris-Sud, France

**Nonlinear Constraints and the Max-Entropy Method** Fabrice Gamboa, Université Paris-Sud, France

## Weak and Strong Convergence and Applications to Stochastic Systems

Weak and almost sure limiting behavior of stochastic processes has enjoyed a continued interest of research activities. Its applications arise, for example, in mathematical physics, statistical mechanics, and stochastic control. One often needs to study the asymptotic properties of some stochastic systems. These systems are normally modeled by stochastic ordinary differential equations, stochastic partial differential equations, or differential equations with random perturbations. Therefore, to examine the limit behavior of the underlying systems modeled by these equations by means of weak and almost sure convergence methods becomes an important matter.

The speakers presented some current research on the convergence of approximate solutions to Zakais equation, wave propagation in randomly-layered media, stochastic equations with small parameters, singular perturbations in manufacturing systems, and various techniques for computing conditional exponential moments and Onsager Machlup functionals. They also discussed various applications.

**Organizers:** K.M. Ramachandran, University of South Florida; George Yin, Wayne State University

**On the Convergence of Approximate Solutions to Zakai's Equation**  Pao-Liu Chow, Wayne State University

**Wave Propagation in Randomly-Layered Media**  Benjamin S. White, Exxon Research and Engineering Company, Annandale

**Convergence Problems in Singularly Perturbed Stochastic Systems**  Gangaram S. Ladde, University of Texas, Arlington

**Weak Convergence of Stochastic Differential Game with a Small Parameter**  K.M. Ramachandran, Organizer

### Suggested Reading

M. Asch, W. Kohler, G. Papanicolaou, M. Postel, and B. White, *Frequency content of randomly scattered signals*, SIAM Rev., 33 (1991), pp. 519-626.

G. M. Baudet, *Asynchronous iterative methods for multiprocessors*, J. Assoc. Comput. Mach., 28 (1975), pp. 226-244.

E. Ben-Jacob, D. J. Bergman, B. J. Matkowsky, and Z. Schuss, *Phys. Rev. A*, 26 (1982), p. 2805.

B. Z. Bobrovsky and Z. Schuss, *A singular perturbation method for the computation of the mean first passage time in a nonlinear filter*, SIAM J. Appl. Math., 42 (1982), p. 174.

V. S. Borkar and M. K. Ghosh, *Stochastic differential games*, 1990, preprint.

M. Buttiker, *Scattering theory of thermal and excess noise in open conductors*, Phys. Rev. Lett., 65 (1990), p. 2901.

D. C. Chin, *Comparative study of several multivariate stochastic approximation algorithms*, Proc. Amer. Statist. Assoc., Statist. Comp. Section, 1990, pp. 223-228.

P. L. Chow, J. L. Jiang, and J. L. Menaldi, *Pathwise convergence of approximate solution to Zakai's equation in a bounded domain*, Proc. Conference on Stochastic PDE's and Applications III, Lecture Notes in Math., Springer-Verlag, New York, Berlin, to appear.

B. Derrida and E. Gardner, *Lyapunov exponent of the one-dimensional Anderson model: Weak disorder expansions*, J. Physique, 45 (1984), p. 1283.

W. F. Eddy and M. J. Schervish, *Asynchronous iteration*, Comput. Science and Statist.: Proc. 20th Symposium on the Interface, 1989, pp. 165-173.

C. W. Gardiner, *Handbook of Stochastic Methods*, Springer-Verlag, New York, 1982.

J. Golec and G. Ladde, *Averaging principle and systems of singularly perturbed stochastic differential equations*, J. Math. Phys., 31 (1991), pp. 1116-1123.

U. Haussmann and Q. Zhang, *Discrete stochastic adaptive control with small observation noise*, Appl. Math. Optim., to appear.

A. Heunis and M. A. Kouritzin, *Convergence in stochastic averaging*, Tech. Report, Department of Electrical and Computer Engineering, University of Waterloo, Canada, 1991.

Y. C. Ho and X. R. Cao, *Perturbation Analysis of Discrete Event Dynamic Systems*, Kluwer Academic, Boston, 1991.

M. M. Klosek, B. M. Hoffman, B. J. Matkowsky, et al., *Diffusion theory of multidimensional activated rate processes: The role of anisotropy*, J. Chem. Phys., 90 (1989), p. 1141.

M. M. Klosek, B. J. Matkowsky, and Z. Schuss, *The Kramers problem in the turnover regime: The role of the stochastic separatrix*, Ber. Bunsenges. Phys. Chem., 95 (1991), p. 331.

M. M. Klosek, A. Nitzan, M. Ratner, and Z. Schuss, *Anisotropic diffusion*, in preparation.

R. Kuske, Z. Schuss, I. Goldhirsch, and S. H. Noskowicz, *Schrodinger's equation on a one-dimensional lattice with weak disorder*, SIAM J. Appl. Math., to appear.

F. Marvasti, *A Uniform Approach to Zero-Crossings and Nonuniform Sampling of Single and Multidimensional Signals and Systems*, Nonuniform Publications, Oak Park, IL.

———, *Advanced Topics in Shannon Sampling and Interpolations Theory*, R. Marks II, ed., Springer-Verlag, New York.

———, *Analysis and recovery of sample-and-hold and linearly interpolated signals with irregular samples*.

F. Marvasti et al., *Recovery of speech signals with lost samples*, 1992, in preparation.

T. Naeh, M. M. Klosek, B. J. Matkowsky, and Z. Schuss, *Uniform solution of Kramers' problem by a direct approach*, Springer-Verlag Lectures Appl. Math., 27 (1991), p. 241.

J. Qiu, S. M. Shahidehpour, and Z. Schuss, *IEEE Trans. Power Systems*, 4 (1989), pp. 197-204.

J. C. Spall, *Multivariate stochastic approximation using a simultaneous perturbation gradient approximation*, IEEE Trans. Automat. Control, 1992, in press.

J. C. Spall and J. A. Cristion, *Neural networks for control of uncertain systems*, Proc. Test Technology Symposium IV, April 1991, sponsored by the U.S. Army Test and Evaluation Command, to appear.

D. J. Thouless, in *Ill-condensed Matter*, R. Balian, R. Maynard, G. Toulouse, eds., North-Holland, Amsterdam, New York, 1979, p. 1.

A. J. Viterbi, *Principles of Coherent Communications*, McGraw-Hill, New York, 1966.

C. Z. Wei, *Nonatomic martingale transform with applications to stochastic approximation*, Tech. Report, Institute of Statistical Science, Academia Sinica, Taiwan, 1991.

# Chapter 19
# CHEMICAL KINETICS AND COMBUSTION

## Mathematical Modeling of Detonations

Detonation models involve the equations of fluid mechanics, augmented to include the effects of chemical reaction. Because the full set of equations is too complicated to be solved exactly, numerical procedures are employed to obtain quantitative solutions. Alternatively, simplified detonation models can be studied analytically with the aid of asymptotic methods to obtain a qualitative understanding of detonation processes.

The qualitative theory of steady, plane detonations with simple chemistry is well known. The speakers presented recent progress in detonation theory for problems that are unsteady, not planar, or have complicated chemistry.

**Organizer:** Glenn W. Ledder, University of Nebraska, Lincoln

**Evolution of a Sharp Thermal Nonuniformity in a Nondiffusive Reactive Medium**  Ashwani K. Kapila, Rensselaer Polytechnic Institute

**Self Similar Accumulation of Shock and Ignition Events**  Ashwani K. Kapila, Rensselaer Polytechnic Institute

**Traveling Waves in Model Reacting Flows**  J. David Logan, University of Nebraska, Lincoln

**A Signaling Problem for Near-Equilibrium Flows in the Fickett Model of Combustion**  Glenn W. Ledder, Organizer; J. David Logan, University of Nebraska

## Nonlinear Dynamics in Mathematical Combustion Theory

As analytical and numerical techniques have become more refined, combustion theory has emerged as a fruitful applications area of nonlinear dynamics. In this session, the speakers presented the results of recent nonlinear analyses of several combustion problems, including the single-phase deflagration of solids, gases, and high-density fluids, as well as the multi-phase burning of solid/liquid propellants and porous solids. The main focus was on primary, secondary, and higher-order bifurcation of various nonsteady and/or nonplanar modes of combustion that can occur beyond the steady, planar neutral stability threshold, and on the transition to chaotic flame propagation.

**Organizer:** Stephen B. Margolis, Sandia National Laboratories, Livermore

**Quasi-Periodic and Chaotic Combustion of Solids and High-Density Fluids Near Points of Strong Resonance**  Stephen B. Margolis, Organizer

**Ginzburg-Landau Type Equations in Combustion**  David O. Olagunju, University of Delaware; Bernard J. Matkowsky, Northwestern University

**The Structure of Supercritical Diffusion Flames with Arrhenius Mass Diffusivities**  Stephen B. Margolis, Organizer; John K. Bechtold, Sandia National Laboratories

**Nonlinear Dynamics in Combustion**  Alvin Bayliss, Northwestern University; Bernard J. Matkowsky, Northwestern University; Gary K. Leaf, Argonne National Laboratory

**Modes of Burning in Filtration Combustion and Their Stability**  Michael R. Booty, Southern Methodist University; Bernard J. Matkowsky, Northwestern University

### Suggested Reading

A. Bayliss and B. J. Matkowsky, *Two routes to chaos in solid fuel combustion*, SIAM J. Appl. Math., 50 (1990), pp. 437–459.

M. R. Booty and B. J. Matkowsky, *Modes of burning in filtration combustion*, European J. Appl. Math., 2 (1991), pp. 17–41.

A. Liñán, *The asymptotic structure of counterflow diffusion flames for large activation energies*, Acta Astronautica, 1 (1974), pp. 1007–1039.

S. B. Margolis, *New routes to quasi-periodic combustion of solids and high-density fluids near resonant Hopf bifurcation points*, SIAM J. Appl. Math., 51 (1991), pp. 693–726.

———, *Chaotic combustion of solids and high-density fluids near points of strong resonance*, Proc. Roy. Soc. Lond. Ser. A., 433 (1991), pp. 131–150.

D. O. Olagunju and B. J. Matkowsky, *Burner-stabilized cellular flames*, Quart. Appl. Math., 48 (1990), pp. 645–664.

# Chapter 20
# COMPUTATIONAL FLUID DYNAMICS

## Development of Navier-Stokes Simulation Methods

This minisymposium is concerned with the development and construction of 3-D simulation methods for compressible and incompressible viscous flows, and their wide range of applications, incorporating both hydraulic flow problems in civil engineering and flows around elements of civil aircraft. In order to cope with the various applications, much attention has been given to the design of tools from informatics to obtain a multi-usable simulation system.

Current research is directed at compact efficient solvers with a high degree of accuracy. This also requires attention for the far field boundary conditions. Present day turbulence models are a weak point in the solvers and need validation to obtain insight in the boundaries of applicability. Only the development of methods based on direct simulation may remedy this situation in the future. The speakers described recent research and developments on modeling and simulation of flows.

**Organizer:** Pieter J. Zandbergen, University of Twente, The Netherlands

**Top-Down Design of CFD Software** Marli E.S. Vogels, National Aerospace Laboratory NLR, The Netherlands

**On Far Field Boundary Conditions and Coupling With Boundary Integral Equations** Wolfgang L. Wendland, G. Warnecke, and T. Sonar, Universität Stuttgart, Germany

**Numerical Analysis of Finite Volume Methods for Navier-Stokes Equations** Bill Morton, Oxford University, United Kingdom

**Applications and Development of a Compressible Navier-Stokes Solver** Hans G.M. Kuerten, University of Twente, The Netherlands

**An Incompressible Navier-Stokes Solver in General Coordinates** Piet Wesseling, Delft University of Technology, The Netherlands

**Validation of Navier-Stokes Computations** Arthur E. Mynett, Delft Hydraulics, The Netherlands

**On the Modeling of Turbulence and Applications** Jacques Laminie and Frédéric Pascal, Université Paris-Sud, France

## Fluid Dynamics Applications on Parallel Computers

Computational fluid dynamics has benefited greatly from advances in numerical analysis and computer hardware. Flow past complete aircraft configurations, fluid and combustion interactions, and turbulence are now candidates for simulation. With parallel computers overtaking conventional computers in both speed and memory, another leap in the complexity of fluids problems capable of numerical treatment stands to be realized. Parallel software design is still in rapid evolution; however, parallel flow simulations underway highlight both strengths and bottlenecks of present technology.

The speakers described diverse applications of fluid dynamics: aerodynamics, combustion, turbulence, and heat transfer, and present a number of different numerical schemes on a diversity of machines: the N cube, Intel cube, Connection Machine, and the Suprenum.

**Organizers:** David E. Keyes, Yale University;
Richard B. Pelz, Rutgers University

**Parallel Spectral Element Methods for the Incompressible Navier-Stokes Equations** Paul Fischer, California Institute of Technology

**3-D Unsteady Navier-Stokes Calculations on the CM-2** Dennis Jespersen, NASA Ames Research Center; Creon Levit, NASA Ames Research Center

**Domain Decomposed Computations of Combustion** William D. Gropp, Argonne National Laboratory; David E. Keyes, Organizer

**Parallel Multigrid Solution of 2-D Incompressible Navier-Stokes Equations** Johannes Linden, Gesellschaft für Mathematik und Datenverarbeitung, Germany

**Vortex Dynamics and Turbulence on Ensemble Architectures** Richard B. Pelz, Organizer

John Ockendon and Charles Holland

## Multilevel Adaptive Methods in Computational Fluid Dynamics

In order to obtain sufficient accuracy, efficiency, and flexibility in grid generation in the computational modeling of flows in complicated domains, discretization methods and solution algorithms have to interact during the course of a computation. Adaptive discretization methods and dependable a-posteriori error estimates are required. Multilevel adaptive methods provide an excellent framework for this, combining accuracy and efficiency. This is a rapidly developing field, with computational fluid dynamics as an important application area. The speakers presented recent developments in research, and discussed applications in gas dynamics and reservoir engineering.

**Organizers:** Steve McCormick, University of Colorado, Denver; Pieter Wesseling, Delft University of Technology, The Netherlands

**Nested Multigrid for Compressible Flow Computations on Solution Adaptive Unstructured Grids**  Herman Deconinck, von Karman Institute for Fluid Dynamics, Belgium; P.W. Hemker, H.T.M. Van der Maarel, and B. Koren, Center for Mathematics and Computer Science, The Netherlands

**Multilevel Adaptive Methods in Computational Fluid Flow**  Steve McCormick, Organizer

**Adaptive Multigrid Methods for the Euler and Navier-Stokes Equations Using Unstructured Meshes**  Dimitri Mavriplis, ICASE NASA Langley Research Center

**Porous Media Flow on Locally Refined Grids**  Geert H. Schmidt, Koninklijke Shell Exploration and Production Laboratory, The Netherlands

### Suggested Reading

R. H. J. Gmelig Meyling, W. A. Mulder, and G. H. Schmidt, *Porous media flow on locally refined grids*, in Proc. Workshop on Numerical Methods for the Simulation of Multiphase and Complex Flows, T. Verheggen, ed., Springer-Verlag, Berlin, 1991.

W. D. Gropp and D. E. Keyes, *Domain decomposition methods in computational fluid dynamics*, Internat. J. Numer. Methods Fluids, 13 (1991).

P. W. Hemker, H. T. M. Van der Maarel, and C. T. H. Everaars, *Basis: A data structure for adaptive multigrid computations*, CWI Report NM-R9014, Centrum voor Wiskunde en Informatica, Amsterdam, 1990.

D. C. Jespersen and C. Levit, *A computational fluid dynamics algorithm on a massively parallel computer*, Internat. J. Supercomput. Appl., 3 (1989), pp. 9–27.

S. McCormick, *Multilevel Adaptive Methods for Partial Differential Equations*, SIAM Frontiers VI, Society for Industrial and Applied Mathematics, Philadelphia, 1989.

———, *Multilevel Projection Methods for Partial Differential Equations*, Society for Industrial and Applied Mathematics, Philadelphia, 1992.

G. H. Schmidt and F. J. Jacobs, *Adaptive local grid refinement and multigrid in numerical reservoir engineering*, J. Comput. Phys., 77 (1988), pp. 140–165.

H. T. M. Van der Maarel, P. W. Hemker, and C. T. H. Everaars, *Euler: An adaptive Euler code*, CWI Report NM-R9014, Centrum voor Wiskunde en Informatica, Amsterdam, 1990.

## Numerical Methods for Viscous Incompressible Flow

Various numerical methods have been developed for viscous incompressible flow, for example, spectral, finite difference, nonlinear Galerkin, and vortex methods. The problems of gaining more precision, calculating flow with high Reynolds number, simulating long-time behavior, and computing singularities are of most interest. The speakers presented new results and discussed their applications.

**Organizer:** Long-an Ying, Peking University, People's Republic of China

**Unsteady Time Asymptotic Flows in the Square Driven Cavity**  John Goodrich, NASA Lewis Research Center

**A Hybrid Spectral Method for the Three-Dimensional Numerical Modeling of Nonlinear Flows in Shallow Seas**  Robin Lardner and Yuahe Song, Simon Fraser University, Canada

**An Optimal Error Analysis for the Projection Method for the Navier-Stokes Equations**  Jie Shen, Indiana University, Bloomington

**Steady Flow Past Bluff Bodies**  Tao Tang, Simon Fraser University, Canada; Derek B. Ingham, University of Leeds, United Kingdom

**Gevrey Class Regularity and Approximate Inertial Manifolds for 3D Bénard Convection in Porous Media**  Edriss Titi, University of California, Irvine and Cornell University; Michael D. Graham and Paul H. Steen, Cornell University

# Chapter 21
# COMPUTER SCIENCE

## Analog Cryptography
The speakers explored the application of encryption in the analog domain to the protection of analog data where digital encryption is not feasible or desirable. Methods of analog data encryption were explored for both private key and public key cryptosystems. The efficiency and security of the methods were evaluated and the difference between these and digital algorithms were discussed.

**Organizers:** George I. Davida, University of Wisconsin, Milwaukee; Gilbert G. Walter, University of Wisconsin, Milwaukee

**Digital and Analog Cryptography**  George I. Davida, Organizer

**Analog Encryption Using Wavelets**  Gilbert G. Walter, Organizer

**Complexity of Ill-Posed Problems**  M. Zuhair Nashed, University of Delaware

**Cryptosystems and Codes Based on Infinite Structures**  R. Blakeley, Texas A&M University

## Continuous Computational Complexity
Problems in science and engineering often have continuous models. However, much of the work in computational complexity to date has applied to discrete or combinatorial problems.

Many continuous problems have been shown to be intractable in the worst case deterministic setting. Their complexity grows exponentially with dimension. This is disturbing because high dimensionality occurs in many mathematical problems such as integration, and in many applications from physical science, engineering, and economics.

Can randomization or settling for an average or probabilistic assurance break intractability? The speakers presented what is known and then proposed new directions for research.

Application of continuous computational complexity to such mathematical problems as linear programming and partial differential equations and to areas such as optimization, control theory, and computer vision were also discussed.

**Organizer:** Joseph F. Traub, Columbia University

**Power and Limitations of Randomization**  Joseph F. Traub, Organizer

**Computational Complexity of Multivariate Problems**  Henryk Wozniakowski, Columbia University

**Quadrature Formulas for Monotone Functions**  Erich Novak, University Erlangen-Nürnberg, Germany

**Ill-Posed Problem Instances**  James M. Renegar, Cornell University

## Fortran 90: How the New Language will Benefit Applied Mathematics
By the time of the meeting, it was anticipated that Fortran 90 will have been adopted as an international (ISO) standard and vendors will be working hard to provide implementations. It is an upward-compatible extension of Fortran 77, which will preserve the vast investment in codes for applied mathematics. The additional features lie in seven major areas: arrays, numerical computation, parameterized data types, user-defined data types, modules, pointers, and language evolution.

The speakers explained the advantages of the new language by case studies in three areas of applied mathematics and considered implementation issues based on the experience of one implementer.

**Organizer:** John K. Reid, Rutherford Appleton Laboratory, United Kingdom

**The Use of Fortran 90 for Numerical Software**  Brian T. Smith, University of New Mexico

**How Optimization Codes will Benefit From Fortran 90**  John K. Reid, Organizer

**The Use of Fortran 90 for Solving Partial Differential Equations**  John R. Rice, Purdue University; Ronald F. Boisvert, National Institute for Standards and Technology, Gaithersburg

**The Implementation of Fortran 90**  Malcom J. Cohen, Numerical Algorithms Group, United Kingdom

## Knowledge-Based Systems for Mathematical Software: Tools, Technologies, Case Studies
The minisymposium addressed the issue of knowledge-based systems (KBS) designed to assist computer users in the solution of numerical and statistical problems. While algorithms in the form of library routines or application-specific packages are available for solving a wide range of problems, it is evident that many users experience difficulties in relating their particular problems to a given set of algorithmic capabilities and in using those capabilities effectively. KBS advocates argue that proper emphasis on knowledge engineering and human-computer interface issues can alleviate many of those difficulties. The speakers highlighted the progress made and obstacles encountered in specific areas of KBS development and addressed the issue of a common, underlying technology for the construction of such systems.

**Organizer:** Brian Ford, Numerical Algorithms Group Ltd., United Kingdom

**Automated Selection of Mathematical Software Using Functional Representation**  Ian Gladwell, Southern Methodist University; Michael Lucks, Johns Hopkins University

**Knowledge-Based Synthesis of Numerical Programs for the Solution of Partial Differential Equations**
A. Daniel Kowalski, Rutgers University, Piscataway

**The ESPRIT FOCUS Project: A Framework for the Development of Knowledge-Based Front Ends**
Stephen J. Hague, Numerical Algorithms Group Ltd., United Kingdom

**A Focus Case Study: The Design and Construction of a Knowledge-Based Library Routine Advisor**   Colin W. Cryer, Westflische Wilhelm-Universität, Germany

## Scientific and Numerical Computing in C

The popularity of the C programming language for writing numerical software, and for use in scientific computing in general, is increasing. The language was recently standardized by ANSI but its numerical properties are still under scrutiny. Examples of the potential problem areas are exception handling, the IEEE floating point issues, variably dimensioned arrays, parallelization, and complex arithmetic.

This minisymposium focused on the advantages and disadvantages of the current C facilities relevant to numerical and scientific computing, and outlined the ongoing research and the standardization work of the ANSI X3J11 associated Numerical C Extensions Group (NCEG). The speakers addressed some of the important problem areas in more detail, and discussed the conversion of FORTRAN to C, the efficiency of the current C compilers, and the use of C in massively parallel systems.

**Organizers:** Stuart I. Feldman, BELLCORE, Morristown; Mladen A. Vouk, North Carolina State University

**An Overview of the C Standardization Work—ANSI X3J11 and Numerical C Extensions Group**   Mladen A. Vouk, Organizer

**Mathematical Software in Standard C — Why Not Standard Results?**   Cleve Moler, Mathworks

**Converting FORTRAN to C**   Stuart I. Feldman, Organizer

**Are Vectorizing C Compilers Helping the Scientific C Programmer?—A Comparative Analysis of C Compilers and a C Usage Analysis**   Lauren L. Smith, Supercomputing Research Center, Bowie

**Programming Massively Parallel Systems in C—C* on the Connection Machine**   James L. Frankel, Thinking Machines Corporation, Cambridge

**Coping with Arrays and Pointers**   Mladen A. Vouk, Organizer

Ha-Jine Kimn of Seoul, South Korea, with his son, Jung-han Kimn.

## SLI Arithmetic: An Alternative to Floating-Point

Symmetric level-index, or SLI, arithmetic is a recently proposed alternative to floating-point (FLP) arithmetic. One particularly awkward problem with FLP is its liability to overflow and underflow. SLI overcomes this difficulty using an iterated logarithmic representation for real numbers.

In this minisymposium, the speakers considered the analysis, implementation, and application of the SLI system. The first speaker provided an overview of the representation and its error analysis, highlighting the closure of the system and comparing it to FLP. The second speaker was concerned with implementations—both existing software implementations and some possible approaches to hardware. The last two speakers concentrated on applications—one on numerical algorithm design with the theme of numerical approaches to singularity, and the other on the advantages of the system for scientific software design.

**Organizer:** Peter R. Turner, United States Naval Academy

**SLI Arithmetic 1: Introduction and Analysis**   Frank W.J. Olver, University of Maryland, College Park; Charles W. Clenshaw, University of Lancaster, United Kingdom

**SLI Arithmetic 2: Implementation**   Peter R. Turner, Organizer; Daniel W. Lozier, National Institute of Standards and Technology

**SLI Arithmetic 3: Impact on Numerical Algorithms**
Charles W. Clenshaw, University of Lancaster, United Kingdom; J.L. Buchanan, United States Naval Academy; Daniel W. Lozier, National Institute of Standards and Technology; Frank W.J. Olver, University of Maryland, College Park; Ian Reid, Numerical Algorithms Group Ltd., United Kingdom; Peter R. Turner, Organizer

**SLI Arithmetic 4: Software Engineering Aspects**
Daniel W. Lozier, National Institute of Standards and Technology; Peter R. Turner, Organizer

## Software for Discrete Mathematics

This minisymposium focused on the issues surrounding the design, implementation, and use of software for discrete mathematics. The need for versatile software oriented toward problems of a discrete character is strongly felt in many applied areas such as communication networks and computer design. Such software can be a valuable research tool.

The major objectives of recent work in creating such software packages are to cover most fundamental needs, including the design of specialized languages, the development of methods for visualizing combinatorial objects, and algorithm animation. Several more specific software systems are developed for symbolic set manipulation and automatization of the discovery process.

The speakers considered how the domain of discrete mathematics uniquely affects software and system issues, described a number of software packages, and discussed such issues as portability, extendibility, and linking of different packages.

**Organizers:** Mark K. Goldberg, Rensselaer Polytechnic Institute; Gregory E. Shannon, Indiana University, Bloomington

**Set Player: A Software System for Symbolic Set Manipulation**   Mark K. Goldberg, Organizer; Dave Berque, Rensselaer Polytechnic Institute; Ronald Cecchini, Rensselaer Polytechnic Institute; Reid Rivenburg, Rensselaer Polytechnic Institute

**OPL — A Notation for Formulating and Solving Discrete Optimization Problems**  Robert Moll, University of Massachusetts, Amherst; Bruce Macleod, University of Maine

**Combinatorics and Graph Theory with Mathematica**  Steven S. Skiena, State University of New York, Stony Brook

**Interactive Software for Graphs**  Gregory E. Shannon, Organizer

**Automated Discovery in Mathematics**  Siemion Fajtlowicz, University of Houston, University Park

**Graphical Methods for Analyzing Networks**  Nathaniel Dean, Bellcore, Morristown

**Software System for Manipulating Graphs and Digraphs**  Mukkai Krishnamoorthy, Rensselaer Polytechnic Institute; Thomas Spencer, Rensselaer Polytechnic Institute; Eric McCaughrin, Rensselaer Polytechnic Institute

## Use of Automatically Derivable Analytic Information in Numerical Computation

Scientific computation deals mainly with differential equations and other analytic problems. Such problems have functions as input data (coefficients, boundary values, objective functions, etc.) which commonly appear as arithmetic expressions derived from the underlying mathematical model. While these data functions are evaluated many times in a solution algorithm, other mathematical information contained in them is not exploited although it could often lead to a more efficient and robust computation. The information would be retrievable from the arithmetic expression form of the data and from transformations applied during solution, but there are no tools for its retrieval within the context of a numerical computation.

The speakers considered directions for progress. They dealt with preliminary ways of retrieving and incorporating analytic information into numerical computation, with internal representations for functions which facilitate this, and with ways to integrate the mathematical and computational aspects in scientific problem solving.

**Organizer:** Hans J. Stetter, Technical University of Vienna, Austria

**Some Ways of Dealing with Function Data in Numerical Computation**  J.A. van Hulzen, Technical University of Twente, The Netherlands

**Preconditioning the Numerical Solution of Singularly Perturbed Nonlinear Systems Using Automatically Derived Analytic Information**  Mark F. Russo, Sterling Drug Inc., Malvern, PA

**Rigorous Recursive Calculation with Functions**  Arnold Neumaier, University of Freiburg, Germany

**Acquiring, Storing, and Using Mathematical Information in Scientific Software**  Stuart I. Feldman, Bell Communications Research, Morristown

### Suggested Reading

W. S. Brainerd, C. H. Goldberg, and J. C. Adams, *Programmer's Guide to Fortran 90*, McGraw–Hill, New York, 1990.

C. W. Clenshaw and F. W. J. Olver, *Beyond floating point*, J. Assoc. Comput. Mach., 31 (1984), pp. 319–328.

C. W. Clenshaw, F. W. J. Olver, and P. R. Turner, *Level-index arithmetic: An introductory survey*, Numerical Analysis and Parallel Processing, P. R. Turner, ed., Lecture Notes in Math., Vol. 1397, Springer-Verlag, New York, Berlin, 1989, pp. 95–168.

N. Dean and G. E. Shannon, eds., *Computational support for discrete mathematics*, Proc. from a workshop to be held in March 1992 at DIMACS, DIMACS Series in Discrete Mathematics and Theoretical Computer Science, AMS/ACM, 1992.

*Fortran*, ISO Publications Department, ISO/IEC 1539:1991, Geneva, Switzerland, 1991.

P. Gaffney and E. Houstis, eds., *Programming Environments for High Level Scientific Problem Solving*, Proc. IFIP WG 2.5 Working Conference 6, North-Holland, Amsterdam, 1992.

A. Griewank and G. F. Corliss, eds., *Automatic Differentiation of Algorithms: Theory, Implementation, and Applications*, Society for Industrial and Applied Mathematics, Philadelphia, 1991.

D. W. Lozier and F. W. J. Olver, *Closure and precision in level-index arithmetic*, SIAM J. Numer. Anal., 27 (1990), pp. 1295–1304.

D. W. Lozier and P. R. Turner, *Robust parallel computation in floating-point and SLI arithmetic*, Computing, 1992, to appear.

M. Metcalf and J. Reid, *Fortran 90 Explained*, Oxford University Press, Oxford, New York, Tokyo, 1990.

A. Neumaier, *Interval Methods for Systems of Equations*, Cambridge University Press, Cambridge, U.K., 1990, Ch. 2 and 5.

E. Novak, *Deterministic and Stochastic Error Bounds in Numerical Analysis*, Lecture Notes in Math., Vol. 1349, Springer-Verlag, Berlin, 1988.

J. M. Renegar, *It is possible to know a problem instance is ill-posed? Some foundations for a general theory of condition numbers*, Tech. Report, Department of Operations Research and Industrial Engineering, Cornell University, Ithaca, NY, 1991.

S. Skiena, *Implementing Discrete Mathematics: Combinatorics and Graph Theory with Mathematica*, Addison-Wesley, Reading, MA, 1990.

J. F. Traub, G. W. Wasilkowski, and H. Wozniakowski, *Information-Based Complexity*, Academic Press, New York, 1988.

P. R. Turner, *Implementation and analysis of extended SLI Operations*, Proc. ARITH10, P. Kornerup and D. W. Matula, eds., IEEE Computer Society, Washington DC, 1991, pp. 118–126.

A. G. Werschulz, *The Computational Complexity of Differential and Integral Equations*, Oxford University Press, Oxford, U.K., 1988.

H. Wozniakowski, *Average case complexity of multivariate integration*, Bull. Amer. Math. Soc. (N.S.), 24 (1991), pp. 185–194.

# Chapter 22
# CONTROL AND SYSTEMS THEORY

## Boundary Control Theory for Partial Differential Equations and Applications

The subject of this minisymposium is boundary control theory of systems described by partial differential equations, where the control function acts on the boundary and where the boundary itself is a control variable. Applications include control of large space flexible structures, robotics, phase transition problems, and dam problems.

Current directions of research in this field call for development of new techniques in the area of partial differential equations. Indeed, these have recently accounted for the solution of a number of questions in control and stabilization. However, inaccuracies of modeling, the presence of uncontrolled perturbations, and inherent nonlinearities present an array of new challenges for future research.

The speakers addressed several issues related to control and stabilization of various models of partial differential equations. Both linear and nonlinear models were discussed.

**Organizer:** Irena Lasiecka, University of Virginia

**Optimal Control for Problems in Partial Differential Equations**  Avner Friedman, Institute for Mathematics and Its Applications, University of Minnesota

**Controllability and Stabilizability of Nonlinear Plates**  John E. Lagnese, Georgetown University

**Riccati Equations for Elastic Systems—Theoretical and Numerical Aspects**  Roberto Triggiani, University of Virginia

**On Control and Stabilization of Nonlinear Rotating Beams**  Günter Leugering, Georgetown University

## $H_\infty$ Optimal Control

The $H_\infty$ optimal control problem has now been studied for more than a decade. Recently, the introduction of state space techniques has influenced substantially the accomplishments in $H_\infty$ control and has led to a deeper understanding of the intricate structure of many problems in this area. To broaden the scope of applications of $H_\infty$ control theory, current research is directed towards applications in robust stabilization problems and multi objective control problems. In this minisymposium, some of the recent theoretical achievements were addressed, and various new developments were discussed.

**Organizer:** S. Weiland, Rice University

**The Singular Minimum Entropy $H_\infty$ Control Problem**  A.A. Stoorvogel, University of Michigan, Ann Arbor

**$H_\infty$ Filtering of Sampled Data Systems**  W. Sun Jr. and P.P. Khargonekar, University of Michigan, Ann Arbor; K.M. Nagpal, University of Illinois, Urbana

**On the Mixed $H_2/H_\infty$ Problem**  H.L. Trentelman, University of Gröningen, The Netherlands

**$H_\infty$ Optimal Control and Dissipative Dynamical Systems**  S. Weiland, Organizer

## Hereditary Systems and Their Applications

Hereditary systems are systems described by functional differential equations, integro-differential equations and time-delay equations. Their applications are found primarily in regulation theory, mechanics, ecology, and medicine. The problems under consideration are: stability, periodic oscillations, optimal control, adaptive control, control with incomplete state information, and optimal state estimate.

The speakers presented current research and discussed unresolved problems.

**Organizer:** Vladimir Kolmanovskii, MIEM, USSR

**Stability of Hereditary Systems and Their Applications**  Vladimir Kolmanovskii, Organizer

**Periodic Oscillations of Hereditary Systems**  Valerii Nosov, MIEM, USSR

**Optimal Control of Hereditary Systems**  E. Bruce Lee, University of Minnesota, Minneapolis

**State Estimation of Hereditary Systems**  Karl Kunisch, Technische Universität Graz, Austria

## HJB Equations, Viscosity Solutions, and Optimal Control of Hydrodynamics

Optimal control theory of viscous flow is a very new subject. However, it has the potential of growing into a major discipline of applied mathematics and engineering science. In this minisymposium, we focused on two of the new and very fundamental results: the Pontryagin maximum principle and dynamic programming. The Hamilton-Jacobi-Bellman equation satisfied by the value function was shown to play a central role. Major mathematical techniques utilized were the viscosity solution concept of Crandall and Lions and the Ekeland variational principle. Related issues for general nonlinear evolution equations were also discussed.

**Organizer:** S.S. Sritharan, University of Southern California

**Dynamic Programming of the Navier-Stokes Equations**  S.S. Sritharan, Organizer

**The Maximum Principle for Flow Control Problems**  Hector O. Fattorini, University of California, Los Angeles

**Applications of Viscosity Solutions to Problems in Optimal Control**  Lawrence Evans, University of California, Berkeley

**Regularity Properties of the Value Function and Their Application to the Optimal Control of Evolutionary Systems**   Helene Frankowska, Université de Paris-Dauphine, France; Piermarco Cannarsa, Università di Pisa, Italy

## Matrix Problems in Control Theory

Two areas of major current interest in the mathematical theory of linear control systems were addressed. The first concerned descriptor systems, where the derivative vector has a singular matrix coefficient. First, a solution was given for the determination of a stabilizing state feedback which decouples the output from external disturbances. Next, for systems containing input derivatives, an inversion procedure was developed using matrix generalized inverses. The second problem area relates to equivalence of representations of linear systems. A new polynomial-based transformation was presented which preserves the finite and infinite frequency system characteristics. Finally, for bilinear strict equivalence of matrix pencils, a new set of invariants was introduced, thereby extending some classical results.

**Organizer:** A.C. Pugh, Loughborough University of Technology, United Kingdom

**On Inversion of Linear Control Systems Containing Input Derivatives**   Victor Lovass-Nagy, Clarkson University; David L. Powers, Clarkson University; R. J. Schilling, Clarkson University

**A Polynomially-Based Transformation for the Generalized Study of Linear Systems**   A.C. Pugh, Organizer; G.E. Hayton, Hull University, United Kingdom

**New Invariants on the Bilinear Strict Equivalence of Matrix Pencils**   Nicholas Karcanias, City University, United Kingdom; Grigoris Kalogeropoulos, University of Athens, Greece

## Numerical Computation of Optimal Robot Trajectories

Numerical optimal control methods for the computation of robot trajectories offer significant advantages for the path-planning problem. Whereas heuristic strategies are often far from being time-, energy- or jerk-optimal, numerical optimal control can significantly improve the performance of a robot manipulator for a given task; it also provides reference movements ideally suited for optimal robot design.

Optimal robot controls have difficult structures — multiple bang-bang and singular arcs occur and local minima exist. Standard indirect (e.g., maximum-principle-based) methods are therefore often too complicated or time consuming for routine industrial application, which calls for easy-to-use and fast direct numerical methods. For applications where computing time is the critical factor (e.g., interactive design and real time optimal control), the development of scalable performance parallel algorithms is important.

The session covered point-to-point and trajectory-prescribed robot control problems, single and multiple manipulator tasks, and new (parallel) direct numerical algorithms.

**Organizer:** Georg H. Bock, University of Augsburg, Germany

**Minimum Time Motions of a Two-Arm Robot**   Arthur E. Bryson, Stanford University; Rhonda Slattery, Stanford University

**Semi-Infinite Programming Techniques in Robotics**   Rainer Hettich, University of Trier, Germany; Elke Haaren-Retagne, University of Trier, Germany

**Physical Interpretation of Time-Optimal Movements of Various Types of Industrial Robots**   Richard W. Longman, Columbia University; George H. Boch, Organizer; Mark Steinbach, Universität Augsburg, Germany

**Efficient Parallel Algorithms for the Numerical Computation of Optimal Point-to-Point Trajectories of Robots**   Marc C. Steinbach, Universität Augsburg, Germany; Georg H. Bock, Organizer

## Numerical Methods for the Hamilton-Jacobi-Bellman Equations of Optimal Control Problems

The numerical solution of optimal control problems is one of the necessary requirements for many applications, e.g., in engineering, economics, and robotics. It is crucial in the majority of nonlinear control problems, where explicit algebraic solutions are missing and the synthesis of feedback optimal controls cannot be formally derived. In the last decade, the rapid development of the theory of generalized (viscosity) solutions for Hamilton-Jacobi-Bellman equations has given a rigorous mathematical basis to the dynamic programming formulation. For many control problems, the value function can now be characterized as the unique viscosity solution of a corresponding Hamilton-Jacobi-Bellman equation.

This approach also has been used to construct several approximation schemes which have interesting features. First of all, they provide approximate value functions in a domain, whereas the existing numerical methods, based on Pontryagins principle, give the approximate value of the solution in one point. Moreover, approximate feedback controls can be obtained in that domain.

The speakers presented several results in this direction for equations related to nonlinear deterministic and stochastic control problems and to differential games.

**Organizer:** Maurizio Falcone, Università di Roma "La Sapienza", Italy

**Numerical Methods for Controlled Reflected Processes and the Skorokhod Problem**   Harold J. Kushner, Brown University; Paul Dupuis, Brown University

**Some Results on Discrete Approximation for Optimal Stochastic Control**   José L. Menaldi, Wayne State University

**On the Numerical Approximation of the Hamilton-Jacobi-Bellman Equation**   Jean-Pierre Quadrat, INRIA, France; A. Sulem, INRIA, France; M. Akian, INRIA, France

**Numerical Solution of Deterministic Optimal Control Problems**   Maurizio Falcone, Organizer

## Optimal Control Methods in Meteorology and Oceanography

Numerical weather forecasting needs a model and data. The problem is to insert the data into the model in such a way that the resulting solution of the model is in good agreement with the observation and the flow has the general properties of atmospheric flow. Optimal control methods have been proposed to verify these requirements; many operational centers in numerical weather prediction plan to carry out these methods especially for using satellite data.

The development of these methods implies numerous improvement in numerical methods in fluid dynamics, understanding dynamic systems, and optimization algorithms for very large systems. The speakers presented an overview of current models, techniques, and problems in this area.

**Organizer:** Francois-Xavier L. Le Dimet,
Université Blaise Pascal, France

**Optimal Control for Data Assimilation in Meteorology** Francois-Xavier L. Le Dimet, Organizer

**Optimal Control Methods and Efficient Minimization Algorithms for Variational Data Assimilation in Meteorology** Michael I. Navon, Florida State University; Xioalei Zou, Florida State University

**Variational Assimilation of Meteorological Observations** Olivier Talagrand, Laboratoire de Meteorologie Dynamique, CNRS, France; Jean-Noel Thepaut, European Center for Medium Range Weather Forecasts, United Kingdom

**Very Large Scale Control Problems in Oceanography** Carl Wunsch, Massachusetts Institute of Technology

## Parallel Computing and Optimization

Parallel computers have given a fresh perspective to the design of algorithms in numerical analysis and mathematical programming. There is interest in efficient parallel implementations of good "serial" algorithms and renewed interest in algorithms which are not usually competitive in a serial computing environment, but which lend themselves well to massively parallel computation.

For nonlinear optimization algorithms, there is often the need to calculate derivatives. Here, parallel automatic differentiation techniques could save the user the trouble of writing efficient parallel code for derivative evaluation.

Speakers in this session discussed issues in parallel optimization such as the design of optimal algorithms for large-scale convex and nonlinear programming, the development of parallel approaches to network optimization, and parallel implementation of automatic differentiation.

**Organizer:** Stephen J. Wright, Argonne National Laboratory

**Adaptive Multilevel Methods for Compact Fixed Point Problems** C. Tim Kelley, North Carolina State University

**An Auction Algorithm for Shortest Paths** Dimitri P. Bertsekas, Massachusetts Institute of Technology

**On the Performance of Optimization Software on High-Performance Computers** Brett Averick, Argonne National Laboratory and University of Minnesota, Minneapolis; Richard Carter, Argonne National Laboratory and University of Minnesota, Minneapolis; Jorge J. Moré, Argonne National Laboratory

**Parallel Function and Derivative Evaluation** Christian H. Bischof, Argonne National Laboratory; Andreas Griewank, Argonne National Laboratory

## Recent Progress in Stochastic Control Theory

Research in the theory of stochastic control is motivated by problems from aerospace guidance and control, computer and communication networks, manufacturing processes, and other areas. For models based on stochastic differential equations, a great deal of progress has been made over the past two decades.

In this minisymposium, we focused on three recent areas of development: (1) the theory of singular stochastic control which arises from the so-called "heavy traffic" diffusion approximation in queueing-system models for communication networks, including analytical techniques and the related free-boundary problems; (2) the extension of traditional applications to systems engineering and to models in financial economics, such as pricing of stock options; and (3) an optimal stochastic control problem with an unobserved state wherein the solution to a linear quadratic control system will be given.

**Organizers:** Pao-Liu Chow, Wayne State University;
José-Luis Menaldi, Wayne State University

**An Overview of Singular Stochastic Control—Methods and Applications** Maurice Robin, INRIA, France

**Stochastic Control Models for Financial Economics** Wendell H. Fleming, Brown University

**A New Representation for the Optimal Stopping Risk and Its Applications** Ioannis Karatzas, Columbia University

**An Optimal Control Depending on the Conditional Density of the Unobserved State** Kurt Helmes, University of Kentucky; Raymond W. Rishel, University of Kentucky

## Shape and Boundary Variations

In many new and challenging applications, the control variable is not only a function defined on the domain or its boundary, but the domain or the boundary itself. Shape optimization provides a broad framework for optimal control problems in which the geometry is an integral part of the control scheme.

Numerical methods are an extremely important part of this activity. Shape sensitivity analysis provides the basic tools for the shape calculus. Its applicability extends to many physical systems modeled by variational equalities and inequalities. Important classes of free boundary problems can now be described by shape variational principles.

The existence of optimal shapes and boundaries is becoming an urgent issue. New efficient tools are needed to predict, describe, and characterize optimal domains and boundaries. The underlying mathematical problems are challenging and of paramount importance for applications and the choice of numerical methods.

**Organizers:** Michel C. Delfour, Université de Montréal, Canada; Jean-Paul Zolésio, Université de Nice, France

**Bounded Perimeter and Total Curvature Sets in Shape Optimization** Jean-Paul Zolésio, Organizer

**Optimization Over Sets of Finite Perimeter** Steven J. Cox, Rice University; Marty Ross, Rice University

**Shape Analysis of Nonsmooth Domains** Michel C. Delfour, Organizer

**Sensitivity Analysis of Shape Optimization Problems** Jan Sokolowski, Polish Academy of Sciences, Poland

**Numerical Computation of Free Boundary Problems** Michel Fortin, Université Laval, Canada

**Dynamical Shape Control of Nonlinear Thin Rods** Juergen Sprekels, Universität-GH Essen, Germany

**Three-Dimensional Shape Optimization of an Elastic Structure Using the Homogenization Method** Noboru Kikuchi, University of Michigan, Ann Arbor

**Optimization of Photocell to Transform Light into Electricity** Olivier Pironneau, INRIA, France; Y. Achdou, Université de Paris VI, France

**Recent Developments in Finite Element Based Shape Sensitivity Analysis** Pierre Becker, Université de Liege, Belgium

**Shape Sensitivity Analysis for Creeping Materials**
Robert B. Haber, University of Illinois, Urbana-Champaign; H.S. Lee, University of Illinois, Urbana-Champaign; C. Vidal, University of Illinois, Urbana-Champaign

**Suggested Reading**

M. Akian, *Analyse de l'algorithme multigrille FMGH de resolution d'equations d'Hamilton-Jacobi-Bellman*, in Analysis and Optimization of Systems, A. Bensoussan and J. L. Lions, eds., Lecture Notes in Control and Inform. Sci., Vol. 144, Springer-Verlag, New York, Berlin, pp. 113–122.

J. P. Aubin and H. Frankowska, *Set Valued Analysis*, Birkhauser Publications, Basel, Switzerland, 1990.

A. V. Balakishnan, *A Continuum Model for Interconnected Lattice Trusses*, Proc. of the Eighth VPII & SU Symposium on Dynamics and Control of Large Structures, Blacksburg, VA, 1991.

M. C. Bancora–Imbert, P. L. Chow, and J. L. Menaldi, *On the numerical approximation of an optimal correction problem*, SIAM J. Sci. Statist. Comput., 9 (1988), pp. 970–991.

V. Barbu, E. N. Barron, and R. Jenson, *Necessary conditions for optimal control in Hilbert spaces*, J. Math. Anal. Appl., 133 (1988), pp. 151–162.

M. Bardi and M. Falcone, *An approximation scheme for the minimum time function*, SIAM J. Control Optim., 28 (1990), pp. 950–965.

D. P. Bertsekas and J. N. Tsitsiklis, *Parallel and Distributed Computing*, Prentice–Hall, Englewood Cliffs, NJ, 1989.

C. H. Bischof, *Issues in parallel automatic differentiation*, MCS—P235—0491, Mathematics and Computer Science Division, Argonne National Laboratory, IL, 1991.

H. G. Bock and K. J. Plitt, *A multiple shooting algorithm for direct solution of optimal control processes*, Proc. 9th IFAC World Congress, Budapest, Pergamon Press, Elmsford, NY, London, 1984.

I. Capuzzo Dolcetta and M. Falcone, *Discrete dynamic programming and viscosity solutions of the Bellman equation*, Ann. Inst. H. Poincare Anal. Non Lineaire, 6 Supp. (1989), pp. 161–184.

P. L. Chow, J. L. Menaldi, and M. Robin, *Additive control of stochastic linear systems with finite horizon*, SIAM J. Control Optim., 23 (1985), pp. 858–899.

J. J. Dongarra, I. S. Duff, D. C. Sorensen, and H. A. van der Vorst, *Solving Linear Systems on Vector and Shared-Memory Computers*, Society for Industrial and Applied Mathematics, Philadelphia, 1991.

J. Doyle, K. Glover, P. P. Khargonekar, and B. A. Francis, *State space solutions to standard $H_2$ and $H_\infty$ control problems*, IEEE Trans. Automat. Control., 34 (1989), pp. 831–847.

P. Dupuis and H. Ishii, *PDE's with oblique reflection on nonsmooth domains*, Ann. Probab., 1990.

N. El Karoui and I. Karatzas, *A new approach to the Skorohod problem and its applications*, Stochastica, 34 (1991), pp. 57–82.

H. Fattorini and S. S. Sritharan, *Optimal control of viscous flow problems*, to be published.

W. H. Fleming and T. Zariphopoulou, *An optimal investment consumption model with borrowing*, Math. Oper. Res., 1991.

B. A. Francis, *A Course in $H_\infty$ Control Theory*, Lecture Notes in Control and Inform. Sci., Vol. 88, Springer-Verlag, Berlin, 1987.

A. Friedman, *Variational Principles and Free Boundary Problems*, John Wiley, New York, 1982.

W. Hackbush and U. Trottenberg, eds., *Multigrid Methods*, Lecture Notes in Math., Vol. 960, Springer-Verlag, New York, Berlin, 1982.

K. Gopalsamy, *Equations, Part I*, Math. Ecol., 1991, preprint.

E. Haaren-Retagne, *Robot trajectory planning with semi-infinite programming*, IEEE Trans. Automat. Control, submitted.

J. K. Hale, *Theory of Functional Differential Equations*, Springer-Verlag, New York, 1977.

K. L. Helmes and R. W. Rishel, *An optimal control depending on the conditional density of the unobserved state*, Proc. 2nd U.S.–French Workshop on Applied Stochastic Analysis, to appear.

P. P. Khargonekar and M. A. Rotea, *Multiple objective optimal control of linear systems: The quadratic norm case*, IEEE Trans. Automat. Control, 36 (1991), pp. 14–24.

V. B. Kolmanovskii and V. R. Nosov, *Stability of Functional Differential Equations*, Academic Press, New York, London, 1986.

Y. Kuang and H. L. Smith, *Global stability in diffusive delay Lotka–Volterra systems*, Differential Integral Equations, 4 (1991), pp. 117–128.

H. J. Kushner, *Numerical methods for stochastic control problems in continuous time*, SIAM J. Control Optim, 28 (1990).

J. Lagnese, *Boundary Stabilization of Thin Plates*, Society for Industrial and Applied Mathematics, Philadelphia, 1989.

J. Lagnese and G. Leugering, *Uniform stabilization of a nonlinear beam by nonlinear boundary feedback*, J. Differential Equations, to appear.

I. Lasiecka and R. Triggiani, *Differential and Algebraic Riccati Equations with Applications to Boundary/Point Control Problems: Continuous Theory and Approximation Theory*, Lecture Notes in Control and Inform. Sci., Vol. 164, Springer-Verlag, New York, Berlin, 1991.

D. C. McFarlane and K. Glover, *Robust controller design using normalized coprime factor plant descriptions*, Lecture Notes in Control and Inform. Sci., Vol. 138, Springer-Verlag, Berlin, 1990.

J. L. Menaldi, *Some estimates for finite difference approximations*, SIAM J. Control Optim., 27 (1989), pp. 579–607.

R. A. Slattery, *Optimal control of closed-chain robotic systems*, Dissertation, Stanford University, Stanford, CA, 1991.

S. S. Sritharan, *An optimal control problem in exterior hydrodynamics*, Proc. Roy. Soc. Edinburgh, to appear.

———, *Dynamic programming of the Navier Stokes equations*, Systems Control Lett., 16, pp. 299–307.

M. C. Steinbach, H. G. Bock, and R. W. Longman, *Time-optimal extension or retraction in polar-coordinate robots: A numerical analysis of the switching structure*, Proc. 1989 AIAA Guidance, Navigation, and Control Conference, Boston, 1989, pp. 895–910. J. Optim. Theory Appl., submitted.

———, *Time optimal control of SCARA robots*, Proc. 1990 AIAA Guidance, Navigation, and Control Conference, Portland, OR, 1990.

A. A. Stoorvogel, *The $H_\infty$ control problem: A state space approach*, Ph. D. Thesis, University of Eindhoven, Eindhoven, the Netherlands, 1990.

S. J. Wright, *Partitioned dynamic programming for optimal control*, SIAM J. Optim., 1 (1991), pp. 620–642.

# Chapter 23
# DISCRETE MATHEMATICS

## Combinatorics Meets (and Joins) Algebra and Topology

Interaction of combinatorialists with algebraists and topologists catalyzes exciting research in combinatorics. The result is an exchange of perspectives, borrowing of basic techniques, collaboration on projects of mutual interest, and occasional application of a theorem in one field to solve an important problem in another.

The speakers surveyed current areas of interaction, diverse in subject and sophistication. These included arrangements and polytopes, the combinatorics and topology of subgroup complexes, combinatorial invariant theory, and representation theory.

**Organizer:** Lynne M. Butler, Princeton University

**Generalized Flags, Poset Embeddings, and Partitions**  Lynne M. Butler, Organizer; Alfred W. Hales, University of California, Los Angeles

**The Combinatorics and Topology of Hyperplane Arrangements**  Paul Edelman, University of Minnesota, Minneapolis

**Recent Combinatorial Results in Representation Theory**  Andrei Zelevinsky, Cornell University

**Free Non-Associative Algebras Arising from Variations on the Lie Bracket**  Sheila Sundaram, Université du Québec à Montréal, Canada

## Constructive Combinatorics

Over the past decade, the area of constructive combinatorics has developed with great impetus, building on the tradition of combinatorial methods for explicit computations and constructions in a variety of mathematical fields such as representation theory and topology. Constructive combinatorics provides insightful proofs which lend themselves to the development of combinatorial algorithms, and reveals new results as by products of the method of proof.

Significant results obtained in connection with the theory of partitions, group representations, and partially ordered sets have enriched and advanced modern combinatorics, and constructive combinatorics is rich in interesting open problems and challenging conjectures. The speakers presented recent results and open problems in the area of constructive combinatorics.

**Organizer:** Rodica Simion, George Washington University

**Substance and Shadow, Combinatorial Proof and Algebraic Proof**  Doron Zeilberger, Temple University, Philadelphia

**Descents and Fixed Points of Permutations**  Michelle Wachs, University of Miami, Coral Gables; Jacques Désarménien, Université Robert-Schumann, France

**Variations on Schensted's Algorithm**  Ira M. Gessel, Brandeis University

**Relationships between Standard Young Tableaux and Rigged Configurations**  Kathleen O'Hara, University of Iowa; Dennis Stanton, University of Minnesota

**A Combinatorial Interpretation of the Inverse Kostka Matrix and Applications**  Jeffrey Remmel, University of California, San Diego

**The Flag Variety and Young's Natural Representation**  Helene Barcelo, University of Michigan, Ann Arbor

## Hypergraphs and Ramsey Theory

Hypergraph theory and Ramsey theory are both dynamically developing, important branches of discrete mathematics. Hypergraphs concern fundamental combinatorial objects of finite sets and Ramsey theory is a study of the underlying unavoidable structures. These two areas are intimately related to each other.

The speakers in this minisymposium presented important work in these areas.

**Organizer:** Peter Frankl, CNRS, France

**Ramsey Theory and Computational Complexity**  Stefan Burr, City University of New York

**My Favorite Problem in Ramsey Theory**  Ronald L. Graham, AT&T Bell Laboratories, Murray Hill

**Rado Numbers**  Sao-Bing Loo, City University of New York

**The Erdös-Ko-Rado Theorem for Small Families**  Richard Duke, Georgia Institute of Technology; Vojtech Rödl, Emory University

## Random Generation and Approximate Counting via Rapidly Mixing Markov Chains

The problem of efficiently computing the permanent of a matrix has intrigued mathematicians ever since Cauchy defined the permanent in 1812. The #P-completeness of this problem established its intractability and led researchers to look for polynomial time approximation algorithms, especially for the case of 0/1 matrices which corresponds to counting the number of perfect matchings in a related bipartite graph. Recently developed techniques have resulted in significant progress on this problem. Approximately counting the number of solutions to a self-reducible problem has been shown to be equivalent to generating a random one. The latter problem is tackled by showing that a certain Markov chain, whose states encode individual solutions, is rapidly mixing and therefore converges in polynomial time to its stationary (uniform) distribution.

This technique has been used for approximating other important #P-complete problems as well, e.g. computing the volume of a convex body, and counting the number of total orders consistent with a given partial order. The speakers presented recent advances in these important classes of problems.

**Organizers:** Vijay Vazirani, Cornell University; Umesh Vazirani, University of California, Berkeley

**Generating Random Spanning Trees**  Andrei Broder, DEC Systems Research Center, Palo Alto

**On the Convergence Rate of Markov Chains**  Milena Mihail, Bellcore, Morristown

**Path Techniques for Bounding Conductance**  Umesh Vazirani, Organizer

**The Mixing Rate of Markov Chains, an Isoperimetric Inequality, and Computing the Volume**  Miklos Simonovits, Hungarian Academy of Sciences, Hungary

## Sequence Analysis in Molecular Biology

There is an evident need for more sophisticated tools for sequence analysis to deal with the ever increasing quantities of protein and DNA sequence data being generated by molecular biologists. Some problems of sequence analysis are familiar to computer scientists, e.g., searching for all occurrences of a fixed pattern in a long DNA sequence. Other problems have interesting combinatorial features, e.g., searching for repeated subsequences, palindromes, inverted complemented palindromes. Still other problems lack clear mathematical definition, e.g., finding intron-exon boundaries in DNA. A common characteristic of all these problems is "dirty" data: the occurrence of random deletions, insertions, and substitutions of letters in sequences.

The speakers presented an overview of this developing area.

**Organizer:** Eugene L. Lawler, University of California, Berkeley

**Linear Approximation of Shortest Superstrings**  Avrim Blum, Massachusetts Institute of Technology; Tao Jiang, McMaster University, Canada; Ming Li, University of Waterloo, Canada; John Tromp, CWI, The Netherlands; Mihalis Yannakakis, AT&T Bell Laboratories, Murray Hill

**Approximate Regular Expression Matching with Concave Gap Penalties**  Gene Meyers, University of Arizona

**Computational Complexity of Inferring Phylogenies**  Tandy J. Warnow, University of California, Berkeley; Sampath Kannan, DIMACS and University of Arizona

## Theoretical Aspects of Parallel Structures

Advances in computer technology have made massively parallel computers a reality rather than just a dream. Myriad problems relating to the design, use, and implementation of such computers can be abstracted to algorithmic or combinatorial problems that are as attractive in their own right as they are relevant to parallel computation.

The speakers presented recent work on a variety of such algorithmic and combinatorial problems. They discussed the effects of machine "topologies" on efficiency of implementation, interprocessor communication, and algorithm development.

**Organizer:** Arnold L. Rosenberg, University of Massachusetts, Amherst

**Coding Theory and Efficient Communication on the Hypercube**  William Aiello, Bell Communications Research, Morristown

**Algorithmic Consequences of Group Structure Underlying Cayley Graphs**  Fred Annexstein, University of Cincinnati

**Theoretical Aspects of VLSI Pin Limitations**  Robert Cypher, IBM Almaden Research Center

**Randomly-Wired Multistage Networks**  Bruce Maggs, NEC Research Institute, Princeton

**Circuits that Sort Most Permutations in Small Depth**  Gregory Plaxton, University of Texas, Austin; Yuan Ma, Massachusetts Institute of Technology; Tom Leighton, Massachusetts Institute of Technology

# Chapter 24
# DYNAMICAL SYSTEMS

## Chaos, Control, and Noise: Deterministic and Stochastic Aspects

Common techniques in the theory of dynamical systems, control systems, and stochastic systems yield fruitful cross fertilization of these areas and lead to important results and applications such as connections between control and chaos, control of complex behavior in nonlinear systems, ergodic and entropy theory of (skew product) flows, applications of Lyapunov exponents in control theory and stochastic systems, stochastic bifurcations, relations between stochastic differential equations and chaotic systems, stability and stabilization.

In this minisymposium, recent approaches towards a unifying theory were presented, together with several applications. The audience was invited to participate in a discussion of complex nonlinear phenomena in the areas indicated above.

**Organizers:** Fritz Colonius, Universität Augsburg, Germany; Wolfgang H. Kliemann, Iowa State University

**Dynamical Systems and Control Systems: An Introduction**   Wolfgang H. Kliemann, Organizer

**Towards a Global Analysis of Control Systems**   Fritz Colonius, Organizer; Wolfgang H. Kliemann, Organizer

**Lyapunov Exponents and Optimal Control**   Russell A. Johnson, University of Southern California

**Applications of Topological Dynamics to Problems of Controllability and Observability**   Mahesh Nerurkar, Rutgers University, New Brunswick

**Controlling Chaos**   Celso Grebogi, University of Maryland, College Park

**Lyapunov Exponents of Stochastic Oscillators—Comparison of Real Noise versus White Noise**   Mark Pinsky, Northwestern University

**Large Stochastic Systems as Dynamical Systems**   Ludwig Arnold, University of Bremen, Germany

**Lyapunov Exponents for Stochastically Perturbed Co-Dimension Two Bifurcations**   Sri Namachchivaya, University of Illinois, Urbana-Champaign: Sanjiv Talwar, University of Illinois, Urbana-Champaign

**Characterization of Stochastic and Chaotic Systems**   Thomas J.S. Taylor, Arizona State University

## Dynamics, Dimensions, Fractals, and Multifractal Decomposition

Fractal dimension and coding space constructions are particularly effective and elegant techniques in the study of dynamical systems. Four examples of these were presented in this minisymposium. The speakers presented a cautionary result that existence of a pointwise limit for the dimension of an ergodic measure does not imply that this pointwise limit is constant almost surely, described an analysis of the geometry of chaotic attractors for cylinder maps using a coding construction based on the dynamics of the map, discussed some rigorous results for multifractals obtained from graph-directed constructions, and described results for the dimension of the graphs of sparse Fourier series, a probabilistic generalization of the classical Weierstrass function.

**Organizer:** Robert Cawley, Naval Surface Warfare Center, Silver Spring

**Dimension Decompositions of Ergodic Dynamical Systems**   Colleen D. Cutler, University of Waterloo, Canada

**Invariant Attracting Continua in Cylinder Maps**   Pat Carter, Naval Surface Warfare Center

**Fractal Dimension of Graphs of Sparse Fourier Series**   Brian R. Hunt, United States Naval Systems Weapons Center, Silver Spring

**Hausforff Dimension and Multifractals**   Gerald A. Edgar, Ohio State University

## Random Eigenvalue Problems and Applications

The modeling of complex (linear) dynamic systems is important in all areas of application. The most important descriptor of such systems is the eigenvalue structure, i.e., eigenvalues/eigenvectors. One approach to modeling complex systems is probabilistic, where the system parameters are interpreted to be random variables. The characteristic equations for such systems are random algebraic polynomials, for which one attempts to estimate the distributions of the random eigenvalues. The speakers discussed theory and application.

**Organizer:** Haym Benaroya, Rutgers University, Piscataway

**Random Eigenvalue Problems of Sturm-Liouville Type**   William E. Boyce, Rensselaer Polytechnic Institute

**Random Eigenvalue Problems and Structural Dynamics**   Haym Benaroya, Organizer

**Random Eigenvalues via Computation for Dynamics**   Mohammed Ettouney, Weidlinger Associates, New York

**Random Flow Induced Vibrations**   K.Y.R. Billah, Princeton University; Masanobu Shinozuka, Princeton University

## Suggested Reading

L. Arnold, H. Crauel, and J.-P. Eckmann, eds., *Lyapunov Exponents*, Springer Lecture Notes in Math., Springer-Verlag, Berlin, New York, 1991.

A. T. Bharucha–Reid and M. Sambandham, *Random Polynomials*, Academic Press, New York, 1986.

F. Colonius and W. Kliemann, *Limit behavior and genericity for nonlinear control systems*, Report No. 305 of the Schwerpunktprogramm der Deutschen Forschungsgemeinschaft 'Anwendungsbezogene Optimierung und Steurung', Universitat Augsburg, 1991.

R. A. Ibrahim, *Structural Dynamics with Parameter Uncertainties*, Appl. Mech. Rev., 40 (1987), pp. 309–328.

S. Namachchivaya and Talwar, *Lyapunov exponents for stochastically perturbed codimension 2 bifurcations*, J. Sound Vibration, submitted.

E. Ott, C. Grebogi, and J. Yorke, *Controlling chaos*, Phys. Rev. Lett., 64 (1990), pp. 1196–1199.

M. Pinsky, *Extremal character of the Lyapunov exponent of the harmonic oscillator*, Ann. Appl. Probab., to appear.

———, *Lyapunov exponents of real-noise driven nilpotent systems and harmonic oscillators*, Stochastics Rep., 35 (1990), pp. 93–110.

M. Pinsky and V. Wihstutz, eds., *Diffusion Processes and Related Problems in Analysis, Vol. 2: Stochastic Flows*, Birkhauser, Basel, Switzerland, 1991.

T. Shinbrot, E. Ott, C. Grebogi, and J. Yorke, *Using chaos to direct trajectories to targets*, Phys. Rev. Lett., 65 (1990), pp. 3215–3218.

T. Taylor, *Observations of chaotic systems and randomness*, in Analysis and Control of Systems, C. Byrnes, C. Martin, and P. Saeks, eds., North–Holland, Amsterdam, New York, 1988.

———, *Systems with fast chaotic components*, in Computation and Control, K. Bowers and J.R. Lund, eds., Birkhauser, Basel, Switzerland, 1989.

J. vom Scheidt and W. Purkert, *Random Eigenvalue Problems*, North–Holland, Amsterdam, New York, 1983.

The CICIAM Committee plans for ICIAM 95.

# Chapter 25
# ELECTROMAGNETICS AND SEMICONDUCTORS

## Advanced Models for Semiconductor Device Simulation

The ongoing miniaturization of semiconductor devices has created a need for new models for charge transport in semiconductors. In VLSI applications, the effects of ballistic and hot carriers as well as quantum mechanical effects have to be accounted for. The classical drift diffusion model does not meet these demands. Among the alternatives are Monte Carlo simulations, kinetic equations, including the semiconductor Boltzmann equation and its quantum mechanical counterpart, the Wigner equation, and the so called hydrodynamic semiconductor model.

The speakers presented an overview of the numerical solution of the Wigner equation, the relation between kinetic and fluid dynamical models, and a new algorithmic approach to semiconductor device modeling.

**Organizer:** Christian Schmeiser, Technische Universität Wien, Austria

**Steady State Solutions of a Simplified Hydrodynamic Model** Christian Schmeiser, Organizer

**Diffusion Approximations for the Semiconductor Boltzmann Equation** Frederic Poupaud, Université de Nice, France

**A "Modified Boltzmann" Algorithm to Model Semiconductor Devices** Thomas I. Seidman, University of Maryland, Baltimore County

**Quantum Steady States: The Block-Poisson Equation** Anton Arnold, Technische Universität Berlin, Germany

## Aspects of the Numerical Solution of the Semiconductor Device Equations

The topic of this minisymposium was new developments in the numerical solution of the semiconductor device equations. The efficient and robust numerical solution of the semiconductor equations is of prime importance for the related industries and recently a number of new developments have been proposed.

This minisymposium emphasized advanced topics such as the use of adaptive spectral methods and the special adaptation of the Scharfetter-Gummel discretization to the avalanche phenomenon. Special attention was also paid to newly developed nonlinear multigrid approaches, in which truly coarse grids are used to accelerate the solution process. These new developments bring together the possibilities of fast solvers and the accurate representation of adaptive discretizations.

**Organizer:** Pieter W. Hemker, CWI, The Netherlands

**A Discretization Method for The Case of Avalanche Generation** Willy H.A. Schilders, Philips Corporation, ISA\CAD Centre, The Netherlands

**Collocation Methods for Hydrodynamic Device Equations** Christian A. Ringhofer, Arizona State University

**Adaptive Multigrid Methods for the Simulation of Semiconductor Devices** Rainer Constapel, University of Duisburg, Germany

**Multigrid and Adaptive Mixed Finite Elements for Two-Dimensional Semiconductor Device Simulation** Hans Molenaar, CWI, The Netherlands

## Computation and Mathematical Modeling in Electromagnetics

The minisymposium addressed mathematical modeling of electromagnetics problems on personal computers to massively parallel computers, unsolved problems, and industry practices and goals.

The speakers presented, for the most part, work on moment methods, which are based on the integral representation of Maxwells equations. Notwithstanding the fact that these methods require solutions to large dense systems and a few unsolved problems, moment methods are the most general and robust computation tools in electromagnetics.

The speakers opened a dialogue regarding computational electromagnetics between applied mathematicians and electrical engineers working on electromagnetics. Input from our mathematical colleagues would be invaluable to the future of computational electromagnetics.

**Organizer:** Elizabeth L. Yip, Boeing Company, Seattle

**Out-of-Core Dense Solver for Electromagnetic Problems on the iPSC/860** David S. Scott, Edward J. Kushner and Enrique Castro-Leon, Intel Corporation

**PC Electromagnetics** John Rockway, J.C. Logan and L.C. Russell, Naval Ocean Systems Center

**Unsolved Problems: The Electrostatics in Thin Dielectric Layers** Robert Olson, Washington State University

**Computational Electromagnetics—Industry Goals** William J. Gray, Boeing Company, Seattle

**Entire Domain Basis Functions for Frequency Selective Surfaces** Elizabeth L. Yip, Organizer; A. Ishimaru, University of Washington

## Fast Computational Methods for Device Design

In the last two decades, computer-aided semiconductor device design has become an interdisciplinary research area with cooperation of physicists, electrical engineers, computer scientists, and applied mathematicians. In particular the evolution of microelectronics towards very small devices is stimulating enhanced and new mathematical methods.

The current directions of mathematical investigation are primarily concentrated on multidimensional nonlinear partial differential equations and algebra with particular emphasis on solution methods implemented on modern computer architectures.

The speakers presented an overview of current work to attract the attention of applied mathematicians to an area which is at present still dominated by engineers. Numerical analysts and engineers should be brought together to foster a fruitful exchange of ideas for the solution of the many outstanding problems.

**Organizer:** Siegfried Selberherr, Technical University of Vienna, Austria

**A Collection of Iterative Algorithms for VLSI-Device Simulation**  Martin Stiftinger, Otto Heinreichsberger and Siegfried Selberherr, Organizer

**Finite Element and Box Method Discretization of the 3D Current Continuity Equations**  Thomas Kerkhoven, University of Illinois, Urbana-Champaign

**The Role of Numerical Algorithms and Visualization in Complex Semiconductor Device Simulation**  William M. Coughran, Jr. and Eric Grosse, AT&T Bell Laboratories, Murray Hill

## Higher-Moment Models for Semiconductor Device Simulation

Semiconductor device simulation has become an important applications area for numerical analysis. Traditionally, the drift-diffusion model, consisting of a Poisson equation for an electrostatic potential and parabolic continuity equations for electron and hole carriers, has been used. Recently, devices have become small enough or include exotic (quantum) effects so that more sophisticated models are needed. Higher-order moments of the Boltzmann transport equation result in so-called hydrodynamic or extended hydrodynamic models. In addition, modifications of drift-diffusion including phenomenological terms or energy-balance equations have been proposed. These newer approaches to device modeling require numerical techniques suitable for hyperbolic or highly convective flow simulations with stiff source terms. The speakers presented an overview of device models and appropriate specialized numerical techniques.

**Organizers:** William M. Coughran, Jr., AT&T Bell Laboratories, Murray Hill; Carl Gardner, Duke University

**Two-Dimensional MESFET Simulations via the Hydrodynamic Model**  Joseph W. Jerome, Northwestern University; S. Osher, University of California, Los Angeles; Chi-Wang Shu, Brown University

**Numerical and Engineering Aspects of the Hydrodynamic Model**  Farouk Odeh, IBM T.J. Watson Research Center

**Electron Shock Waves in Submicron Semiconductor Devices**  Carl Gardner, Duke University

**Fitting Hydrodynamic Parameters and the Limitation of the Model**  William M. Coughran, Jr., Organizer; Wolfgang Fichtner, ETH-Zurich, Switzerland; M.R. Pinto, AT&T Bell Laboratories, Murray Hill

## Numerical Methods for Advanced Semiconductor Device Models

Increased miniaturization in modern semiconductor devices leads to a physical regime in which the assumptions under which the drift diffusion equations (today's standard in device simulation) are derived from the Boltzmann equation are no longer valid. Thus, device simulation will have to rely on more realistic models in the future. These models consist of either macroscopic equations, the so-called hydrodynamic approximation, or the underlying transport equations themselves. This opens up a whole new array of challenging numerical problems including the calculation of shocks and the solution of high dimensional partial differential equations via deterministic and statistical methods. The minisymposium speakers presented an overview of recent advances in numerical methods for the hydrodynamic equations, the Boltzmann and the Wigner transport equations.

**Organizer:** Christian A. Ringhofer, Arizona State University

**Ensemble Monte Carlo Modeling with Simultaneous Molecular Dynamics Simulation**  David K. Ferry, Arizona State University; Alfred M. Kriman, State University of New York, Buffalo

**Numerical Solution of the Wigner Equation Using the Deterministic Particle Method**  Francis Nier, École Polytechnique, France

**Some Modeling Aspects of the Hydromodel for Semiconductor Devices**  Farouk M. Odeh, IBM T.J. Watson Research Center

**Higher Order ENO Schemes for the Hydrodynamic Model**  Stanley J. Osher, University of California, Los Angeles; Emad A. Fatemi, University of Minnesota

## Numerical Prediction of Scattered High Frequency Electromagnetic Waves

The purpose of this minisymposium was to give an overview of the most recent methodologies used in the numerical solution of electromagnetic and acoustic waves illuminating conductive or non-conductive bodies. This problem is crucial for many industrial applications. For example, in aerospace engineering, numerical simulations of the radar cross section of bodies with coated surfaces at high frequencies is encountered.

Academic and industrial experts from several countries presented their computational experience in this field and described efficient iterative algorithms on parallel machines for solving the time harmonic or time domain Maxwells equations focusing on special treatment of different absorbing boundary conditions.

**Organizers:** Gérard A. Meurant, CEA, France; Jacques Periaux, AMD/BA, France

**Lagrange Multipliers and Domain Decomposition Methods for the Maxwell Equations in the Frequential Domain**  Roland Glowinski, University of Houston; Jacques Périaux, Organizer

**RCS Calculations Using Time-Domain Solvers and Adaptive Unstructured Grids**  R. Lohner and B. Petitjea, George Washington University, C. R. DeVore, Naval Research Laboratory, and J. J. Ambrosiano, Lawrence Livermore National Laboratory

**Iterative Solution of Symmetric Complex Systems**  Gérard A. Meurant, Organizer

**Time Domain Electromagnetics Computation for Arbitrarily Shaped Layered 3-D Objects**  Vijaya Shankar, Rockwell International Science Center

**Long Range Effects of Composite Wave Absorbing Paints**  Yves Achdou and Olivier Pironneau, Université Paris VI, France

**A Finite Element Method for Time Harmonic Acoustics and Electromagnetics Problems in Three Space Dimensions**  D.P. Young, Boeing Computer Services, Seattle; F.T. Johnson, Boeing Commercial Airplanes, Seattle; J.E. Bussoletti, Boeing Commercial Airplanes, Seattle; G.X. Sengupta, Boeing Commercial Airplanes, Seattle; R.H. Bukhart, Boeing Computer Services, Seattle; R.G. Melvin, Boeing Computer Services, Seattle; M.B. Bieterman, Boeing Computer Services, Seattle

## Transport Equations for Gases and Semiconductors

The improvements in the production of semiconductor devices are increasingly tied up with improved modeling and better understanding of existing models by means of mathematical and computational analysis. For recent devices, the Boltzmann equation appears to be required in place of the more traditional diffusion model. The study of this equation has recently received a great impetus, because flight in the upper atmosphere is now required. Typical current directions of research are: rigorous derivation of the kinetic equations and of the various fluid dynamic limits, justification of intuitive numerical procedures, and existence and uniqueness of global solutions. The speakers presented some of the current research in these areas.

**Organizer:** Carlo Cercignani, Politecnico di Milano, Italy

**Existence and Uniqueness of Global Classical Solutions in the Wigner-Poisson Equations for Two Dimensions**  Anton Arnold, Purdue University

**On the Boltzmann Equations for Semiconductors**  Frédéric Poupaud, Université de Nice, France

**Incompressible Fluid Dynamic Limits of the Boltzmann Equation**  François Golse, Université Paris VII, France

**Kinetic Equations from Particle Systems**  Mario Pulvirenti, Università "La Sapienza" Roma, Italy

### Suggested Reading

U. M. Ascher, P. A. Markowich, P. Pietra, and C. Schmeiser, *A phase plane analysis of transonic solutions for the hydrodynamic semiconductor model*, Math. Models Meth. Appl. Sci., 1991.

C. Cercignani, *Theory and Application of the Boltzmann Equation*, Springer-Verlag, New York, 1988.

———, *Mathematical Methods in Kinetic Theory*, Second Edition, Plenum Press, New York, 1990.

J. J. Dongarra, I. S. Duff, D. C. Sorensen, and H. Van der Vorst, *Solving Linear Systems on Vector and Shared Memory Computers*, Society for Industrial and Applied Mathematics, Philadelphia, 1991.

P. A. Markowich, C. Ringhofer, and C. Schmeiser, *Semiconductor Equations*, Springer-Verlag, Wien, 1990.

P. Markowich, C. Schmeiser, and C. Ringhofer, *Semiconductor Equations*, Springer-Verlag, New York, 1990

E. Miller, *A selective survey of computational electromagnetics*, IEEE Trans. Antenna and Propagation, 36 (1988), pp. 1281–1305.

R. Mittra, C. H. Chan, and T. Cwik, *Techniques for analyzing frequency selective surfaces—A review*, Proc. IEEE, 76 (1988), pp. 1593–1615.

J. Molenaar and P. W. Hemker, *A multigrid approach for the solution of the 2D semiconductor equations*, IMPACT Comput. Sci. Engrg., 2 (1990), pp. 219–243.

R. G. Olsen and O. Einarsson, *Boundary element methods for weakly 3-dimensional quasi-electrostatic problems*, IEEE Trans. Power Delivery, PWRD-2 (1987), pp. 1276–1284.

S. J. Polak, C. den Heijer, W. H. A. Schilders, and P. Markowich, *Semiconductor device modelling from the numerical point of view*, Internat. J. Numer. Methods Engrg., 24 (1987), pp. 763–838.

F. Poupaud, *On a system of nonlinear Boltzmann equations of semiconductor physics*, SIAM J. Appl. Math., 50 (1990), pp. 1593–1606.

———, *Derivation of a hydrodynamic system hierarchy for semiconductors from the Boltzmann equation*, Appl. Math. Lett., 4 (1991), pp. 75–79.

J. W. Rockway, et al., *The MININEC System: Microcomputer Analysis of Wire Antennas*, Artech House, 1988.

W. Schönauer, *Scientific Computing on Vector Computers*, North–Holland, Amsterdam, 1987.

D. S. Scott, E. Castro–Leon, and E. J. Kushner, *Solving very large dense systems of linear equations on the iPSC/860*, Proc. 5th Distributed Memory Computing, 1990, pp. 286–290.

T. I. Seidman, *A new algorithmic model for the transient semiconductor problem*, Internat. Ser. Numer. Math., 93 (1990), Birkhauser Verlag, Basel, Switzerland.

S. Selberherr, *Analysis and Simulation of Semiconductor Devices*, Springer-Verlag, New York, Berlin, 1984.

Renato Spigler and Paul Davis

## Chapter 26
## ENVIRONMENTAL SCIENCE

### Environmental Modeling

This minisymposium is devoted to a vital application of mathematics that is not receiving the attention it warrants. Applied mathematicians must turn in numbers to problems of the environment. The speakers presented such problems and—in the Environmental Decade—perhaps triggered more applied mathematicians to begin working on environmental problems. It is unfortunate to note the dearth of mathematicians among any group of scientists in a conference that deals with environmental issues. We have much to contribute, and we have a responsibility to contribute.

**Organizer:** B.A. Fusaro, Salisbury State University

**Rainfall, Gumbel Distributions, and the Environment**
Roy Leipnik, University of California, Santa Barbara

**Viability Analysis of Endangered Species: How Few is too Few?** Roland H. Lambertson, Humboldt State University

**A Decision-Theoretic Approach to the Conservation of Biological Diversity** Robert W. McKelvey, University of Montana and EPA Environmental Research Laboratory, Corvallis

**Traffic Congestion and Urban Land Use** Frederic Y.M. Wan, University of Washington

**Ocean Response Times and Greenhouse Forcing**
R.D. Braddock, University of Maryland, Baltimore

## Chapter 27
## FLUID DYNAMICS

### Asymptotic and Exact Solutions for Turbulent Shear Layers

Recent research has shown that analytical and approximate asymptotic solutions can be constructed for the usual model equations that describe turbulent shear layers. These may be thin or thick boundary layers such as occur on aircraft wings or over hills, or they may be free shear layers occurring in separated flows, on jets, or in buoyancy driven flows. Characteristic features of these layers are a very wide range of scales and intense processes occurring at the smallest scales. Purely numerical models usually cannot resolve these regions and therefore can give little guidance regarding the general behavior of the regions as parameters and turbulence models change, whereas analytical methods or a combination of analytical and numerical methods can do so.

This minisymposium focused on the interesting challenges in this new field of applied mathematics, focused on some areas of controversy, and showed how the application of analytical methods to the model equations of turbulence is leading to new insights that may produce improved and computationally faster models.

**Organizer:** Julian C.R. Hunt, University of Cambridge, United Kingdom

**A New Asymptotic Method for Turbulent Flows**
R.E. Melnik, Grumman Aerospace Corporation

**On Turbulent Separation Flow Past a Bluff Body**
F.T. Smith, University College, United Kingdom; A. Neish, University College, United Kingdom

**Turbulent Free Shear Layers** S.K. Lele, NASA Ames Research Center

**Multi-Layer Asymptotic Analysis of Perturbed Turbulent Boundary Layers** Julian C.R. Hunt, Organizer; Stephen Belcher, Stanford University

## Darcy Flow in Composite and Random Porous Media: Numerical and Perturbation Methods

This minisymposium covered combined numerical and perturbation methods for solving Darcy flow and related processes in multidimensional composite, stratified, random porous media and fracture networks. Exact closed form solutions of the governing equations are not known, except for very special cases such as one spatial dimension. Some approximate solution methods considered by the speakers include direct numerical simulations, Monte-Carlo simulations, and perturbation methods for nonrandom as well as random media. The unifying theme was the complementarity of "numerical" versus "perturbation" approaches. The main goal was to elucidate the spatial-statistical structure and effective properties (conductivity, dispersity) of flow systems in highly heterogeneous earth materials such as water-bearing and oil-bearing geologic formations.

**Organizer:** Rachid Ababou, Southwest Research Institute, San Antonio

**Direct Simulations of Stochastic Darcy Flow and Perturbation Analysis of Finite Difference Errors** Rachid Ababou, Organizer

**A Transformation of Percolation of Random Fracture Networks to Percolation on Regular Lattices** Jane C.S. Long, University of California, Berkeley; Kevin Hestir, Lawrence Berkeley Laboratory; Kunal Ghosh, Jackson State University

**Spatial Averaging Laws for Conductivity in Three-Dimensional Heterogeneous Porous Media** Alexandre Desbarats, Geological Survey of Canada, Canada

**Stochastic Homogenization and Effective Conductivity Tensor in Random Anisotropic Media** Rachid Ababou, Organizer

**Lubrication Approximation for Fluid Flow in Rock Fractures** G.S. Bodvarsson, Lawrence Berkeley Laboratory; S. Kumar, Lawrence Berkeley Laboratory; R.W. Zimmerman, Lawrence Berkeley Laboratory

**Homogenization Theory Applied to Darcy Flow and Dispersive Transport in Almost Periodic Porous Media** Chiang C. Mei, Massachusetts Institute of Technology

## Discrete Models in Fluid Dynamics

In continuous fluid mechanics, one writes the Navier-Stokes equations and solves them by the discretization process related to classical numerical methods. This minisymposium was devoted to the works in which the models, not the equations, are discretized.

More precisely, the speakers surveyed kinetic theory with a discrete space velocity (mathematical properties of equations and mechanical behavior and motions of fluids also were modeled). They presented recent work on global existence of solutions of the initial value problem and properties of solutions when one considers not only binary collisions but also multiple collisions.

**Organizers:** Nicolas Bellomo, Politecnico di Torino, Italy; Henri Cabannes, Université Pierre et Marie Curie, France

**Constitutive Laws for a Discrete Velocity Gas** Rene Gatignol, Université Pierre et Marie Curie, France

**On the Fractional Step Method for Discrete Models in a Box** Giuseppe Toscani, University of Ferrara, Italy

**Initial-Boundary Value Problems for the Discrete Boltzmann Equation** Shuichi Kawashima, Kyushu University, Japan

**Large Time Behavior of the Broadwell Model of a Discrete Velocity Gas, with Specularly Reflective Boundary Conditions** Marshall Slemrod, University of Wisconsin, Madison

## Dynamics of Slender Vortices

Slender vortices generate supplementary vortex lift until vortex breakdown prevails over the wing. The complexity of these three-dimensional vortical flows can now be analyzed using supercomputers.

Although inviscid methods were already available for predicting the roll-up process of leading-edge vortices, they cannot simulate detailed flow structures of vortical interaction nor the complicated process of vortex breakdown. On the other hand, numerical solutions of Navier-Stokes equations have shown that vortical interaction and breakdown of viscous slender vortices can be well simulated.

This minisymposium focused on the current state of analytical and numerical solutions for three-dimensional dynamics of vortical flows.

**Organizer:** Chen-Huei Liu, NASA Langley Research Center

**Interaction of Vortex Filaments: Intersection, Cancellation and Reconnection** Lu Ting and Francis Bauer, Courant Institute of Mathematical Sciences, New York University

**Breakdown of Slender Vortices** Egon Krause, Technische Hochschule Aachen, Germany

**Simulation of Three-Dimensional Leading-Edge Vortices** Chung-Hao Hsu, NASA Langley Research Center; Chen-Huei Liu, Organizer

## Fluid Mixing and Spatial Chaos

Fluid mixing is connected with such fundamental problems as transition to turbulent or chaotic flow associated with the Navier-Stokes equations, large time behavior of accelerated interfaces in flows associated with the inviscid Euler equations, and "scale-up" in oil reservoir flows. Prototypical fluid mixing problems include: unstable stratification in fluid density (Rayleigh-Taylor instability) and fluid viscosity (flow in porous media), shear flow (Kelvin-Helmholtz instability), and the interactions of vortices in a background of irrotational or weakly rotational fluid. This minisymposium illustrated the commonality of features of these fluid mixing problems and presented new results emerging from research on 3-dimensional vortex flows, large-time, multiple-mode studies of accelerated interfaces, and mixing behavior in 2-dimensional, heterogeneous porous media.

**Organizers:** James Glimm, State University of New York, Stony Brook; W. Brent Lindquist, State University of New York, Stony Brook

**Validation of the Chaotic Mixing Renormalization Group Fixed Point** Qiang Zhang, State University of New York, Stony Brook

**Vortex Reconnection and Turbulent Intermittency** D. Boratavee, Rutgers University, New Brunswick; Z. Gu, Rutgers University, New Brunswick; R. Pelz, Rutgers University, New Brunswick; D. Silver, Rutgers University; Norman J. Zabusky, Rutgers University, New Brunswick

**Mixing By Interfacial Instabilities** Grétar Tryggvason and S. Ozen Unverdi, University of Michigan, Ann Arbor

**Transition to Turbulence in Complex Geometry Flows** Steven A. Orszag, Princeton University; G. Karniadakis, Princeton University

**A Random Field Model and Fluid Mixing in Heterogeneous Porous Media** W. Brent Lindquist and James Glimm, Organizers; Frederico Furtado and Felipe Pereira, State University of New York, Stony Brook

## Hydrodynamics of Sea Surface Impact

The hydrodynamics of solid body impact on the sea surface is important in ocean engineering. Examples are the wave impact of large structural members; impact and ricochet of missiles; slamming of ship bows in heavy seas; and the periodic impact of wavemakers, buoys, and other floating bodies.

Various methods are being utilized to study these phenomena, including classical hydrodynamics and modern techniques such as matched expansions and nonlinear surface element computations.

The four speakers covered a wide scope—the ricochet results correlated centuries of old data; the splash theory went beyond classical treatments; the bow impact analysis utilized nonlinear computations to provide practical tools for design; and the wavemaking theory explained surprising experimental results.

**Organizer:** Marshall P. Tulin, University of California, Santa Barbara

**Ricochet off Water of Spherical Projectiles** Touvia Miloh, University of Tel-Aviv, Israel

**Free-Surface Flows in the Vicinity of a Surface Piercing Structure** Raymond Cointe, Bassin d'Essais des Carenes, France

**Hydrodynamics of Planing Hulls** Hajime Maruo, Yokahama University, Japan

**Wavemaking by a Large Body in a Narrow Tank** Yitao Yao and Marshall P. Tulin, Organizer

## Low Mach Number Turbulence

Under many conditions, low Mach number flows can be properly described by ignoring compressible effects. In the approximation, sound waves and other compressible effects are decoupled from the flow and are evaluated a posteriori. This technique, nevertheless, does describe very important applications of flows at low Mach numbers in which compressible effects are important: lubrication flows (from disk-storage technology) and MHD flows (for plasma confinement technology).

The methods used for studying this problem are a combination of analytical and computational tools. The computational tools are being used to validate theoretical models describing the low Mach number limiting process. Current simulations are restricted to two-dimensional geometries.

The speakers discussed the relevance of compressible effects at low Mach numbers.

**Organizer:** Luis G. Reyna, IBM T.J. Watson Research Center

**MHD Turbulence and the Dynamo Effect** Satoshi Hamaguchi, IBM T. J. Watson Research Center

**Hyperbolic-Parabolic Problems with Fast Waves** Jens Lorentz, University of New Mexico

**Flows in Disk Drives** Luis G. Reyna, Organizer

## Mathematical Contributions to Membrane Separation Technology

Membrane separation processes have become increasingly important in industry, medicine, and scientific laboratories since their introduction in the 1960s. A membrane may be defined as a thin natural or synthetic sheet that is selectively permeable. The "permselectivity" of a porous membrane is determined by its pore-size distribution, which may lead to retention of particles (microfiltration), or macromolecules (ultrafiltration), according to size. Nonporous membranes may also be permselective because of their intrinsic chemical properties (reverse osmosis).

Many aspects of membrane separation processes are not fully understood, and the appropriate mathematical model for a particular process is often the subject of controversy. Of particular importance is the study of factors which limit membrane permeation rate, such as the buildup of solute concentration near the membrane, precipitation of salts and colloidal suspensions, and pore-blockage by particles. Most membrane modules employ crossflow of the feed solution in the laminar or turbulent regimes to enhance membrane permeation rate.

The speakers discussed numerical calculations of turbulent flow, semi-analytical calculations of laminar flow, the formation of deposits, and transport through networks of charged polymers.

**Organizer:** Mary E. Brewster, University of Colorado, Boulder

**Numerical Prediction of the Hydrodynamics and Turbulent Mass Transfer in a Rectangular Ultrafiltration Duct** Christer Rosen, University of Lund, Sweden; Christian Tragardh, University of Lund, Sweden

**Temporal Description of the Formation (and Removal) of the Deposit on (and from) a Rotating Membrane in a Couette System** Roger Ben Aim, CNRS, Laboratoire de Genie Chemique, France; Minggang Liu, CNRS, France

**Dean Vortices in Spiral Channel Membrane Systems** Mary E. Brewster, Organizer; Kun-Yong Chung, Rensselaer Polytechnic Institute; Georges Belfort, Rensselaer Polytechnic Institute

**Transport of Water and Solutes through Polymer Networks** Christopher G. Phillips, Imperial College of Science and Technology, U. K.

## Mathematical Modelling in Fluid-Structure Interaction

In this minisymposium, the speakers discussed problems in fluid-structure interaction. The problems concern the interaction of special structures, such as a circular or square cylinder and an elastic plate, with a flowing medium, and have applications to dynamics of offshore structures, overhead transmission lines and energy transfer in ships acoustics.

For the study of the interaction of a structure with a flowing medium one may carry out experiments in water channels or tanks and wind tunnels. Analytical and numerical models may be used to describe phenomena and measurements derived from these experiments. Depending on the numerical values of the parameters involved one may arrive at linear or non-linear systems of ordinary and/or partial differential equations as model equations.

Interesting asymptotic, projection and GreenÆs function methods as well as numerical methods may be used to analyze the model equations and to describe a.o. phase shift of the fluid-structure formation, flow separation, (in)stability mechanisms and energy transfer.

**Organizer:** Adriaan H.P. van der Burgh, Delft University of Technology, The Netherlands

**An Extended Approach Towards Mathematical Modelling of the Fluidelastic Lock-In Regime**
O. Mahrenholtz, Technische Universität Hamburg-Harburg, Germany; B. Fago, Technische Universität Hamburg-Harburg, Germany

**Flow Phenomena around an Oscillating Cylinder**
A. Otter, Shell Research B.V., The Netherlands

**Wind-Induced Response of Simple Elastic Structures** Adriaan H.P. van der Burgh, Organizer

## Mathematical Problems in Viscoelastic Liquid Flows

Viscoelastic liquids are of great importance in many technological devices (for instance, shaping processes for polymer materials). A number of theoretical questions need to be resolved before numerical simulation of polymer processing operations can be carried out with confidence at the high stress level typically observed in industry. The equations that must be solved for even the most elementary models of viscoelastic liquids are considerably more complex than the Navier-Stokes equations and many unresolved issues of fundamental nature remain. In particular, extrusion instabilities (e.g., melt fracture) have not yet received a satisfactory explanation and have no counterpart in the theory of Newtonian liquids.

This minisymposium provided state-of-the-art knowledge on mathematical questions arising from the modeling of flows of viscoelastic fluids of technological significance.

**Organizers:** Colette Guillopé, Université Paris-Sud, France; Jean-Claude Saut, Université Paris-Sud, France

**Stability Issues in Viscoelastic Fluid Mechanics**
Colette Guillopé, Organizer; Jean-Claude Saut, Organizer

**Stability of Discontinuous Steady States in Non-Newtonian Fluids** John A. Nohel, University of Wisconsin, Madison

**Shear Flows for KBK-Z Type Models** Hans P. Engler, Georgetown University

**Global Existence of Smooth Shearing Motions in a KBK-Z Fluid** Deborah Brandon, Virginia Polytechnic Institute and State University; William John Hrusa, Carnegie-Mellon University

**A Singular Perturbation Problem in Viscoelastic Flow** Pamela Cook, University of Delaware; Gilberto Schleiniger, University of Delaware

**Interplay Between Mathematical, Numerical, and Physical Issues in Viscoelastic Flow Simulations**
Roland Keunings, Université Catholique de Louvain, Belgium

**Absorbing Boundary Conditions for the Flow of Viscoelastic Fluids** Marc Tajchman, Université Paris-Sud, France

**Finite Time Step Strain Phenomena for Differential Viscoelastic Fluid Models** Robert W. Kolkka, Michigan Technological University

**About Nonlinear Viscoelastic Materials with Singular Kernels** William J. Hrusa, Carnegie-Mellon University

## Transonic Aerodynamics

Transonic aerodynamics involves flows for which the flow speed, $q$, is close to the local speed of sound. (The Mach number $M = q/a \cup 1$). In this regime, the flow is dominated by nonlinear effects. For an airplane flying in a uniform stream, the drag rises precipitously as the Mach number far upstream approaches one. Thus, there is great interest in analyzing the transonic range of flight and designing reduced drag wings.

The equations governing transonic flow are inherently nonlinear and must, ultimately, be solved numerically. Analytic and asymptotic results have greatly simplified and enriched both the theory and the computation of these flows. In this minisymposium, the speakers discussed recent results connecting analytical and numerical understanding of these flows.

**Organizer:** L. Pamela Cook, University of Delaware

**Two-Dimensional Choked Wind Tunnel Flow**
Gilberto Schleiniger, University of Delaware

**An Asymptotic Approach to Stability Problems of the Transonic Boundary Layer** O.S. Ryzhov, USSR Academy of Sciences, USSR

**Combined Asymptotics and Numerics in Transonic and Hypersonic Flow** Norman D. Malmuth, Rockwell International Science Center, Thousand Oaks

**Transonic Flow About a Thin Airfoil with a Parabolic Nose** Z. Rusak, Rensselaer Polytechnic Institute; J.D. Cole, Rensselaer Polytechnic Institute

**Recent Progress on Existence of Mixed Type Flow Past a Profile** Cathleen S. Morawetz, Courant Institute of Mathematical Sciences, New York University

### Suggested Reading

R. Ababou, *Identification of effective conductivity tensor in randomly heterogeneous and stratified aquifers*, Proc. Fifth Canadian-American Conf. on Hydrogeology: Parameter Identification and Estimation for Aquifers and Reservoirs, Calgary, Alberta, NWWA, Dublin, OH, 1990, pp. 155–157.

———, *Three-dimensional flow in random porous media*, Ph. D. Thesis, Department of Civil Engineering, Massachusetts Institute of Technology, Cambridge, MA, 1988.

A. Alves, *Proc. Euromech Colloquium on Discrete Models in Fluid Dynamics*, Figuera da Foz, Portugal, World Scientific, London, ed., 1990.

G. I. Barenblatt, *On the scaling laws (incomplete self-similarity with respect to Reynolds numbers) for the developed turbulent flows in tubes*, C. R. Acad. Sci. Paris Ser. II, 313 (1991), pp. 307–312.

G. Belfort, *Fluid mechanics in membrane filtration: Recent developments*, J. Membrane Sci., 40 (1989), pp. 123–147.

E. Berger, *On a mechanism of vortex excited oscillations of a cylinder*, J. Wind Engrg. Ind. Aero., 28 (1987), pp. 301–310.

A. Bhattacharjee and E. Hameiri, Phys. Ref. Lett., 57 (1986), p. 206.

R. D. Blevins, *Flow-Induced Vibration*, Second Edition, Van Nostrand Reinhold, New York, 1990.

P. R. Brazier-Smith and J. F. Scott, Wave Motion, 6 (1984), p. 547.

E. G. Broadbent and D. W. Moore, Philos. Trans. Roy. Soc. London Ser. A, 290 (1979), p. 353.

J. E. Burrell and D. G. Crighton, J. Fluids Structures, 1989. *Complex Systems*, 1 (1987).

D. G. Crighton and J. S. Williams, J. Fluid Mech., 1989.

M. Denn, *Issues in viscoelastic fluid mechanics*, Ann. Rev. Fluid Mech., 22 (1990), pp. 13–34.

A. J. Desbarats, *Support effects and the spatial averaging of transport properties*, Math. Geol., 21 (1989), pp. 383–390.

C. C. Feng, *The measurement of vortex induced effects in flow past stationary and oscillating circular and D-section cylinders*, M. A. Sc. Thesis, University of British Columbia, 1968.

R. Gatignol, *Théorie Cinetique des Gaz à Répartition Discrète de Vitesses*, Lecture Notes in Phys., Vol. 36, Springer-Verlag, Heidelberg, 1975.

O. M. Griffin and S. F. Ramberg, *The vortex-street wakes of vibrating cylinders*, J. Fluid Mech., 66 (1974), pp. 553–576.

C. Guillopé and J. C. Saut, *Existence results for the flow of viscoelastic fluids with a differential constitutive law*, Nonlinear Anal. Theory Meth. Appl., 15 (1990), pp. 849–869.

P. Hall, *On the stability of the unsteady boundary layer on a cylinder oscillating in a viscous fluid*, J. Fluid Mech., 146 (1984), pp. 347–367.

S. Hamaguchi, Phys. Fluids B, 1 (1989), p. 1416.

W. D. Henshaw, L. G. Reyna, and J. Zufiría, *Compressible Navier–Stokes computations for slider air-bearings*, J. Tribology, 113, pp. 73–79.

H. Honji, *Streaked flow around an oscillating circular cylinder*, J. Fluid Mech., 107 (1981), pp. 509–520.

C.-H. Hsu and C. H. Liu, *Simulation of leading-edge vortex flows*, Theoret. Comput. Fluid Dynamics, 1 (1990), pp. 379–390.

J. C. R. Hunt, S. Leibovich, and K. J. Richards, *Turbulent shear flow over low hills*, Quart. J. Roy. Met. Soc., 114 (1988), pp. 1435–1470.

K. Ishii, F. Hussain, K. Kuwahara, and C. H. Liu, *The dynamics of vortex rings in an unbounded domain*, Advances in Turbulence 2, H.-H. Fernholz and H. E. Fiedler, eds., Springer-Verlag, Berlin, Heidelberg, 1989, pp. 51–56.

R. N. Iyengar and O. Mahrenholtz, *Nonlinear oscillations of a vortex excited cylinder in wind*, Solid Mech. Arch., 7 (1982), pp. 411–432.

K. M. Jansons and C. G. Phillips, *On the application of geometric probability theory to polymer networks and suspensions*, I, J. Colloid Interface Sci., 137 (1990), pp. 75–91.

D. D. Joseph, *Fluid Dynamics of Viscoelastic Liquids*, Springer-Verlag, New York, 1990.

R. Keunings, in *Fundamentals of Computer Modelling in Polymer Processing*, C. L. Tucker, ed., Carl Hansler Verlag, 1987.

S. Klainerman and A. Majda, *Compressible and incompressible fluids*, Comm. Pure Appl. Math., 35 (1982), pp. 629–651.

V. F. Kop'ev and E. A. Leont'ev, Soviet Phys. Acoust., 29 (1983), p. 111.

E. Krause, *The Solution to the Problem of Vortex Breakdown*, Lecture Notes in Phys., Vol. 374, K. W. Morton, ed., Springer-Verlag, Berlin, Heidelberg, 1990.

H.-O. Kreiss and J. Lorenz, *Initial-Boundary Value Problems and the Navier–Stokes Equations*, Academic Press, New York, 1989.

H.-O. Kreiss, J. Lorenz, and M. J. Naughton, *Convergence of the solutions of the compressible to the solutions of the incompressible Navier–Stokes equations*, Adv. Appl. Math., 12 (1991), pp. 187–214.

J. C. S. Long and D. M. Billaux, *From field data to fracture network modeling: An example incorporating spatial structure*, Water Resources Res., 23 (1987), pp. 1201–1216.

O. Mahrenholtz, *Fluidelastische Schwingungen*, Z. Angew. Math. Mech., 66 (1986), pp. 1–22.

C. C. Mei and J.-L. Auriault, *Mechanics of heterogeneous porous media with several spatial scales*, Proc. Roy. Soc. London Ser. A, 426 (1989), pp. 391–423.

R. E. Melnik, *Some applications of asymptotic theory to turbulent flow*, AIAA 91-0220, 29th Aerospace Sciences Meeting, Reno, NV, 1991.

A. Michaels, *New vistas for membrane technology*, Chemtech, March, 1989, pp. 162–172.

R. Monaco, *Proc. Workshop on Discrete Kinetic Theory, Lattice Gas Dynamics, and Foundations of Hydrodynamics*, Torino, Italy, World Scientific, London, ed., 1988.

A. Neish and F. T. Smith, *On turbulent separation in the flow past bluff bodies*, J. Fluid Mech., 1992, in press.

L. A. Ostrovski, S. A. Rybak, and L. Sh. Tsimring, Soviet Phys. Uspekhi, 29 (1986), p. 1040.

A. Otter, *Damping forces on a cylinder oscillating in a viscous fluid*, Appl. Ocean Res., 12 (1990), pp. 153–155.

———, *On a hydroelastic problem in offshore structures. The forces on an oscillating cylinder revisited*, Proc. 1st Internat. Offshore and Polar Engineering Conference, Edinburgh, Vol. 3, 1991, pp. 172–177.

M. Renardy, *Mathematical analysis of viscoelastic flows*, Ann. Rev. Fluid Mech., 21 (1989), pp. 21–36.

M. Renardy, W. J. Hrusa, and J. A. Nohel, *Mathematical Problems in Viscoelasticity*, Longman, J. Wiley & Sons, New York, 1987.

T. Sarpkaya, *Force on a circular cylinder in viscous oscillatory flow at low Keulegan–Carpenter numbers*, J. Fluid Mech., 165 (1985), pp. 61–71.

———, *Vortex-induced oscillations: A selective review*, J. Appl. Mech., 46 (1979), pp. 242–258.

H. R. Strauss, Phys. Fluids, 29 (1986), p. 3668.

L. Ting and R. Klein, *Viscous Vortical Flows*, Lecture Notes in Phys., Vol. 374, Springer-Verlag, Berlin, Heidelberg, 1991.

C. G. A. van der Beek, *Normal forms and periodic solutions in the theory of nonlinear oscillations. Existence and asymptotic theory*, Internat. J. Non-Linear Mech., 25 (1989), pp. 263–279.

C. G. A. van der Beek and A. H. P. van der Burgh, *On the periodic wind induced vibrations of an oscillator with two degrees of freedom*, Nieuw Arch. Wisk, 2 (1987), pp. 207–225.

A. H. P. van der Burgh, *On the modeling of a continuous oscillator by oscillators with a finite number of degrees of freedom*, Proc. Third European Conf. on Math in Industry, Kluwer Academic Publ. & B. G. Teubner, Stuttgart, 1990, pp. 159–170.

G. P. Zank and W. H. Matthaeus, *Nearly incompressible hydrodynamics and head conduction*, Phys. Rev. Lett., 64, pp. 1243–1246.

R. W. Zimmerman, S. Kumar, and G. S. Bodvarsson, *Lubrication theory analysis of the permeability of rough-walled fractures*, Proc. 2nd Ann. Internat. Conf. on High-Level Radioactive Waste Management, Las Vegas, NV, American Nuclear Society, Vol. 1, 1991, pp. 535–541.

# Chapter 28
# GEOMETRIC MODELING, DESIGN, AND COMPUTATION

## Automatic Finite Element Meshing from Solid Models

This minisymposium addressed the problem of generating automatically discrete models (meshes) for finite element (FE) analysis of solid domains defined in a solid modeling system. Automatic meshing is one of the primary areas of application of solid modeling technology. The ability to generate automatically a valid FE model is mandatory for the development of effective automatic CAD/CAE procedures for production design in industry. Automatic FE modeling is also a critical tool for numerical simulation in advanced research areas such as computational fluid dynamics and fracture mechanics. At present, research in automatic meshing is largely focused on algorithms based on domain Delaunay triangulation and recursive spatial decomposition with application to nonmanifold topologies. The speakers presented current research and development in academia and industry in the U.S. and Europe on this topic.

**Organizer:** Renato Perucchio, University of Rochester

**Finite Element Meshing for the Computation of Flow Fields over Complete Aircraft**  Timothy J. Baker, Princeton University

**Automatic Mesh Generation in Non-Manifold Topology Environment**  Mukul Saxena, Peter M. Finnigan, C. Graichen, A. F. Hathaway, V.N. Parthasarathy, General Electric Corporate Research and Development Center

**Topological Validity and Computational Complexity of Algorithms for Delaunay Tetrahedrization of Solids**  Nickolas Sapidis, University of Rochester; Renato Perucchio, Organizer

## Geometric and Physical Modeling for Computer-Aided Design and Visualization

Computer-aided design and visualization poses a rich variety of problems in numerical methods, symbolic computation, graphics, and systems organization. Approaching these problems from an interdisciplinary perspective promises comprehensive solutions that unify partial answers found previously in specialized investigations.

The speakers described these problems from a variety of perspectives, including a detailed exposure of the numerical dimension, a description of innovative approaches to modeling deformable objects, and a study of the interaction between geometric shape and physical behavior.

**Organizer:** Christoph M. Hoffmann, Purdue University

**Numerical Aspects of Simulation of Physical Models in CAD and Visualization**  Christoph M. Hoffmann, Organizer

**Simulation of Physical Systems for Geometric Models**  Christoph M. Hoffmann, Organizer

**Applied Deformable Models**  Demetri Terzopoulus, University of Toronto, Canada

**Alpha 1: Complex Geometric Modeling in Support of Computer Aided Manufacturing**  Richard F. Riesenfeld, University of Utah

*Left and above:*
Attendees enjoy the banquet.

## Introduction to CAGD

Computer Aided Geometric Design (CAGD) is a new branch of mathematics, computer science, and engineering involved with the creation and manipulation of geometric objects. Curves, surfaces, solid modeling, interactive computer graphics, and computational geometry are some of the ingredients contributing to this fertile area. From the beginning, CAGD has shared a symbiotic partnership with the automation of design and manufacturing (CAD/CAM) and computer graphics.

This minisymposium provided an introduction to the field. The speakers looked at CAGD from four different perspectives: historical, theoretical, application, and graphical. They also presented their viewpoints on what will happen in CAGD in the next few years.

**Organizer:** Rosemary E. Chang, Silicon Graphics

**CAGD: A Historical Perspective**  Robert E. Barnhill, Arizona State University

**CAGD: Some Theoretical Foundations**  Alyn P. Rockwood, Arizona State University

**Computer Aided Geometric Design: Theories in Action**  Tracy M. Whelan, CAMAX Systems Inc., Minneapolis

**Computer Aided Geometric Design: Graphical Realization**  Rosemary E. Chang, Organizer

## Leaving Flatland: Higher Dimensional Methods in Computer-Aided Geometric Design

Many geometric problems become simpler to solve when rephrased in a higher-dimensional space. In this minisymposium, the speakers described the use of higher-dimensional spaces in a variety of problems in computer-aided geometric design (CAGD). CAGD is concerned with the construction, manipulation, and analysis of complex objects such as those found in solid modeling systems, CAD/CAM environments, and computer vision systems. Higher-dimensional spaces greatly simplify CAGD applications such as: the representation and analysis of offset curves and surfaces, blend and fillet surfaces, equidistance surfaces, the development of multi-sided generalizations of Bezier surface patches, and the extension of B-spline curves to geometric continuity.

**Organizer:** Tony D. DeRose, University of Washington

**Algorithms for Geometrically Continuous Spline Curves of Arbitrary Degree**  Hans-Peter Seidel, University of Waterloo, Canada

**Two-Surfaces in Higher-Dimensional Space: Why, How, and What For?**  Christoph M. Hoffmann, Purdue University

**S-Patches: A Multi-sided Generalization of Bezier Surfaces**  Tony D. DeRose, Organizer

**Simplicial Approximation of Surfaces and Manifolds**  Kurt Georg, Colorado State University

## Massively Parallel Algorithms for Geometric Computations

Geometric computations are fundamental to a wide variety of applications such as physical modelling, robotics, computer graphics, data visualization, and manufacturing automation. Very large data sets are characteristic of these problems and the complexity of existing serial algorithms is often quadratic or worse. Issues that need to be addressed are data structures for efficient geometric computation on an SIMD hypercube, communications-efficient programming techniques, and the design of massively parallel geometric algorithms.

The speakers presented results on massively parallel solutions to the problems of triangulation, incremental adaptive surface fitting, polygon layout, and self-avoiding random surfaces. They also discussed the attendant practical considerations of implementation of a fine-grained parallel computer.

**Organizer:** Bryant W. York, Boston University

**Incremental Adaptive Surface Fitting of Very Large Data Sets**  Bryant W. York, Organizer

**Efficient Polygon Placement**  Robert N. Moll, University of Massachusetts, Amherst

**Parallel Triangulation and Shelling**  Francis Sullivan and Isabel Beichl, National Institute of Standards and Technology

**Dynamical Triangulations: Theory, Simulations and Applications**  M. Aghistine, Princeton University

## Robustness and Stability in Computational Geometry

In recent years, there has been growing concern about the robustness of geometric algorithms. Since the aim of computational geometry is application of the algorithms to real-world problems, attention must be paid to the actual computational behavior of the methods. Errors due to truncation and roundoff have more dramatic manifestations for geometrical questions than they have in other areas. Finding a slightly inaccurate solution to a linear system may be acceptable, unless the inaccuracy causes a code to report that a given point is on both sides of a fixed plane! Therefore, it is important to devise algorithms that are robust (give a self-consistent answer to a problem related to the actual input) and stable (the problem solved is close to the question asked).

The speakers considered the emerging theory of robustness for computational geometry and the well-developed theory of those numerical methods that ultimately determine the numerical stability of real codes.

**Organizer:** Francis Sullivan, National Institute of Standards and Technology

**Numerical Stability of Geometric Algorithms**  Steven Fortune, AT&T Bell Laboratories, Murray Hill

**Stability of Numerical Primitives for Computational Geometry**  Francis Sullivan, Organizer; Isabel Beichl, National Institute of Standards and Technology

**Robust Solid Modeling Using Internal Methods**  Michael Karasick, Derek Lieber and Lee Nackman, IBM T.J. Watson Research Center

**Some Techniques for Robust Geometry**  Victor Milenkovic, Harvard University

# GEOMETRIC MODELING, DESIGN, AND COMPUTATION

## Surface Modeling

Parametric surfaces have established themselves as an indispensable tool for the design of cars, ships, planes, and other structures. The speakers presented a geometrically oriented introduction to the main concepts and then pointed to more recent developments. They discussed Bernstein-Bezier methods, NURBS, and some nonstandard surface methods.

**Organizer:** Wolfgang Boehm, Technische Universität Braunschweig, Germany

**Rational Curves and Surfaces**   Gerald E. Farin, Arizona State University

**S-Patches and Generalized B-splines of Arbitrary Topology**   Tony D. DeRose, University of Washington

## Visualizing Multidimensional Geometry and Applications

By means of a system of parallel coordinates, relations among $N$ real variables are mapped into (indexed) subsets of $R^2$ whose geometrical properties enable the visualization of the corresponding hypersurfaces in $R^N$. This is the only known methodology for the visualization of relations among an arbitrary number of variables. For $N = 2$ a point $\leftarrow \rightarrow$ line duality leads to a new convexity duality and some optimal convexity algorithms. A line in $R^N$ is represented by $N-1$ points and in general $p$-flats in $R^N$ are represented by $(N-p)\,p$ indexed planar points, enabling the visualization of polyhedra in $R^N$. Algorithms exist for translations, rotations, hyperplane intersections, and point membership queries. Developable hypersurfaces, considered as the envelope of their tangent hyperplanes, are represented by $N-1$ indexed planar curves. An alternate representation of a class of hypersurfaces yields an algorithm for constructing and displaying interior points and provides local curvature information.

The speakers presented applications to statistics (for exploratory data analysis and model testing), computational geometry including the conflict avoidance problem in air traffic control, computer vision, and process control.

**Organizer:** Alfred Inselberg, IBM Scientific Center, Santa Monica and University of Southern California

**Mathematical Foundations of Parallel Coordinates**   Alfred Inselberg, Organizer; B. Dimsdale, IBM Scientific Center, Santa Monica and University of Southern California

**Some Statistical Applications of the Parallel Coordinate, Axis System**   Chris Gennings, Medical College of Virginia, Virginia Commonwealth University; Kathryn Dawson, Medical College of Virginia, Virginia Commonwealth University; Hans W. Carter, Medical College of Virginia, Virginia Commonwealth University

**Parallel Coordinates, $p$-Flats, and Multi-Dimensional Visualization**   John Scott Eickemeyer, Information Technology Institute, Singapore

## Suggested Reading

T. J. Baker, *Automatic mesh generation for complex three-dimensional regions using a constrained Delaunay triangulation*, Engrg. Comput., 5 (1989), pp. 161–175.

R. Barnhill, *Representations and approximation of surfaces*, in Mathematical Software III, J. R. Rice, ed., Academic Press, New York, 1977.

———, *Surfaces in computer aided geometric design: A survey with new results*, Comput. Aided Geom. Design, 2 (1985), pp. 1–17.

R. Barnhill, G. Farin, L. Fayard, and H. Hagen, *Twists, curvatures and surface interrogation*, Comput. Aided Design, 20, pp. 341–346.

R. Barnhill and R. Riesenfeld, eds., *Computer Aided Geometric Design*, Academic Press, New York, 1974.

J. Beck, R. Farouki, and J. Hinds, *Surface analysis tools*, IEEE Comput. Graphics Appl., 6 (1986), pp. 18–36.

M. Bercovier, *Finite Element Methods in Industrial Applications*, ECMI Series, Teubner Publishers, Stuttgart, 1992, to appear.

G. Farin, *Curves and Surfaces for Computer Aided Geometric Design*, Second Edition, Academic Press, New York, 1990.

I. Faux and M. Pratt, *Computational Geometry for Design and Manufacture*, Ellis Horwood, 1979.

C. Hoffman and J. Hopcroft, *Simulation of physical systems from geometric models*, IEEE J. Robotics Automat., RA-3 (1987), pp. 194–206.

C. Lawson, *Software for $C^1$ surface interpolation*, in Mathematical Software III, J. R. Rice, ed., Academic Press, New York, 1977.

R. Perucchio, M. Saxena, and A. Kela, *Automatic mesh generation from solid models based on recursive spatial decompositions*, Internat. J. Numer. Methods Engrg., 28 (1989), pp. 2469–2501.

N. Sapidis and R. Perucchio, *Domain Delaunay tetrahedrization of arbitrarily shaped curved polyhedra defined in a solid modeling system*, Proc. ACM/SIGGRAPH Symposium on Solid Modeling Foundations and CAD/CAM Applications, Austin, TX, 1991, pp. 465–480.

M. Saxena, P. M. Finnigan, C. M. Graichen, et al., *Octree-based automatic mesh generation for nonmanifold domains*, Proc. 1991 ASME International Computers in Engineering Conference, Vol. I, Santa Clara, CA, 1991, pp. 435–443.

D. Terzopoulos and K. Fleischer, *Deformable models*, The Visual Computer, 4 (1988), pp. 306–331.

D. Terzopoulos and K. Waters, *Physical-based facial modeling, analysis, and animation*, J. Visualization Comput. Animation, 1 (1990), pp. 73–80.

Chapter 29
# GEOPHYSICAL SCIENCES

## Efficient Computational Methods in Modeling Porous Media Flow

One of the largest single uses of supercomputing power is the modeling of flow in porous media. The petroleum industry has the main commercial application, but the application to contaminant transport modeling in the analysis of the effects of pollution is gaining significant importance. Information on porous media is often available on a scale which is finer than that being used in the simulation. Therefore, it is important to develop better mathematical and computational tools to solve ever larger problems.

Researchers developed algorithms which take advantage of various computer architectures, such as parallel and vector machines. Others have developed methods relying on the differential equations themselves; among these are better preconditioners and various multigrid methods. Others have used adaptive discretization.

The speakers described the state of the art in the development of the fastest computational methods that are used in modeling flow in porous media, and discussed possible new directions in the search for improvements.

**Organizer:** Thomas W. Fogwell, Ferme Le Bied, Switzerland

**Multigrid on SIMD Machines for Modeling Porous Media Flow**   Joel E. Dendy, Jr., Los Alamos National Laboratory; Jeffrey M. Rutledge, Chevron Oil Field Research Company, La Habra

**The Connection Machine Implementation of a New Method for Porous Media Flow Modeling**   Philip C. Emeagwali, University of Michigan, Ann Arbor

**A New Semicoarsening Multigrid Approach for Interface Estimates of Domain Decomposition in Oil Reservoir Simulation on MIMD Computers**
Rao Bhogeswara, University of Houston; John E. Killough, University of Houston

**Petroleum Reservoir Simulation on Supercomputers**   Larry C. Young, Reservoir Simulation Research Corporation, Tulsa

Chapter 30
# INVERSE PROBLEMS

## Computational Methods for Solving Positron Emission Tomography Problems

One of the main reasons that positron emission tomography (PET) for finding cancer tumors has not moved from the experimental laboratory into the clinical laboratory is that the medical community thinks that the computational costs are too high. However, the most commonly used algorithms for image reconstruction can be speeded up by using common sparse matrix techniques and well-known methods in function optimization. Moreover, these algorithms are very well suited to parallel and vector machines where they can be made economically feasible even on medically reasonable problems of 16,000 variables.

The speakers considered incorporating the statistical nature of the noise of the data into the model, studied the convergence of various proposed algorithms, suggested ideas for accelerating convergence, considered several ideas for approximating the solution and the data on various unusual grids, and showed how to adapt some of the algorithms to parallel and vector machines.

**Organizer:** Linda Kaufman, AT&T Bell Laboratories, Murray Hill

**Optimization Methods in Positron Emission Tomography**  Alvaro R. DePierro, IMECC-UNICAMP, Brazil

**Bayesian Statistical Methods for Scatter and Attenuation Compensation in Emission Tomography**  Donald E. McClure, Brown University; Stuart Geman, Brown University; Kevin M. Manbeck, Brown University; John Mertus, Brown University

**Rotationally-Symmetric Basis Functions for Iterative Image Reconstruction in PET**  Robert M. Lewitt, University of Pennsylvania

**Block-Iterative Algorithms for Feasibility and Optimization Problems in Image Reconstruction**  Yair Censor, Technion-Israel Institute of Technology, Israel

**Implementing and Accelerating Emission Tomography Algorithms**  Linda Kaufman, Organizer

## Electromagnetic Inverse Problems

The problems considered in this minisymposium were electromagnetic inverse problems, where there is a need to reconstruct the electromagnetic state of the interior of a body from measurements made on the exterior. Such problems are important in a variety of fields, ranging from medical imaging to nondestructive testing of materials.

Methods under investigation include time-domain approaches and frequency-domain approaches, some of which involve low frequencies and some high frequencies. In addition, work is being done on the full multidimensional Maxwells equations.

This minisymposium brought together people working on a number of these approaches.

**Organizers:** Margaret Cheney, Rensselaer Polytechnic Institute; David Isaacson, Rensselaer Polytechnic Institute; Fadil Santosa, University of Delaware

**Electromagnetic Imaging at Fixed Frequency**  Margaret Cheney, Organizer; David Isaacson, Organizer

**Inverse Scattering Problems of Electromagnetic Waves in Biisotropic Media Using Time-Domain Techniques**  Gerhard S. Kristensson, Lund Institute of Technology, Sweden; Sten Rikte, Lund University, Sweden

**An Inverse Problem for Maxwells Equations**  Erkki Somersalo, Institute of Technology, University of Helsinki, Finland

**Preconditioning in Electrical Impedance Tomography**  Fadil Santosa, Organizer

**Dual Feasibility Constraints for Electrical Impedance Tomography**  James G. Berryman, Lawrence Livermore National Laboratory

**Iterative Methods for Electromagnetic Profile Inversion**  R.E. Kleinman, University of Delaware; Xinming Jiang, University of Delaware; P.M. van den Berg, Delft University of Technology, The Netherlands

**Impedance Tomography in Geophysical Exploration**  Apo Sezginer, Schlumberger-Doll Research

**A Relaxed Functional of Impedance Tomography**  Michael Vogelius, Rutgers University, New Brunswick

## Inverse Problems Using Spectral Data

A broad selection of inverse problems were represented. Data included eigenvalues, nodal positions, scattering data or impulse response data. Uniqueness theorems, well-posedness results, asymptotic forms, the Heisenberg principle, analysis, and algorithms were presented for these nonlinear problems. Applications are in geophysics, nondestructive testing, identification, and design.

**Organizer:** Joyce R. McLaughlin, Rensselaer Polytechnic Institute

**Solutions of Inverse Nodal Problems**  Ole H. Hald, University of California, Berkeley; Joyce R. McLaughlin, Organizer

**The Free Oscillation Problem for the Earth**  Robert S. Anderssen, CSIRO, Australia

**The Application of Schurs Algorithm to an Inverse Eigenvalue Problem**  Graham M.L. Gladwell, University of Waterloo, Canada

**The Inverse Backscattering Problem**  James V. Ralston, University of California, Los Angeles; Gregory Eskin, University of California, Los Angeles

**The Eigenvalues of the Dirichlet to Neumann Map**  John Sylvester, University of Washington

**Simultaneous Concentration in the Physical and Spectral Domains**  Alberto Grünbaum, University of California, Berkeley

**Identification of Parameters for Beams and Narrow Plates Using Spectral Data**  David L. Russell, Virginia Polytechnic Institute and State University; Luther W. White, University of Oklahoma

**Localization of the Spectrum for a Periodic Schroedinger Operator**  Leonid Friedlander, University of California, Los Angeles

## Inverse Scattering Problems

Inverse scattering theory is concerned with the problem of determining the physical properties of an unknown inhomogeneity in a medium from a knowledge of its effect on a given acoustic, elastic, or electromagnetic wave. Applications of inverse scattering theory occur in medical imaging, geophysical prospecting, and nondestructive testing.

The main mathematical problems in inverse scattering theory arise from the fact that they in general are not only highly nonlinear but also improperly posed. These difficulties are currently being addressed through the use of either optimization schemes or asymptotic analysis coupled with innovative numerical methods. Both approaches lack an adequate error analysis and are often computationally far too expensive, particularly for multi-dimensional problems. The speakers addressed these issues with an emphasis on computational and multi-dimensional problems.

**Organizer:** David L. Colton, University of Delaware

**An Overview of Inverse Scattering**  David L. Colton, Organizer

**Multidimensional Acoustic Inverse Problems for Inhomogeneous Media**  Peter B. Monk, University of Delaware

**Multidimensional Inverse Scattering at Fixed Energy**  Adrian L. Nachman, University of Rochester

**Velocity Inversion in Reflection Seismology**  William W. Symes, Rice University

## Inverse Scattering

The speakers considered inverse scattering problems in acoustics and electromagnetics—the reconstruction of scattering obstacles and inhomogeneous media from far-field measurements. Problems of this type occur in a number of applications such as ultrasonic medicine, seismic imaging, nondestructive testing, etc. They are difficult to solve since they are nonlinear and ill-posed.

Mainstream current research consists of theoretical investigations and numerical approximation for two- and three-dimensional problems. One important aspect is uniqueness results, i.e., how much information is needed to determine the scatterer. In respect to numerical approximation, recent efforts try to replace the more traditional approaches by methods based on a stable reformulation of the inverse problem as a nonlinear optimization problem.

**Organizer:** Rainer Kress, Universität Gottingen, Germany

**An Inverse Problem for the Conductive Boundary Value Problem**  Andreas Kirsch, Universität Erlangen-Nürnberg, Germany

**Three-Dimensional Elastic Wave Inverse Scattering as Applied to Nondestructive Evaluation**  Karl J. Langenberg, University of Kassel, Germany

**The Uniqueness of a Solution to an Inverse Scattering Problem for Electromagnetic Waves**  Lassi Päivarinta, University of Helsinki, Finland; Valeri Serov, Leninie Gory Fakultet BMK, USSR; Erkki Somersalo, University of Helsinki, Finland

**Three-Dimensional Reconstructions in Inverse Obstacle Scattering**  Rainer Kress, Organizer

## Mathematical Methods in Computerized Tomography

The problem of computerized tomography is still very important in many areas of science, medicine, and industry. While the basic problems have been solved in the last twenty years, there are still problems in emission tomography and in incomplete data tomography. In both cases, existing algorithms are not satisfactory

The speakers presented a state-of-the-art review from the standpoint of industry as well as from the standpoint of academic research. Statistical iterative methods for emission tomography and the lambda-method in local tomography were discussed.

**Organizer:** Frank Natterer, Universität Münster, Germany

**State of the Art in Medical Imaging**  Alfred K. Louis, Universität Saarbrucken, Germany

**Medical Reconstructive Imaging—An Ongoing Challenge to Industrial Mathematics**  Guenter Schwierz, Siemens A.G., Germany

**Local Tomography**  Adel Faridani, Oregon State University; E.L. Ritman, Mayo Medical School; Kennan T. Smith, Oregon State University

**Probabilistic Methods in Emission Tomography**  Stuart A. Geman, Brown University; Donald E. McClure, Brown University; Kevin M. Manbeck, Brown University; John Mertus, Brown University

## Parameter Identification from Boundary Measurements with Industrial Applications

In various application fields such as electrical impedance computed tomography and nondestructive testing (NDT), the identification of parameter functions in partial differential equations from boundary measurements has become important in recent years. There have been two different approaches to parameter identification from boundary measurements—identification from knowledge of the "Dirichlet-to-Neumann-map" and from one or finitely many (noisy) boundary measurements. This minisymposium focuses on the latter approach.

The first two presentations dealt with questions of uniqueness (identifiability) and stability. The next two dealt with numerical approaches, namely iterative methods and the application of Tikhonov regularization as a stable method for attaching this (nonlinear) ill-posed problem. The last two dealt with applications to geophysics (where again, Tikhonov regularization is used) and to nondestructive testing. In the latter presentation, a project with a large European company was described, where the aim was to devise a practical, feasible method for locating reinforcement bars in concrete from distortions of a magnetostatic field.

**Organizer:** Heinz W. Engl, Johannes Kepler Universität, Austria

**Inverse Conductivity Problems with One Measurement**  Victor Isakov, Wichita State University

**Iterative Procedures for the Reconstruction of Coefficients from Boundary Data**  William Rundell, Texas A&M University; Michael Pilant, Texas A&M University

**Stability and Convergence (Rates) of Tikhonov Regularization for Parameter Identification from Boundary Measurements**  Otmar Scherzer, Johannes Kepler Universität, Austria

**Parameter Identification from Boundary Measurements: The Electrical Conductivity of the Earth**  Robert S. Anderssen, CSIRO, Australia; Heinz W. Engl, Organizer; Otmar Scherzer, Johannes Kepler Universität, Austria

**Identification of Reinforcement Bars in Concrete from the Distortion of a Magnetostatic Field: Model, Identifiability, Numerical Results**  Heinz W. Engl, Organizer; Victor Isakov, Wichita State University

## Seismic Inverse Methods in Exploration Research

Inverse methods applied to the problem of determining earth structure for seismic exploration is a research topic of intense interest and activity. The implementation of the ideas of the past ten years has significantly enhanced our ability to create improved images of the interior of the earth and to estimate medium parameters. This dramatic growth has occurred almost in lock step with enhanced computer capabilities that make the new methods accessible to almost all users. Nonetheless, many challenges remain because the underlying problem is nonlinear, the available data is incomplete, and the true complexity of the propagation problem is only approximately modeled, even when describing the earth to be as complex as an anisotropic, elastic medium.

There are two major schools of approach to the seismic inverse problem: (1) an exact iterative inversion for determination of parameters at every point in the earth, and (2) a high-frequency recursive procedure for generating a reflector map of the earths interior with a post-processing estimate of impedance coefficients on each reflector for estimation of parameter changes across reflectors. A third approach somewhat hybrid in respect to this classification, is travel time inversion, which is an iterative method applied to the travel time.

The speakers presented a compendium of current capabilities and research trends in this area.

**Organizers:** Norman Bleistein, Colorado School of Mines; William W. Symes, Rice University

**What Are the Goals of Seismic Inversion?**  Samuel H. Gray, Amoco Production Company, Tulsa

**Subsurface Seismic Images and Spatial Statistics**  Wafik Beydoun, ARCO Oil and Gas Company, Plano

**Mathematical Aspects of Seismic Data Inversion (Uniqueness and Stability)**  V.A. Cheverda, Novosibirsk University, USSR

**Depth Migration Tomography**  Doug Hanson, Conoco, Inc., Ponca City

**Seismic Inversion for Wide-Angle Reflection and Refraction Data**  Robert L. Nowack, Purdue University

**Realistic Expectations from Seismic Inversion Methods**  François Chapel, Societé Nationale Elf Aquitaine, CSTCS, France

**Seismic Inversion for the Delineation of Complex Structure: A Kinematic Approach**  Patrick Lailly, Institut Français du Pétrole, France

## Time Domain Techniques for Inverse Scattering Problems

The purpose of this minisymposium was to review recent progress concerning time domain techniques for inverse scattering problems as they appear in acoustic, electromagnetic, and elasto-dynamic problems. Several methods for the solution of such problems in one spatial dimension have been developed (such as invariant imbedding, layer-stripping, etc.). Substantial current effort is concentrated on generalizations of these methods to two and three spatial dimensions. The speakers focused on recent results concerning these aspects. Results about transient scattering problems have applications in nondestructive testing and characterization of materials.

**Organizers:** Gerhard S. Kristensson, Lund Institute of Technology, Sweden; Staffan E. Strom, Royal Institute of Technology, Sweden

**Invariant Imbedding and Wave Splitting in 3D**  Vaughan Weston, Purdue University, West Lafayette

**Computational Aspects for the Time Domain Inverse Scattering Algorithms**  David J. Wall, University of Canterbury, New Zealand

**Time Domain Techniques in Electromagnetic Inverse Profiling**  Anton G. Tijhuis, Delft University of Technology, The Netherlands

**Electromagnetic Inverse Problems in Medical Science**  Richard A. Albanese and Richard L. Medina, United States Air Force, Brooks Air Force Base

**Suggested Reading**

A. Allers and F. Santosa, *Stability and resolution analysis of a linearized problem in electrical impedance tomography*, Tech. Report No. 90-13, Department of Mathematical Sciences, University of Delaware, Newark, DE, 1990.

J. G. Berryman, *Convexity properties of inverse problems with variational constraints*, J. Franklin Inst., 328 (1991), pp. 1–13.

D. Colton and R. Kress, *Inverse Acoustic and Electromagnetic Scattering Theory*, Springer-Verlag, Berlin, 1992.

D. Colton and P. Monk, *A new method for solving the inverse scattering problem for acoustic waves in an inhomogeneous medium*, Inverse Problems, 5 (1989), pp. 1013–1026, 6 (1990), pp. 935–947.

H. W. Engl and V. Isakov, *On the identifiability of steel reinforcement bars in concrete from magnetostatic measurements*, Report 430, Mathematics Department, University of Linz, Austria, 1991, preprint.

H. W. Engl and A. Neubauer, *On an inverse problem from magnetostatics*, in Computational Techniques and Applications, W. L. Hogarth and B. J. Noye, eds., Hemisphere Publishing Company, New York, 1990, pp. 3–15.

G. Eskin and J. Ralston, *Inverse back scattering*, J. Analyse Math., special issue dedicated to Shmuel Agmon, to appear.

L. Friedlander, *On the spectrum of the periodic problem for the Schrödinger operator*, Comm. Partial Differential Equations, 15 (1990), pp. 1631–1647.

A. Friedman and V. Isakov, *On the uniqueness in the inverse conductivity problem with one measurement*, Indiana Univ. Math. J., 38 (1989), pp. 563–579.

S. Geman, K. M. Manbeck, and D. E. McClure, *A comprehensive statistical model for single photon emission tomography*, in Markov Random Fields: Theory and Applications, R. Chellappa and A. Jain, eds., Academic Press, New York, 1991.

S. Geman and D. E. McClure, *Statistical methods for tomographic image reconstruction*, Bull. Inst. Internat. Statist., 52 (1987), pp. 5–21.

G. M. L. Gladwell, *Inverse Problems in Vibrations*, Martinus Nijhoff Publishers, 1986.

F. A. Grunbaum, *The scattering transform and the Heisenberg uncertainty principle*, in Nonlinear World Vol. 1, Proc. IV International Workshop on Nonlinear and Turbulent Processes in Physics, Kiev, 1989, V. G. Bar Yakhtar and V. Grigor, eds., 1990, pp. 121–129.

O. H. Hald and J. R. McLaughlin, *Solutions of inverse nodal problems*, Inverse Problems, 5 (1989), pp. 307–347.

G. T. Herman, *Image Reconstruction from Projections. The Fundamentals of Computerized Tomography*, Academic Press, New York, 1980.

A. C. Kack and M. Slaney, *Principles of Computerized Tomography Imaging*, IEEE Press, New York, 1988.

L. Kaufman, *Implementing and accelerating the EM algorithm for positron emission tomography*, IEEE Trans. Medical Imaging, MI-6 (1987), pp. 37–51.

R. Kleinman and P. van den Berg, *Nonlinearized approach to profile inversion*, Internat. J. Imaging Systems Tech., 2 (1990), pp. 119–126.

G. Kristensson, *Time domain inversion techniques for electromagnetic scattering problems*, Tech. Report LUTEDX/(TEAT-7011)/1-33, Lund Institute of Technology, Sweden, 1990.

R. M. Lewitt, *Multidimensional digital image representations using generalized Kaiser–Bessel window functions*, J. Opt. Soc. Amer. A, 7 (1990), pp. 1834–1846.

B. Lowe, M. S. Pilant, and W. Rundell, *The recovery of potentials from finite spectral data*, SIAM J. Math. Anal., 23 (1992), pp. 482–504.

A. McNabb, R. S. Anderssen, and E. R. Lapwood, *Asymptotic behavior of the eigenvalues of a Sturm-Liouville system with discontinuous coefficients*, J. Math. Anal. Appl., 54 (1976), pp. 741–751.

A. Nachman, *Reconstructions from boundary measurements*, Ann. of Math., 128 (1988), pp. 531–576.

F. Natterer, *The Mathematics of Computerized Tomography*, Wiley-Teubner, New York, 1986.

Physics in Medicine and Biology: Special Issue on Fully Three-Dimensional Image Reconstruction in Nuclear Medicine and Radiology, 37 (1992), to appear.

P. C. Sabatier, ed., *Inverse Methods in Action*, Springer-Verlag, Berlin, New York, 1990.

F. Santosa and W. Symes, *An Analysis of Least-Squares Velocity Inversion*, Society of Exploration Geophysics, Tulsa, OK, 1989.

F. Santosa and M. Vogelius, *A computational algorithm to determine cracks from electrostatic boundary measurements*, Internat. J. Engrg. Sci., 29 (1991), pp. 917–937.

O. Scherzer, *The use of Tikhonov regularization in the identification of electrical conductivities from overdetermined boundary data*, Results Math., to appear.

O. Scherzer, H. W. Engl, and R. S. Anderssen, *Parameter identification from boundary measurements in a parabolic equation arising from geophysics*, Report 428, Mathematics Department, University of Linz, Austria, 1991, preprint.

A. Sezginer, *Resistivity imaging by nonlinear optimization techniques*, Proc. Progress in Electromagmetic Research Symposium (PIERS), 1989, p. 286.

E. J. Somersalo, M. Cheney, D. Isaacson, and E. L. Isaacson, *Layer-stripping: A direct numerical method for the impedance imaging problem*, Inverse Problems, 7 (1991).

E. J. Somersalo, D. Isaacson, and M. Cheney, *An inverse boundary value problem for Maxwell's equations*, in 7th Annual Review of Progress in Applied Computational Electromagnetics, Naval Postgraduate School, 1991.

A. G. Tijhuis, *Born-type reconstruction of material parameters of an inhomogeneous, lossy dielectric slab from reflected-field data*, Wave Motion, 11 (1989), pp. 151–173.

Y. Vardi, L. A. Shepp, and L. Kaufman, *A statistical model for positron emission tomography*, J. Amer. Statist. Assoc., 80 (1985), pp. 8–20, 34–37.

D. J. N. Wall, *On some differential equation equations arising in time domain scattering problems for a dissipative wave equation*, Transport Theory Statist. Phys., 20 (1991), pp. 29–54.

V. H. Weston, *Invariant imbedding for the wave equation in three dimensions and the applications to the direct and inverse problems*, Inverse Problems, 6 (1990), pp. 1075–1105.

# Chapter 31
# MANUFACTURING SYSTEMS

## Engineering and Manufacturing Systems

Manufacturing theories and methods are at the core of manufacturing: for example, geometric modeling is the basis of surface forming. The methods of applied mathematics can be used to develop theoretical solutions to outstanding problems. But, these solutions are of interest only if they are available to those that need them, i.e., only if they are incorporated into the manufacturing design/build process. The solution to this technology transfer problem is complicated and involves developing new computer programs, teaching new techniques and methods, correcting mathematical shortfalls of in-place systems and modifying theoretical solutions to meet real-world constraints. In this symposium there were four talks by mathematicians, physicists and engineers, all of whom have experience in developing tools or systems for industrial use.

**Organizer:** David R. Ferguson, Boeing Computer Services

**Shipping Math to the Shop Floor**  Stephen P. Keeler, Boeing Computer Services, Seattle

**Mathematical Problems in Design and Manufacturing**  Joseph Pegna, Rensselaer Polytechnic Institute

**Sheet Metal Forming Simulation Using Finite Elements**  Thomas B. Stoughton, General Motors Research Laboratories, Warren

**Technology Transfer from University to Industry**  Heinz W. Engl, Johannes-Kepler-Universität, Austria

# Chapter 32
# MATERIAL SCIENCE

## Applications of the Theory of Nonlinear Diffusion Equations in Applied Science

The nonlinear diffusion equation, sometimes called the percolation or porous medium equation, exhibits a rich structure of phenomena including finite wave speeds, waiting times, and blow-up. Extensive analytical results are available, and there is a wide range of applications to problems in applied science, such as infiltration of moisture in a partially saturated medium, combustion in a porous medium, dopant diffusion in silicon, and models for epidemics. A number of these applications may be generalized to nonlinear "diffusion type" problems for which few analytical results are available. It is valuable to compare and contrast these extensions. Work on such problems is in progress at a number of centers in Europe and collaborative links have been established with the help of European Community support.

The speakers presented an overview of these generalized nonlinear "diffusion type" problems with examples.

**Organizer:** Alan B. Tayler, Oxford University, United Kingdom

**Free Boundaries with Cusps in Certain Porous Media Flow Problems**  Hans J. van Duijn, Delft University of Technology, The Netherlands

**Optimal Control Problems for Reaction-Diffusion Systems Modeling Man-Environment-Man**  Vincenzo Capasso, University of Milan and SASIAM, Italy

**Analysis of a Model for Substitutional Diffusion**  John R. King, Nottingham University, United Kingdom

**Threshhold Nonlinearities in Porous Medium Combustion**  John Norbury, Oxford University, United Kingdom

## Dynamics of Patterns and Interfaces

An understanding of interfacial dynamics is a central problem in the study of dynamic pattern formation such as solidification, combustion, chemical reaction, and regional partition in mathematical biology. Recent developments in singular limit approaches to these problems are remarkable both in theoretical and numerical aspects. Examples of these developments include the level set approach and viscosity method for flows by mean curvature and the SLEP method for reaction diffusion systems.

The speakers in this minisymposium reviewed recent progress on modelings, methods, and numerics on interfacial problems and clarified the relation between reaction diffusion models and their singular limit equations with sharp interfaces. The presentations started with an overview on these issues, compareed and contrasted several mathematical approaches and numerics, and identified promising research directions.

**Organizer:** Yasumasa Nishiura, Hiroshima University, Japan

**Mathematical Problems in Phase Separation and Coarsening**   Paul C. Fife, University of Utah

**Pattern Formation of Growing Snow Crystals as Third Boundary Value Problem**   Etsuro Yokoyama, Carnegie Mellon University

**Pattern Dynamics in Reacting and Diffusing Media**   Masayasu Mimura, Hiroshima University, Japan; Ryo Kobayashi, Ryukoku University, Japan

**Numerical Algorithms for Propagating Interfaces: Crystal Growth, Minimal Surfaces, and Mean Curvature Flow**   James A. Sethian, University of California, Berkeley

**Phase Transitions and Generalized Motion by Mean Curvature**   Lawrence C. Evans, University of California, Berkeley

## Geometry of Evolving Interfaces and Singular Layers

Evolving interfaces appear in combustion, phase transitions, hydrodynamical instabilities, and many other situations. To compute these evolving interfaces, it is necessary to understand them mathematically. Conversely, the computation of evolving interfaces has been a source of interesting partial differential equations (PDEs).

There are essentially two approaches. Start with a time dependent PDE describing the phenomenon, show that it generates interfaces as singular layers, and study their evolution. This may be very difficult and says little about singularities that will appear. Alternatively, one may start with the geometrical dynamics of the interface, a nonlinear, usually degenerate set of PDEs involving geometric quantities such as curvature.

The speakers presented an overview of research on multidimensional problems, in which interfaces appear by competition between diffusion and nonlinear effects.

**Organizer:** Michelle V. Schatzman, C.M.A.I. Université Lyon 1, France

**Front Propagation for Reaction-Diffusion Equations**   Lia Bronsard, Princeton University

**Spiral Waves for Reaction-Diffusion Equations in a Fast Reaction/Slow Diffusion Limit**   Andy Bernoff, Northwestern University

**Nonlocal Reaction-Diffusion Equations and Nucleation**   Peter Sternberg, University of Indiana, Bloomington; Jacob Rubinstein, Technion Israel Institute of Technology, Israel

## Interfacial Instabilities and Bifurcations

The subject of this minisymposium was the generation of secondary solutions on account of bifurcation where the nonlinearity is at an interfacial boundary condition. It has applications in the areas of materials processing, such as crystal growth and etching, and in chemical engineering, such as 2-phase flows in bore wells, oil displacement by water in porous media and flame front stability. The length scales in these problems are all different and all highly dependent on the various transport phenomena. Some of the unresolved features are related to the modeling of problems where multiple length scales are involved. Methods of linearized stability yield simple results but such problems are generally non-self adjoint.

The speakers presented an overview of the physical problems and demonstrated some of the difficulties in the analytical and numerical calculations.

**Organizer:** Ranga Narayanan, University of Florida, Gainesville

**Core Annular Flows**   Charles Maldarelli, City College of New York

**Interaction of Taylor Couette Flow and Morphological Instabilities**   Geoffrey B. McFadden, National Institute of Standards and Technology; S. R. Coriell, B. T. Murray, M. E. Glicksman, and M. E. Selleck, NIST

**Hydrodynamic Instabilities in Solidification Processes**   Adam A. Wheeler, University of Bristol, United Kingdom

## Methods and Software for Structural Design of Composites

This minisymposium addressed the development of effective methods and software for the design of composites that is competitive with development by empirical means.

For many years, optimization has prevailed as a major direction of research. It benefitted from the progress in techniques, but was limited by the large number of variables and constraints. Ranking methods are not as complex and have proved their efficiency for designing laminated composites, since their introduction in the mid 1980s. Polar description opened new directions of research in designing with anisotropy and also simplified other methods because it reduces the number of pertinent material parameters. Finally, fast re-analysis methods with finite elements are a good tool for preliminary design, but require experienced users to control the process.

The speakers presented an overview and a comparison of methods and software for structural design of composites.

**Organizer:** Georges Verchery, École des Mines de Saint-Etienne, France

**Fast Re-analysis Methods with Finite Elements for Composite Structures**   Shahram Aivazzadeh, Université de Dijon, France

**Ranking Methods for Design of Composite Structures**   Stephen W. Tsai, Stanford University; Thierry Massard, Commissariat a l'Energie Atomique, France

**Polar Description of Anisotropy and Application to Design of Laminated Composites**   Georges Verchery, Organizer

## Phase Transition Kinetics and Nonlinear Diffusion: Numerical and Renormalization-Group Methods

Coarse-grained mesoscale descriptions of phase transition processes and pattern evolution such as spinodal decomposition crystal growth have been sources of interesting nonlinear partial differential equations (PDEs). In this minisymposium, the speakers discussed derivation of coarse-grained PDE models, approaches for qualitatively solving nonlinear PDEs computationally, and analytic renormalization group approach to PDEs in conjunction with semilinear parabolic PDEs. The equivalence of the theory of intermediate asymptotics and the renormalization group theory was demonstrated, and renormalized perturbation approaches to PDE problems were illustrated.

**Organizer:** Yoshitsugu Oono, University of Illinois, Urbana-Champaign

**Renormalization-Group Theory of Anomalous Dimensions in Nonlinear Diffusion** Nigel D. Goldenfeld, University of Illinois, Urbana; Y. Oono, Organizer; L. Y. Chen, University of Illinois, Urbana; F. Liu, University of Illinois, Urbana; Olivier Martin, CUNY-City College

**Pattern Evolution in Excitable Reaction-Diffusion Systems** Takao Ohta, Ochanomizu University, Japan

**Simulations of Dendritic Crystal Growth** Ryo Kobayashi, Ryukoku University, Japan

**Computationally Efficient Coarse-Grained Modeling and the Renormalization Group** Yoshitsugu Oono, Organizer

## Solidification: Theory and Applications

This minisymposium discussed some ways in which mathematics can be used to understand solidification phenomena and processes. These include the description and analysis of industrial casting, mushy regions and dendritic growth, all of which pose fascinating mathematical challenges. The speakers presented an overview of some of the theoretical techniques which seem most relevant.

**Organizer:** John R. Ockendon, Oxford University, United Kingdom

**Unravelling the Complexities of Industrial Casting Processes with Mathematical Models for Quality** J.K. Brimacombe, University of British Columbia, Canada

**Statistical Dynamics of Mushy Zones** Steven P. Marsh and C.S. Pande, Naval Research Laboratory; M.E. Glicksman, Rensselaer Polytechnic Institute

**Interfacial Wave Theory of Dendritic Growth** J.J. Xu, McGill University, Canada

**Grain Boundary Grooves and the Gibbs-Thomson Term** Geoffrey B. McFadden, S.C. Hardy and S.R. Coriell, National Institute of Standards and Technology, Gaithersburg; P.W. Voorhees, Northwestern University; R.F. Sekerka, Carnegie Mellon University

## Superconductivity, Classical and Modern

One of the most exciting developments in contemporary science and applied mathematics has been the discovery of high temperature superconductivity. This area is important in applications because room temperature superconductivity, if achieved, will have a profound influence on everyday life in such areas as power transmission, train travel, electrical devices, and energy production.

Various mathematically-based models, BCS and the Ginzberg-Landau, are being extended and tested for their relevance to high Tc experiments.

The speakers reviewed the new analytical and numerical approaches known to describe conventional type II superconductors. In addition, they discussed the latest developments in new high Tc superconductivity.

**Organizer:** Melvyn S. Berger, University of Massachusetts, Amherst

**Symmetric Vortices and Type II Superconductors** Melvyn S. Berger, Organizer

**Vortices for the Ginzburg-Landau Equations—The Nonsymmetric Case in Bounded Domain** Yi Ying Chen, Princeton University

**A Unified Description of the Cuprate Superconductors** Robert Soulon, Naval Research Laboratory

**Superconducting Electronics in Supercomputing** Fernand Bedard, National Security Agency

## Mathematical and Computational Modeling of Materials Forming Processes

Materials forming processes (melting, molding, pressing, stamping, cutting,...) represent a major component of product manufacture. In addition to behavior with respect to forming processes, material properties after forming must be appropriate to intended product use. Matching material properties, forming processes and intended use characteristics presents a very substantial challenge involving large expenditures of time, equipment and money. This session presented several efforts to understand materials forming through modeling on scales ranging from molecular structure on up to macroscopic behavior. Actual applications to industrial processes were presented.

**Organizer:** Peter E. Castro, Eastman Kodak Company, Rochester

**Materials Simulation for Industry** Douglas C. Allan, Corning Incorporated; Michael P. Teter, Corning Incorporated, Corning

**Simulation of Injection Molding** Richard Ellson, Eastman Kodak Company, Rochester; Karen Malburne, Eastman Kodak, Rochester

**Metal Processing** Russell L. Mallett, Alcoa Laboratories, Alcoa Center

John Burns, William Kolata, and Mac Hyman plan future SIAM meetings.

**Suggested Reading**

G. Barles, L. Bronsard, and P. E. Souganidis, *Front propagation for reaction-diffusion equations of bistable type*, to appear.

Y. G. Chen, Y. Giga, and S. Goto, *Uniqueness and existence of solutions of generalized mean curvature flow*, J. Differential Geom., 33 (1991), pp. 749–786.

S. H. Davis, *Hydrodynamic interactions in directional solidification*, J. Fluid Mech., 212 (1990), p. 241.

P. de Mottoni and M. Schatzman, *Geometrical evolution of developed interfaces*, Trans. Amer. Math. Soc., accepted for publication.

J. Dewynne, J. R. Ockendon, and P. Wilmott, *On a mathematical model for fiber tapering*, SIAM J. Appl. Math., 49 (1989), p. 987.

L. C. Evans, H. M. Soner, and P. E. Souganidis, *The Allen-Cahn equation and generalized motion by mean curvature*, to appear.

L. C. Evans and J. Spruck, *Motion of level sets by mean curvature I*, J. Differential Geom., 33 (1991), pp. 635–681.

P. Fife, *Models for phase separation and their mathematics*, in Nonlinear Partial Differential Equations and Applications, M. Mimura and T. Nishida, eds., Kinokuniya Publishers, Tokyo, to appear.

———, *Pattern dynamics for parabolic partial differential equations*, preprint.

E. Georgiou, D. T. Papageorgiou, C. Maldarelli, and D. S. Rumschitzki, *The double layer capillary stability of an annular electrolyte film surrounding a dielectric-fluid core in a tube*, J. Fluid Mech., 226 (1991), p. 149.

E. Georgiou, C. Maldarelli, D. T. Papageorgiou, and D. S. Rumschitzki, *An asymptotic theory for the linear stability of a CAF in the thin annular limit*, J. Fluid Mech., submitted.

M. E. Glicksman, S. R. Coriell, and G. B. McFadden, *Interaction of flows with the crystal-melt interface*, Ann. Rev. Fluid Mech., 18 (1986), p. 307.

N. Goldenfeld, O. Martin, and Y. Oono, *Intermediate asymptotics and renormalization group theory*, J. Sci. Comput., 4 (1989), pp. 355–372.

N. Goldenfeld, O. Martin, Y. Oono, and F. Liu, *Anomalous dimension and the renormalization group in a nonlinear diffusion process*, Phys. Rev. Lett., 64 (1990), pp. 1361–1364.

K. H. Hoffman and J. Sprekels, *Free and moving boundary problems*, Pitman Res. Notes, 185, 186 (1991).

R. Kobayashi, *Modeling and numerical simulations of dendritic crystal growth*, Physica D, submitted.

G. B. McFadden, S. R. Coriell, M. E. Glicksman, and M. E. Selleck, *Instability of a Taylor-Couette flow interacting with a crystal melt interface*, PCH Physico Chemical Hydrodyn., 11 (1989), p. 387.

G. B. McFadden, S. R. Coriell, B. T. Murray, et al., *Effect of a crystal-melt interface on Taylor-vortex flow*, Phys. Fluids A, 2 (1990), p. 700.

M. Mimura, Y. Nishiura, and T. Takaishi, *Coupled pattern generator of reaction diffusion systems*, preprint.

Y. Nishiura, *Coexistence of infinitely many stable solutions to reaction diffusion systems in the singular limit*, Dynamics Reported, Springer-Verlag, in press.

T. Ohta, A. Ito, and A. Tetsuka, *Self-organization in an excitable reaction-diffusion system: Synchronization of oscillatory domains in one dimension*, Phys. Rev. A, 42 (1990), pp. 3225–3232.

Y. Oono, *Modeling macroscopic nonlinear space-time phenomena*, IEICE Trans., E74 (1991), pp. 1379–1387.

D. T. Papageorgiou, C. Maldarelli, and D. S. Rumschitzki, *Nonlinear interfacial stability of core-annular flows*, Phys. Fluids A, 2 (1990), p. 340.

L. M. Pismen and J. Rubinstein, *Dynamics of defects*, to appear.

D. S. Riley and S. H. Davis, *Long-wave morphological instabilities in the directional solidification of a dilute binary alloy*, SIAM J. Appl. Math., 50 (1990), p. 420.

J. A. Sethian, *Curvature and evolution of fronts*, Comm. Math. Phys., 101 (1985), p. 487.

———, *Numerical algorithms for propagating interfaces: Hamilton-Jacobi equations and conservation laws*, J. Differential Geom., 31 (1990), pp. 131–161.

A. A. Wheeler, *A strongly nonlinear analysis of the morphological instability of a freezing binary alloy: Solutal convection, density change and nonequilibrium effects*, IMA J. Appl. Math., in press.

E. Yokoyama and T. Kuroda, *Pattern formation in growth of snow crystals occurring in the surface kinetic process and the diffusion process*, Phys. Rev. A, 41 (1990), pp. 2038–2049.

## Chapter 33
# MATHEMATICS EDUCATION

### SIAM Student Paper Competition
Three students presented award winning papers in applied and computational mathematics. The students were chosen from among those submitting papers to the SIAM Student Paper Competition. To qualify, a student had to submit a singly-authored paper to SIAM by February 15, 1991, be a student in good standing, and at the time of submission, must not have received his/her doctorate.

**Chair:** Margaret H. Wright, AT&T Bell Laboratories, Murray Hill

**Uniform Stabilization of the Euler-Bernoulli Plate with Feedback Operator Acting via Bending Moments**   Mary Ann Horn, University of Virginia

**Models and Computations for Interstellar Magnetic Gas Clouds: A Variational Method**   Brian C. Morris, University of Massachusetts

**Nonlinear Feedback Control Law Learning by Neural Networks: A Viability Approach**   Nicholas Seube, University of Paris, France

### The Mathematical Contest in Modeling (MCM)
The MCM encourages undergraduates to work on realistic, open-ended problems in a team effort. It is hoped that overseas visitors who see these students in action might be prompted to start such a contest at home.

The speakers were undergraduate winners of MCM for 1991. Two teams of undergraduates presented their solution to the MCM 1991 problems.

**Organizer:** B.A. Fusaro, Salisbury State University

**Estimating the Flow From a Water Tank**   Anna Baumgartner, University of Alaska, Fairbanks; Anupama Rao, University of Alaska, Fairbanks; Eiluned Roberts, University of Alaska, Fairbanks; R. A. Hollister, advisor

**Minimal Spanning Trees for a Communications Network**   Zvi Margaliot, University of Western Ontario, Canada; Alex Pruss, University of Western Ontario, Canada; Patrick Surry, University of Western Ontario, Canada; H. Rasmussen, advisor

## Chapter 34
# MECHANICS, WAVES, AND SOLIDS

### Diffraction by Connected Wedges
The longstanding problem of diffraction by wedge-shaped obstacles where two or more wave velocities are involved has so far resisted all attempts to obtain an approach that is both mathematically strict and computationally efficient for significant applications in acoustics, radioscience, and geophysics.

The methods being developed to solve these problems include direct numerical methods and reduction to functional equations, in which the functional equations are treated by direct methods or by reduction to integral equations. Direct methods provide numerical results but cannot display the fine structure of wave fields. New versions of reduction to integral equations are strict but they have not been sufficiently examined in practice.

The speakers discussed the mathematical aspects of every approach, compared their efficiencies, and surveyed major applications.

**Organizer:** Bair Vl. Budaev, Steklov Mathematical Institute, USSR

**Diffraction of Elastic Waves By a Rigid Wedge**
Jesper Larsen, Math-Tech APS, Denmark

**Application of Sommerfeld Integrals to the Problems of Diffraction by Cones of Arbitrary Cross-Sections**   Valery P. Smyshlyaev, USSR Academy of Sciences, USSR

**Diffraction of Elastic Waves by a Free Wedge**   Bair Vl. Budaev, Organizer

## Finite Rotations in Plates and Shells—Nonlinear Theories, Computational Methods

A current trend in nonlinear structural mechanics is the formulation of improved theories and numerical methods based on well defined order-of-magnitude considerations for strains and rotations. In recent years, many papers have been devoted to the formulation of new geometrically nonlinear plate and shell theories for small strains accompanied by moderate, large or unrestricted rotations. These investigations have been carried out in the Kirchhoff-Love type theory of thin, elastic, isotropic plates and shells. The results have been extended also to more complicated problems of technical interest, e.g. anisotropic laminated composite plates and shells and elastic-plastic structures.

This session gave an account of recent research in the theory and finite element analysis of finite rotation plate and shell problems. The presentations included the development of new nonlinear theories, variational methods, finite elements, and the numerical analysis of the stability and post-buckling behavior of plates and shells.

**Organizers:** Rudiger Schmidt, University of Wuppertal, Germany; J.N. Reddy, Virginia Polytechnic Institute and State University

**Analysis of Laminated Composite Plates Using First- and Second-Order Moderate Rotation Theories**  J.N. Reddy, Organizer; E. Sacco, University of Rome, Italy

**Moderate Rotations in Elastic-Plastic Shells and Plates**  Dieter Weichert, University of Lille 1, France

**Nonlinear Theory of Laminated Composite Plates and Shells Accounting for the Interlaminar Continuity Conditions**  Liviu Librescu, Virginia Polytechnic Institute and State University

ICIAM attendees enjoyed visiting the software and book exhibits.

INRIA was the host for ICIAM 87.

## Non-Classical Problems of the Theory and Behavior of Structures Exposed to Complex Environmental Conditions

This minisymposium was concerned with the modeling and examination of the strategies to be implemented toward the solution of a number of problems confronting the behavior of structures constructed of composite materials and exposed to severe environmental conditions. It is a well-known fact that the new generations of aeronautical/aerospace vehicles, rocket engines, submarines, space stations, space antennas, and nuclear reactors, etc. are likely to be constructed at least in part of composite materials. However, in contrast to their metallic counterparts, these composite structures cannot be modeled within the classical hypotheses proper to the standard, metallic ones. In addition to the necessity of adequately modeling these structures, a better understanding of their behavior is required.

The speakers examined a variety of problems including modeling and optimization of composite structures exhibiting large deformations, the static and dynamic snap-through behavior of composite structures, buckling and post-buckling behavior of composite structures exhibiting initial geometric imperfections and exposed to a complex system of mechanical and thermal in-plane loadings.

**Organizer:** Liviu Librescu, Virginia Polytechnic Institute and State University

**Postbuckling Behavior of Shear Deformable Composite Panels Exhibiting Initial Geometric Imperfections and Thermal Loadings Exposed to In-Plane Mechanical and Thermal Loadings**  Liviu Librescu, Organizer; Marco Antonio Souza, Pontificia Universidade Catolica do Rio de Janeiro, Brazil

**Non-Classical Theories and Approximations of Shells**  Gerald Wempner, Georgia Institute of Technology

**Improved Modeling of Shear Deformable Composite Laminated Plates and Shells—New Results and Evaluations**  Marco Di Sciuva, Politecnico di Torino, Italy

**The Behavior of Shear Deformable Composite Cylindrical Shells Subjected to Axial Loading, Torsion and an Axisymmetric Temperature Field**  Victor Birman, University of Missouri, Rolla

**Stabilization of Distributed Parameter Structures by Time-Delayed Velocity Feedback Control**  Sarp Adali, University of Natal, South Africa; J.M. Sloss, University of California, Santa Barbara; I.S. Sadek, University of California, Santa Barbara; J.C. Bruch, University of California, Santa Barbara

**Snap-Through Buckling of Composite Cylindrical Shells**  Anthony N. Palazotto, Air Force Institute of Technology

**Stability of Laminated Cylindrical Shells: Thin, Thick, Perfect and Imperfect**  George J. Simitses, University of Cincinnati

**On the Strain Energy of Anisotropic Thin Bodies**  Dewey H. Hodges, Georgia Institute of Technology; A.R. Atilgan, Georgia Institute of Technology

**Aeroelastic Stability Analysis of Anisotropic Composite Wings Based on a Refined Structural Model**  Gabriel Karpouzian, United States Naval Academy; Liviu Librescu, Organizer

# MECHANICS, WAVES, AND SOLIDS

One of the most popular displays from the Joy of Mathematics exhibit.

## Nonlinear Waves in Deformable Solids

With the implementation of higher and higher levels of input energies, the desire to avoid response distortion, and the technical potentialities to benefit from nonlinearities in other cases, a thorough study of nonlinear dynamical effects in deformable solids has become a necessity. This has developed into a highly technical subject of applied mathematics that is somewhat paradigmatic with its emphasis on asymptotic expansions, new methods of treating multidimensional nonlinear evolution problems, shock-wave and soliton theories, and chaos in some cases.

The minisymposium addressed the solution of such difficult nonlinear problems while delineating the various types of practical solution approaches, whether analytical or numerical. Applications abound in solid-state devices, signal-processing, nondestructive testing methods, and seismology; but a common framework emerges which favors a true communication between specialists. Among the methods exhibited are modulation theory, a projection method, the Whitham-Newell theory, various asymptotic approaches, and direct numerical experiments.

**Organizer:** Gérard A. Maugin, Université Pierre et Marie Curie, France

**The Projection Method for Nonlinear Surface Acoustic Waves**   David F. Parker, University of Edinburgh, United Kingdom; Alexei A. Maradudin, University of California, Irvine; Andreas P. Mayer, University of Edinburgh, United Kingdom

**The Dynamic Response of a Stretched Hyperelastic Membrane Subjected to Normal Impact**   J.B. Haddow, University of Victoria, Canada; J.L. Wegner, University of Victoria, Canada; J.L. Jiang, University of Victoria, Canada

**Nonlinear Dynamics of 2D Defects in Discrete Models of Elastic Solids**   Joel Pouget, Université Pierre et Marie Curie, France

**Nonlinear Waves in a Layer with Energy Influx**   Juri Engelbrecht, Estonian Academy of Sciences, USSR

**Nonlinear Wave Interactions in a Micropolar Elastic Medium**   Sadik Dost, University of Victoria, Canada; S. Erbay, University of Victoria, Canada; H.A. Erbay, University of Victoria, Canada

**Surface Solitons on Elastic Surfaces: Analysis and Numerics**   Hichem Hadouaj, Université Pierre et Marie Curie, France; Gérard A. Maugin, Organizer

**Nonlinear Surface Acoustic Waves on a Piezoelectric Solid**   A.P. Harvey, Cambridge University Press, New York; G. E. Tupholme, University of Bradford, England

## Waves in Inhomogeneous Elastic Media

The talks planned for this minisymposium centered around wave propagation in heterogeneous and anisotropic elastic media. The speakers came from diverse backgrounds: they are material scientists, mechanical engineers, geologists, and mathematicians. Their research interests range from Earths stratigraphic mapping to the design of "stealth" submarines. The common thread in their work is the phenomenon of the interaction of waves with inhomogeneities in an elastic medium. They described a diverse set of approaches taken to understand wave interaction including field testing, laboratory experiments, numerical computations, and functional analysis.

**Organizer:** Rouben Rostamian, University of Maryland, Baltimore County

**Dispersive Effective Medium Theory for Wave Propagation in Periodic Composites**   Fadil Santosa, University of Delaware

**One-Component Surface Waves**   Thomas C.T. Ting, University of Illinois, Chicago

**The Stroh Formulation for 2-D Anisotropic Elasticity**   Thomas C.T. Ting, University of Illinois, Chicago

**Field Observations of Seismic Wave Scattering**   Kandiah Balachandran, Consulting Geophysicist, Baltimore

**Waves in Randomly Layered Anisotropic Media**   Andrew Norris, Rutgers University, Piscataway

**Waves in Finely Layered Media**   Robert Burridge, Schlumberger-Doll Research

**Waves in Continuously Stratified Elastic Media**   William Hager, University of Florida, Gainesville

### Suggested Reading

J. Engelbrecht, *An Introduction to Asymmetric Solitary Waves*, Longman, London, 1992.

H. A. Erbay, S. Erbay, and S. Dost, *Nonlinear wave modulation in micropolar elastic media I, II*, Internat. J. Engrg. Sci., 29 (1991), pp. 845–868.

A. P. Harvey and G. E. Tupholme, *Nonlinear mode coupling of two codirectional surface acoustic waves on a piezoelectric solid*, Internat. J. Engrg. Sci., 29 (1991), pp. 987–998.

G. A. Maugin, B. Collet, R. Drouot, and J. Pouget, *Nonlinear electromechanical couplings*, Nonlinear Science Series, Manchester University Press, U.K., 1992.

G. A. Maugin and H. Hadouaj, *Solitary surface transverse wave on an elastic substrate coated with a thin film*, Phys. Rev. B, 44 (1991).

F. Odeh and J. B. Keller, *Partial differential equations with periodic coefficients and Bloch waves in crystals*, J. Math. Phys., 5 (1964), pp. 1499–1504.

D. F. Parker and E. A. David, *Nonlinear piezoelectric surface waves*, Internat. J. Engrg. Sci., 27 (1989), pp. 565–581.

J. Pouget, *Dynamics of patterns in ferroelastic-Martenistic transformations*, I, II, Phys. Rev., B43 (1991), pp. 3575–3592.

F. Santosa and W. W. Symes, *A dispersive effective medium for wave propagation in periodic composites*, SIAM J. Appl. Math., 51 (1991), pp. 984–1005.

# Chapter 35
# MEDICINE AND BIOLOGY

## Competitive Systems

This minisymposium addressed fundamental and applied aspects for competitive systems. These systems have applications in chemical, biological, ecological, economic, and military models. The speakers discussed important trends: The theory of competitive systems, controlling competitive systems (with possible obstacles), nonmonotonicity and chaotic aspects, and efficient numerical schemes. Partial differential equations, ordinary differential equations, and discrete systems were considered.

**Organizers:** Suzanne M. Lenhart, University of Tennessee, Knoxville; Vladimir Protopopescu, Oak Ridge, National Laboratory

**Permanence in Some Reaction-Diffusion Models**
George C. Cosner and Robert Steven Cantrell, University of Miami; Vivian C.L. Hutson, University of Sheffield, United Kingdom

**Non-Monotonicity, Chaos and Combat Models**
James Dewar, James J. Gillogly and Mario L. Juncosa, Rand Corporation, Santa Monica

**Three-Dimensional Competitive Lotka-Volterra Systems**   Mary Lou Zeeman, Massachusetts Institute of Technology

**Steady States for the Gradostat: A Singular Perturbation Approach**   Hal L. Smith, Arizona State University

**Game for Non-Local Competitive Distributed Systems**   Srdjan Stojanovic, University of Cincinnati; S. Lenhart, Organizer; V. Protopopescu, Organizer

Bernard McDonald, deputy director of NSF's Division of Mathematical Sciences and Charles Holland, director for mathematical sciences at AFOSR.

## Delay Differential Equations in Population Dynamics

Ordinary and partial differential equations have long played a central role in the history of theoretical population dynamics, and will no doubt continue to serve as valuable tools in future investigations. However, more realistic models should include some of the past states of the system; that is, ideally, a real system should be modeled by differential equations with time delays. Indeed, the use of delay differential equations in the modeling of population dynamics is currently very active, largely due to the recent rapid progress achieved in the study of these equations.

In this minisymposium, the speakers focused on qualitative and global aspects of some population models involving delay (ordinary and partial) differential equations. Specifically, they were concerned with global stability, stability switching, oscillatory behavior, and Hopf-bifurcation of the feasible steady states of these equations.

**Organizer:** Yang Kuang, Arizona State University

**Stage-Structured Models of Population Growth**
W.G. Aiello and Herbert I. Freedman, University of Alberta, Canada; J. Wu, York University, Canada

**Global Stability for Infinite Delay Lotka-Volterra Type Systems**   Yang Kuang, Organizer; H.L. Smith, Arizona State University

## Mathematical Immunology

The immunological system is a highly complex collection of cells and molecules that works to destroy pathogens on time scales ranging from milliseconds (for chemical reactions) to weeks (lifetimes of components). Understanding this system presents a major challenge to biologists and has been the subject of an enormous amount of experimental work. Mathematical models have been developed at all levels, particularly in attempts to understand the numerous properties of the system that seem "emergent". Novel types of massive computer simulations have been employed; complex new systems of integro-differential equations have been derived and partially analyzed; and novel stochastic approaches have been attempted. The speakers presented an overview of current research in this area.

**Organizers:** Alan S. Perelson, Los Alamos National Laboratory; Lee A. Segel, Weizmann Institute, Israel

**Mathematical and Simulation Models of Immune Networks**   Alan S. Perelson, Organizer

**The Shape Space Approach to Immune Networks**
Lee A. Segel, Organizer

**Affinity Maturation on Fitness Landscapes**   Catherine Macken, Stanford University; Alan S. Perelson, Organizer

**Modeling Aspects of the Interaction of HIV with the Immune System**   Stephen J. Merrill, Marquette University

## Models for Biological Pattern Formation

This minisymposium presented examples of pattern formation mechanisms in biological systems spanning several size scales, starting at the molecular level, and proceeding to the population level. The speakers began with patterns in immune networks and in cellular systems, and continued with tissue interactions such as those in the dermis and epidermis. A final example was given in the context of ecology and the interactions of species. Mathematical methods included linear and nonlinear stability analysis of partial and integro-differential equations. Our case studies illustrated that common mechanisms are often exploited in various guises by systems which seem unrelated, but which are functionally similar. Such topics, which date back to A. Turing, have provided a fertile area for applied mathematical research with impact on developmental and population biology.

**Organizers:** Leah Edelstein-Keshet, University of British Columbia, Canada; David J. Wollkind, Washington State University

**Pattern Formation in Immune Networks** Lee A. Segel, Weizmann Institute, Israel

**Patterns Formed through Cellular Interactions** Leah Edelstein-Keshet, Organizer

**Pattern Formation in Tissues** Valipuram S. Manoranjan, Washington State University

**Diffusive Instability Pattern Formation in a Model System for Mite Interaction on Fruit Trees** David J. Wollkind, Organizer; John B. Collings, Adams State College; Maria Concepcion Briones Barba, University of the Philippines, Philippines

**Pattern Formation in Hydra—A Theoretical Study** Somdatta Sinha, Centre for Cellular and Molecular Biology, Hyderabad, India

## Systems Theory in Mathematical Biology

The underlying theme of this minisymposium was the question of whether particular populations persist or go extinct in a given system. This problem is relevant in both of its aspects, preservation of endangered species on the one hand, and elimination of pest species on the other. A new application, interacting human populations with differing language backgrounds, was described.

The speakers surveyed the techniques and current research for competitive and predator-prey systems in ecological models and chemostats. Each speaker addressed a particular aspect of the models and techniques, with a focus on either systems of ordinary differential equations or discrete maps.

**Organizer:** Herbert I. Freedman, University of Alberta, Canada

**Competition in the Chemostat with Inhibitor** Paul E. Waltman, Emory University; S.B. Hsu, Tsing Hua University, Republic of China

**Competition in the Chemostat** Gail S.K. Wolkowicz, McMaster University, Canada; Lu Zhiqi, Henan Teacher's University, People's Republic of China

**Discrete Models of Mathematical Ecology** Joseph W.H. So, University of Alberta, Canada

**Applications of Persistence Theory: Food Chains, Competition and Bilingualism** Herbert I. Freedman, Organizer

### Suggested Reading

W. G. Aiello and H. I. Freedman, *A time-delay model of single-species growth with stage structure*, Math. Biosci., 101 (1990), pp. 139–153.

R. S. Cantrell and C. Cosner, *On the uniqueness and stability of positive solutions in the Lotka–Volterra competition model with diffusion*, Houston J. Math., 15 (1989), pp. 341–361.

J. M. Cushing, *Integrodifferential Equations and Delay Models in Population Dynamics*, Lecture Notes in Biomath., Vol. 20, Springer-Verlag, New York, 1977.

J. A. Dewar, J. J. Gillogly, and M. L. Juncosa, *Nonmonotonicity, chaos, and combat models*, R-3995-RC, RAND Corporation, Santa Monica, CA, 1991, preprint.

L. Edelstein-Keshet and G. B. Ermentrout, *Models for contact-mediated pattern formation: Cells that form parallel arrays*, J. Math. Biol., 29 (1990), pp. 33–58.

H. I. Freedman, *Deterministic Mathematical Models in Population Ecology*, Marcel Dekker, New York, 1980.

K. Gopalsamy, *Time lags and global stability in two-species competition*, Bull. Math. Biol., 42 (1980), pp. 729–737.

J. Hofbauer and K. Sigmund, *The Theory of Evolution and Dynamical Systems*, Cambridge University Press, Cambridge, U.K., 1988.

Y. Kuang and H. L. Smith, Global stability for infinite delay Lotka–Volterra type systems, J. Differential Equations, to appear.

———, *Convergence in Lotka–Volterra type diffusive delay systems without dominating instantaneous negative feedbacks*, J. Australian Math. Soc. Ser. B, submitted.

S. Lenhart, S. Stojanovic, and V. Protopopescu, *A two-sided game for nonlocal competitive systems with control on the source terms*, Proc. 1991 Workshop on Variational Problems, Institute of Mathematics and Its Applications, MN, Springer-Verlag, New York, to appear.

C. A. Macken, P. Hagan, and A. S. Perelson, *Evolutionary walks on rugged landscapes*, SIAM J. Appl. Math., 51 (1991), pp.

B. N. Nagorcka, V. S. Manoranjan, and J. D. Murray, *Complex spatial patterns from tissue interactions—an illustrative model*, J. Theoret. Biol., 128 (1987), pp. 359–374.

A. S. Perelson, ed., *Theoretical Immunology, Parts One and Two*, Addison-Wesley, Redwood City, CA, 1988.

———, *Modeling the interaction of HIV with the immune system*, in Mathematical and Statistical Approaches to AIDS Epidemiology, C. Castillo-Chavez, ed., Lecture Notes in Biomath., Vol. 83, Springer-Verlag, New York, 1989, pp. 350–370.

A. S. Perelson and S. A. Kauffman, eds., *Evolution on Rugged Landscapes: Proteins, RNA, and the Immune System*, Addison-Wesley, Redwood City, CA, 1988.

L. A. Segel and A. S. Perelson, *Exploiting the diversity of time scales in the immune system: A B-cell antibody model*, J. Statist. Phys., 63 (1991), pp.1113–1131.

———, *A paradoxical instability caused by relatively short range inhibition*, SIAM J. Appl. Math., 50 (1990), pp. 91–107.

S. Sinha, N. V. Joshi, J. S. Rao, and S. Mookerjee, *A four-variable model for the pattern forming mechanism in hydra*, Biosystems, 17 (1984), pp. 15–22.

P. Waltman, *Competition Models in Population Biology*, Society for Industrial and Applied Mathematics, Philadelphia, 1983.

D. J. Wollkind, J. B. Collings, and M. C. B. Barba, *Diffusive instabilities in a one-dimensional temperature-dependent model system for a mite predator-prey interaction on fruit trees: Dispersal motility and aggregative prey-taxis effects*, J. Math. Biol., 29 (1991), pp. 339–362.

# Chapter 36
# NUMERICAL LINEAR ALGEBRA

## Krylov Iterative Methods on Massively Parallel Architectures: Computational and Mathematical Issues

During the last decade, sparse Krylov iterative methods for solving linear systems and finding eigenvalues have gained widespread acceptance. A current challenge is to find ways to efficiently implement these computationally efficient methods on massively parallel machines. This requires attention to the (frequently conflicting) objectives of computational efficiency, exploitable concurrency, partitioning and mapping, and the ability to preprocess to reduce the performance impact of irregular communication and computation patterns.

The speakers addressed these issues by presenting both theoretical and experimental results obtained by implementing a variety of these algorithms on the CM-2, Masspar, and Intel Touchstone prototype machines.

**Organizer:** Dimitri Mauriplis, ICASE, NASA Langley Research Center

**Parti and Optimizations**   Ravi Donnusamy, NASA Langley Research Center

**Padé Approximants for Computation of Eigenvalue of Large Hermitian Matrices**   Nahid Emad, Université Pierre et Marie Curie and Écoles Nationale des Mines de Paris, France

**Sparse Polynomial Preconditioned Conjugate Gradient on Massively Parallel Architectures**   Serge Petiton, Etablissement Technique Central de l'Armement, France and Yale University; Christine Weill, Université Pierre et Marie Curie, France

**Parallel Tool Kit for Efficiently Implementing Unstructured Problems**   Raja Das, NASA Langley Research Center

## LAPACK: A Linear Algebra Library for High-Performance Computers

The aim of the LAPACK project has been to develop a portable linear algebra library for efficient use on as broad a variety of high-performance, general purpose, scientific computers as possible. The library is based on the widely used LINPACK and EISPACK packages for solving linear equations, eigenvalue problems, and linear least squares problems, but extends their functionality in a number of ways and results in a single uniform package. LAPACK is designed to exploit the parallel processing capabilities of many high-performance machines, is freely available, and is intended to become part of the infrastructure of scientific computing.

The speakers described the background and algorithmic basis for LAPACK and reported on the performance and future development of LAPACK.

**Organizer:** Sven J. Hammarling, Numerical Algorithms Group Ltd., United Kingdom

**Introduction and Background to the LAPACK Project**   Sven J. Hammarling, Organizer

**Performance of Block Algorithms in LAPACK**   Edward Anderson, University of Tennessee, Knoxville

**Algorithmic Highlights in the LAPACK Project**   Christian H. Bischof, Argonne National Laboratory

**LAPACK 2 and the Way Forward**   Jack J. Dongarra, University of Tennessee, Knoxville and Oak Ridge National Laboratory

*Left:* Masatake Mori, elected to the board of directors, Japan SIAM, and invited speaker, Masayasu Mimura.

*Above:* Marline Block and Eugene Levner.

## Matrix Methods on Array Processors

Parallel algorithms for linear algebraic methods have become important due to the commercial emergence of parallel computers and the dominance of matrix methods in both off-line scientific computing and dedicated real-time applications.

Matrix algorithms have properties of locality, recursiveness, and regularity that match well the fine-grain parallelism and dense connectivity of array processors. When the dimension of the problem corresponds to the dimension of the systolic array, a matrix operation can have very low overhead in communication and synchronization. The speakers presented an overview of software kernels for scientific computing, methods in signal processing, unitary matrix and transform techniques, and Jacobi-type update methods and applications.

**Organizer:** George J. Miel, Hughes Research Laboratories

**State-of-Art and Design Methodology of Fixed-Size Arrays**  Ed F. Deprettere, Delft University of Technology, The Netherlands

**Massively Parallel Matrix Algebra Kernels for Scientific Computing**  Petter Bjorstad, University of Bergen, Norway

**Jacobi-Type Updating Algorithms and Systolic Arrays**  Marc Moonen, ESAT Katholieke Universiteit Leuven, Belgium; Joos Vandewalle, ESAT Katholieke Universiteit Leuven, Belgium

**Unitary Matrices and Fast Transforms in Array Processing**  George J. Miel, Organizer

## Polynomial Based Iterative Methods for Non-symmetric Linear Systems

The iterative solution of large and sparse linear systems of algebraic equations arises in many scientific applications. During the last decade, there has been an increasing interest in systems with a non-symmetric coefficient matrix. An important example is systems from the discretization of convection-diffusion problems. New iterative schemes have been designed to solve these problems. Polynomial based methods play an important role. They can be used as preconditioners for the conjugate gradient method and as an ingredient for hybrid methods. In addition, polynomial based methods are well suited for parallel and vector architectures.

The speakers surveyed recent research on these methods, discussed their efficient implementation, and presented recent convergence results based on approximation theoretical tools.

**Organizer:** Wilhelm E. Neithammer, Universität Karlsruhe, Germany

**A Polynomial Iteration for Indefinite Linear Systems**  Bernäd Fischer, Universität Hamburg, Germany

**Stable Biconjugate Gradient Type Methods for Non-symmetric Linear Systems**  Roland W. Freund, NASA Ames Research Center

**Lanczos Type Methods for Non-symmetric Linear Systems**  N.M. Nachtigal, ETH-Zentrum, Switzerland

**On a Generalization of Faber Polynomials in Connection with ADI**  Gerhard Starke, Kent State University

**The Parallel Solution of Almost Block Diagonal Systems Arising in Numerical Methods for BVPs for ODEs**  Kenneth R. Jackson and Richard H. Pancer, University of Toronto, Canada

## Solution of Sparse Systems in a Parallel Environment

This minisymposium was concerned with both sparse systems and parallelism, thus addressing two important issues in solving large-scale problems in science and engineering. The talks concentrated on direct methods for solving sparse linear equations since such techniques have the widest range of applicability.

The speakers focused primarily on frontal and multifrontal approaches. They discussed why this class of methods is so powerful, particularly on vector and parallel computers, and described current research that extends the techniques to handle very sparse systems and unsymmetric structures.

They also examined the performance, on several parallel computers, of a robust, portable, and efficient multifrontal code on problems from a wide range of disciplines.

**Organizer:** Iain S. Duff, Rutherford Appleton Laboratory, United Kingdom and CERFACS, France

**The Use of Level 2 and Level 3 BLAS in Sparse Codes**  Iain S. Duff, Organizer

**A Multifrontal Approach to Parallel Sparse Triangular Solutions**  Alex Pothen, Pennsylvania State University, University Park

**An Unsymmetric-Pattern Multifrontal Method**  Timothy A. Davis, University of Florida; Iain Duff, Organizer

**Vectorized LU Factorization of Extremely Sparse Matrices**  Robert F. Lucas, Supercomputing Research Center

## Sparse Problems in Electric Power Networks

Electric power system control, monitoring, and forecasting lead to a variety of challenging mathematical and computational problems. Network topology and the particular character of power measurements give a rich structure to the large sparse systems that are common to these problems. The utility setting offers special challenges including demands for real-time processing in large networks and the best use of efficient architectures.

The speakers presented an overview of a variety of current problem areas, including problems arising in several forms of estimation and optimization.

**Organizers:** Kevin A. Clements, Worcester Polytechnic Institute; Paul W. Davis, Worcester Polytechnic Institute

**An Overview of Sparse Problems in Electric Power Networks**  Kevin A. Clements, Organizer

**Parallel Computation for Sparse Network Problems: Fundamental Ideas**  Fernando L. Alvarado, University of Wisconsin, Madison; Ramón Betancourt, San Diego State University

**Parallel Computation for Sparse Network Problems: Ordering and Organization**  Ramón Betancourt, San Diego State University; Fernando L. Alvarado, University of Wisconsin, Madison

**Sparse Matrix Implementations for Electric Power Networks**  Brian Stott, Power Computer Applications Corporation, Mesa

## The Effect of Ordering on Preconditioned Conjugate Gradients and Ordering Strategies

This minisymposium focused on the effect of ordering on preconditioned conjugate gradient methods and on the ordering algorithms. The use of preconditioners with conjugate gradient, GMRES, ORTHOMIN, or CGs accelerations now form the iterative techniques of choice for solving large sparse systems arising in finite element and finite volume modeling. Recent research has demonstrated that ordering of the equations and variables is highly significant for the effectiveness of these preconditioners. The goal of understanding the effects of ordering and the study of ordering strategies is an active area of current research.

While it is unlikely that algorithms for optimal ordering can be developed for arbitrary general sparse matrices, it is expected that effective strategies can be developed for relatively broad and important classes. The speakers addressed these issues and present current research in this area.

**Organizers:** Peter A. Forsyth, University of Waterloo, Canada; Andrew Sherman, Yale University; Wei P. Tang, University of Waterloo, Canada

**Analysis of Parallel Incomplete Point Factorizations**
Victor Eijkhout, University of Illinois, Urbana-Champaign

**Accelerated Line Iterative Methods for Discrete Convection-Diffusion Problems** Howard Elman, University of Maryland, College Park

**Ordering Methods for PCG Techniques Applied to Incompressible Viscous Flow** Paulina Chin, University of Waterloo, Canada; E. D'Azevedo, University of Waterloo, Canada; Peter A. Forsyth, Organizer; Wei P. Tang, Organizer

**Numerical Experiments with Ordering Strategies for the Preconditioned Conjugate Gradient Method**
Gérard Meurant, Centre d'Etudes de Limeil-Valenton, France

## Suggested Reading

F. L. Alvarado, W. F. Tinney, and M. K. Enns, *Sparsity in large-scale network computations*, in Advances in Electric Power and Energy Conversion, System Dynamics and Control, C. T. Leondes, ed., Academic Press, New York, 1991.

F. L. Alvarado, D. C. Yu, and R. Betancourt, *Partitioned sparse A-inverse methods*, IEEE Trans. Power Systems, 5 (1990) pp. 452–459.

P. R. Amestoy and I. S. Duff, *Vectorization of a multiprocessor multifrontal code*, Internat. J. Supercomput. Appl., 3 (1989), pp. 41–59.

E. Anderson, Z. Bai, C. Bischof, et al., *LAPACK: A portable linear algebra library for high-performance computers*, in Proc. Supercomputer '90, IEEE, New York, 1990, pp. 2–11.

E. Anderson, C. Bischof, J. Demmel, et al., *Prospectus for an extension to LAPACK: A portable linear algebra library for high-performance computers*, LAPACK Working Note 26, Tech. Report CS-90-118, Department of Computer Science, University of Tennessee, Knoxville, TN, 1990.

A. Bose and K. A. Clements, *Real-time modeling of power networks*, Proc. IEEE, 75 (1987), pp. 1607–1622.

T. A. Davis and I. S. Duff, *Unsymmetric-pattern multifrontal methods for parallel sparse LU factorization*, Report TR-91-23, Computer and Information Sciences Department, University of Florida, Gainesville, FL, 1991.

J. J. Dongarra, J. Du Croz, I. S. Duff, and S. Hammarling, *A set of level 3 basic linear algebra subprograms*, ACM Trans. Math. Software, 16 (1990), pp. 1–17.

J. J. Dongarra, J. Du Croz, S. Hammarling, and R. J. Hanson, *An extended set of FORTRAN basic linear algebra subprograms*, ACM Trans. Math. Software, 14 (1988), pp. 1–32.

J. J. Dongarra, I. S. Duff, D. C. Sorensen, and H. A. van der Vorst, *Solving Linear Systems on Vector and Shared Memory Computers*, Society for Industrial and Applied Mathematics, Philadelphia, 1991.

R. W. Freund and N. M. Nachtigal, *QMR: A quasi minimal residual method for non-Hermitian linear systems*, Numer. Math., to appear.

M. H. Gutknecht, *A completed theory of the unsymmetric Lanczos process and related algorithms. Part I*, SIAM J. Matrix Anal. Appl., 13 (1992), pp. 594–639.

A. Pothen and F. L. Alvarado, *A fast reordering algorithm for parallel sparse triangular solution*, SIAM J. Sci. Statist. Comput., 13 (1992), pp. 645–653.

A. Pothen and X. Yuan, *Optimal reordering of sparse Cholesky factors for parallel triangular solution*, 1991, in preparation.

P. Sadayappan and V. Visvanathan, *Efficient sparse matrix factorization for circuit simulation on vector supercomputers*, IEEE TCAD, CAD-8 (1989), pp. 1276–1285.

B. Stott and O. Alsac, *An overview of sparse matrix techniques for on-line network applications*, IFAC Symposium on Power System Operation, Beijing, 1986.

*Above:* Ivar Stakgold and Therese Bricheteau.

*Right:* Reinhard Menniken, Helmut Neunzert and Wolfgang Walter enjoy an invited presentation.

# Chapter 37
# NUMERICAL METHODS IN ORDINARY DIFFERENTIAL EQUATIONS

## Advances in the Computational Treatment of Differential Algebraic Equations

Differential algebraic equations (DAEs) arise naturally in various applications in science and engineering (constrained variational problems, descriptor (or semistate) systems, etc.). Formally, a DAE represents a uniformly singular implicit ordinary differential equation $f(x,x,t) = 0$, providing a vector field on a manifold. DAE solution behavior may differ considerably from that of standard ordinary differential equations, which causes serious numerical difficulties.

The speakers described some significant recent results on the computational treatment, based on a solid analytical knowledge. It was shown how these results are relevant in solving real world problems. In particular, different discretization schemes for higher index DAEs were proposed, shooting methods were considered, and parameter estimation problems were investigated.

**Organizer:** Roswitha März, Humboldt-Universität zu Berlin, Germany

**Projected Runge-Kutta Methods for Higher Index Differential Algebraic Equations** Uri M. Ascher, University of British Columbia, Canada; Linda R. Petzold, University of Minnesota

**Numerical Approximation of Boundary Value Problems for Index 2 Differential Algebraic Equations** Michael Hanke, Humboldt-Universität zu Berlin, Germany

**The Condition of Differential Algebraic Equation Boundary Value Problems and Shooting Methods** Marianela Lentini, Universidad Simon Bolivar, Venezuela; Rene Lamour, Humboldt-Universität zu Berlin, Germany; Roswitha März, Organizer

**Solvability of Boundary Value Problems for Systems of Singular Differential-Algebraic Equations** Roswitha März, Organizer; Ewa Weinmüller, Technische Universität Wien, Austria

I. E. Block and William Boyce enjoy a moment's rest from a busy schedule.

## Continuous Runge-Kutta Methods for Initial Value Problems—Applications and Theory

In recent years there have been several advances in the development, implementation, and use of interpolating schemes for explicit Runge-Kutta formulas. The resulting continuous numerical methods provide approximate solutions which are particularly convenient for users who must solve initial value problems (IVPs) which have special difficulties such as discontinuities, nonstandard stopping criteria, or frequent output requirements.

In this minisymposium, we surveyed some of these theoretical and practical advances which have made possible a new generation of robust and reliable Runge-Kutta software. The topics were chosen to be of interest to researchers active in developing methods for IVPs in ODEs as well as to practitioners who must solve real problems and are interested in how the state-of-the-art is evolving.

**Organizer:** Wayne H. Enright, University of Toronto, Canada

**Developing a Runge-Kutta Code with Interpolants** Ian Gladwell, Southern Methodist University; Lawrence F. Shampine, Southern Methodist University; Richard W. Brankin, Numerical Algorithms Group Ltd., United Kingdom

**Deriving Continuous Extensions of Explicit Runge-Kutta Methods** Jim H. Verner, Queen's University, Canada

**Interpolation Schemes for Explicit Runge-Kutta Nyström Methods** Philip W. Sharp, University of Toronto, Canada

**Strong Stability Properties of Continuous Runge-Kutta—Methods and Applications** Marino Zennaro, Università di Udine, Italy

## Novel Finite-Difference Techniques for the Numerical Integration of Differential Equations

The fundamental problem addressed here was the construction of appropriate finite-difference schemes for the numerical integration of differential equations. The importance of this problem is directly related to questions concerning numerical instabilities that can occur in finite-difference models of differential equations. There are a number of nonstandard "novel" procedures for assuring the elimination of certain types of numerical instabilities and their efficient implementation. A major advantage of certain of these procedures is that they provide accurate numerical results for large step-sizes.

The speakers reviewed some recently derived nonstandard finite-difference schemes and discussed some applications.

**Organizer:** Ronald E. Mickens, Clark Atlanta University

**Nonstandard Finite-Difference Schemes** Ronald E. Mickens, Organizer

**Explicit Numerical Solution of Initial-Value Problems on Parallel Computers** John N. Shoosmith, NASA Langley Research Center

**Marching Methods for Elliptic Problems in Computational Ocean Acoustics** Donald F. St. Mary, University of Massachusetts, Amherst; G.H. Knightly, University of Massachusetts, Amherst

**A Vectorized Parallel Finite-Difference Method for Nonlinear Partial Differential Equations** Suhrit K. Dey, Eastern Illinois University, Charleston

## Parallel Numerical Methods for Initial Value Problems for ODEs

With the advent of "useful" parallel computers, many researchers have begun to explore various approaches to exploit parallelism in the numerical solution of initial value problems (IVPs) for ordinary differential equations (ODEs) with the goal of solving large problems more quickly than is currently possible on sequential machines. Although some promising results have been obtained, this research is still in a preliminary phase: no effective robust reliable general-purpose production codes are yet available.

We began with a short introduction and survey of the area by the organizer followed by a more detailed exploration of a few promising current research directions by the four speakers.

**Organizer:** Kenneth R. Jackson, University of Toronto, Canada

**Parallel Implementation of Waveform Relaxation Method for Ordinary Differential Equations** Alfredo Bellen, Università di Trieste, Italy

**Parallel Methods for the Numerical Solution of Ordinary Differential Equations** Kevin Burrage, University of Auckland, New Zealand

**Recent Results on Picard-Lindelöf Iteration** Olavi Nevanlinna, Helsinki University of Technology, Finland

**Stability of Parallel Explicit ODE Methods** Robert D. Skeel, University of Illinois, Urbana-Champaign; Hon-Wah Tam, University of Illinois, Urbana-Champaign

## Parallel Solution of ODE Boundary Value Problems

Parallel methods for the numerical solution of boundary value problems (BVPs) in ordinary differential equations are important because (1) many large scale scientific problems are modeled by ODE boundary value problems, (2) solving realistic problems results in large-scale time-consuming computations, (3) the "best" sequential techniques clearly have scope for parallelization, yet the resulting algorithms have severe bottlenecks remaining, and (4) new techniques for this problem teach us much about the possibilities for the parallel solution of time dependent and stationary partial differential equations when solved by the method of lines.

Current research concentrates on implementations of the straightforward approaches to parallelization, on the design of parallel algorithms to ease the bottlenecks, and on specially designed methods.

**Organizer:** Ian Gladwell, Southern Methodist University

**PCOLNEW: A Parallel Boundary-Value ODE Code for Shared-Memory Machines** Karin R. Bennett, University of Kentucky; Graeme Fairweather, University of Kentucky

**On Parallel Methods for Boundary Value ODEs** Uri M. Ascher, University of British Columbia, Canada; S.Y. Chan, University of British Columbia, Canada

**Stable Parallel Algorithms for Two-Point Boundary Value Problems** Stephen J. Wright, Argonne National Laboratory and North Carolina State University

**The Parallel Solution of Almost Block Diagonal Systems Arising in Numerical Methods for BVPs for ODEs** Kenneth R. Jackson and Richard N. Pancer, University of Toronto

**Level 3 BLAS Approach to Parallel Solution of ABDs from Separated and Non-separated BCs** Ian Gladwell, Organizer; Marcin Paprzycki, University of Texas, Permian Basin

### Suggested Reading

U. M. Ascher and S. Y. P. Chan, *On parallel methods for boundary value ODEs*, Computing, 46 (1991), pp. 1–17.

K. R. Bennett and G. Fairweather, *PCOLNEW: A parallel boundary-value ODE code for shared-memory machines*, Tech. Report CCS-90-8, Center for Computational Sciences, University of Kentucky, Lexington, KY, 1990.

S. K. Dey, *A massively parallel algorithm to solve nonlinear parabolic PDE's*, available from S. K. Dey, Department of Mathematics, Eastern Illinois University, Charleston, IL 61920.

W. H. Enright, K. R. Jackson, S. P. Nørsett, and P. G. Thomsen, *Interpolants for Runge-Kutta formulas*, ACM Trans. Math. Software, 12 (1986), pp. 193–218.

———, *Effective solution of discontinuous IVPs using a Runge-Kutta formula pair with interpolants*, Appl. Math. Comput., 27 (1988), pp. 313–335.

I. Gladwell, L. F. Shampire, L. S. Baca, and R. W. Brankin, *Practical aspects of interpolation in Runge-Kutta codes*, SIAM J. Sci. Statist. Comput., 8 (1987), pp. 322–341.

K. R. Jackson, *A survey of parallel numerical methods for initial value problems for ordinary differential equations*, IEEE Trans. Magnetics, 1991.

K. R. Jackson and R. N. Pancer, *The parallel solution of ABD systems arising in numerical methods for BVPs for ODEs*, Tech. Report 255/91, Computer Science Department, University of Toronto, Canada, 1991.

G. H. Knightly and D. F. St. Mary, *Marching methods for elliptic models of underwater sound propagation*, in Computational Acoustics: Wave Propagation, D. Lee, R. L. Sternberg, and M. H. Schultz, eds., Elsevier Science Publishers G.V., Amsterdam, the Netherlands, 1988, pp. 397–407.

M. E. Kramer and R. M. M. Mattheij, *Application of global methods in parallel shooting*, Report RANA 91-06, Department of Mathematics and Computing Science, Eindhoven University, 1991.

R. E. Mickens, *Exact solutions to a finite-difference model of a nonlinear reaction-advection equation: Implications for numerical analysis*, Numer. Methods Partial Differential Equations, 3 (1989), pp. 313–325.

B. Owren and M. Zennaro, *Order barriers for continuous explicit Runge-Kutta methods*, Math. Comp., to appear.

M. Paprzycki and I. Gladwell, *Using Level 3 BLAS to solve almost block diagonal systems*, Proc. Fifth SIAM Conference on Parallel Processing for Scientific Computing, D. Sorensen, ed., Society for Industrial and Applied Mathematics, Philadelphia, 1992.

L. F. Shampire, *Interpolation for Runge-Kutta methods*, SIAM J. Numer. Anal., 22 (1985), pp. 1014–1022.

J. N. Shoosmith, *A stable numerical method for the solution of initial value problems on parallel computers*, unpublished manuscript, available from J. N. Shoosmith, NASA, Langley Research Center, Hampton, VA 23665.

S. J. Wright, *Stable parallel elimination for boundary value ODEs*, MCS-P229-0491, Argonne National Laboratory, 1991, preprint.

# Chapter 38
# NUMERICAL METHODS IN PARTIAL DIFFERENTIAL EQUATIONS

## Adaptive Multilevel Finite Element Methods: The Next Generation

In the active field of multilevel methods for PDEs, fully adaptive realizations have not been given enough attention up to now—in the sense of "self-adaptivity," which means successive mesh refinement, possibly highly nonuniform, for nasty real life geometries, and controlled essentially by the successive approximations of the PDE solutions. The minisymposium covered the most recent generation of such methods, with talks given by young scientists only. Thus the title wording "next generation" is understood in a double sense. Special attention was given to time dependent problems and 3-D stationary problems. Fields of applications touched were hyperthermia in medical computing and polyreaction kinetics in chemical computing.

**Organizer:** Peter J. Deuflhard, Konrad-Zuse Center, Germany

**Domain Decomposition and Multilevel Methods for Adaptive Mesh Refinement**   Jinchao Xu, Pennsylvania State University, University Park

**Adaptive Multilevel Discretizations for Time Dependent PDEs - with Application to Cancer Therapy**   Folkmar A. Bornemann, Konrad-Zuse Center, Germany

**Discrete Multilevel Techniques for Countable ODEs—with Application to Polyreaction Kinetics**   Michael Wulkow, Konrad-Zuse Center, Germany

**A Multilevel Domain Decomposition Algorithm for Elliptic Problems in Three Dimensions**   Barry F. Smith, Argonne National Laboratory

## Asymptotics-Induced Numerical Methods and Domain Decomposition

The combination of asymptotic and numerical analyses improves accuracy for obtaining solutions to multiple scale problems. Laminar flow of a slightly viscous fluid or combustion with high activation energy are classical examples of such physical problems.

The asymptotic analysis of such singular perturbation problems brings an analytically motivated decomposition of the domain. Different scalings correspond to different processes which occur in different geometrical zones. Each subproblem within its subdomain is easier to solve than the original problem. Specific techniques connect these subproblems through the boundaries. Further, these numerical methods take advantage of the architecture of parallel computers. The speakers presented some recent results in this area and discussed their application.

**Organizer:** Marc Garbey, Université Lyon I, France and Argonne National Laboratory

**Domain Decomposition: An Instrument of Asymptotic-Numerical Methods**   Ray C.-Y. Chin, Indiana University-Purdue University, Indianapolis; Gerald Hedstrom, Indiana University-Purdue University, Indianapolis

**Adaptive $h$- and $p$- Refinement Finite Element Methods for Singularly Perturbed Parabolic Systems**   Joseph E. Flaherty, Rensselaer Polytechnic Institute; Slimane Adjerid, Rensselaer Polytechnic Institute; Yun Wang, Rensselaer Polytechnic Institute

**Strategies for Adding Asymptotic Information to Programs**   William D. Gropp, Argonne National Laboratory

**Asymptotics-Induced Numerical Methods for Conservation Laws**   Marc Garbey, Organizer

**Asymptotic-Numerical Methods for Sturm-Liouville Systems**   Steven Pruess, Colorado School of Mines

## Boundary Element Methods

The speakers presented recent developments in the mathematical foundations and numerical aspects of boundary element methods. These methods have become important and effective numerical methods for treating many problems in engineering applications. For example, they have proved important in 3-D electromagnetic wave propagation, elastodynamics, and in computational mechanics, in particular for crack and punch problems.

The speakers covered recent advances in the combined methods of boundary and finite elements, applications to 3-D electromagnetic fields, time dependent boundary element methods, and some computational experience with hypersingular integral equations.

**Organizer:** Wolfgang L. Wendland, University of Stuttgart, Germany

**Recent Advances in the Combined Methods of Boundary and Finite Elements**   George C. Hsiao, University of Delaware

**Regularization in 3-D Electromagnetic Wave Propagation**   Jean C. Nedelec, École Polytechnic, France

**Time Dependent Boundary Element Methods**   Martin Costabel, University Nantes, France

**Some Computing Experience with Hypersingular Integral Equations**   Frank J. Rizzo, Gunaseclan Krishnasamy and Yijun Liu, University of Illinois, Urbana-Champaign

## Coupling of Different Equations: Modeling and Numerical Simulation

This minisymposium focused on equations with different types in different subregions of the computational domain. These equations have applications in several physical problems, including viscous-inviscid interaction for boundary layers, external aerodynamics around a body, and the interaction between molecular and continuous flow regime in the upper atmosphere. The speakers discussed recent advances in this field

**Organizer:** Valeri I. Agoshkov, USSR Academy of Sciences, USSR

**Some Imbedding Theorems and the Solvability of the System of Hyperbolic-Parabolic Equations** Valeri I. Agoshkov, Organizer

**Interface Problems in Fluid-Dynamics and Related Numerical Algorithms** Alfio Quarteroni, Politecnico di Milano, Italy

**Numerical Coupling of Boltzmann and Navier-Stokes Equations** Patrick Le Tallec, Université Paris-Dauphine, France

**Asymptotics for Domain Decomposition: Toward Uniform Distribution of Work and Errors** Jeffrey S. Scroggs, North Carolina State University

## Interface Methods in Fluids and Their Applications

Interface methods exactly preserve the boundary between distinct computational regions. Such methods have found wide application in many areas of fluid dynamics such as modeling oil reservoirs, detonation and shock waves, jets and wakes.

Current research in this area requires the understanding of mathematical aspects of two-dimensional front-tracking methods and extension of front-tracking methods to three-dimensional problems, the application of interface methods to the study of heterogeneous reservoirs and Hele-Shaw cells, and the use of boundary-integral methods to study vortex sheet separation from a sharp plate. The speakers described current work in these areas.

**Organizer:** Bruce G. Bukiet, New Jersey Institute of Technology

**The Performance of Parallel Algorithms for Interface Problems on MIMD Message-Passing Machines** Yi Wang, State University of New York, Stony Brook; Yue Fan Deng, State University of New York, Stony Brook; James Glimm, State University of New York, Stony Brook

**Numerical Solution of the Hele-Shaw Equations** Nathaniel Whitaker, University of Massachusetts, Amherst

**Vortex Sheet Roll-Up at a Sharp Edge** Robert Krasny, University of Michigan, Ann Arbor

## Lattice Gas Theory and Applications

Lattice gas methods are new discrete computational methods for approximating many partial differential equations, including Navier-Stokes, Burgers, Poisson, wave, and reaction-diffusion equations. They are efficient and easily incorporate complicated boundaries. Lattice gas methods can be used to model different physical phenomena like hydrodynamics, turbulent flows, flows involving phase transition, multi-phase flows, chemically reacting flows, combustion processes, magnetohydrodynamics, thermohydrodynamics, liquid crystal flows, and biochemical flows.

The speakers discussed different models used to investigate these phenomena and provided an overview of the directions and applications of current research.

**Organizers:** Gary D. Doolen, Los Alamos National Laboratory; Anna T. Lawniczak, University of Guelph, Ontario, Canada

**Lattice Gas Overview and Applications** Gary D. Doolen, Organizer

**Renormalization of the Transport Coefficients in a Diffusive Lattice Gas** Bruce M. Boghosian, Thinking Machines Corporation; C. David Levermore, University of Arizona; Washington Taylor, IV, University of California, Berkeley and Thinking Machines Corporation

**Lattice Gas Automata for Complex Fluids** Shiyi Chen, University of Delaware

**Lattice-Boltzmann Modeling of Immiscible Fluids** Andrew K. Gunstensen and Daniel H. Rothman, Massachusetts Institute of Technology; Stephane Zaleski, CNRS, France; Gianluigi Zanetti, Princeton University

**Lattice Gas Theory and Applications** Anna T. Lawniczak, Organizer

**Lattice Gas Methods Applied to Flows through Porous Media** Gary D. Doolen, Organizer

**Lattice Gas Model for Wave—Applications in Acoustics** Susan K. Numrick, Naval Research Laboratory

**Lattice Gas Methods for Reactive Systems** Anna T. Lawniczak, Organizer

**Lattice Gas Models for Magnetohydrodynamics** William H. Matthaeus, University of Delaware

**Lattice Gas Automata Modeling of Fluid Flows with a Non-ideal Gas** Hudong Chen, Dartmouth College

## Numerical Methods for the Boltzmann Equation

The Boltzmann Equation has regained importance in recent years through new theoretical results and new numerical algorithms as well as through new applications in space and vacuum technology. There are still improvements to and extensions of the traditional Direct Simulation Monte Carlo (DSMC) being made. At the same time, more deterministic methods called Low Discrepancy methods were developed, which compete with DSMC.

This minisymposium focused on numerical methods including extensions to more complicated physical situations, the main topic particle methods for Boltzmann equations, including the problems of transition to the fluid regime (Navier-Stokes) and of boundary values. The speakers presented an overview of current research and applications in this area.

**Organizer:** Helmut Neunzert, University of Kaiserslautern, Germany

**Low Discrepancy Methods: New Ideas and Applications** Jens Struckmeier, University of Kaiserslautern, Germany

**Improvements of the Traditional Direct Simulation Monte Carlo Algorithms and Applications to 3-D Industrial Problems** François Coron, Aerospatiale, France

**Weighted Particle Approximations** Sylvie Mas-Gallic, École Polytechnique, Palaiseau, France

**Problems in Connections with the Transition Boltzmann-Aerodynamics** Frank Gropengiesser, University of Kaiserslautern, Germany

## Numerical Integration Methods for Mechanical Engineering Dynamics

Numerical integration methods for mechanical engineering dynamics are an important simulation tool in computer aided engineering. Applications can be very complex, e.g., in the automotive car industry where complete vehicles are simulated. Probably the most common mechanical model class are multibody systems, i.e., rigid bodies connected by joints. Depending on the problem, the coordinate systems, and the mechanical formalism chosen, one obtains linearly-implicit ODEs, strongly, or loosely coupled differential-algebraic equations, which have a specific structure reflecting the structure of the mechanical system and the variational system.

Because of the importance of industrial applications and the mathematical challenge, the field has a very rapid development. Present research focuses on invariants and stabilization or projection methods, the exploitation of inherent structures, and the treatment of discontinuities in solutions.

The speakers discussed all these aspects with an emphasis on the first two.

**Organizers:** Georg H. Boch, Universität Ausburg, Germany; Germund Dahlquist, Royal Institute of Technology, Sweden; Ben Leimkuhler, University of Kansas; Linda Petzold, University of Minnesota

**On Index and Existence Theories for DAEs and Their Applications** Werner Rheinboldt, University of Pittsburgh

**Efficient Numerical Integration of DAEs for Mechanical Systems with Constraints and Invariants** Edda Eich, Universität Augsburg, Germany

**Stability of Computational Methods for Constrained Dynamics Systems** Linda R. Petzold, Organizer; Uri M. Ascher, University of British Columbia, Canada

**Waveform Relaxation for Multibody Dynamics** Benedict J. Leimkuhler, Organizer

**An Implicit Numerical Integration Algorithm for Differential- Algebraic Equations of Multibody Dynamics** Edward J. Haug, University of Iowa; Jeng Yen, Computer and Design Software Inc.

**Multistep Methods for Solving Constrained Equations of Motion** Florian A. Potra, University of Iowa

**Some Numerical Aspects of the Simulation of Mechanical Systems with Constraints and Invariants** Taifun Alishenas, Royal Institute of Technology, Sweden; Germund Dahlquist, Organizer

**Computation of Consistent Initial Values of Multibody Systems in Descriptor Form** Peter Rentrop, Technical University München, Germany

## Particle Methods

Particle methods are used in incompressible fluid mechanics (vortex methods), in rarefied gas dynamics (Monte Carlo techniques), in plasma physics, in astrophysics, etc. In the study of rarefied gas dynamics, Monte Carlo methods have been the major numerical technique used. In some other areas, a minority of the computational work currently carried out involves particle methods. Great progress has been made in particle methods in the last few years, making them increasingly competitive with other schemes in some of these areas. Recent advances include the fast methods for calculating interparticle interactions and the development of deterministic schemes (as alternatives to the stochastic particle methods). Speakers discussed several areas of application of particle methods, including both theoretical and practical aspects.

**Organizers:** Georges-Henri Cottet, Université de Grenoble, France; Claude A. Greengard, IBM T.J. Watson Research Center

**Numerical Resolution of the Wigner Equation: Using the Deterministic Particle Method** Francis Nier, École Polytechnique, France

**Boltzmann Approximation of the Euler Equations** Benoit Perthame, Université d'Orleans, France

**Vortex Sheets From Vortex Layers** Mario Pulvirenti, Università dell'Aquila, Italy

**Viscous Splitting of the Navier-Stokes Equations with Boundaries** J. Thomas Beale, MSRI, Berkeley

**A Monte-Carlo Analysis of a Vortex Method for Euler and Navier-Stokes Equations with Non-Smooth Data** Yann Brenier, Université de Paris VII, France; G.-H. Cottet, Organizer

**Fast Vortex Methods in Three Dimensions** Philip Colella, University of California, Berkeley

**Hybrid Particle-Grid Methods in Fluid Dynamics** Georges-Henri Cottet, Organizer

**Design of Fast and Accurate Monte Carlo Particle Schemes** Hans Babovsky, IBM Scientific Center, Heidelberg, Germany

## The Nonlinear Galerkin Method in Large Scale Scientific Computing

The increase of computing power generated by supercomputers leads to new challenges in scientific computing. For the numerical solution of large systems such as the Navier-Stokes equations, we must handle a very large number of unknowns and integrate the equations on large intervals of time. Most existing methods were designed at a time of scarce computing resources and are not necessarily well suited for modern computing.

The object of this minisymposium was to present a new algorithm, the Nonlinear Galerkin Method, which stems from dynamical systems theory and is adapted to large-scale computing. The speakers presented the motivations and the numerical analysis of the method, and they described the effective implementation of the method on supercomputers or parallel computers and showed that the method is effective as far as stability, accuracy, and gain in computing time are concerned.

**Organizers:** Martine Marion, École Central de Lyon, France; Roger Temam, Université Paris-Sud, France and Indiana University, Bloomington

**Inertial Manifolds and Inertial Algorithms** Roger Temam, Organizer

**The Nonlinear Galerkin Method** Martine Marion, Organizer

**Approximation of the Two- and- Three Dimensional Navier-Stokes Equations by Spectral Methods** Thierry DuBois, Université Paris-Sud, France

**Numerical Analysis of a One-Dimensional Immersed Boundary Method** D. Funaro, Università di Pavia, Italy

**Implementation of the Nonlinear Galerkin Method on a Parallel Computer** Moshe Israeli, Technion-Israel Institute of Technology, Israel

**Incremental Unknown Method** Min Chen, Indiana University, Bloomington; Roger Temam, Organizer

**Approximate Inertial Manifolds and Effective Viscosity in Turbulent Flows** Oscar Manley, Office of Basic Energy Sciences; Ciprian Foias, Indiana University; Roger Temam, Organizer

**Optimal Approximate Inertial Manifolds** George Sell, University of Minnesota, Minneapolis

## The Problem of Locking in Finite Element Analysis

Locking in the context of the numerical solution of partial differential equations is said to occur when an approximating scheme for a parameter-dependent problem deteriorates for values of the parameter close to a particular limit. It occurs, for example, in plates and shells for thickness close to zero, in elastic materials when the Poisson ratio is near 0.5, and in nonlinear problems like plasticity. Locking can completely ruin accuracy and an inordinate amount of effort may be needed to overcome it in the absence of special strategies.

The speakers addressed how locking is handled in engineering practice (NASTRAN). They also described mathematical approaches to the formulation and treatment of locking.

**Organizers:** Ivo Babuska, University of Maryland, College Park; Manil Suri, University of Maryland, Baltimore County

**Locking in Commercial Finite Element Codes** Richard H. MacNeal, MacNeal-Schwendler Corporation, Los Angeles

**A Mathematical Approach to Locking and Robustness** Manil Suri, Organizer

**Mixed Methods for Locking Problems** Franco Brezzi, Instituto di Analisi Numerica del CNR, Italy

**The Membrane Locking Problem in the Finite Element Analysis of Shells** Juhani Pitkaranta, Helsinki Institute of Technology, Finland

### Suggested Reading

A. S. Alves, ed., *Discrete Models of Fluid Dynamics*, Proc. NATO Workshop, Figueira da Foz, Portugal, 1990, World Scientific, 1991.

J. C. Astier, F. Coron, D. Dupuis, et al., *Aérodynamique d'un véhicule de rentrée en phase transitionnelle de ga rarefiée*, Proc. Roscoff, 1990.

H. Babovsky, F. Gropengieber, H. Neunzert, et al., *Application of well-distributed sequences to the numerical simulation of the Boltzmann equation*, J. Comput. Appl. Math., 31 (1990), pp. 15–22.

I. Babuska and M. Suri, *On locking and robustness in the finite element method*, Tech. Note BN-112, IPST, University of Maryland, College Park, 1990, SIAM J. Numer. Anal., 29 (1992).

A. Bendali, *Numerical analysis of the exterior boundary value problem for the time harmonic Maxwell equation by a boundary finite element method*, Math. Comp., 43 (1984).

F. A. Bornemann, *An adaptive multilevel approach to parabolic equations in two space dimensions*, Dissertation, Freie Universitaet Berlin, Germany, 1991.

———, *An adaptive multilevel approach to parabolic equations II. Variable-order time discretization based on a multiplicative error correction*, IMPACT Comput. Sci. Engrg., 3 (1991), pp. 93–122.

J. H. Bramble, J. E. Pasciak, and J. Xu, *Parallel multilevel preconditioners*, Math. Comp., 55 (1990), pp. 1–22.

F. Brezzi and K. J. Bathe, *Studies of finite element procedures—the inf-sup condition, equivalent forms and applications*, in Reliability of Methods for Engineering Analysis, K. J. Bathe and D. R. J. Owen, eds., Pineridge Press, Swansea, U.K., 1986, pp. 197–219.

M. Chen and R. Temam, *The incremental unknown method*, I, II, Appl Math. Lett., 1991.

———, *The incremental unknowns for solving partial differential equations*, Numer. Math., 59 (1991), pp. 255–271.

F. Coron and J. F. Pallegoix, *Computation of transitional rarefied flow*, Advances in Kinetic Theory and Continuum Mechanics, Springer-Verlag, New York, Berlin, 1991.

M. Costabel, *Boundary integral operators for the heat equation*, Integral Equations Operator Theory, 13 (1990), pp. 498–552.

P. Deuflhard and M. Wulkow, *Computational treatment of polyreaction kinetics by orthogonal polynomials of a discrete variable*, IMPACT Comput. Sci. Engrg., 1 (1989), pp. 269–301.

C. Devulder and M. Marion, *A class of numerical algorithms for large time integration: The nonlinear Galerkin method*, SIAM J. Numer. Anal., 29 (1992), pp. 462–483.

G. D. Doolen, ed., *Lattice Gas Methods for Partial Differential Equations*, Addison-Wesley, New York, 1990.

———, ed., *Lattice gas methods for PDEs: Theory, applications and hardware*, Proc. NATO Advanced Research Workshop, Los Alamos, NM, 1989, Physica D, 47 (1991).

C. Foias, O. Manley, and R. Temam, *Approximate inertial manifolds and effective viscosity in turbulent flows*, Phys. Fluids A, 3 (1991), pp. 898–911.

H. Grad, *Principles of the kinetic theory of gases*, Handbuch der Physik XII, 1958, pp. 205–295.

F. Gropengieber, *Domain decomposition in the transition between kinetic theory and aerodynamic*, Dissertation, University of Kaiserslautern, Germany, 1991, in preparation.

F. Gropengieber, H. Neunzert, and J. Struckmeier, *Computational methods for the Boltzmann equation*, Venice, 1989, The State of Art in Applied and Industrial Mathematics, R. Spigler, ed., Kluwer Acad. Publishers, 1990.

Waiting to hear the winner of the NAG drawing.

T. Ha-Duong, *On the boundary integral equations for the crack opening displacement of flat cracks*, Rapport interne No. 194, Ecole Polytechnique, Palaiseau, France, 1991, Integral Equations Operator Theory, to appear.

G. C. Hsiao, *The coupling of boundary element and finite element methods*, Z. Angew. Math. Mech., 70 (1990), pp. T493–T503.

G. C. Hsiao and W. L. Wendland, *Domain decomposition in boundary element methods*, in Domain Decomposition Methods for Partial Differential Equations, R. Glowinski, Y. A. Kuznetsov, G. Meurant, et al., eds., Society for Industrial and Applied Mathematics, Philadelphia, 1991, pp. 14–49.

G. Krishnasamy, F. J. Rizzo, and T. J. Rudolphi, *Continuity requirements for density functions in the boundary integral equation method*, Comput. Mech., to appear.

G. Krishnasamy, L. W. Schmerr, T. J. Rudolphi, and F. J. Rizzo, *Hypersingular boundary integral equations: Some applications in acoustic and elastic wave scattering*, J. Appl. Mech., 57 (1990), pp. 404–414.

R. MacNeal, *The shape sensitivity of isoparametric elements*, FEM in the Design Process, Proc. Sixth World Congress, J. Robinson, ed., Robinson and Associates, Devon, U.K., 1990.

P. Manneville, N. Boccara, G. Y. Vichniac, and R. Bidaux, eds., *Cellular automata and modeling of complex physical systems*, Proc. of the Winter School, Les Houches, France, 1989, Springer-Verlag, Berlin, 1989.

M. Marion and R. Temam, *Nonlinear Galerkin methods*, SIAM J. Numer. Anal., 26 (1989), pp. 1139–1157.

———, *Nonlinear Galerkin methods: The finite elements case*, Numer. Math., 57 (1990), pp. 205–226.

S. Mas-Gallic, Transport Theory Statist. Phys., 16 (1987), p. 855.

S. Mas-Gallic and F. Poupaud, Transport Theory Statist. Phys.

J. C. Nedelec and F. Starling, *Integral equation methods in a quasi-periodic diffraction problem for the time-harmonic Maxwell's equations*, SIAM J. Math. Anal., 22 (1991), pp. 1679–1701.

N. Nishimura and S. Kobayashi, *An improved boundary integral equation method for crack problems*, Proc. IUTAM Symposium, San Antonio, TX, 1987, in Advances in Boundary Element Method, T. A. Cruse, ed., Springer-Verlag, Berlin, Heidelberg, 1987.

J. Pitkaranta, *The problem of membrane locking in finite element analysis of cylindrical shells*, Tech. Report, Institute of Mathematics, Helsinki University of Technology, Finland, 1990.

B. Smith, *A domain decomposition algorithm for elliptic problems in three dimensions*, Numer. Math., to appear.

———, *A parallel implementation of an iterative substructuring algorithm for problems in three dimensions*, Report MCS-P249-0791, Argonne National Laboratory, Argonne, IL.

———, *An iterative substructuring algorithm for problems in three dimensions*, in Fifth International Symposium on Domain Decomposition Methods for Partial Differential Equations, T. F. Chan, E. Keyes, G. A. Meurant, J. S. Scroggs, and R. G. Voigt, eds., Society for Industrial and Applied Mathematics, Philadelphia, 1992.

R. Temam, *Infinite dimensional dynamical systems in mechanics and physics*, Appl. Math., 68 (1988).

M. Wulkow, *Numerical treatment of countable systems of ordinary differential equations*, Dissertation, Freie Universitaet Berlin, Germany, 1990.

J. Xu, *Theory of multilevel methods*, Report AM 48, Department of Mathematics, Pennsylvania State University, University Park, PA, 1989.

# Chapter 39
# OCEAN AND ATMOSPHERIC SCIENCE

## Global Climate Modeling: Challenges and Opportunities

The Department of Energy's CHAMMP (Computer Hardware, Applied Mathematics, Model Physics) research program aims to further global change research by producing an Advanced Climate Model ten thousand times faster than current Cray-based models. Existing models employ spectral methods to solve the atmospheric flow equations. Other physical processes (such as clouds) are generally handled by simple parametrization schemes. In contrast, the CHAMMP model will require numerical methods susceptible to implementation on tens of thousands of processors. New techniques will be required to model physical processes at fine resolutions.

The speakers addressed the particular numerical and computational problems presented by large-scale parallel methods and the role of applied mathematics in the development of these models. It was intended for scientists, engineers, and mathematicians working in climate modeling or interested in the application of mathematics to climate modeling problems.

**Organizers:** Ian T. Foster, Argonne National Laboratory; Hans G. Kaper, Argonne National Laboratory

**CHAMMP and the Advanced Climate Model**
Ari Patrinos, Department of Energy

**The Development of the Numerical Formulation in the ECMWF Forecast Model**  Adrian Simmons, European Center for Medium-Range Weather Forecasts, United Kingdom

**Concepts of Climate Models at the Center for Climate System Dynamics**  Akimasa Sumi, University of Tokyo, Japan

**Potential Detrimental Interactions between Numerical Methods and Physical Parameterizations in Global Atmospheric Models**  David Williamson, National Center for Atmospheric Research

**Development of Global Climate Models for MIMD Parallel Computers**  John B. Drake, Oak Ridge National Laboratory

**The Role of Advanced Computing Methods in the Development of Parallel Climate Models**  Ian T. Foster, Organizer

## Mathematical Models of the Earth's Atmosphere and Space Environment

Members of the general public are increasingly aware of the changes that mankind is unintentionally making to the Earth's environment. These range from pollution from factory chimneys to the increasing greenhouse effect, the depletion of stratospheric ozone, and nuclear reactors aboard Earth-orbiting satellites.

In order to understand such effects, it is essential to have a proper understanding of the processes which play significant roles in determining the natural situation. This understanding is now being incorporated into analytic and computational models. Man-made changes can then be evaluated with some confidence. The speakers highlighted some areas of concern where recent progress has been made by modeling.

**Organizer:** Michael J. Rycroft, Cranfield Institute of Technology, United Kingdom

**Interfaces between Different Atmospheric and Oceanic Layers**  Jullian C.R. Hunt, University of Cambridge, United Kingdom; D.J. Carruthers, Cambridge Environmental Research Consultants, United Kingdom; H.J.S. Fernando, Arizona State University

**Mathematical Models of Stratospheric Ozone Transport and Chemistry**  Richard B. Rood, NASA Goddard Space Flight Center

**Mathematical Models of the Earth's Magnetosphere**
David P. Stern, NASA Goddard Space Flight Center

### Suggested Reading

I. Foster, W. Gropp, and R. Stevens, *The parallel scalability of the spectral transform method*, Monthly Weather Review, March, 1992.

*Ozone Depletion, Greenhouse Gases and Climate Change*, National Academy Press, 1989.

P. Rasch and D. Williamson, *The sensitivity of a general circulation model to the moisture transport formulation*, J. Geophys. Res., in press.

F. S. Rowland and I. S. A. Isaksen, eds., *The Changing Atmosphere*, John Wiley & Sons, New York, 1988.

A. Simmons and D. Dent, *The ECMWF multi-tasking weather prediction model*, Comput. Phys. Reports, 11 (1989), pp. 165–194.

U.S. Department of Energy, *Building an Advanced Climate Model: Program Plan for the CHAMMP Climate Modeling Program*, Publication DOE/ER-0479T, National Technical Information Service, Washington, DC, 1990.

W. Washington and C. Parkinson, *An Introduction to Three-Dimensional Climate Modeling*, University Science Books, 1986.

# Chapter 40
# OPTIMIZATION

## Cooperative Game Theory Related to Programming

Cooperative game theory is one of the most important tools for considering competition and its corresponding splitting after the players have made lateral payments. The speakers examined various concepts, some of them new, for arriving at suitable solutions. Such concepts are very important in applications because when the players are under competition there is always some kind of negotiations which are reflected in such new concepts. The interdependence of such concepts is also considered. The current directions are the Shapley Value, the weighted and distinguished core, and other concepts. The speakers reported on the latest results in these growing fields.

**Organizer:** Ezio Marchi, Universidad Nacional de San Luis, Argentina

**The Shaping Value of Resale-Proof Trades**
Thomas Quint, United States Naval Academy; Luis Quintas, Pace University

**On the Nucleoli**  Stef Tijs, University of Nijmegen, The Netherlands

**An Alternative Algebraic Definition of the Weighted Shapley Values**  Irinel Dragan, University of Texas, Arlington

**The Weighted Core with Distinguished Coalition**
Magdalena Caritisaui, Universidad Nacional de San Luis, Argentina; Ezio Marchi, Organizer

## Inexact Optimization for Intelligent Manufacturing Systems

Many real-life problems arising in modern manufacturing systems (e.g., planning flexible manufacturing systems, scheduling robots, and optimization of automated factory) are characterized by inexact presentation of input data. This factor in many cases makes solving the problems easy, the latter being especially important when solving instances of the problems very many times, each time with modified criteria, constraints, and input data.

A new mathematical instrument for dealing with inexact data is "inexact optimization"—an approach which, in a sense, is concurrent and reciprocal to multi-criteria optimization.

The aim of this minisymposia was to present the main current directions of research in inexact optimization—fuzzy optimization, interval optimization, $s$-almost optimal-approximation—into intelligent decision-making systems.

**Organizer:** Eugene Levner, Jerusalem, Israel

**New Results for the PUMS Problem**  Monique Guignard-Spielberg and Emmanuel Chajahis, University of Pennsylvania; Hochong Lee, Kyunghee University, Korea

**Inexact Optimization in Flexible Manufacturing Systems**  Laureano F. Escudero, IBM T.J. Watson Research Center

**Almost Optimal Algorithms for Scheduling Robots**
Eugene Levner, Organizer

**Multicriteria Optimal Control in Multi Connected Dynamic Systems**  Yu. Shtessel

## Large Scale Nonlinear Optimization

Large scale optimization problems arise in many areas of science and engineering. Novel algorithms and novel computer architectures have greatly expanded the class of problems that can be effectively solved. In particular, much progress has been made in applying interior-point methods to nonlinear problems. The speakers discussed this and related topics in the context of both scalar and parallel computing.

**Organizer:** Stephen G. Nash, George Mason University

**Parallel Algorithms for Constrained Optimization**
Ariela Sofer and Stephen G. Nash, George Mason University

**Resolving the Shell Dual with a Nonlinear Primal-Dual Algorithm**  Garth P. McCormick, George Washington University

**Higher-Order Methods for Large Linear and Quadratic Programming Problems**  Paul T. Boggs, and Christoph Witzgall, National Institute of Standards and Technology, Gaithersburg; Paul D. Domich and Janet E. Rogers, National Institute of Standards and Technology, Boulder

**Decomposition of Algorithms for Optimization**
S.-P. Han, Johns Hopkins University

## Large Scale Optimization via Simplex and Interior-Point Methods: Algorithms and Applications

During the past five to ten years, a number of significant advances for solving large scale linear programming problems have been made. These advances are powerful enough to permit solutions to problems of a size that were, only recently, considered impractical, for example: Karmarkar's 1984 development of the interior-point method, advances made to the simplex method, and advances in performance and price of the computing hardware.

The ability to solve large LP problems is not just of academic interest. Practical problems, that are relevant to industry, are also in this class.

The speakers detailed algorithmic advances in the simplex and interior-point algorithms. Also addressed was the application of both methods to large scale, practical problems.

**Organizer:** Greg Astfalk, Convex Computer Corporation, Richardson, Texas

**Recent Computational Experience with the Primal-Dual Interior-Point Method for Linear Programming**   Roy E. Marsten, Georgia Institute of Technology; Irvin Lustig, Princeton University; David Shanno, Rutgers University

**Optimization Applications at the Military Airlift Command: Importance and Difficulties**   Robert Roehrkasse, Military Airlift Command; M. Ackley, Military Airlift Command; W. Carter, Military Airlift Command; G. Hughes, Military Airlift Command; J. Litko, Military Airlift Command; K. Ware, Military Airlift Command; A. Whisman, Military Airlift Command

**Dual Algorithms for Large Applications in Linear Programming**   Robert E. Bixby, Rice University

**Optimizing the Airlines Set-Partitioning Problems**   Karla Hoffman, George Mason University; Manfred Padberg, New York University

## Large Scale Optimization

For over forty years, linear programs have been solved by the simplex method. The need to solve large problems was immediate and it is now commonplace to solve problems in more than twenty thousand variables. By contrast, it is relatively rare to solve large nonlinear programming problems; even problems in ten thousand nonlinear variables represent significant challenges.

Many of the most successful techniques for small dense nonlinear optimization are not readily adaptable to large structured problems. Also, while it is true that the advent of faster and more powerful computer architectures has considerably enhanced our computational power, this and sophisticated implementations are not sufficient; algorithms must be designed specifically for large scale nonlinear problems. The speakers reported on recent progress.

**Organizer:** Andrew R. Conn, IBM T.J. Watson Research Center

**SQP Methods for Large-Scale Nonlinear Programming**   Walter Murray, Stanford University

**An Exterior-Point Method for Large Convex Quadratic Programming**   Thomas F. Coleman, Cornell University; Jinguo Liu, Cornell University

**Incorporating Variable Metric Information in Algorithms for Large-Scale Constrained Optimization**   Jorge Nocedal, Northwestern University; Richard Byrd, University of Colorado

**LANCELOT—An Algorithm for the Solution of Large-Scale Nonlinear Programming Problems**   Nicholas I.M. Gould, Rutherford Appleton Laboratory, United Kingdom; Philippe L. Toint, Notre Dame de la Paix, Belgium; Andrew R. Conn, Organizer

**The Current Status of a Software Package for Large-Scale Nonlinear Programming**   Philippe L. Toint, Notre Dame de la Paix, Belgium; Andrew R. Conn, Organizer; N.I.M. Gould, Rutherford Appleton Laboratory, United Kingdom

## Nonlinear Equation Approaches to Optimization and Equilibrium Modeling

Finite-dimensional optimization, complementarity, and variational inequality problems as a system of nonlinear (although sometimes nonsmooth) equations provides interesting and potentially fruitful areas for the development of new algorithms. In particular, recent advances in linear and nonlinear programming can be considered as special cases of these more general approaches to such problems.

In recent years, substantial progress has been made in the development of Newton-like methods for nonsmooth equations, continuation methods, etc. This work has laid the foundation for many advances in computational procedures for complementarity and optimization problems. However, this work has been hampered by issues of nonregularity (singularities) and numerical difficulties.

The speakers presented the basics of the equation approach and described recent advances in the theory and computational aspects of applying various algorithms which arise from this methodology.

**Organizer:** Patrick T. Harker, University of Pennsylvania

**A Nonsmooth Equation-Based SQP Method for Solving the NCP**   Steven A. Gabriel and Jong-Shi Pang, Johns Hopkins University

**Global Convergence of Damped Nonsmooth Newton's Method via the Path Search**   Daniel Ralph, Cornell University

**A Gauss-Newton Method for Nonsmooth Equations**   Michael C. Ferris, University of Wisconsin, Madison

**Smooth and Nonsmooth Equation Methods for Solving Finite-Dimensional Variational Inequalities**   Patrick T. Harker, Organizer

## Nonlinear Equations and Optimization in Infinite Dimensional Spaces

Algorithms for nonlinear equations and optimization in infinite dimensional spaces may differ in both analysis and formulation from conventional algorithms for such problems in finite dimensions. Functional analytic considerations, such as choice of spaces or compactness properties of nonlinear maps, are important in the design and theory of such algorithms. When these algorithms are discretized, the resulting methods for the finite dimensional approximate problems are often new and preserve the underlying functional analytic properties such as the sparsity pattern and symmetry.

The speakers discussed a variety of such algorithms, their properties, and applications including optimal control problems, integral equations, boundary value problems, and parameter identification.

**Organizers:** Carl T. Kelley, North Carolina State University; Ekkehard W. Sachs, Universität Trier, Germany

**Broyden-Like Methods for Integral and Differential Equations**   Carl T. Kelley, Organizer

**Infinite Dimensional Optimization Methods for Optimal Control Problems**   Ekkehard W. Sachs, Organizer

**Inexact Newton and Newton Iterative Methods**   Homer F. Walker, Utah State University

**Global Convergence of a Class of Trust Region Algorithms for Optimization Using Inexact Projections on Convex Constraints**  Andrew R. Conn, IBM T.J. Watson Research Center; N.I.M. Gould, Rutherford Appleton Laboratory, United Kingdom; Annick Saertnaer, Facultés Universitaires Notre-Dame de la Paix, Belgium; Philippe L. Toint, Facultés Universitaires Notre-Dame de la Paix, Belgium

## Trust Region Methods for Nonlinear Programming

This minisymposium consisted of four presentations on robust methods for nonlinear programming, respectively, a brief motivation for trust regions in NLP, a catalog of trust-region subproblems, two representative algorithms, and a general convergence theory.
Organizer: J. E. Dennis, Jr., Rice University

**Robust Trust-Region Algorithm for Nonlinear Programming**  J. E. Dennis, Jr., Organizer; Karen Williamson, Rice University

**Results for a Trust-Region for Nonlinear Programming**  Richard H. Byrd, University of Colorado, Boulder; Emmanuel Omojokun, Virginia State University

**A Trust-Region Methods for Nonlinear Programming**  Karen A. Williamson, Rice University; J. E. Dennis, Jr., Organizer

**Convergence Analysis of NLP Trust Region Methods**  James V. Burke, University of Washington

### Suggested Reading

J. Barutt and T. Hull, *Airline crew scheduling: Supercomputers and algorithms*, SIAM News, 23 (1990).

A. S. Belen'kii and E. V. Lerner, *Scheduling models and methods in optimal freight transportation planning*, Automat. Remote Control, 1 (1989), pp. 1–56.

R. E. Bixby, J. W. Gregory, I. J. Lustig, et al., *Very large-scale linear programmming: A case study in combining interior point and simplex methods*, Report J-91-07, Industrial and Systems Engineering, Georgia Institute of Technology, GA, 1991.

P. T. Boggs, J. W. Tolle, and A. J. Kearsley, *A merit function for inequality constrained nonlinear programming problems*, Report NISTIR 4702, National Institute of Standards and Technology, Gaithersburg, MD, 1991.

T. F. Coleman and L. Hulbert, *A globally and superlinearly convergent algorithm for convex quadratic programs with simple bounds*, Tech. Report, Department of Computer Science, Cornell University, Ithaca, NY, 1990.

A. R. Conn, N. Gould, A. Sartenaer, and P. Toint, *Global convergence of a class of trust region algorithms for optimization using inexact projections on convex constraints*, Report 90/4, Department of Mathematics, FUNDP, Namur, Belgium, 1990.

A. R. Conn, N. I. M. Gould, and P. L. Toint, *A proposal for a standard data input format for large-scale nonlinear programming problems*, Tech. Report CS-89-61, Computer Science Department, University of Waterloo, Ontario, Canada, 1990.

———, *Large-scale nonlinear constrained optimization*, in ICIAM 91, Society for Industrial and Applied Mathematics, Philadelphia, 1992, pp. 51-70.

———, *A comprehensive description of LANCELOT*, Tech. Report 91/10, FUNDP, Namur, Belgium, 1991.

S. C. Eisenstat and H. F. Walker, *Globally convergent inexact Newton methods*, Research Report February/91/51, Mathematics and Statistics Department, Utah State University, 1991, SIAM J. Optim., submitted.

S. K. Eldersveld, *Large-scale sequential quadratic programming algorithms*, Ph.D. Thesis, Department of Operations Research, Stanford University, Stanford, CA, 1991.

L. F. Escudero, *A production planning problem in FMS*, Ann. Oper. Res., 17 (1989), pp. 69–104.

J. W. Friedman, *Game Theory with Application to Economics*, Oxford University Press, Oxford, U.K., 1990.

P. T. Harker and B. Xiao, *Newton's method for the nonlinear complementarity problem: A B-differentiable equation approach*, Math. Programming, 48B (1990), pp. 339–357.

K. Hoffman and M. Padberg, *A polyhedral cutting plane approach to solving set-partitioning problems*, Department of Operations Research and Applied Statistics, George Mason University, 1991.

C. T. Kelley and E. W. Sachs, *Pointwise Broyden methods*, SIAM J. Optim., 3 (1992).

M. Kojima and S. Shindo, *Extensions of Newton and quasi-Newton methods to systems of $PC^1$ equations*, J. Oper. Res. Soc. Japan, 29 (1986), pp. 352–374.

F.-S. Kupfer and E. W. Sachs, *Numerical solution of nonlinear parabolic control problems by a reduced SQP method*, Universitat Trier, 1991, preprint.

A. Kusiak, *Intelligent Manufacturing Systems*, Prentice Hall, Englewood Cliffs, NJ, 1990.

T. Ichiishi, *Game Theory and Applications*, Academic Press, New York, 1990.

E. L. Lawler, J. K. Lenstra, and A. H. G. Rinnooy Kan, *Recent developments in deterministic sequencing and scheduling*, in Deterministic and Stochastic Scheduling, M. A. H. Dempster, J. K. Lenstra, and A. H. G. Rinnooy Kan, eds., D. Reidel Publ. Company, Boston, 1982, pp. 35–73.

I. J. Lustig, R. E. Marsten, and D. F. Shanno, *The interaction of algorithms and architectures for interior point methods*, Ann. Oper. Res., 1992, to appear.

G. P. McCormick, *The superlinear convergence of a nonlinear primal-dual algorithm*, Tech. Paper T-550/91, Department of Operations Research, The George Washington University, Washington, DC, 1991.

R. B. Myerson, *Game Theory*, Harvard University Press, Cambridge, MA, 1991.

S. G. Nash and A. Sofer, *A general-purpose parallel algorithm for unconstrained optimization*, SIAM J. Optim., 1 (1991), pp. 530–547.

J. Nocedal and D. C. Liu, *On the limited memory BFGS method for large-scale optimization*, MPB, 45 (1989), pp. 503–528.

J. S. Pang, *Newton's method for B-differentiable equations*, Math. Oper. Res., 15 (1990), pp. 311–341.

S. Robinson, *An implicit-function theorem for a class of nonsmooth functions*, Math. Oper. Res., 16 (1991), pp. 292–309.

———, *Newton's method for a class of nonsmooth functions*, unpublished manuscript, Department of Industrial Engineering, University of Wisconsin, Madison, WI, 1989.

R. Roehrkasse, *KORBX comes to the Military Airlift Command*, IEEE Potentials, December, 1990, pp. 39–40.

R. Roehrkasse and G. C. Hughes, *Crisis analysis: Operation Desert Shield*, OR/MS Today, December, 1990, pp. 22–27.

K. E. Stecke and R. Suri, eds., *Flexible manufacturing systems: Operations research models and applications*, Ann. Oper. Res., 3 (1985).

# Chapter 41
# ORDINARY DIFFERENTIAL EQUATIONS

## Eigenvalue Problems for Ordinary Differential Equations

Eigenvalue problems for ODEs occur in a wide variety of physical applications, including hydrodynamics, elasticity, acoustics, heat conduction, magnetohydrodynamics and quantum mechanics. They are particularly important in the analysis of stability of solutions of both linear and nonlinear PDEs.

The speakers considered numerical and theoretical aspects of Sturm-Liouville equations and related higher order equations and systems, including both regular and singular problems. Highly accurate and robust shooting methods have been developed, which can aim for the $n$-th eigenvalue without consideration of other eigenvalues and which have a posteriori error estimates.

In the case of Sturm-Liouville problems, the speakers discussed several methods that have been implemented by codes that can be used to approximate continuous spectra and spectral functions as well as eigenvalues and eigenfunctions. For higher order problems and systems, the speakers discussed some new methods (including oscillation and Prufer methods) that are currently being developed.

**Organizer:** Leon Greenberg, University of Maryland, College Park

**Approximation of Sturm-Liouville Eigenvalues and Eigenfunctions with Global Error Control**   Steven A. Pruess, Colorado School of Mines

**Numerical Approximation of Singular Spectral Functions for Sturm-Liouville Problems having Continuous Spectra**   Charles A. Fulton, Florida Institute of Technology

**Computing Eigenvalues of Singular Limit-Circle Sturm-Liouville Problems**   Anton Zettl, Northern Illinois University

**Reliable Computation of Resonances of Schrödinger Problems on the Half-Line**   John D. Pryce, Royal Military College of Science, United Kingdom

**Computing Eigenvalues of Self-adjoint Systems of ODEs**   Leon Greenberg, Organizer

**Eigenvalues of Second Order Linear Systems of ODEs**   Marco Marletta, Royal Military College of Science, United Kingdom

**Eigenvalue Problems for Systems of ODEs**   Frederick V. Atkinson, University of Toronto, Canada

## Mathematical Modeling and Simulation of Multibody Systems

In machine, robot, and vehicle engineering, reliable mathematical models are required for the simulation of motion and the analysis of strength. During the last decade, the method of multibody systems has been developed, resulting in explicit and implicit formulations of the constraint conditions characterizing engineering systems. The explicit formulation results in pure ordinary differential equations while the implicit representation requires algebraic-differential equations.

The speakers described the theory and application of the different approaches in multibody system dynamics. They also discussed the experiences of mathematicians and engineers in industry with real time simulations of motion.

**Organizers:** Werner O. Schiehlen, University of Stuttgart, Germany; G. Leister, University of Stuttgart, Germany

**A Multibody Simulation Method for the Dynamics of Vehicles with Active Suspension Elements**
Ronald J. Anderson, Queens University, Canada

**Modeling the Contact Problem of Multibody Dynamics as a System of Differential Algebraic Equations**   Edda Eich, University of Augsburg, Germany; Claus Führer, German Aerospace Research, Germany; Bernd Simeon, Technical University of Munich, Germany

**Simulation of Mechanical Systems with Impact Occurrences Demonstrated at Electric Rotary Hammers with Pneumatic Impact Mechanism**
Dieter Schramm, Robert Bosch GmbH, Germany; G. Leister, Organizer; Robert Bosch, Robert Bosch GmbH, Germany

**Mathematical Relations of Classical/Recursive Multibody Dynamics Formalisms**   Werner O. Schiehlen, Organizer; G. Leister, Organizer

Robert E. O'Malley and Charles Osgood, manager of the National Security Agency's Mathematical Sciences program.

## Perturbation Methods and Applications in Nonlinear Oscillations

This minisymposium focused on computational, perturbation, and averaging methods for a variety of autonomous and forced nonlinear differential equations. Applications are likely to arise in ship capsize problems, Josephson junction circuits in electronics, and forced oscillations of mechanical systems. There is much current interest in homoclinic bifurcation, period doubling, and chaotic outputs of nonlinear systems. The speakers provided a review of some of the latest perturbation and harmonic balance methods in this context.

**Organizer:** Peter Smith, University of Keele, United Kingdom

### A Dual Perturbation Series Approach to Homoclinic Bifurcation in Nonlinear Systems
Peter Smith, Organizer; Jennifer M.E. Yorke, University of Keele, England

### Instability of Softening Spring Oscillators
Lawrence N. Virgin, Duke University

### Computational Averaging for Nonlinear Oscillators
Jonathan Summers, University of Newcastle-upon-Tyne, United Kingdom

### Harmonic Balance Applied to Strongly Nonlinear Oscillators
T.D. Burton and Ameer Hassan, Washington State University

### Suggested Reading

R. J. Anderson and D. M. Hanna, *Comparison of three vehicle simulation methodologies*, in The Dynamics of Vehicles on Tracs and Roads, R. J. Anderson, ed., Swets & Zeitlinger, Amsterdam, 1989.

F. Atkinson, *Discrete and Continuous Boundary Problems*, Academic Press, New York, 1964.

P. Bailey, M. Gordon, and L. Shampine, *Automatic solution of the Sturm–Liouville problem*, ACM Trans. Math. Software, 4 (1978), pp. 193–208.

J. Brindley, M. D. Savage, and C. M. Taylor, *Nonlinear dynamics of journal bearings*, Philos. Trans. Roy. Soc. London Ser. A, 332 (1990), pp. 107–119.

N. Dunford and J. Schwartz, *Linear Operators, Part II: Spectral Theory*, Interscience, New York, 1963.

E. Eich, C. Fuhrer, B. Leimkuhler, and S. Reich, *Stabilization and projection methods for multibody dynamics*, Research Reports, Institute of Mathematics, University of Technology, Helsinki, 1990.

D. W. Jordan and P. Smith, *Nonlinear Ordinary Differential Equations*, Second Edition, Oxford University Press, Oxford, U.K., 1987.

H. Keller, *Numerical Solution of Two Point Boundary Value Problems*, CBMS-NSF Regional Conf. Ser. in Appl. Math., Vol. 24, Society for Industrial and Applied Mathematics, Philadelphia, 1976.

*Proceedings of the Focused Research Program on Spectral Theory and Bondary Value Problems*, Vols. 1–4, ANL-87-26, Argonne National Laboratory, 1987–1989.

W. Schiehlen, ed., *Multibody Systems Handbook*, Springer-Verlag, Berlin, 1990.

———, *Computational aspects in multibody system dynamics*, in Computer Methods in Applied Mechanics and Engineering 90, Elsevier Science Publishers, Amsterdam, pp. 569–582.

P. Smith and J. M. E. Yorke, *A dual perturbation series analysis of homoclinic bifurcation for automonous systems*, Z. Angew. Math. Mech., 1991.

# Chapter 42
# PARTIAL DIFFERENTIAL EQUATIONS

## Computational Sinc Procedures for Time-Dependent Problems in Mathematical Physics

The Sinc-Galerkin method for the numerical solution of partial differential equations is based on the expansion of a function and its derivatives in terms of Whittaker cardinal functions. There are a number of properties of the Sinc-Galerkin method that distinguish it from spectral methods based on more classical coordinate functions. In order to survey the Sinc methodology, various problems from mathematical physics were used to illustrate approximate quadrature, discretizations on unbounded domains, and the Sinc-Galerkin approach.

The speakers addressed aspects of these techniques which arise from transform methods (Laplace inversion), transport problems (Burgers equation), electromagnetic problems (Maxwells equation), and a closing overview addressing distributed (parallel) computation.

**Organizer:** John R. Lund, Montana State University

**Convergence Acceleration for Laplace Transform Inversion** Sven-Åke Gustafson, University of Utah

**A Fully Sinc-Galerkin Method for Transport Problems** Kenneth L. Bowers, Montana State University

**The Sinc-Galerkin Method in Inhomogeneous Electromagnetic Scattering** Kelly M. McArthur, Colorado State University; P. Monk, University of Delaware; F. Stenger, University of Utah

**A Sinc-Galerkin Method for the Recovery of Material Parameters in Euler-Bernoulli Beam Models** Ralph C. Smith, NASA Langley Research Center

## Mathematical Methods for Scattering Problems in Heterogeneous Media

This minisymposium was concerned with modeling and solution methods for wave propagation (or scattering) problems in heterogeneous media. "Modeling" refers to the constitutive properties of the medium and the fluid-solid wave interaction. The problem area addressed dealt with propagation in unbounded, heterogeneous bodies and scattering by obstacles. A number of methods and approximation procedures have been developed for such problems: specific, albeit sophisticated, methods for one-dimensional structures (e.g., layered media); general, three-dimensional methods such as those based on integral representations and asymptotic series; and schemes (such as Kirchhoff's approximation) for fluid-solid wave interaction.

While current research shows wide activity in all the indicated directions, the scope of the minisymposium was to describe the pertinent methods to provide a deeper understanding of the modeling, the solution techniques, and their applications.

**Organizer:** Angelo C. Morro, University of Genoa, Italy

**Waves in Finely Layered Media** Robert Burridge, Schlumberger-Doll Research

**Wave Propagation in Particulate Composites and in Fluid-Filled Cavities in Elastic or Fluid Media** Guillermo C. Gaunaurd, United States Naval Surface Warfare Center

**Integral Representations for Scattering Problems in Heterogeneous Media** Giacomo F. Caviglia, University of Genoa, Italy

**A Surface Moment Problem in Shape Reconstruction** George Dassios, University of Patras, Greece

## Nonlinear Diffusion Equations

In recent years, significant progress has been made in the analysis of nonlinear diffusion equations. Seemingly simple equations like $u_t = \Delta u + f(u)$ have been found to admit richly structured solution sets when the nonlinearity is a simple polynomial function, even on symmetrical domains. Generalizations to more complicated geometries, more complicated operators than Laplace's operator, and reaction-diffusion equations are now beginning to yield useful information for applications in fluid flow, physics, and biology.

Critical phenomena and multiple solutions abound. Asymptotic methods, phase plane analysis, and classical variational techniques have been combined with tools from computer science—symbolic manipulation and numerical software packages—to gain insight into the behavior of solutions near critical parameter values.

In this minisymposium, we highlighted a few methods and results. The five speakers discussed asymptotic methods and the use of numerical and symbolic manipulation software and highlighted applications in mathematical physics and materials science.

**Organizers:** Frederick V. Atkinson, University of Toronto, Canada; Hans G. Kaper, Argonne National Laboratory

**Manipulation Software in the Analysis of Nonlinear Diffusion Equations** Man Kam Kwong, Argonne National Laboratory

**Dynamics on the Attractor for the Complex Ginzburg-Landau Equation** Peter Takác, Vanderbilt University and Argonne National Laboratory

**Self-similar Solutions for Doping in Semi-conductors** William C. Troy, University of Pittsburgh, Pittsburgh

**Geometric Techniques in Nonlinear Diffusion Problems** Christopher K.R.T. Jones, Brown University

**Untitled** Juan L. Vasquez, Universidad de Madrid, Spain

## Nonlinear Systems of Conservation Laws and Shock Waves

Nonlinear systems of conservation laws, which result from the balance laws of continuum physics and other fields, describe many physical phenomena. Important examples occur in fluid dynamics, solid mechanics, petroleum reservoir engineering, electromagnetism, combustion theory, and control theory.

The speakers in this minisymposium presented a sample of current research efforts to attack the nonlinear systems in these areas. Analytical and numerical strategies as well as methods for solving these problems were described; some new results, methods, and phenomena were presented.

**Organizer:** Gui-Qiang Chen, University of Chicago

**Nonlinear Stability of Undercompressive Shocks**
Kevin Zumbrum, State University of New York, Stony Brook

**An Extension of Glimm's Method to Third Order in Wave Interaction**  Robin Young, University of California, Davis

**Weak Solutions of Hamilton-Jacobi Equations, Conical Diffraction and Numerical Methods**
Eduard Harabetian, University of Michigan, Ann Arbor

**Spectral Viscosity Methods for Multidimensional Nonlinear Conservation Laws**  Gui-Qiang Chen, Organizer; Eitan Tadmor, Tel-Aviv University, Israel; Qiang Du, Michigan State University

**A Stochastic Approach to Conservation Laws**
Helge Holden, University of Trondheim, Norway; Nils Henrik Risebro, University of Oslo, Norway

**The Convergence Rate of Approximate Solutions to Nonlinear Conservation Laws**  Eitan Tadmor, Tel-Aviv University, Israel

**Geometric Framework for Bifurcation of Wave Curves**  Bradley J. Plohr, State University of New York, Stony Brook; Eli L. Isaacson, University of Wyoming; Dan Marchesin, Instituto de Matemática Pura e Aplicada, Brazil; C. Frederico Palmeira, Pontifícia Universidade Católica, Brazil

**Riemann Problem for 2-Dimensional Conservation Laws**  Tong Zhang, Academia Sinica, Peoples Republic of China; Dechun Tan, Academia Sinica, People's Republic of China

**Discontinuous Solutions of the Navier-Stokes Equation with Spherical and Cylindrical Summary**
David Hoff, Indiana University

**Some Results on the Relaxation Theory for Nonlinear Hyperbolic Systems and Related Problems**  Pierangelo A. Marcati, Università di L'Aquila, Italy

**Global Existence of Weak Solutions to the Euler Equations for an Exothermically Reacting Gas**
David Wagner, University of Houston

**Asymptotic Behavior of Solutions to Systems of Hyperbolic Conservation Laws**  Zhouping Xin, Courant Institute of Mathematical Sciences, New York University

**Resonant Nonlinear Conservation Laws**  J. Blake Temple, University of California, Davis

**Oscillations and Conservation Laws**  Michel Rascle, University of Nice, France

**Self-Similar Viscous Limits for the Riemann Problem in Conversation Laws**  Athanassios E. Tzavaras, University of Wisconsin, Madison

**Cauchy Problem for Nonstrictly Hyperbolic Conservation Laws**  Pui Tak Kan, University of Wisconsin, Madison

**Oscillations and Cancellations in Conservation Laws**
James Shearer, University of Michigan, Ann Arbor

**Stability of Steady State Waves for Compressible Navier-Stokes Equations for Van der Waals Fluids**
Mohamed Khodja, University of Indiana, Bloomington

## Recent Developments in the Theory of Conservation Laws

This minisymposium focused on recent advances and problems in the theory of shock waves. Such theory has been vital for the qualitative understanding and quantitative computation of flows involving shocks. Conversely, these problems have driven developments in applied analysis, from the theory of functions of bounded variation (in the 1950s and 1960s) to compensated compactness (in the 1980s), which are useful in many other applied contexts.

Dafermos discussed the theory of ODEs with discontinuous coefficients that give characteristics in shocked flows, which gives new bounds on shock interactions. LeFloch examined problems that arise in perturbation theory for shocked flows. Lucier used results in nonlinear approximation theory to design adaptive schemes that are (not just formally) uniformly second-order accurate. Shearer discussed progress in the method of compensated compactness for conservation laws.

Organizer: Jonathan Goodman, Courant Institute of Mathematical Sciences, New York University

**Generalized Characteristic, Stability and Decay for Conservation Laws**  Constantine M. Dafermos, Brown University

**Nonconservative Products and Shocks for Nonconservation Form Equations**  Phillipe LeFloch, Courant Institute of Mathematical Sciences, New York University

**Nonlinear Approximation Theory and Hyperbolic Conservation Laws**  Bradley J. Lucier, Purdue University; Ronald A. DeVore, University of South Carolina

**Oscillations and Concentrations in Conservation Laws**  James Shearer, University of Michigan

V. Rusinov, Yu. V. Prokhorov, and V. V. Sazonov

**Suggested Reading**

R. Burridge, *Waves in finely layered media*, in Applied and Industrial Mathematics, R. Spigler, ed., Kluwer, Dordrecht, 1991, pp. 267–279.

G. Caviglia and A. Morro, *Scattering problems for acoustic waves*, in Applied and Industrial Mathematics, R. Spigler, ed., Kluwer, Dordrecht, 1991, pp. 325–334.

T. Chang (Zhang) and L. Hsiao (Xiao), *The Riemann problem and interaction of waves in gas dynamics*, Pitman Monographs Surveys Pure Appl. Math., 41 (1989).

D. Colton, *The inverse scattering problem for time-harmonic acoustic waves*, SIAM Rev., 26 (1984), pp. 323–350.

R. Courant and K. O. Friedrichs, *Supersonic Flow and Shock Waves*, Wiley-Interscience, New York, 1949.

G. C. Gaunaurd and W. Wertman, *Comparison of effective medium theories for inhomogeneous continua*, J. Acoust. Soc. Amer., 85 (1989), pp. 541–554.

J. Glimm and A. Majda, *Multidimensional hyperbolic problems and computations*, IMA Vol. Math. Appl., 29 (1991).

S.-A. Gustafson, *Computing inverse Laplace transforms using convergence acceleration*, Computation and Control II: Proc. Second Bozeman Conference, Bozeman, MT, Progress in Systems and Control Theory, Birkhauser, Basel, Switzerland, 1991.

P. Lax, *Hyperbolic systems of conservation laws and the mathematical theory of shock waves*, CBMS-NSF Regional Conf. Ser. in Appl. Math., Vol. 11, Society for Industrial and Applied Mathematics, 1973.

J. Lund and K. L. Bowers, *Sinc Methods: Quadrature and Differential Equations*, Society for Industrial and Applied Mathematics, 1992.

A. Majda, *Compressible fluid flow and systems of conservation laws in several space variables*, Appl. Math. Sci., 53 (1984).

K. McArthur, K. L. Bowers, and J. Lund, *The Sinc method in multiple space dimenstions: Model problems*, Numer. Math., 56 (1990), pp. 789–816.

K. McArthur, P. Monk and F. Stenger, *The Sinc-collocation method applied to Maxwell's equations*, preprint available from K. McArthur at Colorado State University, Boulder, CO.

R. Smith, G. A. Bogar, K. L. Bowers, and J. Lund, *The Sinc-Galerkin method for fourth-order differential equations*, SIAM J. Numer. Anal., 28 (1991), pp. 760–788.

J. Smoller, *Shock Waves and Reaction-Diffusion Equations*, Springer-Verlag, New York, 1982.

# Chapter 43
# QUANTUM AND STATISTICAL MECHANICS

## Free Boundaries of Plasmas

Finding stable equilibria of magnetically confined plasmas is a central problem in fusion-oriented plasma physics. The plasma and material wall must be separated by a vacuum, otherwise the plasma and wall would destroy each other. Since the early 1970s codes have been written for designing fusion devices with desirable interface shapes. In today's tokamaks, the axisymmetric interfaces are shaped by magnetic "diverters" and are nonsmooth. In the cross section, the corresponding interface curves have one or two corners.

The speakers focused on associated unsolved or recently solved mathematical problems, in which the plasma/vacuum interface is mostly treated as a free boundary. Some of the equations treated describe steady vortex rings in an ideal fluid as well.

**Organizer:** Rita Meyer-Spasche, Max-Planck Institut für Plasmaphysik, Garching, Germany

**The Ideal Tokamak Configuration**  Dietrich Lortz, Max-Planck Institut für Plasmaphysik, Garching, Germany

**Plasma Corners**  Rolf Kaiser, D. Lortz and G. O. Spies, Max-Planck Institut für Plasmaphysik, Garching, Germany

**A Finite Difference Procedure to Treat Free Boundaries on Rectangular Grids**  Bengt Fornberg, Exxon Research and Engineering Company, Annandale; Rita Meyer-Spasche, Organizer

**Nonuniqueness of Free Boundary Solutions**  Rita Meyer-Spasche, Organizer

## Nonlinear Analysis in the Kinetic Theory of Collisionless Plasmas

Understanding the collective behavior of low density plasmas in which charged particles interact through self-induced electromagnetic fields is crucial to progress in magnetic fusion and to interpreting magnetospheric and space plasma phenomena. The underlying nonlinear Vlasov kinetic equations coupled to Maxwells equations for the fields are fundamentally nonlinear and possess an infinite manifold of non-hyperbolic fixed points. Further, because the linearized operator has continuous spectrum on the entire imaginary axis, standard methods of bifurcation theory and nonlinear analysis are inappropriate. Thus, existence, explicit exact solutions, and stability are crucial results that have not been easily obtained.

The speakers summarized most of the few such rigorous results that have been obtained for the nonlinear Vlasov-Maxwell equations and the related Vlasov-Poisson equations of gravitational stellar dynamics.

**Organizer:** John J. Dorning, University of Virginia

**Nonlinear Analysis in the Kinetic Theory of Collisionless Plasmas**  John J. Dorning, Organizer

**A Survey of Existence Results for the Vlasov-Maxwell Equations**  Robert T. Glassey, Indiana University, Bloomington

**Exact Equilibrium and Traveling Wave Solutions of the Vlasov-Maxwell Equations**  James P. Holloway, University of Michigan, Ann Arbor

**Lagrangian and Eulerian Descriptions of the Vlasov-Poisson System**  Huanchun Ye, University of Texas, Austin; P.J. Morrison, University of Texas, Austin; Jerrold E. Marsden, University of California, Berkeley

**Nonlinear Stability for Spherically Symmetric Models in Stellar Dynamics**  Yieh-Hei Wan, State University of New York, Buffalo

### Suggested Reading

M. L. Buchanan, J. P. Holloway, and J. J. Dorning, *Recent applications of bifurcation theory to nonlinear waves in collisionless plasmas*, in Research Trends in Nonlinear and Relativistic Effects in Plasmas, V. Stefan, ed., American Institute of Physics, New York, 1991.

A. Eydeland and B. Turkington, *A computational method of solving free-boundary problems in vortex dynamics*, J. Comput. Phys., 78 (1988), pp. 194–214.

R. T. Glassey and J. W. Schaeffer, *Global existence for the relativistic Vlasov–Maxwell system with nearly neutral initial data*, Comm. Math. Phys., 119 (1988), pp. 353–384.

S. P. Hirshman, W. I. Van Rij, and P. Merkel, *Three-dimensional free boundary calculations using a spectral Green's function method*, Comput. Phys. Comm., 43 (1986), p. 143.

J. P. Holloway and J. J. Dorning, *Undamped nonlinear plasma waves*, Phys. Rev. A, to appear.

R. Kaiser, D. Lortz, and G. O. Spies, *Plasma corners—1. Constant current density*, Phys. Fluids, to appear.

D. Lortz, *Plane free-boundary equilibria*, Plasma Phys. Control Fusion, 33 (1991), pp. 77–89.

R. Meyer-Spasche and B. Fornberg, *Discretization errors at free boundaries of the Grad-Schluter-Shafranov equation*, Numer. Math., 59 (1991), pp. 683–710.

Y-H. Wan, *Nonlinear stability of stationary spherically symmetric models in stellar dynamics*, Arch. Rational Mech. Anal., 112 (1990), pp. 83–95.

H. Ye, P. J. Morrison, and J. D. Crawford, *Poisson bracket for the Vlasov equation on a symplectic leaf*, Phys. Lett. A, 156 (1991), pp. 96–100.

# Chapter 44
# REAL AND COMPLEX ANALYSIS

## Applications of Generalized Hypergeometric Functions in Physics

For a long time, and maybe even today, it was believed that the use of computing machines would make the special functions of mathematical physics an obsolete part of mathematics and physics.

The speakers surveyed recent results and applications of generalized hypergeometric functions of one and several variables with respect to physical problems. It is important for mathematicians and physicists to come together to discuss their needs and intentions in elaborating the analytic theory of special functions, not to consider them as just an issue of numerical recipes. Results, conjectures, and problems are welcome regarding algebraic or analytic approaches to special functions and their applications in various physical problems. Comments on the concept of the hypergeometric function from the historical point of view will be appreciated.

**Organizer:** Hans Joachim Haubold, Outer Space Division, United Nations

**Applications of Generalized Special Functions in Reaction Theory** A.M. Mathai, McGill University, Canada

**High Order Differential Equations and Some Physical Examples** Richard B. Paris, Dundee Institute of Technology, United Kingdom; Alastair D. Wood, Dublin City University, Ireland

**Generalized Hypergeometric Functions and Their Applications in the Theory of the Light Changes of Eclipsing Variables** H.M. Srivastava, University of Victoria, Canada

**Generalized Hypergeometric Functions and Their Applications in the Theory of Stellar Structure** Hans Joachim Haubold, Organizer

**The Analytical Solution of the Stellar Interior Equations for Approximating Stellar and Protostellar Structures** J. F. Doorish, City University of New York

Heather Wilkinson and James W. Longley.

ICIAM 95 will be held in Hamburg, Germany.

## Exponential Asymptotics and Spectral Theory for Optical Waveguides

Wave propagation in optical waveguides provides unsolved problems in the spectral theory of the Schrodinger operator. Such problems also occur in quantum mechanics. These theories are, to a large extent, dual. The tunneling concept may be viewed in either context, quantum or optical, and related to the theory of resonances and Titchmarsh-Weyl theory.

The pertinent part of spectral theory concerns eigenvalues of nonself-adjoint problems. The computation of the exponentially small imaginary parts of such eigenvalues uses exponentially-improved asymptotic expansions. These require smoothing techniques for Stokes discontinuities related to the work of M.V. Berry and F.W.J. Olver.

The speakers explored the linkages between the above theories and interpreted them in the context of optical waveguides.

**Organizer:** Alastair D. Wood, Dublin City University, Ireland

**Bend Losses in Optical Fibres and Waveguides** Wanda Henry, Cambridge University, United Kingdom

**Spectral Theory of Optical Waveguides** John K. Shaw, Virginia Polytechnic Institute and State University

**Resonances and Optical Tunneling** Fiona Lawless, Dublin City University, Ireland

**Smoothing of the Stokes Phenomena Using Mellin-Barnes Integrals** Richard Bruce Paris, Dundee Institute of Technology, United Kingdom

## Historical Development of Singular Perturbation Concepts

Singular perturbation ideas arose largely in applied contexts, in a variety of different geographical locations, and for a wide spectrum of motivations. Now that this field has achieved some maturity, it is appropriate to review its development and continuing promise. Special attention was given to groups of researchers whose contributions have been substantial, and to very early use of asymptotic matching concepts in applications.

**Organizer:** Robert E. O'Malley, Jr., University of Washington

**Nineteenth Century Roots of Matching in the Applied Literature**  Milton Van Dyke, Stanford University

**Fundamental Concepts of Matching**  Wiktor Eckhaus, University of Utrecht, The Netherlands

**Asymptotics at Caltech's Guggenheim Aeronautical Laboratory**  Julian D. Cole, Rensselaer Polytechnic Institute

**The Study of Singular Perturbations in the Soviet Union, Especially at Moscow State University**  Adelaida B. Vasil'eva, Moscow State University, USSR

**Asymptotic Analysis and Applications at New York University**  Joseph B. Keller, Stanford University

## Singularly Perturbed Integral and Delay Equations

Singular perturbation methods have played an important role in the study of differential equations. However, the potential usefulness of these methods for other types of functional equations involving small parameters has not received the same attention. The wide occurrence of functional equations in applications underscores the importance of developing analytical methods for treating these classes of equations. By providing viable analytical tools, we hope to encourage the increased use of functional equations where appropriate to model physical and biological problems.

Current research primarily attempts to explain boundary-layer phenomena associated with integral and delay-differential equations. Typically, progress is made by a combination of exact solutions, singular perturbation methods, and numerical computations.

The focus of this minisymposium was on recent advances in the development of singular perturbation techniques for integral and delay-differential equations.

**Organizer:** W. Edward Olmstead, Northwestern University

**Singular Perturbation Analysis of Integral Equations**  W. Edward Olmstead, Organizer

**Fredholm Equations with Boundary Layer Resonance**  W. Edward Olmstead, Organizer

**A Matched Expansion Approach to Singularly Perturbed Problems for Singular Integral Equations**  John R. Willis, University of Bath, United Kingdom

**Singular Perturbation Analysis of Boundary-Value Problems for Differential-Difference Equations**  Robert M. Miura, University of British Columbia, Canada

## Turning Points and Canards in Several Dimensions

Turning points lead to difficult problems in asymptotics. A geometrical approach using nonstandard analysis led to the discovery of the "canards phenomenon," which appears to be fundamental in the analysis of some of these problems. Recent progress has been made in that direction. There is, for example, the systematic study of the canards asymptotic expansions which are "nicely diverging" series (results of Y. Sibuya, M. Canalis-Durand, I. Van den Berg). Another is the relation between canards and some physical phenomena called "delayed bifurcations" (A. Neishtadt, G. Wallet, B. Candelpergher, F. & M. Diener and C. Lobry). Even in degenerate cases, it is now possible (with a point of view similar to the one used by N. Koppel and J.L. Callot for the Ackerberg-O'Malley equation) to study turning point problems (A. Dellecroix, M. Diener). There are other very exciting results in the cases of discrete dynamical systems, where new phenomena, partially explained, take place (A. Fruchard).

**Organizer:** Francine Diener, Université de Nice, France

**Diverging Series and Small Exponentials**  Imme P. Van den Berg, University of Gröningen, The Netherlands

**Method for the Study of Solutions Near a Turning Point Using Appropriate Rescalings and Mathematical Rivers**  Marc Diener, Université Paris VII, France

**Dynamical Bifurcation with Noise**  E. Benoit, Écoles des Mines, France; B. Candpergher, and Claude Lobry, Université de Nice, France

**Delayed Bifurcation in Discrete Dynamical Systems: Gamma-Function-Like Behavior**  Augustin Fruchard, Université de Mulhouse, France

## Wavelets and Fast Numerical Algorithms

Recently developed wavelet and wavelet-packet bases (as well as a number of other constructions with similar properties) fill the void that existed between the Fourier basis and the Haar basis in terms of the time-frequency localization. These new tools significantly reduce the gap between the pure and numerical analysis of problems involving integral and differential equations. Based on these new representations numerical algorithms for solving integral and differential equations and computing functions of operators are rapidly advancing the state of the art in numerical analysis. These algorithms are bound to have a significant impact on large scale computing and therefore, on various subfields of applied mathematics and engineering. This minisymposium focused on important examples of such algorithms.

**Organizer:** Gregory Beylkin, University of Colorado, Boulder

**Adaptive Basis in the Solution of Scattering Problems**  Vladimir Rokhlin, Yale University

**Bases for Sparse LU-Factorization of Integral Operators**  Bradley K. Alpert, University of California, Berkeley

**Representations of Differential Operators and Fast Numerical Algorithms**  Gregory Beylkin, Organizer

**Laplace's Equation and the Dirichlet-Neumann Map in Multiply Connected Domains**  Leslie Greengard, Courant Institute of Mathematical Sciences, New York University; Ann Greenbaum, Courant Institute of Mathematical Sciences, New York University; Geoffrey B. McFadden, National Institute of Standards and Technology

**Signal Compression and Fast Computational Algorithms**   Ronald R. Coifman, Yale University

**Fast Texture Discrimination**   Jean-Michel Morel, Université Paris-Dauphine, France; Georges Koepfler, Université Paris-Dauphine, France Christian Lopez, Université Paris-Dauphine, France

**Wavelet Bases and Discrete Subband Coding Schemes**   Albert Cohen, AT&T Bell Laboratories, Murray Hill

**Tutorial on Adaptive Signal Processing Using Wavelet-Packets**   V. Wickerhauser, Yale University

## Wavelets and Generalized Functions

The development of the subject of wavelets has paralleled that of generalized functions a generation earlier. Both fields began with an attempt to solve problems in engineering before their theory was completely understood. Wavelets began with a problem in seismology but soon were found to apply to problems in areas such as signal processing, data compression, and image processing. Generalized functions began with the Heaviside calculus and have become an integral part of engineering analysis.

Recently the two fields have begun to converge. Some of the wavelet theory is simplified by using generalized functions; there are wavelets of point support; some applications use both theories. New discoveries which combine aspects of both theories was presented.

**Organizer:** Gilbert G. Walter, University of Wisconsin, Milwaukee

**A Wavelet Sampling Theorem**   Gilbert G. Walter, Organizer

**Irregular Sampling and the Theory of Frames** John J. Benedetto and William Heller, University of Maryland, College Park

**Optical Wavelet Networks**   Walter Schempp, Universität Siegen, Germany

**Properties of Solutions to Dilation Equations** David Colella, MITRE Corporation, McLean; Christopher E. Heil, Massachusetts Institute of Technology

**On Irregular Operations on Distributions**   Piotr Antosik, California Polytechnic State University

**Suggested Reading**

B. Alpert, G. Beylkin, R. R. Coifman, and V. Rokhlin, *Wavelets for the fast solution of second-kind integral equations*, Tech. Report, Department of Computer Science, Yale Universty, New Haven, CT, 1990.

P. Antosik, J. Mikusinski, and B. Sikorski, *Theory of Distributions: The Sequential Approach*, Elsevier, Amsterdam, 1973.

G. Beylkin, *On the representation of operators in bases of compactly supported wavelets*, SIAM J. Numer. Anal., 29 (1992).

G. Beylkin, R. R. Coifman, and V. Rokhlin, *Wavelets in numerical analysis*, in Wavelets, G. Beylkin, R. Coifman, I. Daubechies, et al., eds., Jones and Bartlett, 1991.

———, *Fast wavelet transforms and numerical algorithms* I, Comm. Pure Appl. Math., 44 (1991), pp. 141–183.

N. Brazel, F. Lawless, and A. D. Wood, *Exponential asymptotics for an eigenvalue of a problem involving parabolic cylinder functions*, Proc. Amer. Math. Soc., in press.

J. Carrier, L. Greengard, and V. Rokhlin, *A fast adaptive multipole algorithm for particle simulations*, SIAM J. Sci. Statist. Comput., 9 (1988).

R. R. Coifman and Y. Meyer, *Remarque sur l'analyse de Fourier à finêtre, série* I, C.R. Acad. Sci., Paris, 312 (1991), pp. 259–261.

R. R. Coifman, Y. Meyer, and V. Wickerhauser, *Wavelet analysis and signal processing*, in Wavelets, G. Beylkin, R. Coifman, I. Daubechies, et al., eds., Jones and Bartlett, 1991.

I. Daubechies, *Orthonormal bases of compactly supported wavelets*, Comm. Pure Appl. Math., 41 (1988), pp. 909–996.

———, *Ten Lectures on Wavelets*, CBMS-NSF Regional Conf. Ser. in Appl. Math., Vol. 61, Society for Industrial and Applied Mathematics, Philadelphia, 1992.

F. Diener and M. Diener, *A very short survey of nonstandard analysis' contributions to the theory of ordinary differential equations*, in Differential Equations, G. M. Dafermos, G. Ladas, and G. Papanicolaou, eds., Marcel Dekker, New York, Basel, 1989, pp. 191–198.

W. Eckhaus, *Asymptotic Analysis of Singular Perturbations*, North–Holland, Amsterdam, 1979.

S. Gottloeber, H. J. Haubold, J. P. Muecket, and V. Mueller, *Early Evolution of the Universe and Formation of Structure*, Akademie-Verlag, Berlin, 1990.

C. Heil and D. Walnut, *Continuous and discrete wavelet transforms*, SIAM Rev., 31 (1989), pp. 628–666.

W. Kath and G. Kreigsmann, *Optical tunnelling: radiation losses in bent fibre optic waveguides*, IMA J. Appl. Math., 41 (1988), pp. 85–103.

J. Kevorkian and J. D. Cole, *Perturbation Methods in Applied Mathematics*, Springer-Verlag, New York, Berlin, 1981.

R. Lutz and M. Goze, *Nonstandard Analysis: A Practical Guide with Applications*, Lecture Notes in Math., Vol. 881, Springer-Verlag, New York, Berlin, 1982.

A. M. Mathai and H. J. Haubold, *Modern Problems in Nuclear and Neutrino Astrophysics*, Akademie-Verlag, Berlin, 1988.

A. M. Mathai and R. K. Saxena, *Generalized Hypergeometric Functions with Applications in Statistics and Physical Sciences*, Lecture Notes in Math., Vol. 348, Springer-Verlag, Berlin, Heidelberg, New York, 1973.

Y. Meyer, *Ondelettes et Opérateurs*, Vols. 1, 2, and 3, with R. R. Coifman, Hermann, Paris, 1990, 1991.

L. Motz, *Astrophysics and Stellar Structure*, Ginn and Co., Waltham, MA, 1970.

M. A. Nashed and G. Walter, *General sampling theorems for functions in reproducing kernel Hilbert spaces*, Math. Control Signals Systems, 4 (1991), pp. 373–412.

R. E. O'Malley, Jr., *Singular Perturbation Methods for Ordinary Differential Equations*, Springer-Verlag, New York, Berlin, 1991.

R. B. Paris and A. D. Wood, *Asymptotics of High Order Differential Equations*, Pitman Res. Notes Math. Ser., Vol. 129, Longman Scientific and Technical, Essex, 1986.

———, *A model equation for optical tunnelling*, IMA J. Appl. Math., 43 (1989), pp. 273–284.

———, *Exponentially improved asymptotics for the gamma function*, J. Comput. Appl. Math., in press.

W. Schempp and B. Dreseler, *Einführung in die harmonische Analyse*, B. G. Teubner, Stuttgart, 1980.

A. W. Snyder and J. D. Love, *Optical Waveguide Theory*, Chapman and Hall, London, 1983.

I. P. Van den Berg, *Nonstandard Asymptotic Analysis*, Lecture Notes in Math., Vol. 1249, Springer-Verlag, New York, Berlin, 1987.

M. Van Dyke, *Perturbation Methods in Fluid Mechanics*, Parabolic Press, Palo Alto, CA, 1975.

A. B. Vasil'eva and V. F. Butuzov, *Asymptotic Methods in Singular Perturbations*, Vysshaja Shkola, Moscow, 1990.

A. K. Zvonkin and M. A. Slobian, *Nonstandard analysis and singular perturbations of ordinary differential equations*, Russian Math. Surveys, 39 (1984), pp. 69–131.

# Chapter 45
# SCIENTIFIC COMPUTATION

## Applications of Interval Arithmetic Yielding Tight Bounds

Interval techniques have been in use for about 30 years providing automated error analysis for a wide variety of numerical problems, yet doubts persist about their ability to achieve tight bounds. Good conventional algorithms often perform poorly as interval algorithms, but mathematical techniques such as fixed point theorems, contractive iterations, Taylor series, intersections, or problem adaptation form the basis for good interval algorithms. The speakers showed how interval techniques have been applied to some industrial strength problems and yielded bounds which differ by only a few units in the last place.

**Organizer:** George F. Corliss, Marquette University

**Achieving Tight Bounds Using Interval Arithmetic**
George F. Corliss, Organizer

**Application of Variable Precision Interval Arithmetic to Problems in Fluid Dynamics**
Jeffery Ely, Lewis and Clark College

**Interval Nonlinear Equation Software: Recent Improvements and Applicability**   R. Baker Kearfott, University of Southwestern Louisiana

**Improving Error Bounds for Initial Value Problems with Interval Automatic Differentiation**   James Daniel Layne, Martin Marietta Astronautics Group

Richard Tapia and a few of his students who attended from Rice University.

## Automatic Differentiation—Derivative Evaluations are Cheap!

Numerical algorithms for problems of optimization, nonlinear equations, quadrature, and ordinary differential equations require the use of ordinary or partial derivatives of a user-defined function. The necessary gradients, Jacobians, Hessians, and Taylor series can be efficiently and accurately computed by automatic differentiation using recurrence relations. Fortunately, the use of automatic differentiation often requires only minor modifications to existing codes. The results are more reliable than finite difference approximations, much less expensive than symbolic computations, and easier to use than routines which require the user to supply code for derivatives.

The speakers presented an overview of automatic differentiation and discussed the implementation and applications.

**Organizer:** George F. Corliss, Marquette University

**Applications of the Jacobian and Higher Derivatives of the Code List**   Louis B. Rall, University of Wisconsin, Madison

**Automated Partial Derivatives**   David R. Hill, Temple University; Lawrence C. Rich, Temple University

**Computing Multivariable Taylor Series Coefficients to Arbitrary Order**   Richard D. Neidinger, Davidson College

**The Computation of Derivative Matrices Based on the Markowitz Count**   Andreas Griewank, Argonne National Laboratory

C. William Gear, D. Allan Bromley, and Robert E. O'Malley pose prior to Bromley's talk on Science, Technology, and Mathematics in the 1990's. Bromley is the Assistant to the President for Science and Technology and Director of the Office of Science and Technology in Washington, DC.

## Computer Arithmetic and Self-Validating Numerical Methods

This minisymposium was concerned with computer arithmetic with extended capabilities. The computer is assumed to be a vector processor in the mathematical sense. This means that apart from usual floating-point arithmetic, all vector and matrix operations over the real and complex numbers and the corresponding intervals are provided by approximations which differ from the correct answer by at most one rounding. These capabilities allow many applications to technical problems where standard floating-point arithmetic cannot provide a sufficient answer.

FORTRAN-SC (ACRITH-XSC) and PASCAL-XSC are extensions of FORTRAN 77 and PASCAL, respectively, that allow a convenient use of the additional arithmetical capabilities. Using these languages, program packages have been developed for many standard problems of numerics. The computer verifies the existence and uniqueness of the solution within tight bounds.

The speakers briefly introduced these languages and presented two problem-solving packages: one for systems of ordinary differential equations and one for linear and nonlinear Fredholm integral equations of the second kind. Some typical engineering applications were also discussed.

**Organizer:** Ulrich W. Kulisch, Universität Karlsruhe, Germany

**Developments of FORTRAN Towards More Reliability in Engineering/Scientific Computation** Wolfgang V. Walter, Universität Karlsruhe, Germany

**PASCAL-XSC, A Powerful PASCAL Extension for Enhanced Scientific Computation Running Under C** Michael Neaga, Numerik Software GmbH, Baden-Baden, Germany

**A Variety of Engineering Problems Solved by Self-Validating Methods** Gunter Schumacher, Universität Karlsruhe, Germany

**Enclosure Methods for a Wide Range of Problems in Ordinary Differential Equations** Rudolf J. Lohner, Universität Karlsruhe, Germany

**Computing Tight Continuous Bounds for the Solution of Systems of Integral Equations** Wolfram Klein, Siemens-München, Germany

## High Performance ESSL Algorithms for the IBM RS/6000 Workstations

About three hundred high performance FORTRAN subroutines have been developed for the RS/6000 workstation family. They constitute the ESSL library and partially include algorithms for matrix operations, linear algebra, signal processing, sorting and searching. Of particular interest are algorithms for Level 2 and Level 3 BLAS. These FORTRAN programs perform uniformly well at near the peak possible rate for these BLAS. For example, all Level 3 BLAS on the model 550 achieve around 75 MFLOPS.

The speakers provided an overview of the ESSL library and its design philosophy, focused on super scalar performance and how it was obtained, and described the algorithms.

**Organizer:** Fred G. Gustavson, IBM T.J. Watson Research Center

**An Overview of the ESSL RS/6000 FORTRAN Library** Fred G. Gustavson, Organizer; Stanley Schmidt, IBM Corporation, Kingston

**New FFT Algorithms and Their Implementation on RS/600** Ali Mechentel, IBM T.J. Watson Research Center

**New Linear Algebra Algorithms** Ramesh Agarwal, IBM T.J. Watson Research Center; Fred Gustavson, Organizer

**Implementation Notes for the Tridiagonal, Sort, Search and Random Number Subroutines for the ESSL 6000 Library** James B. Shearer, IBM T.J. Watson Research Center

## Large Scale Scientific Computations

High performance computing systems are bound to be massively parallel in the next decade. Such systems will be capable of a performance in excess of 1 Teraflop/s by 1995. With a 1,000 fold increase in computational power, compared to the late 1980s, improved models can be used, and new scientific and engineering problems can be tackled. The technology used in the new generation of high performance computer architectures requires that algorithms be devised that exhibit locality of reference, extensive parallelism, and good load balance. In this minisymposia, some techniques that are appropriate for the new type of high performance architectures and large scale industrial applications in fluid mechanics, acoustics, and solid mechanics were discussed. Results from implementations on state-of-the-art high performance architectures were presented.

**Organizer:** S. Lennart Johnsson, Thinking Machines Corporation

**The Free-Lagrange Method on the Connection Machine** Harold E. Trease, Los Alamos National Laboratory

**Calibration of Large-Eddy Simulation Subgrid Models Associated with Monotone Algorithms** Jay P. Boris, Naval Research Laboratory

**An Implementation of 3D One Pass Depth Migration** Peter Highnam, Schlumberger Laboratory for Computer Science, Texas

## Recent Developments in Methods for Enclosing Solutions of Nonlinear Systems

In the past few years, many methods have been developed to iteratively enclose solutions of nonlinear systems. They are important in practice since they automatically deliver safe error-bounds.

From a theoretical point of view, most of these methods are now rather well understood. However, in order to make them work efficiently, one has to take into account the special properties of the particular underlying problem. In addition, the efficient use of vector and parallel computers also requires specific modifications of these methods.

The speakers presented theoretical results and practical experience in this area.

**Organizer:** Goetz E. Alefeld, University of Karlsruhe, Germany

**Efficient Parallel Methods for Enclosing Zeroes of Systems of Nonlinear Equations** Andreas J. Frommer, University of Karlsruhe, Germany

**Preconditioners for the Newton-Gauss-Seidel Method** R. Baker Kearfott, University of Southwestern Louisiana

**Computable Bounds for the Inverse Eigenvalue Problem** Guenter A. Mayer, University of Karlsruhe, Germany

**Domain Decomposition Enclosure Methods with Applications to Nonlinear PDEs** Hartmut Schwandt, Fachbereich Mathematik, TU Berlin, Germany

## Solving Parameter Dependent Nonlinear Equations by Reduced Basis Methods

Reduced Basis Methods were originally developed by engineers to lessen the computational effort in deflection analysis of elastic structures. Today, the Reduced Basis Method is a (numerical) device to treat general single and multiparameter depending problems in a Banach space setting.

The Reduced Basis approach is a Galerkin-type condensation of large nonlinear systems to similar low-dimensional substitutes, wherein the Galerkin-ansatz is adapted to model the systems local behavior with respect to parameter changes. Parameter dependence can thus be studied within easily solved small systems. Reduced Basis Methods have been developed for several problem areas such as nonlinear algebraic equations, (discretized) elliptic and parabolic PDEs, ODEs, DAEs, and unconstrained minimization.

The speakers presented an overview of Reduced Basis techniques, and discussed some practical applications.

**Organizers:** Wolfgang G. Mackens, Universität Hamburg, Germany; Werner C. Rheinboldt, University of Pittsburgh, Pittsburgh

**On the Theory and Error Estimation of the Reduced Basis Method** Werner C. Rheinboldt, Organizer

**Recent Applications of the Reduced Basis Techniques** Ahmed K. Noor, NASA Langley Research Center

**Adaptive Spectral Condensation Methods** Wolfgang G. Mackens, Organizer

## Variational Methods and Grid Generation

Variational methods are used to develop robust numerical grid-generation algorithms to numerically solve partial differential equations in geometrically complex regions. There is a good intuitive understanding of how the methods work; however, there is not a solid mathematical grid generation theory. The purpose of this subsession was to discuss some of the mathematical aspects of variational grid generation methods. This should give a better understanding of how and why some of the methods work and what their limitations are. Some of the problems discussed were: existence and uniqueness of solutions to the grid generation equations, rational choice of parameters (weights), grid generation on curves and surfaces, effect on the errors when solving partial differential equations, including geometry and solution adaption. Another important aspect of these algorithms is what solver should be used to solve the algebraic equations, SOR, multigrid, conjugate gradient, etc.

**Organizers:** José E. Castillo, San Diego State University; Stanley Steinberg, University of New Mexico

**Variational Methods in Harmonic Grid Generation** S.S. Sritharan, University of Southern California

**Some Recent Results on the Mathematics of Grid Generation** G. Liao, University of Texas, Arlington

**Using Variational Grids on Finite-Difference Calculations** Stanley Steinberg, Organizer; Patrick Roache, Ecodynamics Research Associates, Inc.

**Goodness-of-Grid on the Scale 0-10** Bharat Soni, Mississippi State University

**The Discrete Grid Generation Method on Curves and Surfaces** José E. Castillo, Organizer

**A New Elliptic Grid Generation** Patrick M. Knupp, Ecodynamics Research Associates, Albuquerque

**Grid Generation by Harmonic Mappings** Renzo Arina, Politecnico di Torino, Italy

**Adaptive Grid Generation** Dale Anderson, University of Texas, Arlington

## Suggested Reading

Acrith-XSC, *IBM High Accuracy Arithmetic--Extended Scientific Computation*, Version 1, Release 1, IBM Deutschland GmbH, Boblingen, Germany, 1990.

A. S. Arcilla, J. Hauser, P. R. Eiseman, and J. F. Thompsom, eds., *Numerical Grid Generation in Computational Fluid Dynamics and Related Fields*, Pineridge Press Limited, Swansea, U.K., 1991.

J. E. Castillo, ed., *Mathematical Aspects of Grid Generation*, Frontiers Appl. Math., Vol. 8, Society for Industrial and Applied Mathematics, Philadelphia, 1991.

G. F. Corliss, *Industrial applications of interval techniques*, in Computer Arithmetic and Self-Validating Numerical Methods, C. Ullrich, ed., Academic Press, New York, 1990, pp. 91–113.

J. Ely, *Application of variable precision interval arithmetic to the solution of problems in vortex dynamics*, in Proc. 13th World Congress on Computation and Applied Mathematics, IMACS '91, Dublin, Ireland, 1991, pp. 69–70.

J. P. Fink and W. C. Rheinboldt, *On the error behavior of the reduced basis technique for nonlinear finite element approximations*, Z. Angew. Math. Mech., 63 (1983), pp. 21–28.

———, *Local error estimates for parametrized nonlinear equations*, SIAM J. Numer. Anal., 22 (1985), pp. 729–735.

A. Griewank, *Direct calculation of Newton steps without accumulating Jacobians*, MCS-P132-0290, Mathematics and Computer Science Division, Argonne National Laboratory, 1990, preprint.

D. R. Hill and L. C. Rich, *Automatic Differentiation in MATLAB*, Appl. Numer. Math., to appear.

H. Jarausch and W. Mackens, *Numerical treatment of bifurcation by adaptive condensation*, in Numerical Methods for Bifurcation Problems, T. Kuepper, H. D. Mittelmann, and H. Weber, eds., ISNM Vol. 70, Birkhauser-Verlag, Basel, Switzerland, 1984, pp. 296–309.

———, *Solving large nonlinear systems of equations by an adaptive condensation process*, Numer. Math., 50 (1987), pp. 633–653.

R. B. Kearfott, *Algorithm 681, INTBIS, a portable interval Newton/bisection package*, ACM Trans. Math. Software, 16 (1990), pp. 152–157.

R. Klatte, U. Kulisch, M. Neaga, et al., *PASCAL-XSC, Sprachbeschreibung mit Beispielen*, Springer-Verlag, Berlin, Heidelberg, New York, 1991.

U. Kulisch, ed., *Wissenschaftliches Rechnen mit Ergebnisverifikation. Eine Einfhrung*, Akademie Verlag, Berlin, und Vieweg Verlagsgesellschaft, Wiesbaden, 1989.

U. Kulisch and W. L. Miranker, *Computer Arithmetic in Theory and Practice*, Academic Press, New York, 1981.

———, eds., *A New Approach to Scientific Computation*, Academic Press, New York, 1983.

———, *The arithmetic of the digital computer: A new approach*, IBM Research Center Report RC 10580, 1984, SIAM Rev., 28 (1986), pp. 1–40.

J. D. Layne, *Applying automatic differentiation and self-validating numerical methods in satellite simulations*, in Automatic Differentiation of Algorithms: Theory, Implementation, and Application, A. Griewank and G. F. Corliss, eds., Society for Industrial and Applied Mathematics, Philadelphia, 1991.

R. Lohner, *Enclosing the solutions of ordinary initial- and boundary-value problems*, in Computerarithmetic, E. Kaucher, U. Kulisch, and C. Ullrich, eds., Teubner, Stuttgart, 1987, pp. 255–286.

W. Mackens, *Numerical differentiation of implicitly defined space curves*, Computing, 41 (1989), pp. 237–260.

R. D. Neidinger, *An efficient method for the numerical evaluation of partial derivatives of arbitrary order*, ACM Trans. Math. Software, to appear.

A. K. Noor, *Recent advances in reduction methods for nonlinear problems*, Comput. & Structures, 13 (1981), pp. 31–44.

———, *On making large nonlinear problems small*, Comput. Methods Appl. Mech. Engrg., 34 (1982), pp. 955–985.

A. K. Noor, C. M. Andersen, and J. A. Tanner, *Exploiting symmetries in the modeling and analysis of tires*, Comput. Methods Appl. Mech. Engrg., 63 (1987), pp. 37–81.

A. K. Noor, C. D. Balch, and M. A. Shibut, *Reduction methods for nonlinear steady-state thermal analysis*, Internat. J. Numer Methods Engrg., 20 (1984), pp. 1323–1348.

A. K. Noor and J. M. Peters, *Reduced basis technique for nonlinear analysis of structures*, AIAA J., 18 (1980), pp. 455–462.

———, *Recent advances in reduction methods for instability analysis of structures*, Comput. & Structures, 16 (1983), pp. 67–80.

A. K. Noor, J. M. Peters, and C. M. Andersen, *Mixed models and reduction techniques for large-rotation nonlinear problems*, Comput. Methods Appl. Mech. Engrg., 44 (1984), pp. 67–89.

L. B. Rall, *Differentiation arithmetics*, in Computer Arithmetic and Self-Validating Numerical Methods, C. Ullrich, ed., Academic Press, New York, 1990, pp. 73–90.

———, *Automatic Differentiation: Techniques and Applications*, Lecture Notes in Computer Science, Vol. 120, Springer-Verlag, Berlin, 1981.

S. Steinberg and P. J. Roache, *Discretizing symmetric operators in general coordinates*, in preparation.

C. Ullrich, ed., *Computer Arithmetic and Self-Validating Numerical Methods*, Acedemic Press, New York, 1990.

———, ed., *Proceedings of SCAN-89*, IMACS Ann. Comput. Appl. Math., 1990.

C. Ullrich and J. Wolff v. Gudenberg, eds.., *Accurate Numerical Algorithms, A Collection of DIAMOND Research Papers*, Springer-Verlag, Berlin, New York, 1989.

M. Vinokur, *An analysis of finite-difference and finite-volume formulations of conservation laws*, J. Comput. Phys., 81 (1989), pp. 1-52.

# Chapter 46
# SIMULATION AND MODELING

## Applied Mathematics in Latin America

This minisymposium was devoted to applied mathematics in Latin America. The speakers, who are Latin-American mathematicians, presented case studies of mathematical modeling of problems taken from their countries. Applied mathematics has demonstrated much appeal to mathematicians in Latin America, where universities have a strong orientation towards the outside world. Moreover, the fact that mathematical techniques can help in solving real world problems is a way of using their knowledge in the effort for their countries development. This minisymposium offered Latin-American mathematicians an opportunity to break their isolation and to establish contacts with mathematicians from other countries.

**Organizers:** Ellis Cumberbatch, Claremont Graduate School; Victor Latorre, Multiciencias, Peru; Roberto M. Semenzato, CNR, Italy

**Tangent Continuous Algebraic Splines** Marco Paluszny, Universidad Central de Venezuela, Venezuela; Richard R. Patterson, Indiana University-Purdue University at Indianapolis

**A Mathematical Model for Planning in the Medium-Term of Electrical Power Generation** Wilfredo Sosa, Universidad Nacional de Ingenieria, Peru

**Simulation of Ground Transportation of Gasoline** Rafael A. Pastoriza, Universidad Central de Venezuela, Venezuela; Adrian J. Bottini, Universidad Central de Venezuela, Venezuela; Jorge Serebrisky, Universidad Central de Venezuela, Venezuela

**Scheduling of Tanker Fleet** Marianela Lentini, Universidad Simon Bolivar, Venezuela; A. Reinoza, Universidad Simon Bolivar, Venezuela; J.E. Smith, Universidad Simon Bolivar, Venezuela

**Support Decision System for Programming Sea Transportation of Oil Products** Marcelo Salvador, Escuela Politecnica Nacional, Ecuador

**Projection Matrix and Optimal Designs** Edith Seier, Universidad Nacional Mayor de San Marcos, Peru

## A View of Chemical Industrial and Applied Mathematics

This minisymposium dealt with the solution of mathematical and computational problems that arise in chemical engineering and chemistry, in particular those involving process modeling and simulation.

Systems of algebraic equations, ordinary differential equations (ODEs), and partial differential equations (PDEs) can be used to describe a fundamental problem in chemical kinetics or a large simulation of a plant comprised of several processing units. Complex processes such as film coating (as in the manufacturing of laminated materials) lie in this class of problems. The methods in current use include numerical methods for: optimization, solving nonlinear systems of algebraic equations, ODEs, PDEs, DAEs (differential-algebraic equations), and linear algebraic equations. Tools from graphics, artificial intelligence, and data base management are also being used. Often, as in the case of the more powerful simulation tools, these elements are combined.

The speakers covered some of this important area in an understandable way and presented an opportunity to learn about an area where applied mathematics and computation have played and do play an important role.

**Organizer:** George D. Byrne, Exxon Research and Engineering Company

**When Simulations Die** Arthur W. Westerberg, Carnegie Mellon University

**Pitfalls in Dynamic Simulation** Roger W.H. Sargent, Imperial College of Science Technology and Medicine, United Kingdom

**A Solution is Not Enough! A Cry from the Trenches** L.E. Scriven, University of Minnesota, Minneapolis

**Finding All Solutions to Systems of Nonlinear Equations with Applications to Chemical Engineering** J.D. Seader, University of Utah

**Optimizing and Solving DAEs Simultaneously in Chemical Engineering Applications** Lorenz T. Biegler, Carnegie Mellon University

**Pragmatic Approaches to Incorporate "Real Chemistry" into Kinetic Models** Anthony M. Dean, Exxon Research and Engineering Company

**The Challenges of Process Modeling in the Petrochemical Industry** R. M. Furzeland, Shell International Petroleum Company

**Solving Differential Equation Models in Chemical Industry Applications** George D. Byrne, Organizer

## Mathematical Modelling of Etching Processes

This minisymposium was concerned with the analytical and numerical treatment of hyperbolic and parabolic partial differential equations and moving boundaries arising in micromechanics and in the development of microchip and laser technology.

The speakers addressed the numerical treatment of convection-diffusion equations and with moving boundaries; asymptotic methods for moving boundary problems involving diffusion in two and three dimensions, ray methods for hyperbolic moving boundary problems, and the numerical simulation of plasma etching reactors.

**Organizer:** Hendrik K. Kuiken, Eindhoven University of Technology and Philips Research Laboratories, The Netherlands

**Numerical Simulation of the Shape Evolution of Cavities During Wet Chemical Etching** Demetre J. Economou, University of Houston

**Etching Profiles Near Mesh Edges for Diffusion Controlled Etching Using Stationary Etchants** Hendrik K. Kuiken, Organizer

**Moving Boundary Problems for Hyperbolic Conservation Laws: Applications to Ion Etching** David S. Ross, Kodak Research Laboratories, Rochester

**Numerical Simulation of Transport and Reaction in Plasma Etching Reactors** Demetre J. Economou, University of Houston

## Mathematics in Photographic Science

This minisymposium addressed the application of mathematics to various problems in photographic science.

From the perspective of the applied mathematician, photographic science is important because it is a fecund and virtually untapped source of good mathematical and computational problems. The speakers in this minisymposium concentrated on applications of differential equation models and stochastic process models to problems in photographic science. They touched on many basic areas of photographic science: making film (diffusion in swelling gelatin), taking pictures (latent image formation), and sending it to the drug store (development, inhibition).

**Organizer:** David S. Ross, Kodak Research Laboratories, Rochester

**Nonlinear Problems in Electrophotography** Avner Friedman, Institute for Mathematics and Its Applications, University of Minnesota

**The Image Transfer Function in Diffusion Transfer Film Systems** Alan E. Ames, Polaroid Corporation, Cambridge; Bior Holtgren, Polaroid Corporation, Cambridge

**A Reaction-Diffusion Model of Color Negative Film Development** Kam-Chuen Ng and David Stewart Ross, Kodak Research Laboratories, Rochester

**A Stochastic Model of Latent Image Formation** Hans Weinberger, University of Minnesota, Minneapolis; David Sattinger, Yale University

## University-Industry Collaboration in Applied Mathematics

Meaningful interaction between academic mathematicians and mathematicians, scientists, and engineers in industry provide valuable stimuli for the development of applied mathematics. Several formal programs have been established which facilitate such interactions. Each of the speakers described current problems that have involved university-industry interactions. The speakers in this minisymposium are actively involved in university programs that promote industrial mathematics. The speakers also described both the opportunities and difficulties involved in initiating and maintaining direct interactions between academic and industrial mathematics.

**Organizer:** Stavros N. Busenberg, Harvey Mudd College

**Mathematical Models from Industry** Alan B. Tayler, Oxford University, United Kingdom

**Heat Conduction from a VLSI Chip** Ellis Cumberbatch, Claremont Graduate School

**Mathematical Problems in Airspinning Processes** Manfred Baecker, University of Kaiserslautern, Germany; Helmut Neunzert, University of Kaiserslautern, Germany

**Dynamics of Fluidizing Agents in Coal-Water Slurries** Antonio Fasano and Mario Primicerio, University of Florence, Italy; Carla Manni, University of Pisa, Italy

### Suggested Reading

B. E. Bayer and J. F. Hamilton, *Computer investigation of a latent-image model*, J. Optical Soc. Amer., 55 (1965).

E. Cumberbatch, *Heat transfer from a VLSI chip*, Mathematics Clinic Report, Claremont Graduate School, 1991.

J. Dewynne, J. Ockendon, and P. Wilmott, *On a mathematical model for fiber tapering*, SIAM J. Appl. Math., 49 (1989), pp. 983–990.

A. Fasano and M. Primicerio, *Modelling the rheology of a coal-water slurry*, Proc. 4th Symposium in Mathematics in Industry, ECMI 89, H. Wacker, ed., Teubner-Kluwer, Stuttgart, 1991.

A. Friedman, *Mathematics in Industrial Problems*, Springer-Verlag, New York, 1988, chap. 4.

———, *Mathematics in Industrial Problems*, Part 2, Springer-Verlag, New York, 1989, chaps. 10, 14, and 17.

T. H. James, *The Theory of the Photographic Process*, Macmillan, New York, 1977.

H. K. Kuiken, *Etching: A two-dimensional mathematical approach*, Proc. Roy. Soc. London Ser. A, 392 (1984), pp. 199–225.

———, *Mathematical modelling of etching processes*, in Free Boundary Problems: Theory and Applications, Vol. 1, K. H. Hoffmann and J. Sprekels, eds., Pitman Res. Notes Math. Ser., Vol. 185, Longman Scientific and Technical, Essex, U.K., 1990, pp. 89–109.

K. C. Ng and D. S. Ross, *Diffusion in swelling gelatin*, J. Imaging Sci., 35 (1991).

S.-K. Park and D. J. Economou, *Analysis of low pressure rf glow discharges using a continuum model*, J. Appl. Phys., 68 (1990), pp. 3904–3915.

D. S. Ross, *Two new moving boundary problems for scalar conservation laws*, Comm. Pure Appl. Math., 41 (1988), pp. 725–737.

C. B. Shin and D. J. Economou, *Forced and natural convection effects on the shape evolution of cavities during wet chemical etching*, J. Electrochemical Soc., 138 (1991), pp. 527–538.

# AUTHOR INDEX

Ababou, Rachid, 300
Achdou, Y., 290, 298
Ackley, M., 333
Adali, Sarp, 317
Adjerid, S., 326
Agarwal, Ramesh, 345
Aghistine, M., 305
Agoshkov, Valeri I., 327
Aiello, William, 293
Aivazzadeh, Shahram, 313
Akian, M., 289
Albanese, Richard A., 310
Alefeld, Goetz E., 345
Alishenas, Taifun, 328
Allan, D.C., 314
Alpert, Bradley K., 342
Alvarado, Fernando L., 322
Ambrosiano, J., 297
Ames, Alan E., 349
Anderson, Dale, 346
Anderson, Edward, 321
Anderson, Ronald J., 335
Anderssen, Robert S., 308, 309
Annexstein, Fred, 293
Antosik, Piotr, 343
Arina, Renzo, 346
Arnold, A., 296, 298
Arnold, Ludwig, 294
Ascher, Uri M., 324, 325, 328
Astfalk, Greg, 332
Atilgan, A. R., 317
Atkinson, Frederick V., 335, 337
Averick, Brett, 290

Babovsky, Hans, 328
Babuska, Ivo, 329
Baecker, Manfred, 349
Baker, Timothy J., 304
Balachandran, Kandiah, 318
Ball, John M., 3
Barba, M.C.B., 320
Barcelo, Helene, 292
Barenblatt, Grigorii I., 15
Barnhill, Robert E., 305
Bauer, Frances, 300
Baumgartner, A., 316
Bayliss, Alvin, 282
Beale, J. Thomas, 328
Bechtold, J.K., 282
Becker, Pierre, 290

Bedard, Fernand, 314
Beichl, I., 305
Belfort, Georges, 301
Bellen, Alfredo, 325
Bellomo, Nicolas, 300
Benaroya, Haym, 294
Benedetto, J.J., 343
Bennett, K.R., 325
Benoit, E., 342
Berger, Melvyn S., 314
Bernoff, A., 313
Berque, D., 286
Berryman, James G., 308
Bertsekas, D.P., 290
Betancourt, Ramon, 322
Beydoun, Wafik, 310
Beylkin, Gregory, 342
Bhogeswara, R., 307
Biegler, Lorenz T., 348
Bieterman, Michael B., 298
Billah, K.Y.R., 294
Birman, Victor, 317
Bischof, Christian H., 290, 321
Bixby, Robert E., 333
Bjorstad, Petter, 322
Blakley, R., 285
Bleistein, Norman, 310
Blum, Avrim, 293
Boch, Georg H., 289, 328
Bodvarsson, G.S., 300
Boehm, Wolfgang, 306
Boggs, Paul T., 332
Boghosian, Bruce M., 327
Boisvert, R.F., 285
Booty, Michael R., 282
Boratavee, D., 300
Boris, Jay P., 345
Bornemann, Folkmar A., 326
Bosch, R., 335
Bottini, A.J., 348
Bowers, Kenneth L., 337
Boyce, William E., 294
Braddock, R. D., 299
Brady, Michael, 30
Brandon, Deborah, 302
Brankin, R.W., 324
Brenier, Yann, 328
Brewster, M.E., 301
Brezzi, Franco, 329
Brimacombe, J.K., 314

Broder, Andrei, 293
Bronsard, Lia, 313
Bruch, Jr., J.C., 317
Bryson, A.E., 289
Buchanan, J.L., 286
Budaev, Bair Vl., 316
Bukhart, R.H., 298
Bukiet, Bruce G., 327
Burke, James V., 334
Burr, Stefan, 292
Burrage, Kevin, 325
Burridge, Robert, 318, 337
Burton, T.D., 336
Busenberg, Stavros N., 349
Bussoletti, John E., 298
Butler, Lynne M., 292
Buttiker, M., 279
Byrd, Richard H., 333, 334
Byrne, George D., 348

Cabannes, Henri, 300
Candelpergher, B., 342
Cannarsa, P., 289
Cantrell, R.S., 319
Capasso, Vincenzo, 312
Caritisaui, Magdalena, 332
Carruthers, D.J., 331
Carter, H.W., 306
Carter, Pat, 294
Carter, Richard, 290
Carter, W., 333
Castillo, José E., 346
Castro-Leon, E., 296
Castro, Peter E., 314
Caviglia, Giacomo F., 337
Cawley, Robert, 294
Cecchini, R., 286
Censor, Yair, 308
Cercignani, Carlo, 298
Chajahis, E., 332
Chalot, Frédéric L., 146
Chan, S.Y.P., 325
Chang, Rosemary E., 305
Chapel, Francois, 310
Chen, Gui-Qiang, 338
Chen, Hudong, 327
Chen, L.-Y., 284, 314
Chen, Min, 329
Chen, Shiyi, 327
Chen, Yi Ying, 314

Cheney, Margaret, 308
Cheverda, V.A., 310
Chin, P., 323
Chin, Ray C.-Y., 326
Chow, Pao-Liu, 281, 290
Chung, K.Y., 301
Clements, Kevin A., 322
Clenshaw, Charles W., 286
Cohen, Albert, 343
Cohen, Malcom J., 285
Coifman, Ronald R., 41, 343
Cointe, Raymond, 301
Cole, Julian D., 302, 342
Colella, David, 343
Colella, Philip, 328
Coleman, Thomas F., 333
Collings, J.B., 320
Colonius, Fritz, 294
Colton, David L., 308
Conn, Andrew R., 51, 333, 334
Constapel, Rainer, 296
Cook, L. Pamela, 302
Coriell, S.R., 314, 322
Corliss, George F., 344
Coron, Francois, 328
Cosner, G. C., 319
Costabel, Martin, 326
Cottet, Georges-Henri, 328
Coughran, Jr., W.M., 297
Cox, S.J., 290
Cryer, Colin W., 286
Cumberbatch, Ellis, 348
Cutler, Colleen D., 294
Cypher, Robert, 293

Dacunha-Castelle, Didier, 280
Dafermos, Constantine M., 338
Dahlquist, Germund, 328
Das, Raja, 321
Dassios, George, 337
Daum, F.E., 280
Davida, G.I., 285
Davis, Paul W., 322
Davis, Timothy A., 322
Dawson, Clint N., 71
Dawson, K.S., 306
D'Azevedo, E., 323
Dean, Anthony M., 348
Dean, Nathaniel, 287
Deconinck, Herman, 284
Delfour, Michel C., 290
Dendy, Joel E., Jr., 307
Deng, Youfan, 327
Dennis, Jr., J.E., 334
DePierro, Alvaro R., 308
Deprettere, E.F., 322
DeRose, Tony D., 305, 306
Désarménien, J., 292
Desbarats, Alexandre, 300
Deuflhard, Peter J., 326
DeVore, C.R., 297
DeVore, R.A., 338
Dewar, James, 322
Dey, S.K., 325

Diener, Francine, 342
Diener, Marc, 342
Dimsdale, B., 306
Di Sciuva, Marco, 317
Domich, P.D., 332
Dongarra, Jack J., 321
Donnusamy, Ravi, 321
Donoho, David L., 280
Doolen, Gary D., 327
Doorish, J. F., 341
Dorning, John J., 340
Dost, Sadik, 318
Dragan, Irinel, 332
Drake, John, 331
Du, Qiang, 338
DuBois, Thierry, 329
Duff, Iain S., 322
Duke, Richard, 292
Dupuis, P., 289

Eckhaus, Wiktor, 83, 342
Economou, Demetre J., 349
Eddy, William F., 280
Edelman, Paul, 292
Edelstein-Keshet, Leah, 320
Edgar, G.A., 294
Eich, E., 328, 335
Eickemeyer, John Scott, 306
Eijkhout, Victor, 323
Ellson, R., 314
Elman, Howard, 323
Ely, Jeffery, 344
Emad, Nahid, 321
Emeagwali, Philip C., 307
Engelbrecht, Juri, 318
Engl, Heinz W., 309, 312
Engler, Hans P., 302
Enright, Wayne H., 324
Erbay, H.A., 318
Erbay, S., 318
Escudero, L.F., 332
Eskin, G., 308
Ettouney, Mohammed, 294
Evans, Lawrence C., 288, 313

Fago, B., 302
Fairweather, Graeme, 325
Fajtlowicz, Siemion, 287
Fakultet, Leninie Gory, 309
Falcone, Maurizio, 289
Faridani, A., 309
Farin, Gerald E., 306
Fasano, Antonio, 99, 349
Fatemi, Emad A., 297
Fattorini, Hector O., 288
Feldman, Stuart I., 286, 287
Ferguson, David R., 312
Fernando, H.J.S., 331
Ferris, Michael C., 333
Ferry, David K., 297
Fichtner, Wolfgang, 297
Fife, Paul C., 313
Finnigan, Peter M., 304
Fischer, Bernd, 322

Fischer, Paul, 283
Flaherty, Joseph E., 326
Fleming, W.H., 290
Fogwell, Thomas W., 307
Foias, Ciprian, 329
Ford, Brian, 285
Fornberg, Bengt, 340
Forsyth, P.A., 323
Fortin, Michel, 290
Fortune, Steven, 305
Frankel, James L., 286
Frankl, Peter, 292
Frankowska, Helene, 289
Freedman, Herbert I., 319, 320
Freund, R.W., 322
Friedlander, Leonid, 308
Friedman, Avner, 288, 349
Frommer, Andreas J., 345
Fruchard, Augustin, 342
Führer, C., 335
Fulton, C.A., 335
Funaro, D., 329
Furtado, F., 300
Furzeland, Ronald M., 348
Fusaro, B. A., 299, 316

Gabriel, S.A., 333
Gamboa, Fabrice, 280
Garbey, Marc, 326
Gardner, Carl, 297
Gassiat, E., 280
Gatignol, Rene, 300
Gaunaurd, Guillermo C., 337
Geman, Stuart A., 308, 309
Gennings, C., 306
Georg, Kurt, 305
Gessel, Ira, 292
Ghosh, Kunal, 300
Gillogly, James J., 319
Gladwell, Graham M.L., 284, 308
Gladwell, Ian, 285, 324, 325
Glassey, Robert T., 340
Glicksman, M.E., 314
Glimm, James, 300, 327
Glowinski, Roland, 297
Goldberg, Mark K., 286
Goldenfeld, Nigel D., 314
Goldhirsch, I., 279
Golse, Francois, 298
Goodman, Jonathan, 338
Goodrich, J.W., 284
Gould, N.I.M., 51, 333, 334
Graham, Ronald L., 292
Graichen, C., 304
Gray, Samuel H., 310
Gray, William J., 296
Grebogi, Celso, 294
Greenbaum, A., 342
Greenberg, Leon, 335
Greengard, Claude A., 328
Greengard, Leslie F., 342
Griewank, Andreas, 290, 344
Gropengiesser, Frank, 328

# AUTHOR INDEX

Gropp, William D., 283, 326
Grosse, Eric, 297
Grötschel, Martin, 119
Grünbaum, Alberto, 308
Gu, Z., 300
Guignard-Spielberg, M., 332
Guillopé, Colette H., 302
Gunstensen, Andrew K., 327
Gustafson, S., 337
Gustavson, Fred G., 345

Haaren Retagne, E., 289
Haber, Robert B., 291
Haddow, J.B., 318
Hadouaj, Hichem, 318
Hager, William W., 318
Hague, Stephen J., 286
Hald, Ole H., 308
Hales, Alfred W., 292
Hamaguchi, S., 301
Hammarling, Sven J., 321
Han, S.-P., 332
Hanke, Michael, 324
Hanson, Doug, 310
Harabetian, Eduard, 338
Hardy, S.C., 314
Harker, Patrick T., 333
Harvey, A.P., 318
Hassan, A., 336
Hathaway, A.F., 304
Haubold, H.J., 341
Haug, Edward J., 328
Hedstrom, Gerald, 326
Heil, Christopher E., 343
Hein, S., 280
Heinreichsberger, O., 297
Heller, W., 343
Helmes, K., 290
Hemker, P.W., 284, 296
Henry, Wanda, 341
Hernandez, Diego Bricio, 280
Hestir, Kevin, 300
Hettich, Rainer, 289
Heunis, A., 279
Highnam, Peter, 345
Hill, David R., 344
Ho, Y.C., 279
Hodges, Dewey H., 317
Hoff, David, 338
Hoffman, Karla, 333
Hoffmann, C.M., 304, 305
Holden, Helge, 338
Hollister, R. A., 316
Holloway, James P., 340
Holtgren, Bior, 349
Horn, M.A., 316
Hrusa, William J., 302
Hsiao, George C., 326
Hsu, Chung-Hao, 300
Hsu, S.B., 320
Hughes, G., 333
Hughes, Thomas J.R., 146
Hunt, Brian R., 294
Hunt, J.C.R., 299, 331

Hutson, V.C.L., 319

Ingham, D.B., 284
Inselberg, A., 306
Isaacson, David, 308
Isaacson, E.L., 338
Isakov, V., 309
Ishimaru, A., 296
Israeli, Moshe, 329

Jackson, Kenneth R., 322, 325
Jerome, Joseph W., 297
Jespersen, Dennis, 283
Jiang, J.L., 318
Jiang, Tao, 293
Jiang, X., 308
Johnson, Forrester T., 298
Johnson, Russell A., 294
Johnsson, S. Lennart, 345
Jones, C.K.R.T., 337
Juncosa, M.L., 319

Kaiser, Rolf, 340
Kalogeropoulos, G., 289
Kan, Pui Tak, 338
Kannan, Sampath, 293
Kaper, Hans G., 331, 337
Kapila, Ashwani K., 282
Karasick, Michael, 305
Karatzas, Ioannis, 290
Karcanias, Nicholas, 289
Karmarkar, Narendra K., 160
Karniadakis, G., 300
Karpouzian, Gabriel, 317
Kaufman, Linda, 308
Kawashima, Shuichi, 300
Kearfott, R. Baker, 344, 345
Keeler, S.P., 312
Keller, Joseph B., 342
Kelley, C.T., 290, 333
Kerkhoven, Thomas, 297
Keunings, Roland, 302
Keyes, David E., 283
Khargonekar, P.P., 288
Khodja, Mohamed, 338
Kikuchi, Noboru, 290
Killough, J.E., 307
King, John R., 312
Kirsch, Andreas, 309
Klein, Wolfram, 345
Kleinman, R., 308
Kliemann, Wolfgang H., 294
Kloeden, Peter E., 280
Klosek, M.M., 279
Knightly, G.H., 325
Knupp, Patrick, 346
Kobayashi, Ryo, 314, 322
Koepfler, G., 343
Kolkka, Robert W., 302
Kolmanovskii, Vladimir, 288
Koren, B., 284
Kowalski, A.D., 286
Krasny, Robert, 327
Krause, Egon, 300

Kress, Rainer, 309
Kriman, A.M., 297
Krishnamoorthy, Mukkai, 287
Krishnasamy, G., 326
Kristensson, Gerhard S., 308, 310
Kuang, Yang, 319
Kuerten, Hans G.M., 283
Kuiken, Hendrik K., 349
Kulisch, Ulrich W., 345
Kumar, S., 300
Kunisch, K., 288
Kushner, E.J., 296
Kushner, Harold J., 289
Kuske, Rachel, 279
Kwong, Man Kam, 337

Ladde, Gangaram S., 281
Lagnese, John E., 288
Lailly, Patrick, 310
Lambertson, Roland H., 299
Laminie, Jacques, 283
Lamour, Rene, 324
Langenberg, Karl J., 309
Lardner, Robin, 284, 290
Larsen, Jesper, 316
Lasiecka, Irena, 288
Latorre, Victor, 348
Lawler, Eugene L., 293
Lawless, Fiona, 341
Lawniczak, Anna T., 327
Layne, James Daniel, 344
Le Dimet, F.X.L., 290
Le Tallec, P., 327
Leaf, G.K., 282
Ledder, Glenn W., 282
Lee, E. Bruce, 288
Lee, H., 332
Lee, H.S., 291
LeFloch, P., 338
Leighton, T., 293
Leimkuhler, B.J., 328
Leipnik, Roy, 299
Leister, G., 335
Lele, S.K., 299
Lenhart, Suzanne M., 319
Lentini, Marianela, 324, 348
Leugering, Gunther, 288
Levermore, C.D., 327
Levit, Creon, 283
Levner, E., 332
Lewitt, Robert M., 308
Li, Ming, 293
Lia, C., 280
Liao, G., 346
Librescu, Liviu, 317
Lieber, D., 305
Linden, Johannes, 283
Lindquist, W. Brent, 300
Lions, Pierre-Louis, 182
Litko, J., 333
Liu, Chen-Huei, 300
Liu, F., 314
Liu, J., 333
Liu, Y., 326

# AUTHOR INDEX

Lobry, Claude, 342
Logan, J. David, 282
Logan, J.C., 296
Lohner, R.J., 297, 345
Long, Jane C.S., 300
Longman, Richard W., 289
Loo, Sao-Bing, 292
Lopez, C., 343
Lorentz, Jens, 301
Lortz, Dietrich, 340
Louis, A.K., 309
Lovass-Nagy, Victor, 289
Lozier, Daniel W., 286
Lucas, Robert F., 322
Lucier, Bradley J., 338
Lucks, M., 285
Lund, John R., 337
Lustig, I., 333

Ma, Y., 293
Macken, Catherine, 319
Mackens, Wolfgang G., 346
Macleod, Bruce, 287
MacNeal, Richard H., 329
Maggs, Bruce, 293
Mahrenholtz, O., 302
Malburne, K., 314
Maldarelli, Charles, 313
Mallett, R.L., 314
Malmuth, N.D., 302
Manbeck, Kevin M., 308, 309
Manley, Oscar, 329
Manni, Carla, 349
Manoranjan, Valipuram S., 320
Maradudin, Alexei A., 318
Marcati, Pierangelo, 338
Marchesin, D., 338
Marchi, Ezio, 332
Margaliot, Z., 316
Margolis, Stephen B., 282
Marion, Martine, 328
Marletta, Marco, 335
Marsden, Jerrold E., 340
Marsh, Steven P., 314
Marsten, Roy E., 333
Martin, O., 314
Maruo, Hajime, 301
Marvasti, Farokh A., 280
Maryak, John L., 280
März, Roswitha, 324
Mas-Gallic, Sylvie, 328
Massard, Thierry, 313
Mathai, A.M., 341
Matkowsky, Bernard J., 279, 282
Matthaeus, William H., 327
Maugin, Gerard A., 318
Mavriplis, Dimitri, 284, 321
Mayer, Andreas P., 318
Mayer, Guenter A., 345
McArthur, Kelly M., 337
McCaughrin, E., 287
McClure, Donald E., 308, 309
McCormick, G.P., 332
McCormick, S., 284

McFadden, G.B., 314, 322, 342
McKelvey, Robert W., 299
McLaughlin, Joyce R., 308
Mechentel, Ali, 345
Medina, R.L., 310
Mei, Chiang C., 300
Melnik, R.E., 299
Melvin, Robin G., 298
Menaldi, Jose L., 289, 290
Merrill, Stephen J., 319
Mertus, John, 308, 309
Meurant, Gérard A., 297, 323
Meyer-Spasche, Rita, 340
Meyer, Y., 41
Meyers, Gene, 293
Mickens, Ronald E., 324, 325
Miel, George J., 322
Mihail, Milena, 293
Milenkovic, Victor, 305
Miloh, Touvia, 301
Mimura, M., 196, 313
Miura, Robert M., 342
Molenaar, H., 296
Moler, Cleve, 286
Moll, R.N., 287, 305
Monk, P.B., 308, 337
Moonen, Marc, 322
Morro, Angelo C., 337
Morawetz, C.S., 302
Moré, Jorge J., 290
Morel, J.M., 343
Morris, B.C., 316
Morrison, P.J., 340
Morton, Bill, 283
Murray, B.T., 313
Murray, J.D., 212
Murray, Walter, 333
Mynett, Arthur E., 283

Nachman, Adrian L., 308
Nachtigal, N. M., 322
Nackman, L., 305
Nagpal, K.M., 288
Namachchivaya, Sri, 294
Narayanan, Ranga, 313
Nash, Stephen G., 332
Nashed, M.Z., 280, 285
Natterer, Frank, 309
Navon, Michael I., 290
Neaga, Michael, 345
Nedelec, Jean C., 326
Neidinger, R.D., 344
Neish, A., 299
Neithammer, Wilhelm E., 322
Nerurkar, Mahesh, 294
Neumaier, Arnold, 287
Neunzert, Helmut, 327, 349
Nevanlinna, O., 325
Ng, Kam-Chuen, 349
Nier, Francis, 297, 328
Nishiura, Yasumasa, 313
Nocedal, Jorge, 333
Nohel, John A., 302
Noor, Ahmed K., 346

Norbury, John, 312
Norris, Andrew, 318
Noskowicz, S.H., 279
Nosov, Valerii, 288
Novak, Erich, 285
Nowack, Robert L., 310
Numrick, Susan K., 327

Ockendon, John R., 314
O'Hara, Kathleen, 292
Odeh, F.M., 297
Ohta, Takao, 314
Olagunju, David O., 282
Olmstead, W.E., 342
Olson, R., 296
Olver, Frank W.J., 286
O'Malley, Robert E., Jr., 342
Omojokun, E., 334
Oono, Yoshitsugu, 314
Orszag, Steven A., 300
Osher, S., 297
Otter, A., 302
Ozen, S., 300

Padberg, M., 333
Päivarinta, Lassi, 309
Palazotto, Anthony N., 317
Palmeira, C.F., 338
Paluszny, M., 348
Pancer, R., 322, 325
Pande, C.S., 314
Pang, Jong-Shi, 333
Paprzycki, Marcia, 325
Paris, R.B., 341
Parker, David F., 318
Parthasarathy, V.N., 304
Pascal, F., 283
Pastoriza, R.A., 348
Patrinos, Ari, 331
Patterson, R.R., 348
Pegna, J., 312
Pelz, Richard B., 283, 300
Pereira, Felipe, 300
Perelson, Alan S., 319
Périaux, Jacques, 297
Perthame, Benoit, 328
Perucchio, Renato, 304
Petersen, Wesley P., 280
Petitjea, B., 297
Petiton, Serge, 321
Petzold, L.R., 324, 328
Phillips, C.G., 301
Pilant, Michael, 309
Pinsky, Mark, 294
Pinto, M.R., 297
Pironneau, Olivier, 290, 298
Pitkaranta, Juhani, 329
Platen, E., 280
Plaxton, Gregory, 293
Plohr, Bradley, 338
Pothen, Alex, 322
Potra, Florian A., 328
Pouget, Joel, 318
Poupaud, Frederic, 296, 298

# AUTHOR INDEX 355

Powers, D.L., 289
Primicerio, Mario, 349
Protopopescu, V., 319
Pruess, Steven A., 326, 335
Pruss, A., 316
Pryce, John D., 335
Pugh, A.C., 289
Pulvirenti, Mario, 298, 328

Quadrat, Jean-Pierre, 289
Quarteroni, Alfio, 327
Quint, Thomas, 332
Quintas, L., 332

Rall, Louis B., 344
Ralph, Daniel, 333
Ralston, James V., 308
Ramachandran, K. M., 281
Rao, A., 316
Rascle, Michel, 338
Rasmussen, H., 316
Reddy, J.N., 317
Reid, Ian, 286
Reid, John K., 285
Reinoza, A., 348
Remmel, Jeffrey, 292
Renegar, James M., 285
Rentrop, Peter, 328
Reyna, Luis G., 301
Rheinboldt, Werner C., 328, 346
Rice, John R., 285
Rich, Lawrence C., 344
Riesenfeld, Richard F., 304
Rikte, S., 308
Ringhofer, Christian A., 296, 297
Risebro, Nils Henrik, 338
Rishel, Raymond W., 290
Ritman, E.L., 309
Rivenburg, R., 286
Rizzo, Frank J., 326
Roache, P., 346
Roberts, E., 316
Rockway, J.W., 296
Rockwood, Alyn P., 305
Rödl, Vojtech, 292
Roehrkasse, Robert, 333
Rogers, J., 332
Rokhlin, Vladimir, 342
Rood, R.B., 331
Rosen, Christer, 301
Rosenberg, Arnold L., 293
Ross, David S., 349
Ross, M., 290
Rostamian, Rouben, 318
Rothman, D.H., 327
Rubinstein, Jacob, 313
Ruget, Gabriel, 227
Rundell, William, 309
Rusak, Z., 302
Russell, David L., 308
Russell, L.C., 296
Russo, M.F., 287
Rutledge, J.M., 307
Rycroft, Michael J., 331

Ryzhov, O.S., 302

Sacco, E., 317
Sachs, Ekkehard W., 333
Sadek, Ibrahim S., 317
Sallas, William M., 280
Salvador, M., 348
Santosa, Fadil, 308, 318
Sapidis, Nickolas, 304
Sargent, Roger W.H., 348
Saertenaer, Annick, 334
Sattinger, David, 349
Saut, Jean-Claude, 302
Saxena, Mukul, 304
Schatzman, Michelle V., 313
Schempp, Walter, 343
Scherzer, Otmar, 309
Schiehlen, Werner O., 335
Schilders, Willy H.A., 296
Schleiniger, Gilberto, 302
Schmeiser, Christian, 296
Schmidt, Geert J., 284
Schmidt, Rudiger, 317
Schmidt, Stanley, 345
Schramm, Dieter, 335
Schumacher, Gunter, 345
Schuss, Zeev, 279
Schwandt, Hartmut, 346
Schwierz, G., 309
Scott, David S., 296
Scriven, L.E., 348
Scroggs, Jeffrey S., 327
Seader, J.D., 348
Segel, L.A., 319
Seidel, Hans-Peter, 305
Seidman, Thomas I., 296
Seier, E., 348
Sekerka, R.F., 314
Selberherr, Siegfried, 297
Sell, George R., 329
Selleck, M.E., 313
Semenzato, Roberto M., 348
Sengupta, G.X., 298
Serebrisky, J., 348
Serov, Valeri Sergeevich, 309
Sethian, James A., 313
Seube, N., 316
Sezginer, Apo, 308
Shampine, Lawrence F., 324
Shankar, Vijaya, 297
Shanno, D., 333
Shannon, Gregory E., 286, 287
Sharp, Philip W., 324
Shaw, J.K., 341
Shearer, J., 345
Shearer, James, 338
Shen, Jie, 284
Sherman, Andrew, 323
Shinozuka, Masanobu, 294
Shoosmith, John N., 325
Shtessel, Yu., 332
Shu, Chi-Wang, 297
Silver, D., 300
Simeon, B., 335

Simion, Rodica, 292
Simitses, George J., 317
Simmons, Adrian, 331
Simonovits, Miklos, 293
Sinha, Somdatta, 320
Skeel, Robert D., 325
Skiena, Steven S., 287
Slattery, Rhonda, 289
Slemrod, M., 300
Sloss, J.M., 317
Smith, B.F., 326
Smith, Brian T., 285
Smith, F.T., 299
Smith, H.L., 319
Smith, J.E., 348
Smith, K.T., 309
Smith, Lauren L., 286
Smith, Peter, 336
Smith, R.C., 337
Smyshlyaev, Valery P., 316
So, J.W.H., 320
Sofer, Ariela, 332
Sokolowski, Jan, 290
Somersalo, Erkki, 308, 309
Sonar, T., 283
Song, Y., 284
Soni, Bharat, 346
Sosa, W., 348
Soulon, Robert, 314
Souza, Marco Antonio, 317
Spall, James C., 280
Spencer, Thomas, 287
Spies, G.O., 340
Spigler, R.G.C., 280, 304
Sprekels, J., 290
Sritharan, S.S., 288, 346
Srivastava, H.M., 341
St. Mary, D.F., 325
Stanton, D., 292
Starke, Gerhard, 322
Steen, P.H., 284
Steinbach, Marc C., 289
Steinberg, Stanly, 346
Stenger, Frank, 337
Stern, David P., 331
Sternberg, P., 313
Stetter, Hans J., 287
Stiftinger, Martin, 297
Stojanovic, Srdjan, 319
Stoorvogel, A.A., 288
Stott, Brian, 322
Stoughton, T., 312
Strickland, S., 279
Strom, Staffan E., 310
Struckmeier, Jens, 328
Sulem, A., 289
Sullivan, Francis, 305
Summers, Jonathan, 336
Sun, Jr., W., 288
Sundaram, Sheila, 292
Suri, Manil, 329
Surry, P., 316
Sylvester, John, 308
Symes, William W., 308, 310

Tadmor, Eitan, 338
Tajchman, Marc, 302
Takác, Peter, 337
Talagrand, O., 290
Talay, Denis, 280
Talwar, S., 294
Tam, H., 325
Tan, Dechun, 338
Tang, Tao, 284
Tang, Wei P., 323
Tayler, Alan B., 312, 349
Taylor, W., IV, 327
Taylor, Thomas J.S., 294
Temam, R., 328, 329
Temple, J. Blake, 338
Terzopoulos, Demetri, 304
Teter, M.P., 314
Thepaut, J., 290
Tijhuis, Anton G., 310
Tijs, Stef, 332
Ting, Lu, 300
Ting, T.C.T., 318
Titi, E., 284
Toint, Philippe L., 51, 333, 334
Tong, Zhang, 338
Toscani, Giuseppe, 300
Tragardh, Christian, 301
Traub, Joseph F., 285
Trease, Harold E., 345
Trentelman, H.L., 288
Triggiani, Roberto, 288
Tromp, John, 293
Troy, William C., 337
Tryggvason, Gretar, 300
Tsai, Stephen W., 313
Tubaro, L., 280
Tulin, Marshall P., 301
Tupholme, G.E., 318
Turner, Peter R., 286
Tzavaras, Athanassios E., 338

Van den Berg, I.P., 342
van den Berg, P.M., 308
van der Burgh, A.H.P., 302
van der Maarel, H.T.M., 284
van Duijn, H.J., 312
Van Dyke, M., 342
Van Hulzen, Hans A., 287
Vandewalle, J., 322
Vasil'eva, Adelaida B., 342
Vasquez, Juan L., 337
Vazirani, Umesh, 293
Vazirani, Vijay, 293
Verchery, Georges, 313
Verner, Jim H., 324
Vidal, C., 291
Virgin, Lawrence N., 336
Vogelius, Michael, 308
Vogels, M.E.S., 283
Voorhees, P.W., 314
Vouk, Mladen A., 286

Wachs, Michelle, 292

Wagner, David, 338
Walker, Homer F., 333
Wall, David J., 310
Wallace, David J., 241
Walter, G.G., 280, 285, 343
Walter, Wolfgang V., 345
Waltman, Paul E., 320
Wan, Fred, 299
Wan, Yieh-Hei, 340
Wang, Y., 326, 327
Ware, K., 333
Warnow, Tandy J., 293
Wegner, J.L., 318
Wei, C.Z., 279
Weichert, Dieter, 317
Weiland, S., 288
Weill, C., 321
Weinberger, Hans, 349
Weinmüller, Ewa, 324
Wempner, Gerald, 317
Wendland, W.L., 283, 326
Wesseling, Pieter, 283, 284
Westerberg, Arthur W., 348
Weston, Vaughan, 310
Wheeler, A.A., 313
Wheeler, Mary F., 71
Whelan, Tracy M., 305
Whisman, A., 333
Whitaker, Nathaniel, 327
White, Benjamin S., 281
White, L.W., 308
Wickerhauser, V., 41, 343
Williamson, David, 331
Williamson, Karen A., 334
Willis, J.R., 342
Witzgall, C., 332
Wolkowicz, Gail S.K., 320
Wollkind, David J., 320
Wood, A.D., 341
Wozniakowski, H., 285
Wright, Margaret H., 316
Wright, S.J., 290, 325
Wu, J., 319
Wulkow, Michael, 326
Wunsch, Carl, 290

Xin, Zhouping, 338
Xu, J.J., 314
Xu, Jinchao, 326

Yannakakis, M., 293
Yao, Yitao, 301
Ye, H., 340
Yen, Jeng, 328
Yin, George, 279, 281
Yip, Elizabeth L., 296
Yokoyama, E., 313
York, Bryant W., 305
Yorke, Jennifer M.E., 336
Young, D.P., 298
Young, Lawrence, 307
Young, Robin, 338
Yserentant, Harry, 256

Zabusky, Norman J., 300

Zakhuor, A., 280
Zaleski, S., 327
Zandbergen, Pieter J., 283
Zanetti, G., 327
Zayed, A.I., 280
Zeeman, M.L., 319
Zeilberger, Doron, 292
Zelevinsky, Andrei, 292
Zennaro, Marino, 324
Zettl, Anton, 335
Zhang, Qiang, 300
Zhang, Qing, 279
Zhiqi, Lu, 320
Zimmerman, R.W., 300
Zolésio, Jean-Paul, 290
Zou, X., 290
Zumbrum, Kevin, 338

# ATTENDEE LIST

Aavatsmark, Ivar
Norsk Hydro
Norway

Ababou, Rachid
Southwest Research Institute
United States

Abbott, John S.
Corning Incorporated
United States

Abdel-Aziz, Mohammedi, R.
Rice University
United States

Abdullah, Abdul R.
National University of Malaysia
Malaysia

Abell, Martha L.
Georgia Southern University
United States

Abou El-Seoud, Mohamed Samir
Sultan Qaboos University
Germany

Abrahams, Julia
United States

Aceves, Alejandro B.
University of New Mexico
United States

Achdou, Yves
Université Pierre et Marie Curie
France

Ache, Gerardo A.
Venezuela

Adali, Sarp
University of Natal
Central African Republic

Adam, John A.
Old Dominion University
United States

Adjerid, Slimane
Algeria

Afenya, Evans, K.
United States

Agarwal, Ramesh C.
United States

Agoshkov, Valeri
USSR Academy of Sciences
Russia

Ahluwalia, Daljit S.
New Jersey Institute of Technology
United States

Aiello, William
Bellcore
United States

Aitsahlia, Farid
Stanford University
United States

Akian, Marianne
INRIA
France

Akimou, Sadikou
University Center of CUNY
United States

Alabiso, Carlo
Universita di Parma
Italy

Al-Athel, Saleh Abdulrahman
Saudi Arabia

Albanese, Richard A.
United States

Aldama, Alvaro A.
Princeton University
United States

Aldridge, Christopher John
University of Oxford
Great Britain

Aldroubi, Akram
United States

Alefeld, Goetz E.
Germany

Alessandrini, Giovanni
Università de Trieste
Italy

Alevras, Dimitris
United States

Alexandrov, Natalia
Rice University
United States

Alhoori, Amatalelah A.
Howard University
United States

Al-Humadi, Ala
Embry-Riddle Aeronautical University
United States

Alishenas, Taifun
Royal Institute of Technology
Sweden

Allen, Douglas C.
Corning Incorporated
United States

Allen, Richard C., Jr
Sandia National Laboratories
United States

Almgren, Ann Stewart
University of California, Berkeley
United States

Alpert, Bradley K.
University of California, Berkeley
United States

Altman, Thomas
University of Colorado
United States

Altshuller, Dmitry A.
United States

Alvarado, Fernando L.
University of Wisconsin
United States

Amaziane, Brahim
University de Pau/de Pays de l'Adour
France

Ames, Allan E.
United States

An, Lianjun
McMaster University
Canada

Anderson, Dale A.
University of Texas
United States

Anderson, Edward
Cray Research
United States

Anderson, Johannes
Brunel University
Great Britain

Anderson, Paul B.
United States

Anderson, Ronald J.
Queen's University
Canada

Anderssen, Robert S.
CSIRO
Australia

Andrushkiw, Roman I.
New Jersey Institute of Technology
United States

Aneja, Y. P.
University of Windsor
United States

Angelos, James R.
Central Michigan University
United States

Annexstein, Fred S.
University of Cincinnati
United States

Ansorge, R.
Universität Hamburg
Germany

Anterion, Frederic
Saint Gobain Recherche
France

Antosik, Piotr
California Polytechnic State University
Poland

Argabright, Loren N.
Drexel University
United States

Arganbright, Deane E.
Whitworth College
United States

Arina, Renzo
Politecnico di Torino
Italy

Arnold, Anton
Purdue University
United States

Arnold, Ludwig
Universität Bremem
Germany

Arnow, David
Brooklyn College
United States

Arsham, Hossein
University of Baltimore
United States

Artola, Michel
CESTA
France

Asaithambi, N. S.
Mississippi State University
United States

Ascenzi, Maria-Grazia
United States

Ascher, Uri M.
University of British Columbia
Canada

Ashby, Steven F.
Lawrence Livermore National Laboratory
United States

Asic, Miroslav D.
Ohio State University
United States

Astfalk, Greg
Convex Computer Corporation
United States

Atallah, George C.
United States

Atkison, Frederick V.
University of Toronto
Canada

Atkinson, Kendall E.
University of Iowa
United States

Auer, Bruce M.
NASA Lewis Research Center
United States

Augenbaum, Jeffrey M.
United States

Avdeev, A.
Soviet Academy of Sciences
Russia

Averack, Brett M.
Argonne National Laboratory
United States

Averbuch, Amir
Tel Aviv University
Israel

Avila, John H., Jr.
United States

Babovsky, Hans
IBM Germany
Germany

Babuska, Ivo M.
University of Maryland
United States

Bachrach, Benjamin
United States

Badrinath, Vivek
INRS-Telecommunications
Canada

Bagchi, Arunabha
University of Twente
Netherlands

Bahar, Leon Y.
Drexel University
United States

Baheti, Kishan
National Science Foundation
United States

Bailey, John H.
United States

Bajzer, Zeljko
Mayo Clinic
United States

Baker, Timothy John
United States

Baker, William P.
United States

Ball, John
Heriot-Watt University
Scotland

Balla, Katalin
Hungarian Academy of Sciences
Hungary

Banerjee, Bappaditya
Purdue University
United States

Baral, Suresh Chandra
South Carolina State College
United States

Baraniecki, Anna
George Mason University
United States

Barcelo, Helene
University of Michigan
United States

Barcilon, Victor
University of Chicago
United States

Bardi, Martino
Università di Padova
Italy

Barenblatt, Grigorii
USSR Academy of Sciences
Russia

Barles, Guy Raymond
Université de Tours
France

Barlow, Jesse L.
Pennsylvania State University
United States

Barnes, David C.
Washington State University
United States

Barnhill, Robert E.
Arizona State University
United States

Barre, Guy Jerome
Elf Aquitaine
France

Barrera, Pablo
Mexico

Barrett, Andrew V.
United States

Barry, Marian R.
United States

Barshinger, Richard N.
Pennsylvania State University
United States

Barton, Russell R.
Pennsylvania State University
United States

Bassanini, Piero
Universita di Roma "La Sapienza"
Italy

Baumgartner, Anna
University of Alaska, Fairbanks
United States

Beale, J. Thomas
Duke University
United States

Becache, Eliane
CMAP, École Polytechnique
France

Bechtold, John K.
Princeton University
United States

Becker, Leif Eric
Harvard University
United States

Beckermann, Bernhard G.
University of Hannover
Germany

Beckers, Pierre Marie
University of Liege
Belgium

Bedard, Fernand D.
Department of Defense
United States

Bednar, J. Bee
Amerada Hess Corporation
United States

Beezley, Randell S.
United States

Beghelli, S.
University of Bologna
Italy

Beghi, Luigi
University of Padua
Italy

Behrman, William J.
United States

Beichl, Isabel M.
United States

Beiy, Valeriu
Polytechnic Institute of Bucharest
Romania

Belfort, Georges
Rensselaer Polytechnic Institute
United States

Bell, Carl
Philadelphia College of Science & Textiles
United States

Bellen, Alfredo
Italy

Belyaev, Alex
Prime Computers Inc.
United States

Benaroya, Haym
Rutgers University
United States

Benedetto, John J.
University of Maryland
United States

Bengu, Golgen
New Jersey Institute of Technology
United States

Benke, George
Mitre Corporation
United States

Bennett, Karin R.
University of Minnesota
United States

Benoit, Eric
École des Mines de Paris
Paris

BenSaad, Sihem
AT&T Bell Laboratories
United States

Bentil, Daniel E.
University of Washington
United States

Benzi, Michele
North Carolina State University
United States

Beretta, Elena
Rutgers University
United States

Berger, Marsha J.
RIACS/NASA Ames Research Center
United States

Berger, Mel S.
United States

Berger, Neil E.
University of Illinois
United States

Beris, Antony N.
University of Delaware
United States

Berkovich, Simon
Allied Signal ATC
United States

Berman, Zeev
United States

Bermejo, Rodolfo
Atmospheric Environment Service
Canada

Bernardi, Christine
Paris

Bernadou, Michel J.
INRIA
France

Bernoff, Andrew J.
Northwestern University
United States

Berry, Michael W.
University of Tennessee
United States

Berryman, Harry Scott
Yale University
United States

Berryman, James G.
Lawrence Livermore National Laboratory
United States

Bertsekas, Dimitri P.
Massachusetts Institute of Technology
United States

Best, John Philip
University of Birmingham
Great Britain

Betancourt, Ramon
San Diego State University
United States

Betten, Josef
Technical University Aachen
Germany

Bevis, Jean H.
Georgia State University
United States

Beydoun, Wafik B.
United States

Beylkin, Gregory
University of Colorado
United States

Bhattacharayya, P. K.
Polytechnic of Central London
Great Britain

Bhogeswara, Rao
United States

Biegler, L. T.
Carnegie-Mellon University
United States

Bieterman, Michael B.
United States

Billah, Khondakar Y. R.
Princeton University
United States

Birman, Victor
University of Missouri, Rolla
United States

Bischof, Christian H.
Argonne National Laboratories
United States

Bishop, E. R.
Acadia University
Canada

Bissett, Edward J.
General Motors Research Laboratories
United States

Bitman, William R.
United States

Bixby, Robert E.
Rice University
United States

Bjorstad, Petter E.
University of Bergen
Norway

Blackmore, Denis
New Jersey Institute of Technology
United States

Blake, John R.
University of Birmingham, Edgbaston
Great Britain

Blakley, Bob
Texas A&M University
United States

Bleher, Siegfried
McMaster University
Canada

Bleistein, Norman
United States

Bloemendaal, Mark A.
Technical University Delft
Netherlands

Bloom, Frederick
Northern Illinois University
United States

Blowey, James F.
University of Sussex
Great Britain

## 360 ATTENDEE LIST

Blum, Avrim
Carnegie-Mellon University
United States

Bock, Hans Georg
University of Augsburg
Germany

Bodenschatz, John S.
United States

Boffi, Vinicio C.
Italy

Boggs, Paul T.
National Institute of Standards & Technology
United States

Bohigian, Haig E.
United States

Boisvert, Ronald F.
National Institute of Standards & Technology
United States

Boland, Natashia Lesley
University of Western Australia
Australia

Bond, Dave Michael
United States

Booty, Michael R.
Southern Methodist University
United States

Borgers, Christoph
University of Michigan
United States

Boriek, Aladin M.
Baylor College of Medicine
United States

Boris, Jay P.
Naval Research Laboratory
United States

Bornemann, Folkmar
Konrad-Zuse-Zentrum
Germany

Bose, Sujit K.
University of Twente
Netherlands

Bosman, Cheryl L.
Exxon Production Research
United States

Botha, Johan
University of South Africa
United States

Boucher, Mike
United States

Boujot, Jean Paul
United States

Bouras, Abdelaziz
University Claude Bernard Lyon 1
France

Bourgin, Richard
National Bureau of Standards
United States

Bourguignon, Jean Pierre
École Polytechnique
France

Bourjolly, Jean Marie
Concordia University
Canada

Bowers, John C.
Food and Drug Administration
United States

Bowers, Kenneth L.
Montana State University
United States

Boyce, William E.
Rensselaer Polytechnic Institute
United States

Boyd, John P.
Institute for Marine & Costal Sciences
United States

Boyett, Barbara
Mobil Research & Development
United States

Braddock, Roger Davis
University of Maryland-Baltimore County
United States

Brady, Michael
Oxford University
Great Britain

Brand, Clemens
United States

Brandon, Deborah
Carnegie-Mellon University
United States

Brasic, James Robert
United States

Brenier, Yann
DMI-ENS
France

Brewer, Dennis W.
University of Arkansas
United States

Brewster, Mary E.
University of Colorado
United States

Bricheteau, Therese
INRIA
France

Bridgland, Michael F.
Supercomputing Research Center
United States

Brimacombe, J. Keith
University of British Columbia
Canada

Broder, Andrei Z.
DEC Systems Research Center
United States

Brodsky, Mikhail
University of California
United States

Bromley, D. Allen
Yale University
United States

Bronsard, Lia
Carnegie-Mellon University
United States

Brophy, John F.
United States

Brown, Melvin
United States

Brown, Peter N.
Lawrence Livermore National Laboratory
United States

Brown, Robert N.
Food and Drug Administration
United States

Bruno, Oscar
University of Minnesota
United States

Buaquina, Vaughn S.
Ohio University
United States

Bube, Kenneth P.
University of Washington
United States

Buchanan, Mark L.
United States

Budaev, B. V.
Steklov Mathematical Institute
United States

Buettiker, Markus
IBM, T. J. Watson Research Center
United States

Bugarin, Mauricio Soares
University of Brasilia
Brazil

Bukiet, Bruce G.
New Jersey Institute of Technology
United States

Bulens, Pierre Victor
Laborelec
Belgium

Bulgarelli, Jlderico
INSEAN
Italy

Bullivant, David P.
University of Auckland
New Zealand

Bulot, Christophe
CEA - CESTA
France

Bumiller, Carl L.
United States

Bungartz, Hans Joachim
Technical University Muenchen
Germany

Burke, David P.
United States

Burke, James V.
University of Washington
United States

Burley, David M.
University of Sheffield
Great Britain

ATTENDEE LIST  **361**

Burns, John A.
Virginia Polytechnic Institute and State University
United States

Burns, Timothy J.
National Institute of Standards and Technology
United States

Burr, Stefan A.
City College CUNY
United States

Burrage, Kevin
University of Queensland
Australia

Burridge, Robert
Schlumberger-Doll Research
United States

Burton, Thomas
Washington State University
United States

Busenberg, Stavros N.
Harvey Mudd College
United States

Bush, George C.
Middle East Technical University
Turkey

Butcher, John C.
University of Auckland
New Zealand

Butler, Lynne M.
Haverford College
United States

Button, Leslie J.
Corning Incorporated
United States

Byrd, Richard H.
University of Colorado
United States

Byrne, George D.
Exxon Research & Engineering Corporation
United States

Byrne, Helen
University of Oxford
Great Britain

Cabannes, Henri
University of Paris 6
France

Caden, Martin James
University of Technology/Sydney
Australia

Cady, Ralph
United States

Cai, Damu
University of Wisconsin/Madison
United States

Cai, Wei
University of North Carolina
United States

Caldas, Ibere Luiz
Universidade de Sao Palo
Brazil

Caldwell, Ginger A.
NCAR/SCD
United States

Camassa, Roberto
United States

Cambel, Ali
United States

Campana, Emilio
INSEAN
Italy

Campos, Luis M. Braga da Costa
Instituto Superior Tecnico
Portugal

Canic, Suncica
SUNY at Stony Brook
United States

Canning, Francis X.
Rockwell Science Center
United States

Canright, David R.
Naval Postgraduate School
United States

Cao, Jianzhong
United States

Capasso, Vincenzo
Universita Di Milano
Italy

Carabelli, S.
Centro Ricerche Fiat
Italy

Cardoso, Domingos M.
Universidad de Aveiro
Portugal

Carr, Danielle D.
Duke University
United States

Carrere, Frederic
CEA/CESTA
France

Carrington, Walter A.
University of Massachusetts Medical School
United States

Carter, Bob
Boston University
United States

Carter, Patricia H.
United States

Carter, Richard G.
Argonne National Laboratory
United States

Casal, Alfonso C.
Universita Politecnica de Madrid
Spain

Casartelli, Mario
Università di Parma
Italy

Case, James H.
United States

Casey, Stephen D.
The American University
United States

Casper, Jay
United States

Castillo, José E.
San Diego State University
United States

Castro, Peter E.
United States

Castro-Leon, Enrique
Intel Scientific Computers
United States

Cavallaro, Joseph R.
Rice University
United States

Cavaterra, Cecilia
Università di Milano
Italy

Caviglia, Giacomo Franco
University of Genoa
Italy

Cawley, Robert
Naval Surface Warfare Center
United States

Censor, Yair
University of Haifa
Israel

Cercigniani, Carlo
Politecnico di Milano
Italy

Cerda de Groote, Carlos L.
United States

Chadam, John
McMaster University
Canada

Chahine, George
United States

Chajakis, Emmanuel D.
University of Pennsylvania
United States

Chamberland, Marc A.
University of Waterloo
Canada

Chan, Robert P. K.
University of Toronto
Canada

Chan, Tony F. C.
University of California
United States

Chan, Wai Leung
Chinese University Hong Kong
Hong Kong

Chandra, Jagdish
U.S. Army Research Office
United States

Chandrasekaran, Shivkumar
Yale University
United States

Chang, Chih Chen
United States

Chang, Ching Lung
Cleveland State University
United States

## 362 ATTENDEE LIST

Chang, I-Lok
American University
United States

Chang, Mou-Hsiung
University of Alabama/Huntsville
United States

Chang, Rosemary E.
Silicon Graphics Computer Systems
United States

Chang, Yu-Chung
United States

Chapel, Francois
SNEA/CSTCS
France

Chaplain, Mark A. J.
University of Bath
Great Britain

Charron, Richard J.
Memorial University of Newfoundland
Canada

Chattergy, Rahul
United States

Chau, Siu Cheung
University of Lethbridge
Canada

Chaussalet, Thierry J.
North Carolina State University
United States

Chavent, Guy
Maisons-Laffitte
France

Chen, Bin
Clemson University
United States

Chen, Gui-Qiang
University of Chicago
United States

Chen, Hong-Gang
University of Washington
United States

Chen, Hsin-Chu
University of Illinois
United States

Chen, Hudong
Dartmouth College
United States

Chen, K. C.
Texaco Corporation
United States

Cheney, Margaret
Rensselaer Polytechnic Institute
United States

Chen, Mei-Qui
The Citadel
United States

Chen, Min
Indiana University
United States

Chen, Pei-Li
University of Illinois/Carbondale
United States

Chen, Pey-Chun
Rutgers University
United States

Chen, Qi
University of Maryland
United States

Chen, Shiyi
Los Alamos National Laboratories
United States

Chen, Tsu-Fen
University of Texas at Arlington
United States

Chen, Xinfu
University of Pittsburgh
United States

Chen, Xiaojun
Ehime University
Japan

Chen, Yu
Institute for Advanced Study
United States

Cheng, Raymond S.
David W. Taylor Research Center
United States

Chengxian, Xu
Xi'an Jiaotong University
People's Republic of China

Chern, I-Liang
Argonne National Laboratories
United States

Chernesky, Michael P.
University of Maryland
United States

Cheverda, V.A.
Soviet Academy of Sciences
Russia

Chevrier, Pierre
Merlin Gerin
France

Chiareli, Alessandra O.P.
Northwestern University
United States

Chikwendu, Sunday C.
United States

Chin, Ray C.Y.
Indiana University-Purdue University
United States

Chin, Wai
University of Maryland
United States

Chisena, Frank L.
SUNY at Stony Brook
United States

Chiu, Chichia
Michigan State University
United States

Choe, Hi
United States

Choi, U. Jin
Korea Advanced Institute Science & Technology
South Korea

Chou, Pang-Chieh
Rice University
United States

Choudhury, Sudipto Roy
University of Central Florida
United States

Chouikha, Mohamed F.
United States

Chow, Pao-Liu
Wayne State University
United States

Christiansen, Soren
Technical University of Denmark
Denmark

Chung, Fan R. K.
Bell Communications Research
United States

Chung, Kun Yong
Rensselaer Polytechnic Institute
United States

Chu, C. K.
Columbia Univesity
United States

Chu, Moody T.
North Carolina State University
United States

Chu, Sydney C.K.
University of Hong Kong
Hong Kong

Ciment, Melvyn
United States

Claeyssen, Julio Cesar Ruiz
Federal University of Rio Grande do Sul
Brazil

Clark, Kenneth D.
United States

Claudio, Dalcidio M.
Federal University of Rio Grande do Sul
Brazil

Cleghorn, William L.
University of Toronto
Canada

Clements, John C.
Dalhousie University
Canada

Clements, Kevin A.
National Science Foundation
United States

Clemons, Curtis B.
University of Akron
United States

Cockburn, J. Bernardo
University of Minnesota
United States

Coffey, Mark W.
Iowa State University
United States

Cohen, Alan M.
University of Wales
College of Cardiff
Great Britain

Cohen, Gary C.
INRIA
France

Cohen, Herbert E.
United States

Cohen, Hirsh
Alfred P. Sloan Foundation
United States

Cohen, Malcolm
Numerical Algorithms Group Ltd.
Great Britain

Cohen, Michael P.
United States

Cointe, Raymond
Bassin D'Essais des Carenes
France

Colaneri, Patrizio
Politecnico di Milano
Italy

Cole, Julian D.
Rensselaer Polytechnic Institute
United States

Cole, Susan L.
Rensselaer Polytechnic Institute
United States

Colella, David
Mitre Corporation
United States

Coleman, Thomas F.
Cornell University
United States

Colonius, Fritz
Universität Augsburg
Germany

Colombo, Vittorio
Politecnico di Torino
Italy

Colton, David L.
University of Delaware
United States

Comiskey, Catherine M.
Dublin City University
Ireland

Conn, Andrew R.
IBM, Thomas J. Watson Research Center
United States

Conroy, John M.
Supercomputing Research Center
United States

Constapel, Rainer
University of Duisburg
Germany

Contreras, Martha P.
University of California
United States

Cook-Ioannidis, L. P.
University of Delaware
United States

Cooper, Jeffrey M.
University of Maryland
United States

Coppa, Gianni
Politecnico di Torino
Italy

Corduneanu, Constantin C.
University of Texas at Arlington
United States

Cores, Debora
Rice University
United States

Coriell, Sam R.
NIST
United States

Corliss, George F.
Marquette University
United States

Coron, Francois
Aerospatiale
France

Corrias, Lucilla
Citta Universitaria
Italy

Corwin, Thomas L.
Metron
United States

Cosner, George Christopher
University of Miami
United States

Costa, Oswaldo Luiz do Valle
Universidade De Sao Paulo
Brazil

Costa, Peter J.
United States

Costabel, Martin
Université Bourdeaux 1
France

Cottet, Georges-Henri
University of Grenoble
France

Coughran, William M., Jr.
AT&T Bell Laboratories
United States

Countryman, William Mark
Louisiana Technical University
United States

Courtney, Jack
United States

Cowsar, Lawrence C.
Rice University
United States

Cox, Steven J.
Rice University
United States

Crabtree, James B.
United States

Crane, Roger L.
David Sarnoff Research Center
United States

Crawford, Carol G.
U.S. Naval Academy
United States

Crispi, Guido
Osservatorio Geofisico Sperimentale
Italy

Croft, A. J.
Coventry Polytechnic
Great Britain

Crosby, Frank J.
NIST
United States

Crosta, Giovanni
State University of Milan
Italy

Crowley, James M.
United States

Crownover, Richard M.
University of Missouri
United States

Cryer, Colin W.
Westflische Wilhelm-Universitat
Germany

Cuer, Michel
Universite de Montpelier
France

Cumberbatch, Ellis
Claremont Graduate School
United States

Cuminato, Jose Alberto
ICMSC-USP
Brazil

Cutler, Colleen D.m
University of Waterloo
Canada

Cyphers, Robert
IBM Research Division
United States

Dabaghi, Fadi El
INRIA
France

Dabaghi, Zakia B. T.
Institute François du Pétrole
France

Dacunha-Castelle, Sidier
Université of Paris-Sud
France

Dadmehr, Shireen S.
United States

Dafermos, Constantine M.
Brown University
United States

Dahleh, Marie Dillon
University of California
United States

Danial, Edward J.
Kearney State College
United State

Danielson, Donald A.
Naval Postgraduate School
United States

Daripa, Prabir K.
Texas A&M University
United States

# ATTENDEE LIST

Das, Raja
ICASE
United States

Dassios, George
University of Patras
Greece

Daum, Frederick E.
United States

Davida, George
University of Wisconsin
United States

Davidson, Stuart
United States

Davis, Anthony M. J.
University of Alabama
United States

Davis, Chandler
University of Toronto
Canada

Davis, Paul W.
Worcester Polytechnic Institute
United States

Davis, Stephen F.
United States

Davis, Timothy A.
University of Florida
United States

D'Azevedo, Eduardo F.
Oak Ridge National Laboratory
United States

Dean, Anthony M.
Exxon Research & Engineering Company
United States

Dean, Nathaniel
Bellcore
United States

Debnath, Joyati
Winona State University
United States

Debnath, N. C.
Winona State University
United States

Debussche, Arnaud
Indiana University
United States

Decker, William J.
University of Virginia
United States

De Jong, Bartele
University of Twente
Netherlands

Delfour, Michel C.
Université de Montréal
Canada

DeLillo, Thomas K.
Wichita State University
United States

De Long, Michael A.
University of Virginia
United States

Dendy, Joel E., Jr.
Los Alamos National Laboratories
United States

Deng, Yuefan
SUNY at Stony Brook
United States

Denham, Steve C.
U.S. Food & Drug Administration
United States

Dennis, J. E., Jr.
Rice University
United States

Denny, Diane L.
United States

De Pierro, Alvaro R.
IMECC-UNICAMP
Brazil

de Pillis, Lisette G.
University of California, Los Angeles
United States

Deprettere, Ed F.
Delft University of Technology
Netherlands

Desbarats, Alexander J.
Geological Survey of Canada
Canada

Desch, Kristian
Indiana University
United States

Deuflhard, Peter J.
Konrad-Zuse-Zentrum fur Informationstechnik Berlin
Germany

Dewar, James A.
RAND
United States

Dey, S. K.
Eastern Illinois University
United States

Dhavakodi, Salai
University of Florida
United States

Dias, Frederic L.
Université de Nice
France

Diaz, Julio Cesar
University of Tulsa
United States

Diaz, Miguel Martin
University Complutense Madrid
Spain

Diaz, Nicole
United States

Dibos, Francoise
University of Paris/Dauphine
France

Dickey, J. Michael
United States

Diemer, Karen
Los Alamos National Laboratory
United States

Diener, Francine
University of Nice
France

Diener, Marc
University of Nice
France

Dijksetra, Douwe
University of Twente
Netherlands

Djikerman, Robert W.
INRS Telecommunications
Canada

Dikansky, Arnold
St. Johns University
United States

Dimsdale, Bernard
IBM Corporation
United States

Ding, Mingzhou
University of Maryland
United States

Dobson, David C.
University of Minnesota
United States

Doelman, Arjen
Cornell University
United States

Doll, Ferdinand A., Jr.
United States

Donaldson, James A.
Howard University
United States

Donato, June M.
United States

Dongarra, Jack J.
University of Tennessee
United States

Donoho, David Leigh
University of California
United States

Doolen, Gary Dean
United States

Doorish, John
United States

Dorning, John J.
University of Virginia
United States

Dorninger, Dietmar
University of Technology
Austria

Dorr, Milo R.
Lawrence Livermore National Laboratory
United States

Dost, Sadik
University of Victoria
Canada

Dougalis, Vassilios A.
National Technical University
Greece

Dougherty, David E.
University of Vermont
United States

Dougherty, Robert P.
United States

Douglis, Avron
University of Maryland
United States

Dozier, Lewis B.
SAIC
United States

Dragan, Irinel
University of Texas-Arlington
United States

Draghicescu, Cristina I.
University of Massachusetts
United States

Drake, John B.
United States

Drake, Richard
Washington State University
United States

Draper, Richard
United States

Drumheller, David Mark
Naval Research Laboratory
United States

Du, Qiang
Michigan State University
United States

Duan, Jin Qiao
Cornell University
United States

Dubois, Thierry
Université Paris
France

Duck, Kenneth I.
MRJ Inc.
United States

Duff, Iain S.
Rutherford Appleton Laboratory
Great Britain

Duke, Richard A.
Georgia Institute of Technology
United States

Dunn, Jeffrey H.
Naval Research Laboratory
United States

Dupois, Paul
University of Massachusetts
United States

Durack, Donald L.
New Mexico State University
United States

Duran, Marco
Exxon Research & Engineering Company
United States

Durany, Jose
Universidad de Vigo
Spain

Duraiswami, Ramani
Johns Hopkins University
United States

Dutt, Alok
Yale University
United States

Dutta, Nader
British Petroleum Company
United States

Dutto, Laura Cecilia
Universite Laval
Canada

Dwarka, Pankaj R.
Howard University
United States

Eargle, G. Marvin
United States

Easwaran, Chirakkal V.
SUNY College New Paltz
United States

Eberlein, Patricia J.
State University of New York
United States

Ecder, Ali
Yale University
United States

Eckert, Robert P.
United States

Eckhaus, Wiktor
University of Utrecht
Netherlands

Economou J. Demetre
University of Houston
United States

Eddy, William F.
Carnegie-Mellon University
United States

Edelman, Paul H.
University of Minnesota
United States

Edelstein-Keshet, Leah
University of British Columbia
Canada

Edgar, Gerald A.
Ohio State University
United States

Edge, Chad
Indiana University
United States

Edsberg, Lennart
Royal Institute of Technology
Sweden

Eena, Jako
Semmelwets University Medical School
Hungary

Eggert, Kenneth G.
Los Alamos National Laboratories
United States

Ehrenmark, U.T.
City of London Polytechnic
United States

Ehrlich, Louis W.
Johns Hopkins University
United States

Eich, Edda
Universitat Augsburg
Germany

Eickemeyer, John Scott
Information Technology Institute
Republic of Singapore

Eijkhout, Victor L.
University of Tennessee/Knoxville
United States

El-Bakry, Amr Saad
Rice University
United States

Eldridge, Benjamin
Alliant Techsystems
United States

Eleshaky, Mohamed E.
Old Dominion University
United States

Elizondo, Rodolfo
Rice University
United States

El Jai, Abdelhaq
IMP-University of Perpignan
France

Ellacott, Stephen
Brighton Polytechnic
Great Britain

Ellermeier, Wolfgang F.
Technische Hochschule
Germany

Ellson, Richard
Eastman Kodak Company
United States

Elman, Howard C.
University of Maryland
United States

El-Owaidy, Hassan Mostafa
Al-Azhar University
Egypt

Elster, Anne C.
United States

Ely, Jeffrey S.
Lewis and Clark College
United States

Emad, Nahid
École Nationale de Telecoms
France

Emeagwali, Philip C.
United States

Emmerling, F. A.
University Federal Armed Forces/Munich
Germany

Enenkel, Robert Frederick
University of Toronto
Canada

Engel, Mark
University of Regina
Canada

Engl, Heinz W.
Johannes-Kepler-Universitaet
Austria

Engler, Hans P.
Georgetown University
United States

Enright, Wayne H.
University of Toronto
Canada

Epperson, James F.
University of Alabama-Huntsville
United States

Eschenauer, Hans Alex
University of Siegen
Germany

Escudero, Laureano F.
Spain

Ettouney, Mohammed
Weidlinger Associates
United States

Evans, Lawrence
University of California at Berkeley
United States

Evans, Richard B.
United States

Eyre, David Jay
University of Utah
United States

Fabrie, Pierre
Universite Bordeaux I
France

Fadimba, Koffi B.
University of South Carolina
United States

Fahy, Edward J.
United States

Fajtlowicz, Siemion
University of Houston
United States

Fakhroo, Fariba
University of Southern California
United States

Falcone, Maurizio
Universida di Roma "La Sapienza"
Italy

Falgout, Robert D.
Lawrence Livermore National Laboratories
United States

Falk, Richard S.
Rutgers University
United States

Falkner, Julie Carolyn
University of Waterloo
Canada

Falzarano, Jeffrey M.
United States

Fan, Kaisheng
Iowa State University
United States

Fang, Ming
University of Washington
United States

Faridani, Adel
Oregon State University
United States

Farin, Gerald E.
Arizona State University
United States

Fasano, Antonio
University of Florence
Italy

Fatemi, Emad A.
University of Minnesota
United States

Faybusovich, Leonid
University of Notre Dame
United States

Feehan, Paul M. N.
Columbia University
United States

Feldman, Stuart I.
Bellcore
United States

Feng, Wei
University of North Carolina, Wilmington
United States

Ferguson, David R.
United States

Ferreira, Ubirajara R.
UNESP-FEG-DPD
Brazil

Ferretti, Roberto
University di Roma La Sapienza
Italy

Ferris, Michael C.
University of Wisconsin
United States

Fibich, Gadi
New York University
United States

Fife, Paul C.
University of Utah
United States

Figotin, Alexander
University of North Carolina-Charlotte
United States

Fiiedlander, Leonid
University of California, Los Angeles
United States

Filipowski, Sharon K.
United States

Filippopoulos, Nikos
Harvard University
United States

Finn, John McMaster
University of Maryland
United States

Fischer, Bernd
University of Hamburg
Germany

Fischer, I.
Naval Postgraduate School
United States

Fischer, Paul F.
Brown University
United States

Fishman, Louis
United States

Flaherty, Joseph E.
Rensselaer Polytechnic Institute
United States

Fleming, Henry E.
United States

Fleming, Wendell H.
Brown University
United States

Fletcher, Charles W.
United States

Fletcher, John E.
United States

Florens, Danielle
University Paris IX
France

Flores, Jose D.
University of South Dakota
United States

Fogwell, T. W.
G M D
United States

Foka, Rigobert
Thomson-SINTRA-ASM
France

Folk, Robert T.
Lehigh University
United States

Foote, Harlan
Battelle Pacific Northwest Laboratory
United States

Forbes, Alistair B.
National Physical Laboratory
Great Britain

Ford, Brian
NAG Ltd.
Great Britain

Fornberg, Bengt
Exxon Research & Engineering Company
United States

Forney, Glenn P.
United States

Forsyth, Peter A.
University of Waterloo
Canada

Fortin, M.
Université Laval
Canada

Fortune, Steven
AT&T Bell Laboratories
United States

Foster, Clint
SDSMT
United States

Foster, D. J.
Arco Oil & Gas Company
United States

Foster, Ian
Argonne National Laboratory
United States

Frank, Paul D.
United States

Frankel, James L.
Thinking Machines Corporation
United States

Frankel, Paul Henry
National Institute of Health
United States

Frankowska, Helene
University de Paris-Dauphine
France

Freedman, Herbert I.
University of Alberta
Canada

Frenkel, Alexander L.
University of Alabama
United States

Frenzen, Christopher L.
Naval Postgraduate School
United States

French, Donald A.
University of Cincinnati
United States

Freund, Roland W.
NASA Ames Research Center
United States

Friedman, Avner
University of Minnesota
United States

Friesz, Terry
George Mason University
United States

Frigessi, Arnoldo
Rutgers University
United States

Friske, Mel
Wisconsin Lutheran College
United States

Frommer, Andreas
Universität Karlsruhe
Germany

Frosali, Giovanni
University of Ancona
Italy

Fruchard, Augustin
France

Fuehrer, Claus
German Aerospace Research
Germany

Fujino, Seiji
Institute of Computational Fluid Dynamics
Japan

Fulton, Charles T.
Florida Inst of Technology
United States

Funaro, Daniele
University of Pavia
Italy

Furzeland, Ronald
Shell Research B V
Netherlands

Fusaro, B. A.
Salisbury State University
United States

Fusco, Domenico
University of Messina
Italy

Galeone, Luciano
University of Bari
Italy

Galperin, Anatoly
Israel Aircraft Industries
Israel

Gamboa, Fabrice
Université of Paris-Sud
France

Gao, Linda Q.
University of Iowa
United States

Gao, Yang
Harvard University
United States

Garbey, Marc
Hameau De Cromey
St. Sernin Du Plain
France

Garcia, Sonia M. F.
United States

Gardner, Carl L.
Duke University
United States

Gardner, Robert B.
University of North Carolina
United States

Garrett, Tylene S.
Transylvania University
United States

Gartland, Eugene C., Jr
United States

Gary, John M.
NIST
United States

Gassiat, Elizabeth Granier
Université of Paris-Sud
France

Gates, Peter
Sterling Drug Inc.
United States

Gatignol, Renee
University Pierre et Marie Curie
France

Gaunaurd, Bill
Naval Surface Warfare Center
United States

Gavage, Sylvie
University Bernard Lyon 1
France

Gear, C.W.
NEC Research Institute
United States

Geist, Bruce
United States

Geman, Stuart
Brown University
United States

Gennings, Chris
Medical College of Virginia
United States

Georg, Kurt
Colorado State University
United States

Gerber, P. Dean
IBM Corporation
United States

Gervasio, Camille
Rutgers University
United States

Gessel, Ira
Brandeis University
United States

Gheri, Giovanni
University of Pisa
Italy

Ghoreishi, Afshin
Bowdoin College
United States

Giga, Mariko
Nippon Medical School
Japan

Giga, Yoshikazu
Hokkaido University
Japan

Gilbert, Jean Charles
INRIA
France

Gilbert, John R.
Xerox
United States

Gilbert, Robert P.
University of Delaware
United States

Giles, James E.
United States

Gill, Tepper L.
Howard University
United States

Gillis, James T.
United States

Gimse, Tore
University of Oslo
Norway

Gladwell, Graham M. L.
University of Waterloo
Canada

Gladwell, Ian
Southern Methodist University
United States

Glassman, Neal
United States

Glassey, Robert T.
Indiana University
United States

Glimm, James
SUNY at Stony Brook
United States

Glowinski, Roland
University of Houston
United States

Goheen, Christopher H.
United States

Goldberg, Irwin S.
Saint Mary's University
United States

Goldak, John
Carleton University
Canada

Goldberg, Mark K.
Los Alamos National Laboratory
United States

Goldenfeld, Nigel
University of Illinois
United States

Goldring, Tom
N.C. Supercomputing Center
United States

Golin, Simon
Wissensch Nordrhein-Westfalen
Germany

Golub, Gene H.
Stanford University
United States

Gonzalez-Lima, Maria
Rice University
United States

Gooden, Donald K.
Pennsylvania State University
United States

Goodman, Jonathan B.
New York University
United States

Goodrich, John
NASA Lewis Research Center
United States

Gorenstein, Marc V.
Millipore Corporation
United States

Gottlieb, Oded
United States

Gowda, Muddappa S.
University of Maryland
United States

Graham, Ronald L.
AT&T Bell Laboratories
United States

Grasselli, Maurizio
Rutgers University
United States

Gravel, Pierre
College Militaire de St. Jean
Canada

Gray, Samuel H.
Amoco Production Corporation
United States

Grebogi, Celso
University of Maryland
United States

Greenberg, James M.
University of Maryland-Baltimore County
United States

Greenberg, Leon
University of Maryland
United States

Greengard, Claude A.
IBM T.J. Watson Research Center
United States

Greengard, Leslie F.
Courant Institute of Mathematical Sciences
United States

Gressang, Randall V.
United States

Griebel, Michael
Technical University Muenchen
United States

Griewank, Andreas
United States

Griffiths, J. D.
University of Wales
Great Britain

Groenenboom, Albert
Hengelo
The Netherlands

Gropengiesser, Frank
University of Kaiserlautern
Germany

Gropp, William D.
Argonne National Laboratory
United States

Grosse, Eric
AT&T Bell Laboratories
United States

Grossman, George
Central Michigan University
United States

Grötschel, Martin
Konrad-Zuse-Zentrum fur Informationstechnik Berlin
Germany

Grove, John W.
SUNY at Stony Brook
United States

Gruis, Leslie N.
United States

Grunau, Daryl
United States

Grunbaum, F. Alberto
University of California at Berkeley
United States

Gueron, Shay
Technion Israel Institute of Technology
Israel

Guess, Oliver
IAC - CNR
Italy

Guidorzi, Roberto P.
University of Bologna
Italy

Guillope, Colette
Université Paris Sud
France

Gunstensen, Andrew Knut
Massachusetts Institute of Technology
United States

Gunzburger, Max D.
Virginia Technical Institute
United States

Guo, Benqi
University of Manitoba
Canada

Guo, Ben-Yu
Shanghai University of Science and Technology
United States

Guo, Xian-Zhou
University of Maryland
United States

Guptill, James D.
United States

Gurwitz, Chaya Bleich
Brooklyn College
United States

Gustafson, Sven-Ake
HSR
Norway

Ha, Cu D.
AT&T Bell Laboratories
United States

Haaren-Retagne, Elke
Universität Trier
Germany

Haario, Heikki
University of Helsinki
Finland

Haber, Robert B.
University of Illinois
United States

Haberman, Richard
Southern Methodist University
United States

Hackl, Klaus
University of Delaware
United States

Haddow, James Baird
University of Victoria
Canada

Haddard, Ziad
United States

Haffad, A.
Georgia Institute of Technology
United States

Hagan, Patrick S.
Los Alamos National Laboratory
United States

Hager, William W.
University of Florida
United States

# ATTENDEE LIST

Haghighi, Aliakbar Montazer
United States

Hagstrom, Thomas M.
University of New Mexico
United States

Hague, Steve
The Numerical Algorithms Group Ltd.
Great Britain

Hald, Ole H.
University of California
United States

Hales, Alfred W.
University of California
United States

Hallahan, Charles B.
United States

Halvorsen, Svenn A.
ELKEM
Norway

Hamaguchi, Satoshi
IBM Corporation
United States

Hamina, Martti Aulis
University of Oulu
Finland

Hammad, Pierre
Université Aix-Marseille III
France

Hammarling, Sven J.
The Numerical Algorithms Group Ltd.
Great Britain

Hammel, Stephen M.
United States

Hammer, Patricia W.
Hollins College
United States

Hanke, Michael
Humboldt Universität du Berlin
Germany

Hanmouzet, Bernard
University Bordeauxt
France

Hanson, Douglas
Conoco Inc.
United States

Hanson, Richard J.
International Mathematics and
Statistical Libraries
United States

Han, Weimin
University of Iowa
United States

Harabetian, E.
University of Michigan
United States

Harker, Patrick T.
United States

Harnevo, Linda E.
Weizmann Institute of Science
Israel

Harrison, Gary W.
College of Charleston
United States

Haskell, Karen
United States

Hassanzadeh, Siamak
Fujitsu America Incorporated
United States

Hattori, Harumi
West Virginia University
United States

Haubold, Hans Joachim
United Nations
United States

Haug, Edward J.
University of Iowa
United States

Hauge, Sharon K.
University of District of Columbia
United States

Haynes, Todd M.
United States

He, Sailing
Royal Institute of Technology
Sweden

Healy, Dennis M.
Dartmouth College
United States

Heard, Melvin L.
University of Illinois-Chicago
United States

Hegland, Markus
Australian National University
Australia

Heifetz, Daniel
Supercomputing Research Center
United States

Heilio, Matti Ilmari
Lappeenranto University of
Technology
Finland

Hein, Soren
University of California at Berkeley
United States

Heinkenschloss, Matthias
Universität Trier
Germany

Heinricher, Arthur
University of Kentucky
United States

Heller, William Harold
University of Maryland/College Park
United States

Hemker, Pieter W.
The Netherlands

Henry, Wanda Marge
Cambridge University
Great Britain

Herdman, Terry L.
Virginia Polytechnic Institute & State
University
United States

Hereman, Willy A. M.
Colorado School of Mines
United States

Herman, Gerard C.
Delft University
The Netherlands

Hernandez, Diego B.
CIMAT
Mexico

Herrmann, Joseph M.
Texas A & M University
United States

Herron, Isom H.
Howard University
United States

Hesterberg, Tim C.
Franklin & Marshall College
United States

Hestir, Kevin F.
Lawrence Berkeley Lab
United States

Heydon, Clark Allan
Carnegie Mellon University
United States

Heywood, John G.
University of British Columbia
Canada

Hicks, Craig Pinder
Tokyo Institute of Technology
Japan

Highnam, Peter
Schlumberger Laboratory for
Computer Science
United States

Hill, David R.
United States

Hillebrand, Kari Tapio
Technical Research Center of
Finland
Finland

Hilnes, Craig L.
Boeing Computer Service
United States

Hindmarsh, Alan C.
University of California
United States

Hinnestroza, Doris
University of Cincinnati
United States

Hitt, Darren
Johns Hopkins University
United States

Ho, Lop Fat
Wright State University
United States

Hodges, Dewey H.
Georgia Institute of Technology
United States

Hoff, David C.
Indiana University
United States

Hoffend, Thomas R., Jr.
University of Minnesota
United States

Hoffman, Karla Leigh
George Mason University
United States

Hoffmann, Christoph M.
Purdue University
United States

Hoisie, Adolfy
Cornell University
United States

Holden, Helge
University of Trondheim
Norway

Holden, Lisa J.
Kalamazoo College
United States

Hollister, Robert A.
University of Alaska
United States

Hollis, Selwyn L.
Armstrong State College
United States

Hollister, Robert A.
University of Alaska
United States

Holloway, James P.
United States

Holmay, Kathleen
Kathleen Holmay & Associates
United States

Holst, Michael
University of Illinois-Urbana
United States

Holt, Fred B.
Boeing Computer Services
United States

Holway, Lowell
Raytheon Company
United States

Hong, Bin
University of Pittsburgh
United States

Hong, Jianguo
University of Hanover
Germany

Hong, Soon
Grand Valley State University
United States

Horn, Mary Ann
University of Virginia
United States

Horvath, John M.
University of Maryland
United States

Hoskins, Janet
University of Manitoba
Canada

Hou, Lisheng
York University
Great Britain

Houbak, Niels
Technical University of Denmark
Denmark

Howell, Gary W.
Florida Institute of Technology
United States

Howell, Louis H.
Lawrence Livermore National Laboratory
United States

Howes, Frederick A.
Department of Energy
United States

Hromadka, Ted Vincent
United States

Hrusa, William J.
Carnegie-Mellon University
United States

Hsiao, George C.
University of Delaware
United States

Hsieh, Din-Yu Y.
Hong Kong University of Science and Technology
Hong Kong

Hsu, Chung-Hao
NASA Langley Research Center
United States

Hu, Chenyi
University of Houston/Downtown
United States

Huang, Aixiang
Xian Jiatong University
Peoples' Republic of China

Huang, Huaxiong
Johns Hopkins University
United States

Huang, Xun-cheng
Shanghai Institute Comp Technology
Peoples' Republic of China

Huang, Ying Sue
Brown University
United States

Hudson, Andrea Kay
United States

Hughes, Charles Edward
Canada

Hughes, Merritt R.
United States

Hughes, Thomas J. R.
Stanford University
United States

Hughey, Richard P.
Department of Defense
United States

Hulbert, Gregory M.
University of Michigan
United States

Humenik, Keith Edward
University of Maryland-Baltimore County
United States

Hunis, Andrew
University of Waterloo
Canada

Hunt, Brian R.
Naval Surface Warfare Center
United States

Hunt, Fern Y.
Howard University
United States

Hunt, J. C. R.
University of Cambridge
Great Britain

Hurley, Byron
SDSMT
United States

Huschens, Juergen L.
Germany

Hwang, Din-Chih
Naval Surface Warfare Center
United States

Hyde, Peter D.
United States

Hyman, James M.
Los Alamos National Laboratory
United States

Ibrahim, Arsmah
Mara Institute of Technology
Malaysia

Igarashi, Masao
Japan

Ikeda, Tsutomu
Ruykoku University
Japan

Inayat-Hussain, Anis Ahmad
BHP Research
Australia

Indovina, Ronni R.
United States

Infeld, Eric
University of Montreal
Canada

Ing, Susan
The Numerical Algorithms Group Ltd.
Great Britain

Inselberg, Alfred
IBM Scientific Center
United States

Ipsen, Ilse C.F.
Yale University
United States

Isaacson, David
United States

Isakov, Victor
Wichita State University
United States

Ishii, Katsuya
ICFD
Japan

Ito, Masayoshi
Tokushima University
Japan

Izen, Steven H.
Case Western Reserve University
United States

Jackiewicz, Zdzislaw
Arizona State University
United States

Jackson, Kenneth R.
University of Toronto
Canada

Jaco, William H.
American Mathematical Society
United States

Jaffre, Jerome
INRIA
France

Jami, Adrien
Electricite de France
France

Jankins, Mark
Aerospace Corporation
United States

Jankovich, Etienne
Hutchinson S.A.
France

Jannesari, Saeid
United States

Jekot, Tomasz
University of Zululand
Republic of South Africa

Jen, Juif Frank
State University of New York at Stony Brook
United States

Jensen, Soren S.
United States

Jerome, Joseph W.
Northwestern University
United States

Jespersen, Dennis
NASA Ames Research Center
United States

Jewell, James G.
United States

Jiang, Xinming
University of Delaware
United States

Jiang, Yahong
Northwestern University
United States

Jodorkovsky, Mario
Israel

Johnson, Daniel W.
Martin Marietta Aero/Naval Systems
United States

Johnson, Russell A.
Università di Firenze
Italy

Johnsson, S. Lennart
Harvard University
United States

Jolly, Michael S.
Indiana University
United States

Joly, Patrick
INRIA
France

Jones, Alan Frederick
University of Manchester
Great Britain

Jones, Christopher K.R.T
Brown University
United States

Jones, Douglas S.
University of Dundee
Scotland

Jordan, Kirk E.
Thinking Machines Corporation
United States

Joshi, Santosh B.
Government Colony
India

Jou, Emery D.
University of Maryland
United States

Juric, Damir
United States

Kaiser, Ralf
MPI fur Plasmaphxsik
Germany

Kako, Takashi
University of Electro-Communication
Japan

Kalachev, Leonid Viktorovick
University of Washington
United States

Kallrath, Josef
BASF - AG
Germany

Kamberova, Gerda L.
University of Pennsylvania
United States

Kametaka, Yoshinori
Tokushima University
Japan

Kammeyer, Peter C.
U.S. Naval Observatory
United States

Kaneko, Hideaki
Old Dominion University
United States

Kaper, Hans G.
Argonne National Laboratory
United States

Kapila, Ashwani
Rensselaer Polytechnic Institute
United States

Kappos, Efthimios
University of Sheffield
Great Britain

Karasick, Michael
IBM, Thomas J. Watson Research Center
United States

Karatzas, Ioannis
Columbia University
United States

Karim, Salahadin
George Mason University
United States

Karl, Steger
Siemens AG
Germany

Karlsson, Anders Rolf
Royal Institute of Technology
Sweden

Karni, Smadar
University of Michigan
United States

Karowski, Andrzej Joachim
West Virginia University
United States

Karpik, Stephen R.
Canada

Karpouzian, Gabriel N.
U.S. Naval Academy
United States

Kassoy, David R.
University of Colorado
United States

Kastner-Maresch, Alois Erwin
University of Bayreuth
Germany

Katz, I. Norman
Washington University
United States

Katz, Simeon
University of Southern California
United States

Kauffmann, Christian
Delft University of Technlogy
The Netherlands

Kaufman, Linda C.
AT&T Bell Laboratories
United States

Kaul, Raj K.
State University of New York
United States

Kavian, Otared
Université de Nancy I
France

Kawashima, Shuichi
Kyushu University
Japan

Kazarinoff, Nicholas D.
United States

Kearfott, Baker
University of Southern Louisiana
United States

Kearney, Charulata K.
Softech Inc.
United States

Kearsley, Anthony J.
Rice University
United States

Keeler, Stephen P.
United States

Keener, James Paul
University of Utah
United States

Kelley, Carl T.
North Carolina State University
United States

Kerkhoven, Thomas
University of Illinois
United States

Kern, Daniel L.
United States

Kern, Michel E.
Rice University
United States

Keunings, Roland C.
University Catholique de Louvain
Belgium

Kevorkian, Aram K.
United States

Kevorkian, Jirair K.
University of Washington
United States

Keyes, David E.
Yale University
United States

Khaliq, Abdul Qayyum Masud
Western Illinois University
United States

Khodja, Mohamed
Indiana University
United States

Kichenassamy, Satyanad
University of Minnesota
United States

Kikuchi, Noboru
University of Michigan
United States

Kimn, Ha Jine
Ajou University
South Korea

Kimn, Jung-han
Yonsei University
South Korea

Kimura, Ryusuke
University of Tokyo
Japan

Kimura, Yoshifumi
NCAR
United States

Kim, Chul
Kwang Woon University
South Korea

Kim, Haklin
University of Tennessee at Martin
United States

Kim, Jeong-Hoon
United States

Kim, Koonchan
United States

Kim, Kwang H.
Mitre Corporation
United States

Kim, Mi Young
Purdue University
United States

Kincaid, David R.
University of Texas at Austin
United States

Kinch, Dennis F.
United States

Kinderlehrer, David
Carnegie-Mellon University
United States

King, Belinda B.
United States

King, John Robert
University of Nottingham
Great Britain

King, Paul
U.S. Bureau of Mines
United States

Kinnas, Spyros A.
Massachusetts Institute of Technology
United States

Kirchgaessner, K. W.
Germany

Kirkegaard, Peter
Riso National Laboratory
Denmark

Kirsch, Andreas
Universitaet Erlangen
Germany

Klein, Wolfram
Siemens A G
Germany

Kleinman, Ralph E.
University of Delaware
United States

Kliemann, Wolfgang H.
Iowa State University
United States

Klincewicz, John G.
United States

Klosek, Malgorzata M.
University of Wisconsin
United States

Knupp, Patrick
United States

Kobayashi, Ryo
Ryukoku University
Japan

Koh, Eusebio L.
University of Regina
Canada

Koike, Shigeaki
Tokyo Metropolitan University
Japan

Kolkka, Robert W.
Michigan Technological University
United States

Kolmanovsky, V.
Moscow Institute of Electrical Machinery
Russia

Komkov, Vadim
Air Force Institute of Technology/ENC
United States

Konno, Chisato
Hitachi Ltd.
Japan

Kopala, Conrad
United States

Kowalski, Anthony D.
Rutgers University
United States

Kraay, Tom
M.R.J. Incorporated
United States

Kraemer, Walter
University of Karlsruhe
Germany

Krasnoshchekov, Pavel
USSR Academy of Science
Russia

Krasny, Robert
University of Michigan
United States

Kratzer, Steven G.
United States

Krause, Egon
NASA Langley Research Center
United States

Kress, Rainer
University of Gottingen
Germany

Kriman, Alfred M.
State University of New York at Buffalo
United States

Krishnamoorthy, Mukkai S.
Renssalaer Polytechnic Institute
United States

Krishnasamy, Guna
University of Illinois
United States

Kroan, William T.
United States

Krogh, Fred T.
United States

Krohn, Stuart
United States

Krutar, Rudolph Allen
U.S. Naval Research Laboratory
United States

Krystynak, John
NASA-Ames Research Center
United States

# ATTENDEE LIST

Ku, Takang
University of Southern California
United States

Kuang, Yang
Arizona State University
United States

Kuchment, Peter
Wichita State University
United States

Kuerten, Hans
University of Twente
The Netherlands

Kuiken, H.K.
Technical University
The Netherlands

Kulisch, Urlich
University of Karlsruhe
Germany

Kumagai, Teruo
Science University of Tokyo
Japan

Kunoth, Angela
University of South Carolina
United States

Kurss, Herbert
Adelphi University
United States

Kuske, Rachel A.
Northwestern University
United States

Kuz, Ihor
Lvov State University
Ukraine

Kvamsdal, Trond
Norway

Kvriazis, George C.
University of South Carolina
United States

Kwon, Yonghoon
Pohang Institute of Science & Technology
South Korea

Kwong, Man K.
Argonne National Laboratory
United States

Lacomblez, Chantal
University de Bordeaux II
France

Lacroix, Norbert H.
Université Laval
Canada

Ladde, Gangaram S.
University of Texas-Arlington
United States

Laderman, Julian D.
Lehman College
United States

Laffitte, Jean
INSA
France

Lafon, Andre
NASA Ames
United States

Lagnese, John E.
Georgetown University
United States

Lahtinen, Aatos O.
University of Helsinki
Finland

Lai, Yingcheng
University of Maryland
United States

Lailly, Patrick
Institut Francais Du Petrole
France

Lair, Alan V.
AFIT/ENC
United States

Lam, Sau-Hai
Princeton University
United States

Lamberson, Roland H.
Humboldt State University
United States

Lambert, Michael H.
Pittsburgh Supercomputing Center
United States

Lambrakos, Samuel G.
Naval Research Laboratory
United States

Lamour, Rene
Humboldt-University of Berlin
Germany

Lancia, Maria Rosaria
University of Rome "La Sapienza"
Italy

Landowne, Elliott Isaac
CUNY
United States

Landvogt, Markus
J.G. University of Mainz
Germany

Langenberg, Karl Jorg
University of Kassel
Germany

Langford, William F.
University of Guelph
Canada

Langlois, Philippe
University Paul Sabatier-Toulouse
France

Lanzkron, Paul
United States

Lardy, Lawrence J.
Syracuse University
United States

LaRoche, Humberto J.
United States

Larsen, Jesper
Math Tech
Denmark

Larson, Dale A.
United States

Lasseigne, D. Glenn
United States

Lau, Richard Lew
Office of Naval Research
United States

Laurent, Pierre Jean
University Joseph Fourier, Grenoble
France

Lauro, Giuliana
Université Firenze
Italy

Lavery, John E.
NRC-BMS
United States

Lawless, Fiona R.
Dublin City University
Ireland

Lawniczak, Anna T.
University of Guelph
Canada

Lawson, Charles L.
United States

Layne, J. Daniel
United States

Lea, George K.
National Science Foundation
United States

Leap, Wayne Scott
University of Delaware
United States

Leary, Robert H.
San Diego Supercomputer Center
United States

Leaver, Sylvia G.
U.S. Dept of Labor
United States

LeBorne, Richard C.
United States

Ledder, Glenn W.
University of Nebraska-Lincoln
United States

Le Dimet, Francois Xavier
Université Blowe Pascor 1
France

Ledzewicz-Kowalewska, Urszula
Southern Illinois University
United States

Lee, Chung N.
Pohang Institute of Science & Technology
South Korea

Lee, Eva K.
Rice University
United States

Lee, E. B.
University of Minnesota
United States

Lee, Jon
Wright-Patterson Air Force Base
United States

Lee, Jyh-Hao
Academia Sinica, Taipei
Republic of China

Lee, Li
University of Maryland
United States

Lee, Seong H.
Chevron Oil Field Research
United States

Lee, Steven L.
United States

Le Floch, Philippe G.
New York University
United States

Lefton, Lew
University of New Orleans
United States

Leigh, James Ronald
Polytechnic of Central London
Great Britain

Leimkuhler, Benedict J.
University of Kansas
United States

Leister, Guenter
University of Stuttgart
Germany

Leko, Toma D.
United States

Leneveu, Marie-Christine
Beiersdorf AG
Germany

Lenhart, Suzanne M.
University of Tennessee
United States

Lentini, Marianela
University Simon Bolivar
Venezuala

Leon, Steven J.
University of Massachusetts
United States

LeTallec, Patrick
INRIA
France

LeTallec, Xavier
LEPT - ENSAM
France

Leugering, Guenter R.
Georgetown University
United States

Leung, Ming-Ying
University of Texas @ San Antonio
United States

Levin, Jennifer A.
Northwestern University
United States

Levine, William S.
United States

Levner, Eugene V.
Israel

Lewis, Gilbert N.
Michigan Technological University
United States

Lewis, Mark Alun
University of Washington
United States

Lewis, Robert M.
Rice University
United States

Lewitt, Robert M.
University of Pennsylvania
United States

Li, Chen
Wuhan University
Peoples' Republic of China

Li, Duan
University of Virginia
United States

Li, Guangye
Rice University
United States

Li, Jibin
Georgia Institute of Technology
United States

Li, Shidong
University of Maryland-Baltimore County
United States

Li, Xuefeng
United States

Li, Zi-Cai
CRIM
Canada

Liao, Guojun
United States

Librescu, Liviu
Virginia Polytechnic Institute & State University
United States

Lih, Keh Wei
Rutgers University
United States

Lindquist, Brent
SUNY-Stony Brook
United States

Linden, Johannes F.
GMD F1/T
Germany

Lin, Chujen
United States

Lin, Jian-Tong
University of Texas, Arlington
United States

Lin, Patrick Po-Yen
Courant Institute of Math Sciences
United States

Lions, Pierre-Louis
University Paris-Dauphine
France

Lipman, Marc J.
Office of Naval Research
United States

Liron, Nadav
Technion IIT
Israel

Liu, Bing
College of St. Scholastica
United States

Liu, Chen-Huei
NASA Langley Research Center
United States

Liu, Jinn-Liang
National Chiao Tung University
Republic of China

Liu, Jun
Institute National de Recherche en Informatique et Automatique
France

Liu, Lixin
Simon Fraser University
Canada

Liu, Qijin
Wayne State University
United States

Lobo-Pereira, Fernando M F
DEEC-FEUP
Portugal

Lockhart, Deborah Frank
National Science Foundation
United States

Loeve, W.
National Aerospace Laboratory
The Netherlands

Logan, James C.
Naval Ocean Systems Center
United States

Logan, J. David
University of Nebraska
United States

Logan, Roger
College of Charleston
United States

Loheac, Jean-Pierre
École Centrale de Lyon
France

Lohmann, Thomas W.
Germany

Lohner, Rudolf
University of Karlsruhe
Germany

Lohofer, George
DFVLR
Germany

Lome, Louis S.
United States

Longley, James Wilden
United States

Loo, Saoping
CUNY
United States

Lopez, Christian
Université Paris II Dauphire
France

Lorenz, Dan H.
Israel Institute of Technology
Israel

Lorenz, Jens
University of New Mexico
United States

Lortz, Dietrich
Max-Planck Institute
Germany

Louis, Alfred K.
Universitaet Saarbruecken
Germany

Lovass-Nagy, Victor
Clarkson University
United States

Loveday, Dennis Leslie
Loughborough University
Great Britain

Lowes, Leslie Lynne
Jet Propulsion Laboratory
United States

Lozier, Daniel W.
United States

Lube, Gert
Madgebury University of Technology
Germany

Lucas, Robert F.
Supercomputing Research Center
United States

Lucian, Miriam
Boeing Commerical Airplane Company
United States

Lucier, Bradley J.
Purdue University
United States

Luczak, Richard
The Numerical Algorithms Group Ltd.
Great Britain

Ludescher, William H.
General Electric Co.
United States

Luke, Jonathan
United States

Lund, John R.
Montana State University
United States

Luo, Jenn-Ching
Columbia University
United States

Luo, Li-Shi
Los Alamos National Laboratory
United States

Lutoborski, Adam
Syracuse University
United States

Lutz, Melanie P.
University of California
United States

Lyle, Svetlana
Kingston Polytechnic
Great Britain

Lynch, Peter Michael
ICI Explosives
Scotland

Lynn, Yen-Mow
University of Maryland-Baltimore County
United States

MacCluer, Charles R.
Michigan State University
United States

MacCuish, John
Indiana University
United States

MacDonald, Norman
University of Glasgow
Great Britain

MacGillivray, Archibald D.
SUNY/Buffalo
United States

Maciel, Maria Cristina
Rice University
United States

Macken, Catherine A.
Los Alamos National Laboratory
United States

Mackens, Wolfgang
University of Hamburg
Germany

MacNeal, Richard H.
The MacNeal-Schwendler Corporation
United States

Madsen, Kaj
The Technical University of Denmark
Denmark

Maggelakis, Sophia A.
United States

Maggs, Bruce
Massachusetts Institute of Technology
United States

Mahaffy, Joseph M.
San Diego State University
United States

Mahrenholtz, Oskar H.
Technical University of Hamburg
Germany

Mai, Tsunzee
University of Alabama
United States

Mair, Bernard A.
University of Florida
United States

Major, Diana
University of Southwestern Louisiana
United States

Majumdar, Samir R.
University of Calgary
Canada

Maldarelli, Charles
City College of New York
United States

Malek-Madani, Reza
United States Naval Academy
United States

Mallett, Russell L.
United States

Malmuth, Norman D.
Rockwell International
United States

Mandhyan, Indur B.
North American Phillips Corporation
United States

Manitius, Andrzej Z.
United States

Manley, Oscar P.
U.S. Department of Energy
United States

Manoranjan, Valipuram S.
Washington State University
United States

Mansutti, Daniela
Consiglio Nazionale delle Ricerche
Italy

Mantell, Abraham S.
State University of New York
United States

Manteuffel, Thomas A.
University of Colorado at Denver
United States

Marcati, Pierangelo A.
University of L'Aquila
Italy

Marchi, Ezio
University Nactional de San Luis
Argentina

Margaliot, Zvi
University of Western Ontario
Canada

Margerum, Eugene A.
United States

Margolis, Stephen B.
Sandia National Laboratories
United States

Marini, Luisa Donatella
Università di Genova
Italy

Marion, Martine
École Centrale de Lyon
France

Marino, Zennaro
University of L'Aquila
Italy

Markel, Scott A.
David Sarnoff Research Center
United States

Marletta, Marco
Royal Military College of Science
Great Britain

Marron, Christopher S.
University of Virginia
United States

# ATTENDEE LIST

Marsh, Steven
Naval Research Laboratory
United States

Martin, Cynthia Elaine
Naval Research Laboratory
United States

Martin, Douglas R.
National Security Agency
United States

Martinez, Daniel O.
University of Delaware
United States

Martins, Luiz Felipe S.
Brown University
United States

Maruo, Hauime
Japan

Marusic, Miljenko
Mayo Foundation
United States

Maryak, John L.
John Hopkins University
United States

Marzulli, Pietro
University of Pisa
Italy

Marz, Roswitha
Humboldt Universität zu Berlin
Germany

Mascagni, Michael V.
Supercomputing Research Center
United States

Mas-Gallic, Sylvie
Universite Pierre et Marie Curie
France

Mason, David Paul
Witwatersrand University
Republic of South Africa

Massimo, Casciola Carlo
INSEAN
Italy

Mastin, Wayne
Mississippi State University
United States

Masters, Wen C.
Jet Propulsion Laboratory
United States

Masumura, Robert A.
Naval Research Laboratory
United States

Mateus, Geraldo Robson
University of Ottawa
Canada

Mathieson, Alec
Lamar University
United States

Mathai, Arak M.
McGill University
United States

Mathis, Frank H.
Baylor University
United States

Matkowsky, Bernard J.
Northwestern University
United States

Matthaeus, Bill
University of Delaware
United States

Mattheyses, Robert M.
General Electric Research &
Development Center
United States

Mattheij, Robert M.
University of Technology
The Netherlands

Maugin, Gerard A.
Paris, France

Maulino, Consuelo
Universidad Central Venezuela
Venezuela

Mavriplis, Dimitri J.
Institut for Computer Applications in
Science and Engineering
United States

Ma, Baoming
University of Maryland
United States

Ma, Jin Hong
Yale University
United States

Mayer, Gunter A.
Germany

Mazzia, Francesca
Università di Bari
Italy

McArthur, Kelly M.
Colorado State University
United States

McArthur, Raymond P.
SUNY at Buffalo
United States

McCartin, Brian, J.
United States

McClain, Marjorie A.
United States

McClure, Donald E.
Brown University
United States

McCord, Sarah M.
Jet Propulsion Lab
United States

McCormick, Garth
George Washington University
United States

McCormick, Steve
University of Colorado at Denver
United States

McCoy, Paul F.
United States

McCullough, Robert Norman
Ferris State University
United States

McDonald, Bernard R.
National Science Foundation
United States

McDonald, David
University of Ottawa
Canada

McDonnell, John H.
Great Britain

McDonough, Joseph M.
United States

McFadden, Geoffrey B.
National Institute of Standards
United States

McGlynn, Michael James
Department of Defense
United States

McGrath, Joseph F.
Mechanical Dynamics Inc.
United States

McKee, Sean
University of Strathclyde
Scotland

McKelvey, Robert W.
United States

McKendall, Raymond
University of Pennsylvania
United States

McKenna, James
Bellcore
United States

McKinney, William R.
North Carolina State University
United States

McLaughlin, Joyce R.
Rensselaer Polytechnic Institute
United States

McMillan, Christine A.
United States

McNulty, Nieves Austria
United States

McQuain, William D.
United States

Mechentel, Ali
IBM
United States

Medhin, Negash G.
Clark Atlanta University
United States

Meeker, Loren D.
University of New Hampshire
United States

Mehrotra, Sanjay
Northwestern University
United States

Mei, Chiang C.
Massachusetts Institute of
Technology
United States

Meish, Andrew
University College of London
Great Britain

Mejia, Raymond
United States

Melnick, Robert E.
Grumman Corporation
United States

Melrose, Gordon
Old Dominion University
United States

Melville, Robert
Bell Laboratories
United States

Menaldi, Jose Luis
Wayne State University
United States

Mennicken, Reinhard
Universität Regensburg
Germany

Merchant, Gregory J.
Stanford University
United States

Merrill, Stephen J.
Marquette University
United States

Merz, Paul H.
Chevron Research Company
United States

Mesina, George L.
EG&G Idaho Inc.
United States

Messina, Paul C.
California Institute of Technology
United States

Meurant, Gerard A.
Centre d'Etudes de Limeil
France

Meyer-Spasche, Rita
MPI fur Plasmaphysik
Germany

Meyer, Yves F.
Universite Paris IX
France

Miciano, Agnes Razon
Mississippi State University
United States

Mickens, Ronald E.
Clark Atlanta University
United States

Miel, George J.
University of Nevada
United States

Mikhailov, Gennadii Alexeevich
USSR Academy of Sciences
Russia

Milac, Thomas I.
United States

Milenkovic, Victor J.
United States

Miller, Keith
National Security Agency
United States

Miller, Kristin L.
United States

Mills, Patrick L.
United States

Millane, R.P.
Purdue University
United States

Miloh, Touvia
University of Tel-Aviv
Israel

Mimura, Masayasu
Hiroshima University
Japan

Minkoff, Susan
Rice University
United States

Miranda, Guillermo
Ciudad Universitaria
Venezuela

Mitsumura, Ieyoshi
University of Tokyo
Japan

Mitsui, Taketomo
Nagoya University
Japan

Miura, Robert M.
University of British Columbia
Canada

Mokole, Eric
United States

Moler, Cleve B.
United States

Molenaar, J.
Centrum voor Wiskunde en Info
The Netherlands

Molina, Brigida Coromoto
Universita Central de Venezuela
Venezuela

Moll, Robert
University of Massachusetts
United States

Molyneux, John E.
Widener University
United States

Monis, Ann
United States

Monk, Peter B.
University of Delaware
United States

Moonen, Marc
Katholiecke Universiteit
Belgium

Moore, Peter K.
United States

Moore, Robert H.
University of Wisconsin
United States

Moraites, Steve
United States

Morales-Perez, Jose L.
Great Britain

Morawetz, Cathleen S.
New York University
United States

Moré, Jorge J.
Argonne National Laboratory
United States

Mori, Masatake
University of Tokyo
Japan

Moriarty, Julie A.
Maths Institute
Great Britain

Morita, Yoshihisa
Ryukoku University
Japan

Morris, Brian C.
Occidental College
United States

Morrison, John A.
AT&T Bell Laboratories
United States

Morro, Angelo C.
University of Genoa
Italy

Morton, K. W.
Great Britain

Moseley, James L.
West Virginia University
United States

Moskowitz, Ira
Naval Research Laboratory
United States

Moulin, Pierre
Bellcore
United States

Moura, Monique R.
Courant Institute of Mathematical Sciences
United States

Moussouros, Minos
United States

Mucha, Greg
United States

Muehlbach, Guenter W.
University of Hanover
Germany

Mueller, Winfried B.
Universität Klagenfurt
Austria

Muir, Paul H.
St. Mary's University
Canada

Mundrane, Michael R.
Rutgers University
United States

Munger, Fredrick S.
United States

Murakami, Koichi
Tokushima University
United States

Murray, Alan G.
United States

Murray, Bruce
NIST
United States

Murray, James D.
University of Washington
United States

Murray, John
Martin Marietta Aerospace and Naval Systems
United States

Muszynski, Jerzy
Politechnika Warszawska
Poland

Myers, Eugene W.
University of Arizona
United States

Myers, Matthew R.
United States

Mynett, Arthur E.
Delft Hydraulics
United States

Nachman, Adrian
University of Rochester
United States

Nachman, Arje
AFOSR/NM
United States

Nachtigal, Noel M.
United States

Nacul, Evandro Correa
Stanford University
United States

Nadiga, Balasubramanya T.
California Institute of Technology
United States

Nagurney, Anna B.
University of Massachusetts
United States

Naik, Naomi
United States

Nairn, Dean C.
United States

Najarian, John
United States

Nakayama, Maki
Witwatersrand University
Republic of South Africa

Nambu, Takao
Kumamoto University
Japan

Narayanan, Ranganathan
University of Florida
United States

Nardi, Jerry S.
United States

Nash, John C.
Canada

Nash, Stephen G.
George Mason University
United States

Nashed, M. Zuhair
University of Delaware
United States

Nasser, Mahmond G.
United States

Natalini, Roberto
Instituto per Applilcazioni Calculo
Italy

Natterer, Frank
University of Muenster
Germany

Navon, Ionel M.
Florida State University
United States

Nayar, Narinder
University of Virginia
United States

Neaga, Michael
Numerik Software GMBH
Germany

Neal, Leslie Robert
Brunnel University
Great Britain

Negron, Pablo V.
United States

Neidinger, Richard
Davidson College
United States

Nelson, David A.
United States

Nelson, Gary G.
United States

Nerurkar, Mahesh
Rutgers University
United States

Nesbitt, Martha M.
University of Colorado
United States

Neumaier, Arnold
Universitaet Freiburg
Germany

Neunzert, Helmut
University of Kaiserlautern
Germany

Nevanlinna, Olavi
Helsinki University of Technology
Finland

Newton, Paul K.
University of Illinois
United States

Ng, Kam Chuen
United States

Nguyen, Kim Dan
University de Lille 1
France

Nguyen, Mai Khuong
CNRS-ESE-UPS
France

Nicolaides, Roy A.
Carnegie-Mellon University
United States

Niedinger, Richard
Davidson College
United States

Nieland, Hendrik Maarten
CWI
The Netherlands

Nier, Francis
École Polytechnic
France

Niessner, Herbert
ABB Informatik AG
Switzerland

Niethammer, Wilhelm
University Karlsruhe
Germany

Nishiura, Yasumasa
Hiroshima University
Japan

Nocedal, Jorge
Northwestern University
United States

Nochetto, Ricardo
University of Maryland
United States

Noda, Matu-Tarow
Ehime University
Japan

Noel, Claire
GERDSM DCN Toulen France
France

Nohel, John A.
University of Wisconsin-Madison
United States

Noren, Richard D.
Old Dominion University
United States

Norris, A. N.
Rutgers University
United States

Nosov, V.
MIEM
Russia

Nouri-Moghadam, M.
Penn State University
United States

Novak, Erich
Universitat Erlangen
Germany

Nowack, Robert
Purdue University
United States

Nowakowski, Richard S.
Robert Wood Johnson Medical School
United States

Numrich, Susan K.
Naval Research Laboratory
United States

O'Brien, Francis J., Jr.
Naval Underwater Systems Center
United States

Ochs, Robert L., Jr.
University of Toledo
United States

## ATTENDEE LIST

Ockendon, Hilary
Oxford University
Great Britain

Ockendon, John R.
Oxford University
Great Britain

O'Connor, Michael A.
IBM Corporation
United States

Odeh, Farouk
IBM Research Center
United States

Ogawa, Hidemitsu
Tokyo Institute of Technology
Japan

Ogielski, Andrew T.
Bellcore
United States

Ogryzak, Wloozimierz
Marshall University
United States

O'Hara, Kathleen M.
University of Iowa
United States

Ohta, Takao
Ochanomizu University
Japan

Okada, Mari
Yamaguchi University
Japan

Olagunju, David O.
University of Delaware
United States

O'Leary, Dianne P.
University of Maryland
United States

Oliger, Joseph
Stanford University
United States

Oliveira, Aurelio Ribeiro L.
Rice University
United States

Oliver, Carl Edward
Oak Ridge National Laboratory
United States

Oliveira, Pedro
University of Strathclyde
Great Britain

Olmstead, William E.
Northwestern University
United States

Olsen, Robert G.
Washington State University
United States

Olsen, Robert J.
University of Minnesota
United States

Olson, Timothy Edward
Dartmouth College
United States

Olver, Frank W. J.
University of Maryland
United States

O'Malley, R. E., Jr.
University of Washington
United States

Ong, Maria Elizabeth G.
University of California
United States

Oono, Yoshitsugin
University of Illinois
United States

Oosterlef, Kees
Delft University of Technology
The Netherlands

Ordman, Edward T.
Memphis State University
United States

Ortega, Angel Felipe
Universidad Compu Madrid
Spain

Osborn, John E.
University of Maryland
United States

Oser, Hans J.
United States

Osgood, Charles F.
National Security Agency
United States

O'Sullivan, M.J.
University of Auckland
New Zealand

Ott, Edward
University of Maryland
United States

Otter, Andries
Twente University
The Netherlands

Otterheim, Henrik
Royal Institute of Technology
Sweden

Ou, Yuh-Roung
Virginia Tech
United States

Oulton, David B.
United States

Overman, Edward A., II
Ohio State University
United States

Owren, Brynjulf
University of Toronto
Canada

Oxley, Mark Edwin
Air Force Institute of Technology
United States

Oya, Takao
University of North Carolina at Chapel Hill
United States

Paappanen, Teuvo Antero
Technical Research Center of Finland
Finland

Pagani, Carlo
Politecnico di Milano
Italy

Painter, Jeffrey F.
Lawrence Livemore Lab
United States

Pais, John
McDonnell Douglas Research Laboratories
United States

Paivarinta, Lassi
University of Helsinki
Finland

Palacios, Jose L.
United States

Palazotto, Anthony N.
Air Force Institute of Technology
United States

Paluszny, Marco
Universita Central de Venezuela
Venezuela

Pamer, Richard Norman
University of Toronto
Toronto

Pan, Enlin
United States

Pan, Jie
University of Massachusetts
United States

Pan, Jiqin
University of Maryland
United States

Pan, Weixiang
United States

Pandey, Bishun D.
United States

Pantelis, Garry
Australian Nuclear Science & Technology Organization
Australia

Pao, Chia-Ven
North Carolina State University
United States

Paolini, Maurizio
Istituto di Analisi Numerica del C.N.R.

Paprzycki, Marcin
University of Texas Permian Basin
United States

Parikh, Balmukund P.
Wolchand College of Engineering
India

Paris, Richard Bruce
Dundee Institute of Technology
Scotland

Park, Eun-Jae
Purdue University
United States

Parra, Ignacio E.
Universita Politec de Madrid
Spain

Parsons, Lee
United States

Parter, Seymour V.
University of Wisconsin
United States

Pastoriza, Rafiel Alberto
Universidad Central Venezuela
Venezuela

Patrick, Merrel L.
Duke University
United States

Patrinos, Ari
Department of Energy
United States

Payzer, Rochelle
Vitro Corporation
United States

Pearson, John Evan
Los Alamos National Laboratory
United States

Peart, Paul
Howard University
United States

Pecora, Louis M.
Naval Research Laboratory
United States

Pego, Robert L.
United States

Pelissier, Marie-Claude
Université De Toulon
France

Pelz, Richard B.
Rutgers University
United States

Perelson, Alan S.
Los Alamos National Laboratory
United States

Perera, Nawagamuwage B.
Mayfair Center
Pakistan

Peridier, Val J.
Temple University
United States

Perlik, Andrew T.
United States

Pernarowski, Mark C.
University of British Columbia
Canada

Perthame, Benoit
Université D'Orleans
France

Peruchio, Renato
University of Rochester
United States

Peters, Alexander
IBM
United States

Petersen, Wesley P.
IPS
Switerland

Petiton, Serge G.
Yale University
United States

Petrila, Titus
University of Cluj
Romania

Petzold, Linda R.
University of Minnesota
United States

Phillips, Christopher George
Imperial College of Science &
Technology
Great Britain

Phillips, James L.
United States

Phillips, P. Jonathon
Rutgers University

Pickering, William Morgan
University of Sheffield
Great Britain

Piepho, Melvin G.
United States

Pierce, John F.
United States Naval Academy
United States

Pinsky, Mark A.
Northwestern University
United States

Pinsky, Mark
University of Nevada at Reno
United States

Pintar, Anton J.
Michigan Technological University
United States

Pinter, Joseph K.
Canada

Pitkaranta, Juhana
Helsinki University of Technology
Finland

Platen, Eckhard
Karl-Weierstrass-Institut fur Math
Germany

Platenkamp, Gerrit A.
University of California
United States

Plohr, Bradley J.
State University of New York
United States

Pojman, John A.
University of Southern Mississippi
United States

Poliqnone, Debbie
University of Virginia
United States

Pollock, Ray
United States

Pomeranz, Shirley B.
The University of Tulsa
United States

Ponce-Dawson, Silvina
University of Maryland
United States

Ponce-Nunez, Enrique
ETSI Industriales
Spain

Ponnusamy, Ravi
Insitute for Computer Applications in
Science and Engineering
United States

Ponte-Castaneda, Pedro
University of Pennsylvania
United States

Pool, James C. T.
United States

Popel, Aleksander S.
The Johns Hopkins University
United States

Porter, Michael B.
New Jersey Institute of Technology
United States

Pothen, Alex
Pennsylvania State University
United States

Portor-Lockhart, Deborah
Pembroke State University
United States

Potra, Florian A.
University of Iowa
United States

Pouget, J.
Université de Pierre et Marie Curie
France

Poupaud, Frederic
CNRS/Universitaet de Nice
France

Pousin, Jerone
École Polytechnique Lausanne
Switzerland

Power, Desmond T.
Memorial University of
Newfoundland
Canada

Powers, Robert K.
University of Arkansas
United States

Premuda, Francesco Maria
Università Bologna
Italy

Pressburger, Yoram
University of Rochester
United States

Prokhorov, Yu V.
USSR Academy of Sciences
Russia

Promislow, Keith Steven
Indiana University
United States

Prothero, Steve K.
Willamette University
United States

Protopopescu, Vladimir A.
Oak Ridge National Laboratory
United States

Pruess, Steven A.
Colorado School of Mines
United States

Pruss, Alexander R.
University of British Columbia
Canada

Pryce, John Derwent
Royal Military College of Science
Great Britain

Pryor, Daniel V.
United States

Pugh, Ashley Clive
Loughborough University of Technology
Great Britain

Pulvirenti, Mario
Università Dell'Aquila
Italy

Purdy, Carla
University of Cincinnati
United States

Pureza, Vitoria Miranda
Ottawa University
Canada

Purtill, Mark R.
United States

Qian, Maijian
University of Washington
United States

Qian, Shixian
United States

Qiao, Liyuan
University of Waterloo
United States

Qiao, Sanzheng
McMaster University
Canada

Quarteroni, Alfio M.
Politecnico di Milano
Italy

Quinn, Dennis W.
United States

Quinn, Joseph E.
North Carolina Supercomputing Center
United States

Quint, Thomas
United States Naval Academy
United States

Rabe, Ellen H.
Department of Defense
United States

Radomanov, Sergey I.
NIKA
Russia

Raghavan, Padma
United States

Rajala, Mikko Matti
Neste Oy
Finland

Rakesh
Central Michigan University
United States

Rakowska, Joanna
Virginia Polytechnic Institute & State University
United States

Rall, Louis B.
University of Wisconsin-Madison
United States

Ralph, Daniel
Cornell University
United States

Ralston, James
University of California, Los Angeles
United States

Ramachandran, Kandethody M.
University of South Florida
United States

Ramadurai, Kumbakonam S.
St. Francis Xavier University
Canada

Ramanujan, K. S.
University of Medicine and Denistry, New Jersey
United States

Rao, Anupama M. N.
United States

Rascle, M.
Université de Nice
France

Rasmussen, Henning
University of Western Ontario
Canada

Raugh, Michael R.
United States

Rauh, Marianne
Germany

Ravel, Jean Christophe
Université Paris 6
Fance

Ravetto, Piero
Politecnico di Torino
Italy

Rawlings, Robert
United States

Ray, Michael B.
United States

Ray, Richard D.
United States

Raydan, Marcos
University of Kentucky
United States

Reddien, George W., Jr.
Southern Methodist University
United States

Reddy, Satish Chandra
Courant Institute of Mathematical Sciences
United States

Rees, David
SUN-Y-MOR
Great Britain

Reid, John K.
Rutherford Appleton Laboratory
Great Britain

Reid, M. Erik
Carnegie-Mellon University
United States

Reiff, Andrea M.
Rice University
United States

Reissell, Martin
Germany

Reitich, Fernando L.
Carnegie-Mellon University
United States

Remmel, Jeffrey
University of California, San Diego
United States

Ren, Yuhe
Simon Fraser University
Germany

Renegar, James M.
Cornell University
United States

Rentrop, P.
Technical Universität Munchen
Germany

Rheinboldt, Werner C.
University of Pittsburgh
United States

Ribbens, Calvin J
Virginia Polytechnic Institute and State University
United States

Rich, Lawrence C.
United States

Richmond, Owen
ALCOA Laboratories
United States

Riedel, K. S.
New York University
United States

Rienstra, Sjoerd Willem
University of Technology Eindhoven
The Netherlands

Riesenfeld, Richard F.
University of Utah
United States

Rimbey, Scott E.
United States

Ringhofer, Christian A.
Arizona State University
United States

Risebro, Nils Henrik
University of Oslo
Norway

# ATTENDEE LIST

Rishel, Raymond W.
University of Kentucky
United States

Rivas, Damian
Universidad Politec de Madrid
Spain

Rivera, Jaime Edilberto Munoz
National Laboratory for Scientific Computing
Brazil

Rizk, Tony A.
TVA Engineering Laboratory
United States

Rizzo, Frank
University of Illinois
United States

Roberson, Kyle R.
Battelle Pacific N W Lab
United States

Roberts, Catherine A.
Northwestern University
United States

Roberts, Jean E.
Insitut National de Recherche en Informatique et en Automatique
France

Roberts, Lila F.
Georgia Southern University
United States

Robey, Thomas H.
United States

Roche, Jean Roldolphe
Universite de Nancy I-INRIA
France

Rodin, Ervin Y.
Washington University
United States

Rodman, Leiba
College of William & Mary
United States

Rodriguez, Domingo
University of Puerto Rico
United States

Rodriguez, Enrique
Institute Technologico Venezolana
Venezuela

Rodriguez, Jesus Enrique
INTEVEP SA
Venezuela

Rofman, Edmundo
INRIA
France

Rognlie, Dale M.
South Dakota School of Mines & Technology
United States

Rokhlin, Vladimir
Yale University
United States

Romate, Johan E.
The Netherlands

Rood, Richard B.
NASA/Goddard Space Flight Center
United States

Roose, Dirk
Katholieke Universiteit Leuven
Belgium

Root, Robert G.
Lafayette College
United States

Rose, Donald J.
Duke University
United States

Rosen, Christor
University of Lund
Sweden

Rosenau, Philip
Los Alamos National Laboratory
United States

Rosencrans, Steven I.
Tulane University
United States

Ross, David
Eastman Kodak Company
United States

Ross, Nicholas P.
University of Twente
The Netherlands

Rostamian, Rouben
University of Maryland
United States

Rothman, Ernest E.
Cornell University
United States

Roxin, Emilio O.
University of Rhode Island
United States

Ruan, Weihua
Purdue University, Calumet
United States

Ruas, Vitoriano
Université de Saint Etienne
France

Rueda, Norma
Merrimack College
United States

Ruget, Gabriel
Thomson-CSF
France

Rundell, William
Texas A&M University
United States

Rusak, Zvi
United States

Rusanov, V. V.
USSR Academy of Sciences
Russia

Russell, David L.
Virginia Polytechnic Institute & State University
United States

Russell, John M.
United States

Russo, Alessandro
Istituto di Analisi Numerica
Italy

Russo, Mark F.
United States

Rust, Bert W.
National Institute of Standards & Technology
United States

Rybka, Piotr
Warsaw University
Poland

Rye, Ralph
NSP
United States

Ryff, John V.
National Science Foundation
United States

Ryzhov, Oleg S.
USSR Academy of Science
Russia

Saavedra, Jorge I.
United States

Sachs, Ekkehard W.
University of Trier
Germany

Sachs, Jeffrey R.
National Institute of Standards and Technology
United States

Saiac, Jacques Herve
INRIA
France

Saito, Yoshihiro
Nagoya University
Japan

Sajo, Erno
Louisiana State University
United States

Salamonowicz, Paul H.
United States

Salane, Douglas E.
United States

Sallas, William M.
IMSL Incorporated
United States

Salvador, Marcelo P.
Escuela Politecnica Nacionale
Ecuador

Samuelsen, Catherine M.
Rice University
United States

Sandgren, Linda
SDSMT
United States

Sanders, Richard
University of Houston
United States

Santi, L. Michael
Christian Brothers University
United States

# ATTENDEE LIST

Santos, Newton Ribeiro
Universidade Fed Minas Gerais
Brazil

Santosa, Fadil
University of Delaware
United States

Sanugi, Bahrom B.
University Teknologi Malaysia
Malaysia

Saperstone, Stephen H.
George Mason University
United States

Sapidis, Nickolas
GM Research Laboratories
United States

Saranen, Jukka
University of Oulu
Finland

Sargent, Roger W. H.
Imperial College
Great Britain

Sarkar, Kausik
Johns Hopkins University
United States

Sarkis, Marcus
Courant Institute of Mathematical Sciences
United States

Sarvas, Jukka Olavi
University of Helsinki
Finland

Satrape, James V.
Johns Hopkins University
United States

Satterthwaite, James C.
United States

Saunders, Bonita V.
United States

Saxena, Mukul
University of Rochester
United States

Saxton, Katarzyna
Loyola University
United States

Saxton, Ralph
University of New Orleans
United States

Saylor, Paul E.
University of Illinois
United States

Sazegari, Ali
United States

Sazonov, V. V.
USSR Academy of Sciences
Russia

Scales, Lawrence Edwin
Shell Research Limited
Great Britian

Scandrett, Clyde L.
United States

Scapolla, Terenzio
Universita de Pavia
Italy

Schaffers, Wilhelmus J.
DuPont Company
United States

Schatzman, Michelle V.
Universite Lyon 1
France

Scheffe, Molly
United States

Scherzer, Otmar
Universitat Linz
Austria

Schempp, Walter
University of Siegen
Germany

Schleiniger, Gilberto F.
University of Delaware
United States

Schloeder, Johannes P
University of Augsburg
Germany

Schmeiser, Christian
Austria

Schmidt, Geert H.
Shell Oil Company
Houston

Schmidt, Stan
IBM
United States

Schneider, Wilhelm
Technical University of Vienna
Austria

Schot, Steven
The American University
United States

Schovanec, Lawrence E.
Texas Tech University
United States

Schreier, Richard
Oregon State University
United States

Schultz, Cherie L.
United States

Schulz, Volker Hubertus
IWR Universitat Heidelberg
Germany

Schumacher, Guenter
University of Karlsruhe
Germany

Schwartz, Benj L.
TASC
United States

Schwab, Christoph
University of Maryland at Baltimore County
United States

Schwandt, Hartmut M.
Technical University of Berlin
Germany

Schwartz, Ira B.
United States

Schweitzer, Paul
University of Rochester
United States

Schwendeman, Donald W.
Rensselaer Polytechnic Institute
United States

Schwierz, Guenter
United States

Schwetlick, Hubert
Martin-Luther University Halle
Germany

Scriven, L. E.
United States

Scriven, Ronald A.
Central Electricty Research Lab
Great Britain

Scroggs, Jeffrey S.
North Carolina State University
United States

Seader, J.D.
University of Utah
United States

Segall, Richard S.
United States

Segel, Lee A.
Weizmann Institute of Science
Israel

Seidel, Hans-Peter
University of Waterloo
Canada

Seidman, Thomas I.
University of Maryland
United States

Seidel, Wilfried
Universitat der Bundeswehr
United States

Seier, Edith Helfgott
Peru

Seitelman, Leon H.
United States

Selberherr, Siegfried
Austria

Selgrade, James F.
North Carolina State University
United States

Sell, George R.
University of Minnesota
United States

Sembuche, G.C.D.
University of Zimbabwe
Zimbabwe

Semenzato, Roberto
University of Padova
Italy

Sudhanshu, Semwal
University of Colorado
United States

Serbin, Steven M.
University of Tennessee
United States

Seriani, Geza
Osservatorio Geofisico
Italy

Sesay, Mohamed
University of British Columbia
United States

Sethian, James
University of California
United States

Seube, Nicolas
France

Seward, Wendy L.
University of Waterloo
Canada

Sezginer, Apo
Schlumberger-Doll Research
United States

Shah, Jayant M.
Northeastern University
United States

Shannon, Gregory E.
Indiana University
United States

Shao, Sai Lai Sally
Cleveland State University
United States

Shapiro, Louis W.
United States

Shariff, Halim M. B. M.
Coventry Polytechnic
Great Britain

Sharp, Philip W.
Queen's University
Canada

Shashkov, Mikhail
National Center of Math Modeling
USSR Academy of Sciences
Russia

Shaw, John K.
Virginia Technical University
United States

Shearer, James B.
IBM
United States

Shearer, James W.
University of Michigan
United States

Sheen, Dongwee
Purdue University
United States

Sheikh, Qasim M.
Cray Research Incorporated
United States

Shen, Jie
Pennsylvania State University
United States

Shen, Mei Chang
Univesity of Wisconsin
United States

Shene, Ching-Kuang
Johns Hopkins University
United States

Sherman, Arthur S.
NIH-NIDDK-MRB.
United States

Shi, Peter
Oakland University
United States

Shi, Zhong-Ci
Academia Sinica
People's Republic of China

Shibberu, Yosi
United States

Shih, Shagi-Di
University of Wyoming
United States

Shillor, Meir
Oakland University
United States

Shinbrot, Troy
University of Maryland
United States

Shivakumar, Pappur N.
University of Manitoba
Canada

Shoosmith, John N.
United States

Short, Kevin
United States

Showalter, R.E.
University of Texas
United States

Shrager, Richard I.
National Institute of Health
United States

Shrier, Stefan
United States

Shu, Chi-Wang
Brown University
United States

Shubin, Gregory R.
United States

Shutie, Xiao
Tsinghua University
Republic of China

Shvedoff, Vladimir M.
United States

Sidwell, John Mears
Polytechnic of Central London
Great Britain

Sigurdsson, Sven T.
University of Iceland
Iceland

Silva, Pantaleao A.
SUNY at Stony Brook
United States

Silvestri, Vincent
Ecole Polytechnique
Canada

Simion, Rodica E.
George Washington University
United States

Simitses, George J.
University of Cincinnati
United States

Simmons, Adrian John
European Centre for Medium Range Weather Forecasts
Great Britain

Simmonds, James G.
University of Virginia
United States

Simonsen, Kristine M.
Indiana University
United States

Simonovits, Miklos
Hungarian Academy of Sciences
Hungary

Simons, William H.
University of West Virginia
United States

Sincovec, Richard F.
Oak Ridge National Laboratory
United States

Sinha, Prakash C.
Indian Institute of Technology
India

Sinha, Somdatta
Center for Cellular and Molecular Biology
India

Sinsheimer, Janet Suzanne
UCLA
United States

Siripanich, Pachitjanut
NIDA
Thailand

Sjogren, Jon A.
AFOSR/NM
United States

Skeel, Robert D.
Univesity of Illinois
United States

Skiena, Steven
SUNY at Stony Brook
United States

Skorupka, Clements W.
Susquehanna University
United States

Slattery, Rhonda
Stanford University
United States

Slemrod, Marshall
University of Wisconsin
United States

St. Mary, Donald F.
University of Massachusetts
United States

Smedstad, Ole Martin
Sverdrup Technology Inc.
United States

Smit, Jacobus Hendrik
University of Stellenbosch
Republic of South Africa

Smith, Anne C.
University of Virginia
United States

Smith, Barry F.
Argonne National Laboratory
United States

Smith, Brian T.
University of New Mexico
United States

Smith, Hal L.
Arizona State University
United States

Smith, Lauren L.
United States

Smith, Peter
University of Keele
United States

Smith, Ralph C.
Institute for Computer Application in Science and Engineering
United States

Snyder, William V.
Jet Propulsion Laboratories
United States

So, Joseph Wai Hung
University of Alberta
Canada

Sobczyk, Krystyna Maria
University of Mining & Metallurgy
Poland

Sofer, Ariela
George Mason University
United States

Sokolowski, Jan
Systems Research Institute
Poland

Solomatine, Dmitri
IHE Delft
Netherlands

Somersolo, Erkki
Et Hesperiankatu
Finland

Sommerer, John C.
United States

Somwaru, Agapi L.
ERS/DSC
United States

Soner, Halil Mete
Carnegie-Mellon University
United States

Song, Wusheng
University of California/Santa Barbara
United States

Song, Yuhe
Simon Fraser University
Canada

Soni, Bharat K.
Mississippi State University
United States

Sonnad, Vijay
IBM/Austin
United States

Soravia, Pierpaolo
University of Padova
Italy

Sosa, Horacio A.
Drexel University
United States

Sosa, Wilfredo
Peru

Soulen, Robert J.
Naval Research Lab
United States

Spall, James C.
Johns Hopkins University
United States

Spencer, Thomas H.
University of Nebraska
United States

Spies, Ruben D.
Virginia Polytechnic Institute and State University
United States

Spigler, Renato G. C.
Universita di Padova
Italy

Spinner, Stuart
United States

Spoonamore, Janet
United States

Sprekels, Jurgen
Universitat - GHS Essen
Germany

Sritharan, Sivaguru
University of Colorado
United States

Srivastav, Sadanand
Bowie State College
United States

Staffans, Olof
Virginia Polytechnic Institute & State University
United States

Stainthorpe, Julian
Sheffield City Polytechnic
Great Britain

Stakgold, Ivar
University of Delaware
United States

Starke, Gerhard C.
Stanford University
United States

Starr, Page
United States

Steadman, Vernise
Howard University
United States

Stech, Harlan W.
University of Minnesota
United States

Steen, Paul H.
Cornell University
United States

Steffens, Jean-Claude
Laborelec
Belgium

Steihaug, Trond
University of Bergen
Norway

Stein, Melanie E.
United States

Steinbach, Mark C.
Universitat Heidelberg
Germany

Steinbart, Enid
University of New Orleans
United States

Steinberg, Stanley
University of New Mexico
United States

Stern, Julio Michael
United States

Sternberg, Peter
Indiana University
United States

Stetter, Hans J.
Technical University of Vienna
Austria

Steuerwalt, Michael H.
National Science Foundation
United States

Stevens, Rick L.
United States

Stewart, G.W.
University of Maryland
United States

Stewart, Kris W.
San Diego State University
United States

Stewart, Rodney D.
Royal Institute of Technology
Sweden

Stiftinger, Martin K.
Technical University Vienna
Austria

Still, Charles Herbert
Lawrence Livermore National Laboratory
United States

Stoecker, Michael G.
United States

Stoorvogel, Anton A.
Eindhoven University of Technology
Netherlands

Stotland, Stephanie A.
University of Virginia
United States

Stott, Brian
P C A Corporation
United States

Stoughton, Thomas
United States

Strang, Gilbert
Massachuttes Institute of Technology
United States

Strikwerda, John C.
University of Wisconsin
United States

Strom, Staffan E. G.
Royal Institute of Technology
Sweden

Struckmeier, Jens
University of Kaiserlautern
Germany

Styrylska, Teresa
Krakow
Poland

Stuben, Klaus
GMD - F1/T
Germany

Sturtz, Kirk E
United States

Su, Yeong Tzay
United States

Suhartano, Heru
University of Indonesia
Indonesia

Sullivan, Francis E.
National Institute of Standards & Technology
United States

Sumi, Akimasa
University of Tokyo
Japan

Summers, Jonathan Lewis
University of Newcastle Upon Tyne
Great Britain

Sumner, Brian
United States

Sun, Chunguang
Cornell University
United States

Sun, Xiaobai
Argonne National Laboratory
United States

Sundaram, Sheila
University of Quebec
Canada

Sunley, Judith S.
National Science Foundation
United States

Suri, Manil
University of Maryland-Baltimore County
United States

Surry, Patrick
University of Western Ontario
Canada

Svendsen, John A.
Norsk Hydro A/S
Norway

St. Vincent, Michael
United States

Svobodny, Thomas P.
United States

Svoboda, Wolfram D.
International Congress Centrum Belinin
Germany

Swailes, David
University of Strathclyde
Great Britain

Swartzendruber, Budrow A.
United States

Swarztrauber, Paul N.
National Center for Atmospheric Research
United States

Symes, William W.
United States

Szepessy, Anders
Royal Institute of Technology
Sweden

Szymczak, William G.
United States

Taber, Norma J.
United States

Taber, William L.
Jet Propulsion Laboratory
United States

Tabib, Claudette
Outremont
Canada

Taboada, Mario
University of Southern California
United States

Tael, Steven
SUNY-Stony Brook
United States

Taha, Rauf I.
May & Speh, Incorporated
United States

Tajchman, Marc
Université de Paris-Sud
France

Tajdari, Mohammad
Embry-Riddle Aeronautical University
United States

Takac, Peter
Vanderbilt University
United States

Talay, Denis
INRIA
France

Tan, Roger Choon-Ee
National University of Singapore
Republic of Singapore

Tang, Betty
Arizona State University
United States

Tang, Tao
Simon Fraser University
Canada

Tang, Wei Pai
University of Waterloo
Canada

Tapia, Richard A.
Rice University
United States

Taranco, Armand
Aix en Provence University
France

Tardos, Eva
Cornell University
United States

Tarjan, Robert E.
Princeton University
United States

Tarnawsky, Roman
United States

Tarvainen, Kyosti
Rolf Nevanlinna Institute
Finland

Taswell, Carl
United States

Tataru, D.
University of Virginia
United States

Tatalias, Kosmo D.
United States

Tautu, Peter
German Cancer Research Center
Germany

Tavantzis, John
New Jersey Institute of Technology
United States

Taylor, Thomas J. S.
Arizona State University
United States

Taylor, Valerie
University of California at Berkeley
United States

Tekolste, Elton K.
United States

Temam, Roger
University de Paris Sud
France

Temple, Blake
University of California at Davis
United States

Terman, David H.
Ohio State University
United States

Terrill, Louise
Oxford University
Great Britain

Terzopoulos, Demetri
University of Toronto
Canada

Tesei, Alberto
Inst Applicazioni del Calcolo "Mauro Picone"
Italy

Tewarson, Reginald P.
United States

Thepaut, Jean Noel
ECMWF
Great Britain

Thomas, Jean-Marie
Mathematiques Appliquees
France

Thompson, Jon H.
University of New Brunswick
Canada

Thompson, Lisa
Office of Government & Public Affairs
United States

Thompson, Richard
United States

Thompson, Russell C.
Utah State University
United States

Thron, C. Dennis
Dartmouth Medical School
United States

Tier, Charles
University of Illinois at Chicago
United States

Tijhuis, Anton G.
Delft University of Technology
Netherlands

Ting, Lu
Courant Institute of Mathematical Sciences
United States

Ting, Thomas C. T.
University of Illinois at Chicago
United States

Tippett, James M.
United States

Tirmizi, Ikram Abbas
Sultan Qaboos University
Oman Sultante

Titi, Edriss S.
University of California
United States

Tobin, Frank
Smith Kline Beecham
United States

Todd, Michael J.
Cornell University
United States

Toint, Philippe L.
Notre Dame de la Paix
Belgium

Toll, Charles H.
United States

Tome, Murilo Francisco
University of Strathclyde
Great Britain

Torczon, Virginia J.
Rice University
United States

Torokhtij, Anatoli P.
Leningrad Institute of Transport Engineers
Russia

Tran, Victor Ngoc
University of California at Irvine
United States

Trapp, George E., Jr.
United States

Traub, Joseph F.
Columbia University
United States

Traversoni, Leonardo
University Autonoma Metropolitana
Mexico

Trease, Harold E.
Los Alamos National Laboratories
United States

Trefethen, Lloyd N.
Cornell University
United States

Trentelman, H. L.
University of Groningen
Netherlands

Triandai, Ioana
University of Minnesota
United States

Tribbey, William P.
CTI PET Systems Incorporated
United States

Trompert, R. A.
Center for Math/Computer Science
Netherlands

Tron, Andrew W.
Princeton University
United States

Troy, William C.
University of Pittsburgh
United States

Trubuil, Alain
INRIA
France

Truneh, Yohannes
Mesa College
United States

Tryggvason, Gretar
University of Michigan
United States

Tso, Tai-Yih
Iowa State University
United States

Tsuchiya, Takuya
Ehime University
Japan

Tsujikawa, Tohru
Hiroshima Denki Institute of Technology
Japan

Tucsnak, Marius
Universite d'Orleans
France

Tulin, Marshall P.
University of California at Santa Barbara
United States

Tuncel, Levent
Cornell University
United States

Tuong, Ha Duong
CMAP
France

Tupholme, Geoffrey Edward
University of Bradford
Great Britain

Turner, James C., Jr.
United States

Turner, Peter R.
US Naval Academy
United States

Tygar, Justin Douglas
Carnegie-Mellon University
United States

Tzavaras, Athanasios
University of Wisconsin
United States

Unser, M.
National Institutes of Health
United States

Ushijima, Teruo
University Electro-Communications
Japan

Valdes, Teofilo
University Complutense Madrid
Spain

Valdivia, Rebekah N.
United States

Valente, V.
Consiglio Nazionale della Ricerche
Italy

Van de Geijn, Robert A.
University of Texas at Austin
United States

Van Den Berg, Imme Pieter
University of Groningen
Netherlands

Van der Burgh, Adriaan H. P.
University of Technology
Netherlands

Van Duijn, Hans
Delft University of Technology
Netherlands

Van Dyke, Milton D.
Stanford University
United States

Van Harten, Aart
University of Twente
Netherlands

Van Horssen, W. T.
Delft University of Technology
Netherlands

van Hulzen, J. A.
University of Twente
Netherlands

## 388 ATTENDEE LIST

Vargas, John D.
United States

Varosi, Frank
University of Maryland
United States

Vasileva, Adelaida Borisovna
Moscow State University
Russia

Vassilakis, Kostas
Yale University
United States

Vayo, H. W.
University of Toledo
United States

Vazirani, Umesh V.
University of California, Berkeley
United States

Vazquez, Carlos
Universidad de Vigo
Spain

Vega, Jose Manuel
Universidad Politec de Madrid
Spain

Vega-Torres, Yliris
NIST
United States

Venturino, Ezio G.
University of Iowa
United States

Vera, Jorge R.
Cornell University
United States

Verde, Cristina
Instituto de Ingenieria-Universidad Nacional de Mexico
Mexico

Verde-Star, Luis
University Autonoma Metropolitana
Mexico

Verhulst, Ferdinand
University of Utrecht
Netherlands

Vermeer, Pamela
United States

Vermiglio, Rossana
University of Udine
Italy

Verner, James H.
Queens University
Canada

Verri, Maurizio
Politecnico di Milano
Italy

Verwer, J. G.
Stichting Mathematisch Centrum
Netherlands

Vestal, Eric W.
United States

Vidrascu, Marina
INRIA
France

Villamizar, Vianey
Venezuela

Vinayagamoorthy, M.
Arkansas College
United States

Virgin, Lawrence N.
Duke University
United States

Virk, Gurvinder Singii
University of Sheffield
Great Britain

Vishnubhatla, S.
Bell Communications Research
United States

Vistro, Maryann
Ateneo de Manila University
Phillippines

Vogels, Marli
National Aerospace Laboratories
Netherlands

Vogelius, Michael S.
Rutgers University
United States

Voigt, Robert G.
National Science Foundation
United States

Von Schwerin, Reinhold L.
Germany

Voss, David A.
Western Illinois University
United States

Voth, Eric J.
Brown University
United States

Vouk, Mladen A.
North Carolina State University
United States

Vrdoljak, Bozo
University of Split
Yugoslavia

Vuorinen, Matti K. K.
Finland

Wachs, Michelle L.
University of Miami
United States

Wagner, Barbara A.
Los Alamos National Laboratory
United States

Wagner, Claus Albrecht
United States

Wagner, David H.
University of Houston
United States

Wagner, Don
ONR
United States

Walker, Homer F.
Utah State University
United States

Walkington, Noel J.
United States

Wall, David John
University of Canterbury
New Zealand

Wallace, David J.
University of Edinburgh
Scotland

Walter, Gilbert G.
University of Wisconsin
United States

Walter, Wolfgang L.
Universitat Karlsruhe
Germany

Walter, Wolfgang V.
University of Karlsruhe
Germany

Walton, Jay R.
Texas A & M University
United States

Waltman, Paul E.
Emory University
United States

Wan, Frederic Y. M.
University of Washington
United States

Wan, Honghui
Huazhong University of Science
People's Republic of China

Wan, Yieh-Hei
SUNY at Buffalo
United States

Wang, Ching-An
National Chung Cheng University
Republic of China

Wang, Daoliu
Konrad Zuse Center
Germany

Wang, Feng
SUNY at Stony Brook
United States

Wang, Hu
SUNY at Stony Brook
United States

Wang, Hwai-chiuan
National Tsing Hua University
Peoples Republic of China

Wang, Jane I.
Los Alamos National Laboratory
United States

Wang, Samuel
United States

Wang, Shingmin
Northeast Missouri State University
United States

Wang, Shin-Hwa
National Tsing-Hua University
Republic of China

Wang, Tseng-Chan
Jet Propulsion Laboratory
United States

Wang, Xuefeng
Tulane University
United States

Wang, Yuanming
Southeast University
Peoples Republic of China

Wang, Zhongde
University of Houston
United States

Warhola, Gregory T.
Air Force Institute of Technology
United States

Warne, Paul G.
United States

Warnow, Tandy
University of California, Berkeley
United States

Watanabe, Masao
Johns Hopkins University
United States

Watkins, David S.
Washington State University
United States

Watson, Layne T.
VPI & SU
United States

Wazwaz, Abdul-Majid A.
Saint Xavier University
United States

Weaver, Jim
United States

Wedin, Per Ake
University of Umea
Sweden

Weeks, Dennis
MasPar Computer Corp.
United States

Wegman, Edward J.
George Mason University
United States

Wei, Ching Zong
Academia Sinica
People's Republic of China

Wei, Dongming
University of New Orleans
United States

Weichert, Dieter
University of Lille (Eudil)
France

Weidman, Scott T.
National Research Council
United States

Weiland, Siep
Rice University
United States

Weimar, Joerg R.
Belgium

Weinacht, R. J.
University of Delaware
United States

Weinmuller, Ewa B.
Technische Universitat Wien
Austria

Weinberger, Hans F.
University of Minnesota
United States

Weinstein, Mordechai M.
Ministry of Defense
Israel

Weiss, Michael D.
United States

Weiss, Rudiger
Rechenzentrum University Karlsruhe
Germany

Wendland, Wolfgang L.
University Stuttgart
Germany

Weng, Shipei
University of Minnesota
United States

Wesseling, P.
Delft University Technology
Netherlands

Weston, James Stannard
University of Ulster at Coleraine
Northern Ireland

Weston, Vaughan H.
Purdue University
United States

Wheeler, Adam A.
National Institute of Standard & Technology
United States

Wheeler, David H.
Carnegie-Mellon University
United States

Wheeler, Mary Fannett
Rice University
United States

Wheeler, Robert L.
Virginia Polytechnic Institute and State University
United States

Whelan, Tracy
Camac Systems, Inc
United States

White, Benjamin S.
Exxon Research & Engineering
United States

White, Brian A.
RMCS Cranfield
Great Britain

White, Jacob K.
Massachusetts Institute of Technology
United States

White, Robert E.
North Carolina State University
United States

Whitaker, Nathaniel
United States

Whitman, Peter C.
United States

Wick, Joachim
University of Kaiserlautern
Germany

Wickerhauser, Mladen V.
Yale University
United States

Widlund, Olof B.
Courant Institute of Mathematical Sciences
United States

Wiecek, Malgorzata
Clemson University
United States

Wielgosz, Christian
ENSM
France

Wilks, Graham
University of Keele
Great Britain

Williamson, David L.
NCAR
United States

Williams, Joseph J.
University of Manitoba
Canada

Williams, J.E.
Cambridge University
Great Britain

Williamson, Karen A.
Rice University
United States

Willis, John
University of Bath
Great Britain

Willms, N. Bradley
University of Waterloo
Canada

Wilson, David G.
IBM
United States

Wilson, George V.
United States

Winarsky, Norman D.
David Sarnoff Research Center
United States

Winkler, Peter M.
Bellcore
United States

Winther, Ragnar
University of Oslo
Norway

Wlodarski, Krzysztof M.
United States

Wolfe, Peter
University of Maryland
United States

Wolf, Sylvie
IFP
France

Wolkowicz, Gail S.K.
McMaster University
Canada

Wolkowicz, Henry
University of Waterloo
Canada

Wollkind, David J.
United States

Wong, Roderick
University of Manitoba
Canada

Wood, Alastair D.
National Institute for Higher Education
Ireland

Woodard, Joseph W.
United States

Woodbury, Max A.
Duke University
United States

Woodward, Diana E.
Southern Methodist University
United States

Woodward, Michael E.
Loughborough University
Great Britain

Worley, Patrick H.
United States

Wozniakowski, Henryk
Columbia University
United States

Wright, Colin J.
University of Witwatersrand
Republic of South Africas

Wright, Margaret H.
AT&T Bell Laboratories
United States

Wright, Nigel George
University of Leeds
Great Britain

Wright, Stephen J.
Argonne National Laboratory
United States

Wulkow, Michael
Konrad-Zuse-Zentrum
Germany

Wunsch, Carl
Massachusetts Institute of Technology
United States

Wu, Julian J.
U.S. Army Research Office
United States

Wu, Zhijun
Cornell University
United States

Xia, Hong Xing
George Mason University
United States

Xiang, Longwan
Shanghai Jiaotong University
People's Republic of China

Xiao, Ding
University of Connecticut
United States

Xie, Hong
University of Alberta
Canada

Xin, Zhouping
New York University/Courant Institute of Mathematical Sciences
United States

Xiong, Chuyu
Ohio State University
United States

Xu, Jinchao
Pennsylvania State University
United States

Xu, Jian Jun
McGill University
Canada

Xu, Yao-Huan
Grove City College
United States

Xu, Yongzhi
University of Minnesota
United States

Yamaguti, Masaya
Rynkoku University
Japan

Yan, Jia-Jue Joseph
University of North Carolina
United States

Yan, Huan
University of Kentucky
United States

Yan, Xiaopu
University of Pittsburgh
United States

Yan, Yi
University of Kentucky
United States

Yang, Hu
University of Washington
United States

Yang, Jichuan
Brown University
United States

Yang, Yadong
North Carolina State University
United States

Yang, Yisong
University of New Mexico
United States

Yao, Yitao
United States

Yarbrough, Lynn D.
Digital Equipment Corporation
United States

Ye, Huanchun
University of Texas @ Austin
United States

Ye, Qi-xiao
Beijing Institute of Technolog
People's Republic of China

Yen, Jeng
United States

Yerikalapuoy, Vasodena Rao
Andhra University
India

Yih, Tachung
Florida International University
United States

Yin, George
Wayne State University
United States

Yin, Hong Ming
University of Toronto
Canada

Ying, Long-an
Peking University
People's Republic of China

Yip, Elizabeth L.
United States

Yokoyama, Etsuro
Carnegie-Mellon University
United States

York, Bryant W.
United States

You, Yuncheng
University of South Florida
United States

Young, David M., Jr.
United States

Young, David P.
Boeing Computer Services
United States

Young, Larry C.
United States

Young, Robin C.
Courant Institute of Mathematical Sciences
United States

Yserentant, Harry
Universitat Tubingen
Germany

Yuan, He
The Johns Hopkins University
United States

Yuan, Xue Feng
University of Cambridge
Great Britain

Yuan, Ya-Xiang
Northwestern University
United States

Yuri, Flerov
Computing Centre USSR Academy of Science
Russia

Yuskewich, John David
United States

Yu, Lei
University of Maryland
United States

Yves, Perreal
Thomson - CSF/LCR
France

Zachary, Woodford W.
Howard University
United States

Zaghrout, Afaf Abou-Elfotouh S.
Al-Azhar University
Egypt

Zak, Eugene
MAJIQ Incorporated
United States

Zalesak, Rudy
United States

Zamansky, Leo
Ultra Max Corporation
United States

Zandbergen, Pieter
University of Twente
Netherlands

Zasadil, Scott
United States

Zauderer, Erich
United States

Zayed, Ahmed I.
California Polytechnic State University
United States

Zeeman, Mary Lou
University of Texas
United States

Zegeling, Paul A.
Center for Math & Computer Science
Netherlands

Zeitoun, David G.
Hydrological Service
Israel

Zelevinsky, Andrei V.
Cornell University
United States

Zettl, Anton J.
Northern Illinois University
United States

Zhang, Detong
Jilia University
People's Republic of China

Zhang, Duan Zhong
Johns Hopkins University
United States

Zhang, Guodong
University of Maryland
United States

Zhang, Hongbin
University of Virginia
United States

Zhang, Lei
University of Maryland
United States

Zhang, Ning
Pacific Bell Organization
United States

Zhang, Qiang
New York University
United States

Zhang, Qing
University of Toronto
Canada

Zhang, Shangyou
University of Delaware
United States

Zhang, Tong
Germany

Zhang, Wen Wz
United States

Zhang, Weijiang
Shanghai Jiao Tong University
People's Republic of China

Zhang, Yin
University of Maryland
United States

Zhang, Yue
University of Kentucky
United States

Zhang, Sheguang
University of Maryland
United States

Zhong, Xiaohui
United States

Zhou, Jian
University of Maryland
United States

Zhu, Dantong
University of Washington
United States

Zhu, Hang
Brown University
United States

Zhu, Jianping
Mississippi State University
United States

Zhu, Jingyi
University of Utah
United States

Zhu, Mei
University of Washington
United States

Zhu, Yimin
Washington University
United States

Ziegler, Zvi
Israel

Zilli, Giovanni
Universität di Padova
Italy

Zimmerman, Robert W.
United States

Zion, Gary Randall
National Institute of Health & Dental Research
United States

Zito, Jennifer S.
Supercomputing Research Center
United States

Zolesio, Jean-Paul
Centre National de la Recherche Scientifique/USTL
France

Zou, Qisu
Kansas State University
United States

Zubelli, Jorge P.
United States

Zumbrun, Kevin
SUNY at Stony Brook
United States

Zuo, Wei
Rice University
United States

Zurn, Rena M.
United States

Zwillinger, Daniel
United States